Climate Change and Global Warming

Climate Change and Global Warming

Edited by Vivian Moritz

SYRAWOOD
PUBLISHING HOUSE
New York

Published by Syrawood Publishing House,
750 Third Avenue, 9th Floor,
New York, NY 10017, USA
www.syrawoodpublishinghouse.com

Climate Change and Global Warming
Edited by Vivian Moritz

International Standard Book Number: 978-1-68286-446-3 (Hardback)

Cataloging-in-publication Data

Climate change and global warming / edited by Vivian Moritz.
p. cm.
Includes bibliographical references and index.
ISBN 978-1-68286-446-3
1. Climatic changes. 2. Global warming. 3. Climate change mitigation. 4. Climatic changes--Risk management.
I. Moritz, Vivian.
QC903 .C55 2017
551.6--dc23

Printed in China.

TABLE OF CONTENTS

PREFACE

Climate change and global warming are two most crucial, environmental concerns of our times. Researchers across the globe are devising methods and strategies to understand these phenomena and deal with them. Different approaches, evaluations, methodologies and advanced studies in this field have been included in this book. It highlights the different strategies and techniques of dealing with climate change and global warming through different case studies and extensive use of examples. This book is a vital tool for all researching or studying climate change and global warming as it gives incredible insights into emerging trends and concepts. It is an essential guide for both academicians and those who wish to pursue this discipline further.

After months of intensive research and writing, this book is the end result of all who devoted their time and efforts in the initiation and progress of this book. It will surely be a source of reference in enhancing the required knowledge of the new developments in the area. During the course of developing this book, certain measures such as accuracy, authenticity and research focused analytical studies were given preference in order to produce a comprehensive book in the area of study.

This book would not have been possible without the efforts of the authors and the publisher. I extend my sincere thanks to them. Secondly, I express my gratitude to my family and well-wishers. And most importantly, I thank my students for constantly expressing their willingness and curiosity in enhancing their knowledge in the field, which encourages me to take up further research projects for the advancement of the area.

Editor

The Effect of Simulating Different Intermediate Host Snail Species on the Link between Water Temperature and Schistosomiasis Risk

Nicky McCreesh*, Mark Booth

School of Medicine, Pharmacy and Health, Durham University, Durham, United Kingdom

Abstract

Introduction: A number of studies have attempted to predict the effects of climate change on schistosomiasis risk. The importance of considering different species of intermediate host snails separately has never previously been explored.

Methods: An agent-based model of water temperature and *Biomphalaria pfeifferi* population dynamics and *Schistosoma mansoni* transmission was parameterised to two additional species of snail: *B. glabrata* and *B. alexandrina*.

Results: Simulated *B. alexandrina* populations had lower minimum and maximum temperatures for survival than *B. pfeifferi* populations (12.5–29.5°C vs. 14.0–31.5°C). *B. glabrata* populations survived over a smaller range of temperatures than either *B. pfeifferi* or *B. alexandrina* (17.0°C–29.5°C). Infection risk peaked at 16.5°C, 25.0°C and 19.0°C respectively when *B. pfeifferi*, *B. glabrata* and *B. alexandrina* were simulated. For all species, infection risk increased sharply once a minimum temperature was reached.

Conclusions: The results from all three species suggest that infection risk may increase dramatically with small increases in temperature in areas at or near the currents limits of schistosome transmission. The effect of small increases in temperature in areas where schistosomiasis is currently found will depend both on current temperatures and on the species of snail acting as intermediate host(s) in the area. In most areas where *B. pfeifferi* is the host, infection risk is likely to decrease. In cooler areas where *B. glabrata* is the host, infection risk may increase slightly. In cooler areas where *B. alexandrina* is the host, infection risk may more than double with only 2°C increase in temperature. Our results show that it is crucial to consider the species of intermediate host when attempting to predict the effects of climate change on schistosomiasis.

Editor: Ulrike Gertrud Munderloh, University of Minnesota, United States of America

Funding: The research leading to these results has received funding from the European Union's Seventh Framework Programme (FP7/2007-2013) under grant agreement no. 266327. The funders had no role in study design, data collection and analysis, decision to publish, or preparation of the manuscript.

Competing Interests: The authors have declared that no competing interests exist.

* Email: Nicky.mccreesh@durham.ac.uk

Introduction

Little is known about likely effects of climate change on schistosomiasis transmission[1], however neglecting the issue may have a negative effect on control programs and elimination goals by concentrating resources in the wrong areas. Some empirical studies suggest that increasing temperatures may already be expanding the range of schistosomiasis in Uganda, allowing transmission to occur at increasing altitudes[2–4]. Control programs that do not adapt to the changing distribution of schistosomiasis, and changing intensities of infection in different areas, are unlikely to make efficient use of the limited resources available to them.

Different species of *Biomphalaria* and *Bulinus*, the intermediate hosts of *Schistosoma mansoni* and *S. haematobium* respectively, have very different distributions[5], habitats[6], and temperature requirements[7]. Despite this, previous attempts to develop models of schistosomiasis and temperature and/or predict the effects of climate change on schistosomiasis have largely ignored the issue of different intermediate host species. One model of *S. japonicum* transmission and climate change in China used data from a single host species only (*Oncomelania hupensis*)[8], however most previous models of *S. mansoni* and *S. haematobium* have been parameterised using data from multiple species, or even genera, of snail[9–12]. To date, the only model parameterised to a single species of *Biomphalaria* or *Bulinus* snail is an agent-based model of water temperature and *Biomphalaria pfeifferi* population dynamics and *S. mansoni* transmission[13].

In this paper, we modify our previously published *B. pfeifferi* model[13] to allow us to simulate two additional species of *S. mansoni* intermediate host snails: *Biomphalaria glabrata* and *Biomphalaria alexandrina*. These species were selected due to the very limited amount of empirical data available on other *Biomphalaria* species[6]. *B. pfeifferi* is the most widespread intermediate host of *S. mansoni* in Africa[5], and can be found in a range of different types of water body including streams[14], lakes[15], reservoirs[16], irrigated areas[17–19], and rice-paddies[20]. *B. alexandrina* is found in North Africa only, in Egypt, north Sudan, and north-west Libya[5,21]. It is very common in the water supply and drainage networks of the Nile Delta[21,22], and can also be found in springs, streams and the edges of swamps[21]. *B. glabrata* has a widespread distribution in South America and the Caribbean[23], and can be found in pools, marshes, and streams[24].

This study is the first to explore how the population ecology of different species of snail may respond to increasing temperatures, and how this could impact on future schistosomiasis infection risk. It addresses a crucial issue that has been largely ignored in previous modelling efforts, but which nonetheless sits at the forefront of issues to be investigated if adaptation to climate change is to be achieved.

Methods

Model description

The model used is described in full in McCreesh and Booth[13]; a brief description is provided here. The model is an agent-based model of snail population dynamics and infection with schistosomes. All temperature sensitive stages of the snail and schistosome lifecycles are represented by agents (snail eggs, juveniles and adults, miracidia, and cercariae), and the model has a time step of one hour. Figure 1 shows a diagram of the model structure. Snails are born into the model as eggs, and develop at a temperature-dependent rate. When snail egg development is complete, they hatch into juvenile snails. Juvenile snails, in turn, develop at a temperature-dependent rate, becoming adult snails when development is complete. Adult snails produce snail eggs at a rate which is dependent on temperature and the number of snails in the model (to account for reduced fecundity at high snail densities). Snail eggs, juveniles and adults die and are removed from the model at a rate which is dependent on temperature and the number of snails in the model.

Miracidia are introduced into the model at a constant rate. They gain biological age at a temperature-dependent rate and die at a rate that depends on their biological age. They infect juvenile and adult snails at a rate that depends on their biological age and the water temperature. Once infected, juvenile and adult snails become prepatent. Their schistosome infections develop at a temperature-dependent rate, and once development is complete the snails become infectious. Adult snails cease to produce eggs halfway through schistosome development, and infectious snails have increased mortality rates. Once infectious, juvenile and adult snails produce cercariae at a rate which depends on the temperature and the hour of the day. Cercariae gain biological age at a temperature-dependent rate, and die at a biological age dependent rate.

The main output of the model is 'infection risk'. This is equal to the number of cercariae in the model, adjusted for their decreasing infectiousness with increasing biological age. Humans and adult worms are not simulated, and the rate of miracidia introduction is not linked to the number of cercariae or infection risk. This is because the 'correct' linking function would depend on localised factors such as sanitation practices, water contact behaviour, and immunity. Not simulating this link allows the findings of the model to be applied to any area where the relevant intermediate host snail species are found. This issue is discussed in detail in McCreesh and Booth[13].

Model parameterisation

The model was parameterised separately for three different species of *Biomphalaria*: *B. pfeifferi*, *B. glabrata* and *B. alexandrina*, using data from a range of different laboratory experiments and field studies. Details of the model parameterisation to *B. pfeifferi* are given in McCreesh and Booth[13], and details of the model parameterisation to *B. glabrata* and *B. alexandrina* are given in Figures S1–S6, and Tables S1–S3 in Model Parameterisation S1. Figure 2 illustrates values taken by the parameters which vary between snail species in the model: a) the juvenile development rate, b) the egg production rate, c) the egg development rate, d) the egg mortality rate, e) the uninfected and prepatent snail mortality rates, and f) the schistosome development rate.

Experiments

The model was run separately for *B. pfeifferi*, *B. glabrata* and *B. alexandrina* at all water temperatures at which the simulated snail

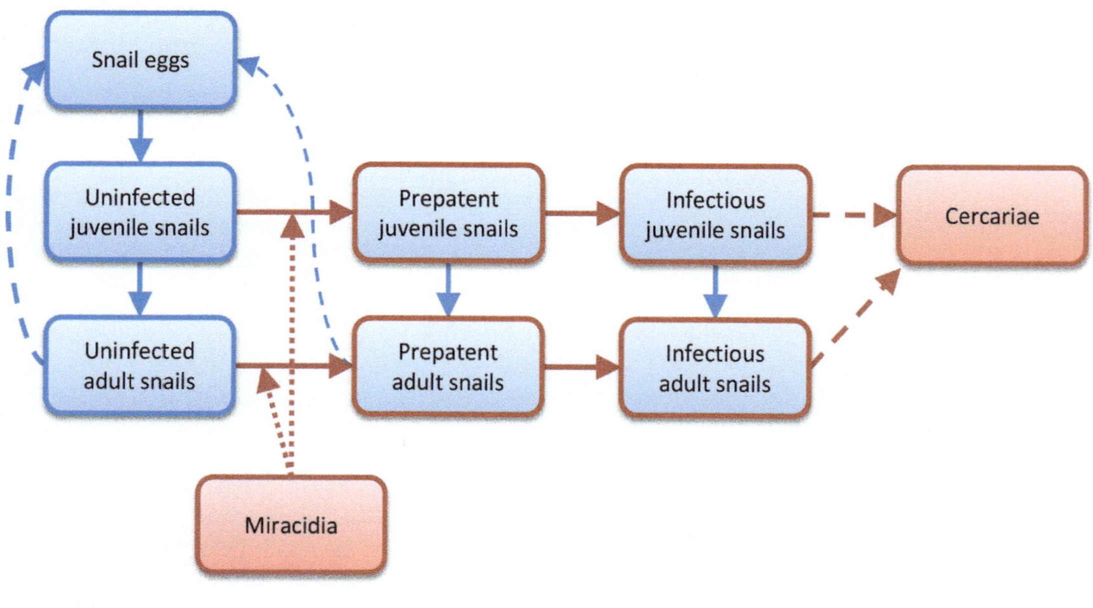

Figure 1. Diagram of the model structure. Boxes show types of agents. Solid arrows show where agents can change from one type into another. Dashed lines show the production of one type of agent by another. Dotted lines indicate infection. Red outlines and arrows indicate the presence of schistosomes. Agents of all types can die and be removed from the model.

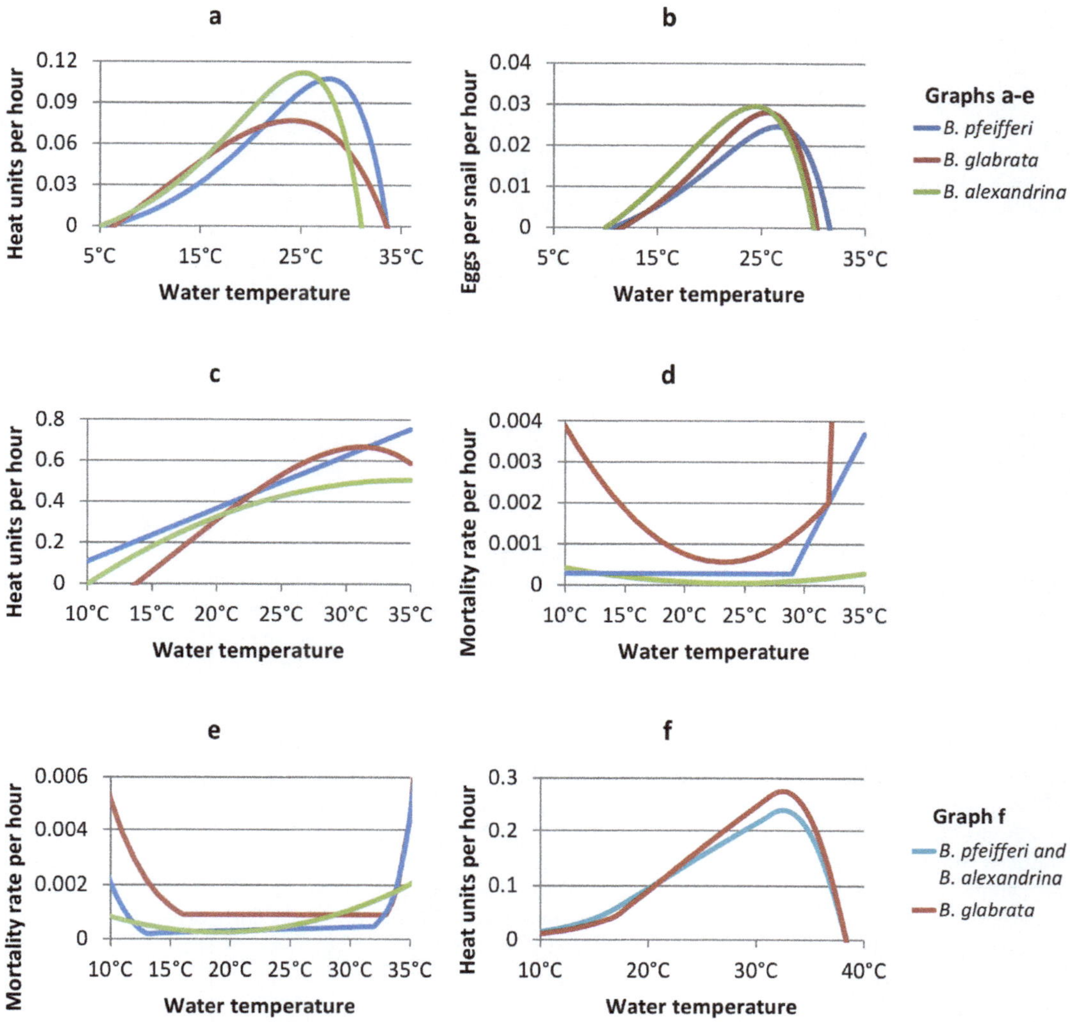

Figure 2. Comparison of parameters used in models of *B. pfeifferi, B. glabrata* and *B. alexandrina*. a) Juvenile development. Heat units gained per hour. Snails become adults and start producing eggs once they have gained 100 heat units. b) Egg production. Eggs per snail per hour. c) Egg development. Heat units gained per hour. Egg hatch once they have gained 100 heat units. d) Egg mortality rate per hour. e) Uninfected and prepatent snail mortality rate per hour. f) Parasite development within the snail. Heat units gained per hour. Snails become infectious once they have gained 100 heat units. Blue lines show parameters for *B. pfeifferi*, red for *B. glabrata*, green for *B. alexandrina*, and turquoise for both *B. pfeifferi* and *B. alexandrina*.

populations could survive indefinitely, with temperature increasing in 0.5°C increments. Outputs were averaged over a minimum of one year and 200 runs. Outputs were averaged over larger numbers of longer runs when necessary to reduce stochasticity. The number of snails in the model was calculated as the total number of uninfected, prepatent and infectious juvenile and adult snails.

In the models for all three snail species, mortality rates estimated from laboratory data were multiplied by 1.35 to account for increased mortality in natural conditions (see Model Parameterisation S1 for details). Estimates of mortality rates in wild snail populations suggest that mortality rates can be much higher than this in some circumstances however[22,24,25]. The effect of this was explored by further doubling mortality rates for all snails (juvenile and adult; uninfected, prepatent and infectious).

A number of intermediate results were also calculated by temperature for each snail species. These results were calculated from the equations used in the model only, and were designed to help improve understanding of the overall model results. They were:

1) The mean proportion of eggs that hatch, calculated from egg development rates and egg mortality rates
2) The mean proportion of juvenile snails that survive to become adults, calculated from juvenile development rates and the uninfected/prepatent mortality rates.
3) The mean proportion of infected snails that survive to become infectious, calculated from parasite development rates and the prepatent mortality rates.
4) The median lifetime cercaria production of infectious snails, calculated from the cercariae production rates and the infectious snail mortality rates.

As these intermediate results were calculated from the input parameters only, and not from model output, 2) to 4) assume that there is no density-dependent increase in mortality rates. 2) also assumes that no juvenile snails develop patent infections before becoming adults (which would increase their mortality rate).

Results

Snail population dynamics

When the lower mortality rates were simulated, the mean total number of *B. glabrata* in the model was lower than the mean total number of *B. pfeifferi* at all water temperatures, and lower than the number of *B. alexandrina* at all temperatures below 29.5°C (figure 3a). The mean total number of *B. alexandrina* was higher than the number of *B. pfeifferi* at all temperatures below 27.0°C. The maximum mean total number of *B. pfeifferi*, *B. glabrata* and *B. alexandrina* were 1015, 850 and 1284 respectively, at 24.5°C, 23.5°C and 25.0°C. Snail numbers were above 90% of their maximum between 17.0°C and 29.5°C for *B. pfeifferi*, 20.0°C and 28.5°C for *B. glabrata*, and 17.0°C and 25.5°C for *B. alexandrina*. Differences were greater between the maximum mean total number of adult snails in each low mortality rate model, with the *B. pfeifferi* model peaking at 506 adult snails at 26.5°C, the *B. glabrata* model at 145 snails at 24°C, and the *B. alexandrina* model at 661 snails at 20.5°C (figure 3b).

Simulating the higher mortality rates reduced both mean snail and adult snail numbers at all temperatures for all snail species, with the exception of the mean number of *B. alexandrina* at 20.5°C to 21.5°C, where snail numbers were 0.09% to 2.03% higher in the higher mortality scenario. This occurred as a result of the density dependence functions used in the model, and would not necessarily occur in wild snail populations. The presence and magnitude of the dip depends on the density functions used. Reductions in the mean number of snails, over the range of temperatures at which the simulated snail populations could survive indefinitely, varied between 4% and 28% for *B. pfeifferi*, 29% and 36% for *B. glabrata*, and 8% and 28% for *B. alexandrina*. Reductions in the mean number of adult snails ranged between 47% and 56% for *B. pfeifferi*, 63% and 66% for *B. glabrata*, and 42% and 59% for *B. alexandrina*.

The range of temperatures between which simulated snail populations could survive indefinitely was smallest for *B. glabrata*, with snail populations unable to survive at temperatures outside 17.0°C to 29.5°C, compared to 14.0°C to 31.5°C for *B. pfeifferi*. Simulated *B. alexandrina* populations survived at lower temperatures than either *B. glabrata* or *B. pfeifferi*, with a minimum temperature for survival of 12.5°C, but did not survive as well as *B. pfeifferi* at high temperatures, dying out at temperatures above 29.5°C. Doubling snail mortality rates slightly reduced the range of temperatures at which simulated *B. pfeifferi* and *B. alexandrina* populations could survive indefinitely to 15.0°C to 31.0°C and 14°C to 28.5°C respectively. Doubling mortality rates had a much greater effect on simulated *B. glabrata* populations, reducing the range of temperatures at which they could survive indefinitely to only 23.0 to 26.5°C.

Infection risk

With the lower mortality rates, infection risk was highest at 16.5°C when *B. pfeifferi* was simulated, at 19.0°C when *B. alexandrina* was simulated, and at 25.0°C when *B. glabrata* was simulated (figure 3c). Either side of these temperatures infection risk fell when simulating all snail species, however infection risk remained high over a wide range of temperatures when *B. glabrata* was the intermediate host snail species. Infection risk remained above 80% of its maximum value at all temperatures between 18°C and 26.5°C in the *B. glabrata* model, between 15°C and 20.5°C in the *B. pfeifferi* model, and between 18°C and 21°C in the *B. alexandrina* model. There was a risk of infection at all temperatures at which simulated *B. glabrata* populations could survive indefinitely. There was no infection risk below 15.0°C in any model, while simulated *B. pfeifferi* and *B. alexandrina* populations

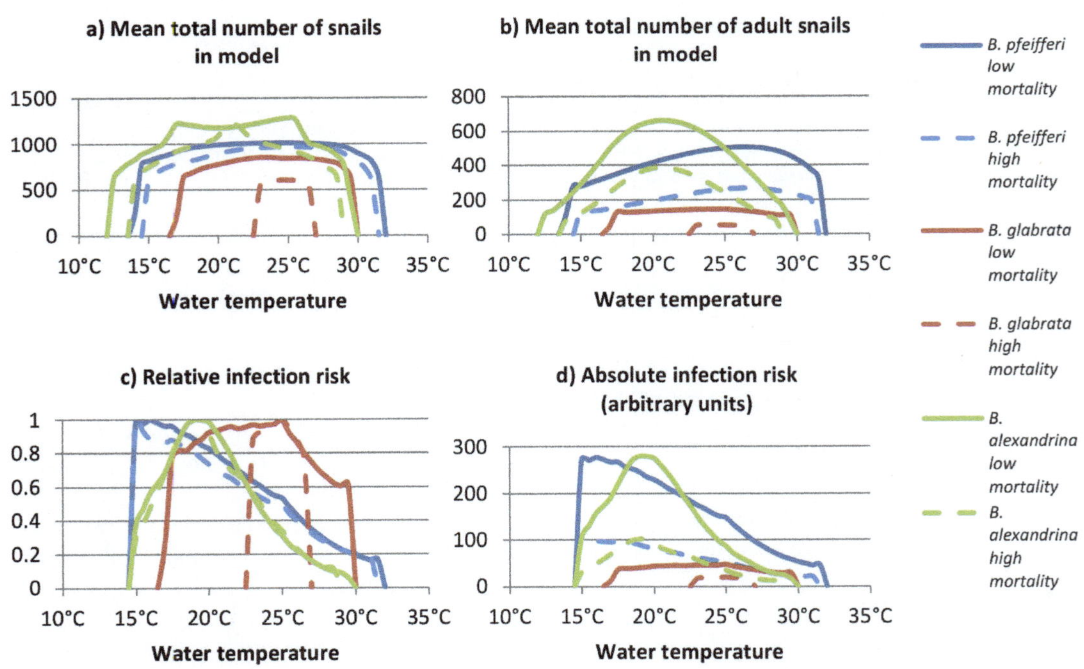

Figure 3. Number of snails and relative infection risk, with high and low mortality rates. a) Mean total number of snails in the model. b) Mean total number of adult snails in the model. c) Infection risk, relative to the maximum risk for the same snail species with the same mortality rates. d) Absolute infection risk (arbitrary units). Blue lines show *B. pfeifferi*, red lines show *B. glabrata* and green lines show *B. alexandrina*. Solid lines show results with a low mortality rate and dashed lines with a high mortality rate.

could survive indefinitely at temperatures as low as 14.0°C and 12.5°C respectively.

Absolute infection risk was highest when *B. pfeifferi* was simulated at all temperatures except between 18.5°C and 22.0°C, when absolute infection risk was highest when *B. alexandrina* was simulated (Figure 3d). Absolute infection risk was lowest when *B. glabrata* was simulated at all temperatures below 28.5°C. Above this temperature, absolute infection risk was lowest when *B. alexandrina* was simulated. Compared to when *B. pfeifferi* was simulated, maximum absolute infection risk was 0.6% higher when *B. alexandrina* was simulated and 83% lower when *B. glabrata* was simulated.

Doubling snail mortality effects had little effect on the relationship between water temperature and infection risk (relative to maximum infection risk in the same scenario), beyond eliminating infection risk at temperatures at which the snail populations were unable to survive indefinitely. It did however reduce absolute infection risk by 53–65% for *B. pfeifferi*, 58–71% for *B. alexandrina* and 58–63% for *B. glabrata* (within the range of temperatures that simulated snail populations were able to survive indefinitely).

Snail survival by stage, and median cercariae production

The proportion of simulated snail eggs that hatch is high for both *B. pfeifferi* and *B. alexandrina* at all moderate temperatures (greater than 80% between 11.0°C and 31.0°C for *B. pfeifferi* and 13.5°C and >35°C *B. alexandrina*) (figure 4a). The proportion is lower for *B. glabrata* at all temperatures, with less than 60% of eggs hatching at temperatures below 18.0°C and above 30.0°C. Assuming that mortality rates do not increase as a result of high snail densities or patent infections, over 50% of simulated *B. pfeifferi* and *B. alexandrina* juveniles will survive to become adults at temperatures between 15.5°C–31.5°C and 15.5°C-27.5°C respectively (figure 4b). Survival to adulthood is much lower for *B. glabrata*, with a maximum of 31% of snails surviving (at 23.0°C to 25.0°C).

The proportion of infected snails that survive to develop patent infections increases gradually with increasing temperature for both *B. pfeifferi* and *B. glabrata* (figure 4c), however the proportion surviving is higher for *B. pfeifferi* at all temperatures. Assuming no density dependent increase in mortality rates, a maximum of 82% of *B. pfeifferi* survive the prepatent period (at 32.0°C), compared to 60% of *B. glabrata* (at 32.5°C). The proportion of *B. alexandrina* surviving prepatency increases with increasing temperature up to 78% at 21.0°C, before falling again.

Figure 4d shows the median number of cercariae produced by each infectious snail during its lifetime, assuming no density dependent mortality. The median number per infectious snail peaks at 26,300 at 18.0°C for *B. pfeifferi* and at 31,300 at 19.5°C for *B. alexandrina*. It remains above 80% of its maximum between 14.5°C and 23.5°C for *B. pfeifferi* and 17.5°C and 22.0°C for *B. alexandrina*. Median numbers are much lower for *B. glabrata*, and vary less with temperature, peaking at 8,400 at 21.0°C, and remaining above 80% of their maximum between 16.0°C and 29.0°C.

Discussion

In areas currently at risk for schistosomiasis, temperatures are projected to increase by 2°C to 5°C by 2090–2099, relative to temperatures in 1980–1999[26]. There is some evidence that

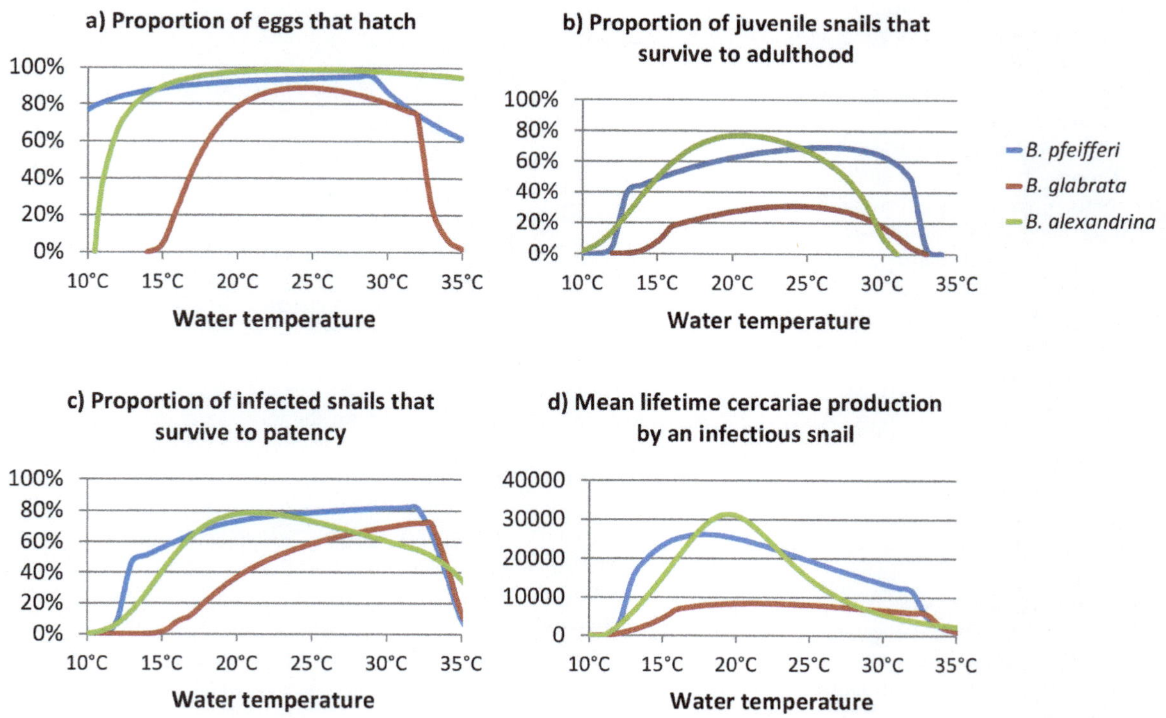

Figure 4. Proportions of eggs and snail surviving each stage, and cercariae production by infected snails. a) Proportion of eggs laid that hatch. b) Proportion of juvenile snails that survive to adulthood (assuming no additional density dependent mortality, and that no juvenile snails develop patent infections before they become adults). c) Proportion of infected snails that survive to patency (assuming no additional density dependent mortality). d) Median lifetime cercariae production by an infectious snail (assuming no additional density dependent mortality). Results are shown for the lower mortality scenarios. Blue lines show *B. pfeifferi*, red lines show *B. glabrata* and green lines show *B. alexandrina*.

rising temperatures may already be affecting the distribution of schistosomiasis[2–4]. Previous attempts to predict the effects of increasing temperatures on schistosomiasis have largely neglected the issue of multiple species of intermediate host snail[9–12]. This is the first study to look at the effects of parameterising a dynamical model of schistosomiasis and temperature to different species of snail.

The results suggest that *B. alexandrina* may be better adapted to slightly cooler water temperatures than *B. pfeifferi*, as both the maximum and minimum temperatures at which the simulated populations could survive indefinitely were lower. Simulated *B. glabrata* populations could survive indefinitely at a smaller range of temperatures than either *B. pfeifferi* or *B. alexandrina* populations, and died out at constant temperatures outside 17.0°C to 29.5°C. Doubling snail mortality rates slightly reduced the range of temperatures at which *B. pfeifferi* and *B. alexandrina* populations could survive, and slightly reduced mean population size at most temperatures. Doubling mortality rates had a greater effect on *B. glabrata*, reducing the range of temperatures at which the population could survive to only 23.0°C to 26.5°C and reducing maximum population size by 29%.

Compared to when *B. pfeifferi* were simulated, infection risk peaked at a slightly higher temperature when *B. alexandrina* were simulated (19.0°C compared to 16.5°C), and at a higher temperature still when *B. glabrata* were simulated (25.0°C). Doubling mortality rates reduced the temperature at which infection risk peaked to 15.0°C when *B. pfeifferi* were simulated, but had no effect on the temperature at which infection risk peaked when *B. glabrata* and *B. alexandrina* were simulated. Absolute infection risk was lower at most temperatures when *B. alexandrina* were simulated, compared to when *B. pfeifferi* were simulated, and much lower at all temperatures when *B. glabrata* were simulated. Doubling snail mortality rates more than halved absolute infection risk for all three species at all temperatures.

Our findings suggest that in most areas where *B. pfeifferi* is the main intermediate host for *S. mansoni*, infection risk will fall as temperatures increase. Infection risk may increase dramatically at the cooler limits of *B. pfeifferi*'s range however. In general, the results suggest that infection risk will also fall in many areas where *B. alexandrina* is the main intermediate host, although it may increase substantially in areas and/or seasons where mean temperatures are currently below around 19.0°C. Where *B. glabrata* is the main intermediate host, infection risk may increase in a much higher proportion of areas and/or seasons, however increases in risk may be smaller. The exception to this is if *B. glabrata* spreads to areas that are currently too cold for it, as infection risk is relatively high even at the coolest temperatures suitable for *B. glabrata*. We will explore this further in future work, running the models using water temperatures from climate projections in regions where the snail species are found. This will also enable us to validate the models by comparing model output for the present day with empirical data from *S. mansoni* prevalence surveys.

The models also suggest that, all else being equal, infection risk may be lower in areas where *B. glabrata* is the intermediate host than in areas where *B. pfeifferi* or *B. alexandrina* are the intermediate host. These results should be interpreted with caution, however, for two reasons. The first is that much of the difference in absolute infection risk is due to the fact that at moderate temperatures simulated *B. glabrata* have much higher mortality rates than simulated *B. pfeifferi* and *B. alexandrina* (figure 2e), resulting in fewer snails surviving the prepatent period (figure 4c), and in lower median lifetime cercariae production by infectious snails (figure 4d). Absolute infection risks in the *B. glabrata* model with standard mortality rates are much closer to absolute infection risks in the *B. pfeifferi* and *B. alexandrina* models with mortality rates doubled. Mortality rates in the model are estimated from mortality rates in laboratory experiments however, and it is therefore possible that the much higher mortality rates of *B. glabrata* were due to different experimental conditions only, and not due to genuine differences between the species.

Secondly, the model does not take into account differences between snail species in susceptibility to *S. mansoni* infection or in cercariae production rates. This is because susceptibility appears to vary as greatly within snail species as it does between species, and there is large variation between individual snails of the same species in cercariae production rates (see Model Parameterisation S1). This should have no effect on the relationship between water temperature and relative infection risk, and very little effect on numbers of snails, but could have had a large effect on absolute infection risk in the models.

The model was parameterised using the best available data for each snail species. We consider there are two main limitations of the data. First, in the vast majority of the experiments that informed model parameterisation, snails were identified using morphological methods only. This means that it is possible that not all of the data were from the species of snail that they were reported to be from, as there can be discrepancies between the molecular and morphological classification of *Biomphalaria* specimens[27]. Second, experimental conditions (e.g. snail densities and feeding strategy) will have varied between the different experiments that produced the data used to parameterise the model. The effect of these differences has been minimised by using data from one experiment only to fit each model parameter wherever possible. This means that the overall relationship between water temperature and snail numbers and relative infection risk should not have been affected greatly by the different experimental conditions. It is important to note that experimental differences may have had a larger effect on absolute snail numbers and infection risk.

The results for all three species suggest that areas currently too cold for substantial schistosome transmission need to be monitored to prevent future epidemics occurring. Uninfected *B. pfeifferi* and *B. alexandrina* may be found in these areas already, whereas *B. glabrata* may need to spread to them from nearby, warmer areas. In areas where schistosomiasis is currently found, the species of intermediate host snail is likely to be important in determining the effects of increasing temperatures on schistosomiasis transmission. Schistosomiasis risk is likely to decrease in the majority of areas where *B. pfeifferi* is the intermediate host. Where *B. glabrata* is the host, risk may increase in some areas, but the increase in risk is unlikely to be large. In some areas where *B. alexandrina* is the intermediate host however, risk may more than double with only a 2°C increase in temperature. These findings highlight the importance of taking a more detailed and nuanced approach to modelling the association between climate change and schistosomiasis. Our approach can also be used as a framework for similar types of investigation into other parasites with life-cycles that include temperature sensitive stages and/or intermediate hosts.

Acknowledgments

The authors would like to thank Healthy Futures project partners who commented on a draft of the manuscript, and Marc Savy for writing an averaging script.

Author Contributions

Conceived and designed the experiments: NM. Performed the experiments: NM. Analyzed the data: NM. Wrote the paper: NM. Provided editorial revisions on manuscript: MB.

References

1. McCreesh N, Booth M (2013) Challenges in predicting the effects of climate change on *Schistosoma mansoni* and *Schistosoma haematobium* transmission potential. Trends Parasitol 29: 548–555.
2. Kabatereine NB, Brooker S, Tukahebwa EM, Kazibwe F, Onapa AW (2004) Epidemiology and geography of *Schistosoma mansoni* in Uganda: implications for planning control. Trop Med Int Health 9: 372–380.
3. Rubaihayo J, Moghusu E, Clouds P, Abaasa A (2008) Schistosomiasis transmission at high altitude crater lakes in Western Uganda. BMC Infect Dis 8: 110.
4. Lachish T, Tandlich M, Grossman T, Schwartz E (2013) High rate of schistosomiasis in travelers after a brief exposure to the high altitude Nyinambuga Crater Lake (1630 m), Uganda. Clin Infect Dis.
5. Stensgaard A-S, Utzinger J, Vounatsou P, Hürlimann E, Schur N, et al. (in press) Large-scale determinants of intestinal schistosomiasis and intermediate host snail distribution across Africa: Does climate matter? Acta Trop.
6. Brown D (1994) Freshwater Snails Of Africa And Their Medical Importance. London, UK: Taylor & Francis.
7. Appleton CC (1978) Review of literature on abiotic factors influencing the distribution and life cycle of bilharzia intermediate host snails. Malacological Review 11: 1–25.
8. Zhou XN, Yang GJ, Yang K, Wang XH, Hong QB, et al. (2008) Potential impact of climate change on schistosomiasis transmission in China. Am J Trop Med Hyg 78: 188–194.
9. Mangal TD, Paterson S, Fenton A (2008) Predicting the impact of long-term temperature changes on the epidemiology and control of schistosomiasis: a mechanistic model. PLoS One 3: e1438.
10. Mangal TD (2009) Developing spatio-temporal models of schistosomiasis transmission with climate change [PhD]: University of Liverpool.
11. Martens WJM, Jetten TH, Rotmans J, Niessen LW (1995) Climate change and vector-borne diseases: A global modelling perspective. Global Environ Change 5: 195–209.
12. Martens WJM, Jetten TH, Focks DA (1997) Sensitivity of malaria, schistosomiasis and dengue to global warming. Clim Change 35: 145–156.
13. McCreesh N, Booth M (2014) The Effect of Increasing Water Temperatures on *Schistosoma mansoni* Transmission and *Biomphalaria pfeifferi* Population Dynamics: An Agent-Based Modelling Study. PLoS One 9(7): e101462. doi:10.1371/journal.pone.0101462
14. Thomas JD, Tait AI (1984) Control of the snail hosts of schistosomiasis by environmental manipulation: A field and laboratory appraisal in the Ibadan area, Nigeria. Philosophical Transactions of the Royal Society of London B, Biological Sciences 305: 201–253.
15. Appleton C (1977) The freshwater Mollusca of Tongaland, with a note on molluscan distribution in Lake Sibaya. Annals of the Natal Museum 23: 129–144.
16. Dupouy J, Mimpfoundi R (1986) Cycle biologique de *Biomphalaria pfeifferi* (Krauss) dans les milieux anthropisés du district de Yaoundé (Cameroun). Compte rendu des séances de la société de biogéographie 62: 47–60.
17. Wibaux-Charlois M, Yelnik A, Ibrahima H, Same-Ekobo A, Ripert C (1982) Etude épidémiologique de la bilharziose à *Schistosoma haematobium* dans le périmètre rizicole de Yagoua (nord Cameroun). Distribution et écologie des hôtes intermédiaires. Bull Soc Pathol Exot Filiales 75: 72–93.
18. Gryseels B (1985) La répartition de *Biomphalaria* et la transmission de *Schistosoma* dans la plaine de la Ruzizi, Burundi: étude préliminaire. Ann Soc belge Méd trop 65: 49–58.
19. Madsen H, Daffalla A, Karoum K, Frandsen F (1988) Distribution of freshwater snails in irrigation schemes in the Sudan. J Appl Ecol: 853–866.
20. Dennis E, Vorkpor P, Holzer B, Hanson A, Saladin B, et al. (1983) Studies on the epidemiology of schistosomiasis in Liberia: the prevalence and intensity of schistosomal infections in Bong County and the bionomics of the snail intermediate hosts. Acta Trop 40: 205.
21. Doumenge J, Mott K, Cheung C, Villenave D (1987) Atlas of the global distribution of schistosomiasis. Bordeaux, France: Presses Universitaires de Bordeaux.
22. Dazo BC, Hairston NG, Dawood IK (1966) The ecology of *Bulinus truncatus* and *Biomphalaria alexandrina* and its implications for the control of bilharziasis in the Egypt-49 project area. Bull World Health Organ 35: 339–356.
23. Paraense WL (2001) The schistosome vectors in the Americas. Mem Inst Oswaldo Cruz 96 Suppl: 7–16.
24. Sturrock RF (1973) Field studies on the transmission of *Schistosoma mansoni* and on the bionomics of its intermediate host, *Biomphalaria glabrata*, on St. Lucia, West Indies. Int J Parasitol 3: 175–194.
25. Jobin WR (1970) Population dynamics of aquatic snails in three farm ponds of Puerto Rico. Am J Trop Med Hyg 19: 1038–1048.
26. IPCC (2007) Summary for Policymakers. Cambridge, United Kingdom and New York, NY, USA.
27. Plam M, Jørgensen A, Kristensen TK, Madsen H (2008) Sympatric *Biomphalaria* species (Gastropoda: Planorbidae) in Lake Albert, Uganda, show homoplasies in shell morphology. Afr Zool 43: 34–44.

2

Vulnerability of Breeding Waterbirds to Climate Change in the Prairie Pothole Region, U.S.A.

Valerie Steen[1,2,3]*, Susan K. Skagen[1], Barry R. Noon[2,3]

1 United States Geological Survey, Fort Collins Science Center, Fort Collins, Colorado, United States of America, **2** Department of Fish, Wildlife, and Conservation Biology, Colorado State University, Fort Collins, Colorado, United States of America, **3** Graduate Degree Program in Ecology, Colorado State University, Fort Collins, Colorado, United States of America

Abstract

The Prairie Pothole Region (PPR) of the north-central U.S. and south-central Canada contains millions of small prairie wetlands that provide critical habitat to many migrating and breeding waterbirds. Due to their small size and the relatively dry climate of the region, these wetlands are considered at high risk for negative climate change effects as temperatures increase. To estimate the potential impacts of climate change on breeding waterbirds, we predicted current and future distributions of species common in the PPR using species distribution models (SDMs). We created regional-scale SDMs for the U.S. PPR using Breeding Bird Survey occurrence records for 1971–2011 and wetland, upland, and climate variables. For each species, we predicted current distribution based on climate records for 1981–2000 and projected future distributions to climate scenarios for 2040–2049. Species were projected to, on average, lose almost half their current habitat (-46%). However, individual species projections varied widely, from +8% (Upland Sandpiper) to -100% (Wilson's Snipe). Variable importance ranks indicated that land cover (wetland and upland) variables were generally more important than climate variables in predicting species distributions. However, climate variables were relatively more important during a drought period. Projected distributions of species responses to climate change contracted within current areas of distribution rather than shifting. Given the large variation in species-level impacts, we suggest that climate change mitigation efforts focus on species projected to be the most vulnerable by enacting targeted wetland management, easement acquisition, and restoration efforts.

Editor: R. Mark Brigham, University of Regina, Canada

Funding: We thank the U.S. Geological Survey, National Climate Change and Wildlife Science Center (NCCWSC) and the Plains and Prairie Pothole Landscape Cooperative (PPPLCC) for research funding. The funders had no role in study design, data collection and analysis, decision to publish, or preparation of the manuscript.

Competing Interests: The authors have declared that no competing interests exist.

* E-mail: valerie.steen@gmail.com

Introduction

The Prairie Pothole Region of north-central North America (central Iowa, U.S.A. to central Alberta, Canada; 900,000 km^2) contains one of the largest wetland areas (40,000 km^2) in the world [1]. Historically, most conservation activities have focused on sustaining extensive, high quality duck habitat because of the associated recreational value of duck-hunting across the U.S. [2]. Increasingly, emphasis is being placed on the diversity of ecosystem services offered by prairie pothole wetlands, including carbon sequestration, flood control, groundwater recharge, water quality improvement, and biodiversity [2]. This includes increasing attention to all 115 species of breeding or migrating waterbirds that depend on the region [3].

Successful management of species requires knowledge of habitat preferences. Strategic management of species also requires identifying those species most vulnerable to future threats. Land conversion continues to be a direct threat to waterbird habitat, but climate change will likely exacerbate loss and interact with changes in land cover. Climate models for the Prairie Pothole Region project increasing temperatures and slight or no increases in precipitation, indicating drier conditions affecting hydroperiods, and the extent and quality of wetland habitat [4,5].

Prairie pothole wetlands are susceptible to climatic variation through impacts on wetland hydroperiod, vegetative condition, and water depth in combination with static factors such as basin size [5]. Well-documented causal relations between past variability in wetland condition and extent and waterbird numbers provide insights to future change in waterbird populations under climate change. In dry years, with fewer wet basins, breeding populations of waterbirds are significantly reduced [6,7]. Building on these causal relations, Sorenson et al. [8] projected population changes for waterfowl under future warming scenarios. Their projections indicated that by 2060 duck populations would be half of their current level. Johnson et al. [5] used mechanistic models relating climate to marsh vegetation dynamics, and projected that the Prairie Potholes in North and South Dakota will be too dry to produce suitable wetland vegetative conditions for breeding ducks in the future.

To address how climate change may impact waterbirds in the Prairie Pothole Region, we created empirically-based species distribution models for a focal group of breeding wetland-associated birds. We related bird occurrence (presence/absence) to climate and land cover predictors. As a species' occurrence varies from year to year in response to dynamic wetland conditions, we used multiple years of bird survey data across 41

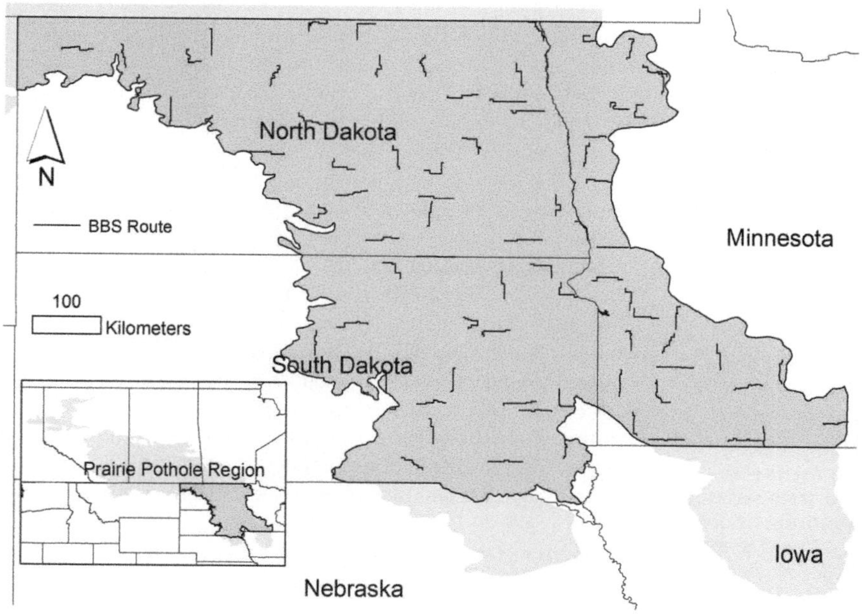

Figure 1. Bird occurrence data were obtained from 77 Breeding Bird Survey (BBS) routes throughout the Prairie Pothole Region (PPR) of North Dakota, South Dakota, and Minnesota. Climate-based projections were also made to the PPR of Iowa.

years, a period that included years of drought and years of heavy precipitation. Although we did not explicitly model wetland condition, we used Random Forests, an ensemble decision tree approach which can capture the interactions between climate variability and the state of wetland basins that drive wetland condition [9]. We projected future waterbird occurrence using species distribution models and future climate projections. To assess how climate change may reduce or expand current suitable habitat, for each species we compared the projections of future distribution to their predicted current distribution, and produced a quantitative estimate of how much habitat would be lost or gained under various climate change scenarios. Additionally, we compared our future projections of waterbird species response to a historic dry period.

Methods

Study Area

The study area (320,000 km^2) was the 45% of the Prairie Pothole Region within four U.S. states (North Dakota, South Dakota, Minnesota, Iowa, Figure 1). The study was restricted to the four states because of available and consistent land cover and downscaled climate data. We excluded Iowa from training the model because too few wetlands remain there to usefully inform the species distribution model, although we did include it in model predictions.

Water-filled glacial depressions termed *potholes* are characteristic of this region and can reach densities greater than 40 per km^2 [10]. Since European settlement, these wetlands have been extensively converted to cropland, with wetland losses greatest in

Table 1. Thirty-one climate and land cover variables used in species distribution models.

Climate		Land Cover	
Temperature	Precipitation	Wetland	Upland
Spring (spr)	Spring	Temporary (temp)	Cropland (crop)
Winter (wint)	Winter	Seasonal (seas)	Grassland (grass)
Fall	Fall	Semipermanent (semi)	Developed (devel)
Summer (sum)	Summer	Lake	Tree
Yearly (1yr)	Yearly	River	
5-year (5yr)	5-year	Shrub	
10-year (10yr)	10-year	Forested (forest)	
5-year std. dev. (5yr_sd)	5-year std. dev.	Total palustrine (pal)	
10-year std. dev. (10yr_sd)	10-year std. dev.	Total	

Temperature calculations were based on averages, while precipitation calculations were based on totals. Land cover variables were based on composition (proportion of total) of that cover type in the landscape. Wetland land cover was apportioned by wetland regime. Total palustrine wetland summed temporary, seasonal, and semipermanent wetlands. Total wetland summed all wetland regimes. Cropland described land planted with crops or fallowed. Grassland included native prairie, conservation reserve program (CRP) land, and hayland.

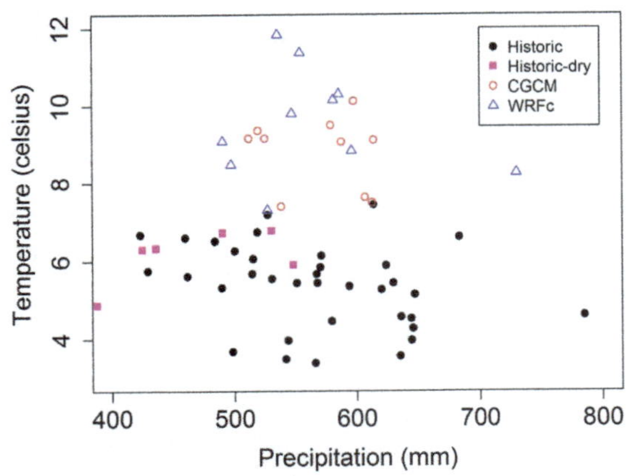

Figure 2. Total precipitation versus average temperature by "bird year" (June of year x-1 to May of year x) for the study area (see Figure 1) for the time periods used to train species distribution models and project future distributions. Historic points showed the years and locations from 1971–2011 used to train the species distribution models with six years withheld. The six years were a dry period from 1987–1992 shown as 'historic-dry'. CGCM and WRFc show two sets of climate projections for the ten year period 2040-49.

the eastern portion of the Prairie Pothole Region: Minnesota (85%), Iowa (95%) and North (49%) and South Dakota (35%) [11,12]. Losses of surrounding upland prairie habitats follow a similar geographic gradient (greatest in the eastern portion of the Prairie Pothole Region) but have been even more severe than wetland losses [3].

Species Occurrence

We obtained species occurrence (presence/absence) data from the North American Breeding Bird Survey (BBS) [13] for waterbirds species with a prevalence of ≥ 0.05. The BBS consists of >3000 routes on secondary roads throughout the continental U.S. and southern Canada. Routes are surveyed once annually during June between 04.45 AM and 10.00 AM. Route locations generally remain the same year after year, although not all routes are surveyed each year and there is variation in the year when a route is initiated. BBS routes are 39.4-km long with 50 stops spaced 0.8 km apart. Three-minute point-count surveys are conducted at each stop. BBS survey data are available for each species and summarized at route totals or 10-stop route segments (https://www.pwrc.usgs.gov/bbs/).

In our study area, BBS surveys took place from late May to early July. This interval extensively overlapped the breeding season (nest-building through brood rearing) for the majority of wetland-dependent species we evaluated. Ten species usually nest during this period and three species occupy brood-rearing habitats. The remainder of the species are engaged in behaviors ranging from incubation to brood-rearing. In addition, seven waterfowl species may be molting body or primary feathers near the end of the survey period.

Even though the breeding cycles of wetland-dependent birds in the Prairie Potholes are not completely synchronous, we believe the BBS survey methods accurately document the presence of all regularly occurring species. Our confidence is based on the overlap between the geographic extent of our survey data, the distribution of our focal species during the breeding season, and the timing of the surveys. The result, we believe, is that the likelihood of correctly documenting the presence of a species was comparable across species, routes, and survey years.

We used data (1971–2011) from high-quality surveys (reported by the BBS as "run type 1") for 77 routes: these were conducted within the correct survey window and not during poor weather. Due to the potential for extensive variation along a route in habitat types, we chose one 10-stop section to model habitat associations rather than use data from the entire route. To accommodate different timing of peak detectability by species, we chose either the first or third section for a species depending which section had

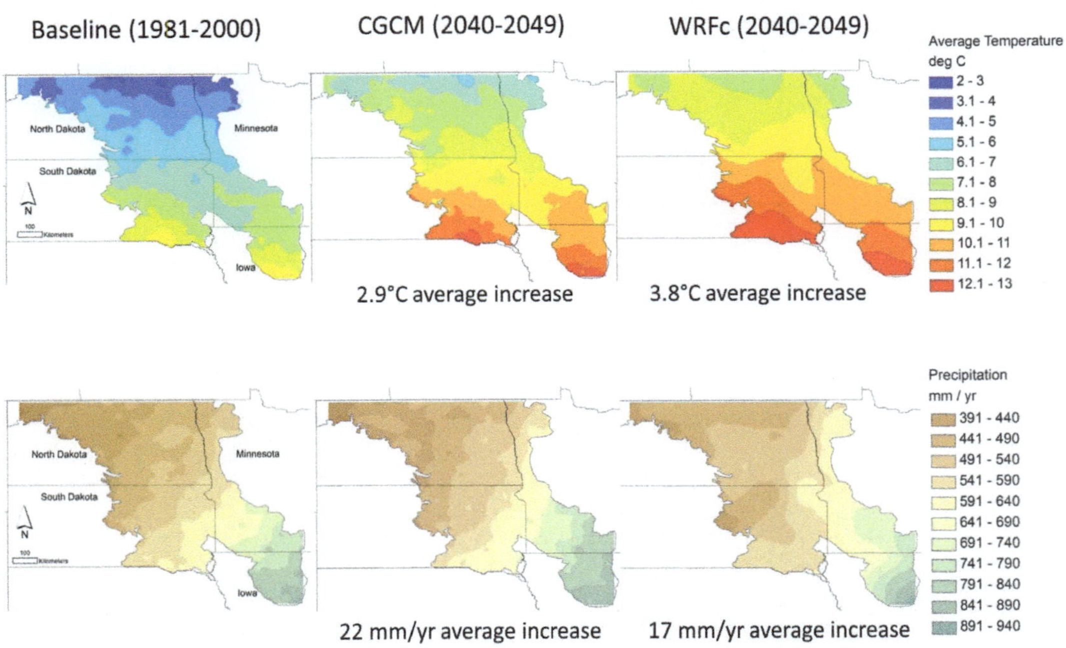

Figure 3. Temperature and precipitation for baseline and two future climate projections for the prairie potholes of North Dakota, South Dakota, Minnesota, and Iowa.

Table 2. Model evaluation showing species prevalence (proportion of data points with species present) for each dataset, and predictive accuracy using the classification matrix and area under the curve (AUC) values.

		Prevalence			To withheld data						To withheld dry years	
Common Name	**Scientific Name**	**Training data (n = 974)**	**Withheld data (n = 817)**	**Dry years data (n = 139)**	**True positive (n)**	**False positive (n)**	**True negative (n)**	**False negative (n)**	**Overall accuracy (%)**	**AUC**	**Overall accuracy (%)**	**AUC**
Canada Goose	Branta canadensis	0.18	0.18	0.09	92	211	457	60	67	0.71	76	0.61
Wood Duck	Aix sponsa	0.06	0.07	0.04	24	187	576	30	73	0.70	84	0.81
Gadwall	Anas strepera	0.11	0.12	0.00	82	183	534	18	75	0.89	74	0.87
American Wigeon	Anas americana	0.05	0.07	0.04	36	129	632	20	82	0.84	71	0.66
Mallard	Anas platyrhynchos	0.60	0.59	0.55	343	101	231	142	70	0.77	57	0.65
Blue-winged Teal	Anas discors	0.36	0.38	0.24	259	139	368	51	77	0.86	63	0.74
Northern Shoveler	Anas clypeata	0.17	0.18	0.09	117	189	481	30	73	0.84	68	0.81
Northern Pintail	Anas acuta	0.21	0.24	0.10	153	161	463	40	75	0.85	69	0.85
Green-winged Teal	Anas crecca	0.05	0.05	0.03	29	180	592	19	77	0.85	72	0.58
Redhead	Aythya americana	0.15	0.12	0.03	87	126	593	11	83	0.91	81	0.95
Ruddy Duck	Oxyura jamaicensis	0.13	0.11	0.06	83	115	612	7	85	0.94	83	0.92
Pied-billed Grebe	Podilymbus podiceps	0.22	0.24	0.11	166	150	474	27	78	0.90	73	0.81
Double-crest. Cormorant	Phalacrocorax auritus	0.09	0.09	0.05	64	201	542	10	74	0.84	68	0.75
American Bittern	Botaurus lentiginosus	0.24	0.27	0.17	190	173	421	33	75	0.84	64	0.76
Great Blue Heron	Ardea herodias	0.07	0.06	0.04	29	180	592	19	76	0.69	88	0.67
Sora	Porzana carolina	0.28	0.27	0.16	178	154	443	42	76	0.86	63	0.71
American Coot	Fulica americana	0.28	0.29	0.15	201	117	467	32	82	0.90	72	0.86
Killdeer	Charadrius vociferus	0.88	0.92	0.86	507	27	41	242	67	0.69	69	0.59
Upland Sandpiper	Bartramia longicauda	0.49	0.48	0.53	308	128	300	81	74	0.82	81	0.84
Willet	Tringa semipalmata	0.16	0.15	0.13	111	163	532	11	79	0.91	70	0.90
Marbled Godwit	Limosa fedoa	0.19	0.18	0.24	121	166	505	28	76	0.88	75	0.90
Wilson's Snipe	Gallinago delicata	0.19	0.20	0.14	140	163	494	23	77	0.90	72	0.80
Wilson's Phalarope	Phalaropus tricolor	0.10	0.12	0.04	79	165	550	23	77	0.86	66	0.84
Franklin's Gull	Leucophaeus pipixcan	0.10	0.11	0.09	64	189	538	26	74	0.81	73	0.86
Ring-billed Gull	Larus delawarensis	0.12	0.16	0.12	103	179	510	25	75	0.84	66	0.78
Black Tern	Chlidonias niger	0.17	0.17	0.07	112	172	505	28	76	0.85	72	0.89
Sedge Wren	Cistothorus platensis	0.27	0.26	0.17	142	168	434	73	71	0.76	81	0.71
Marsh Wren	Cistothorus palustris	0.23	0.24	0.17	166	130	493	28	81	0.89	73	0.78
Common Yellowthroat	Geothlypis trichas	0.83	0.84	0.81	444	41	91	241	65	0.74	60	0.76
Song Sparrow	Melospiza melodia	0.66	0.66	0.57	425	105	172	115	73	0.78	74	0.85

Table 2. Cont.

Common Name	Scientific Name	Prevalence			To withheld data						To withheld dry years	
		Training data (n = 974)	Withheld data (n = 817)	Dry years data (n = 139)	True positive (n)	False positive (n)	True negative (n)	False negative (n)	Overall accuracy (%)	AUC	Overall accuracy (%)	AUC
Yellow-headed Blackbird	Xanthocephalus xanthocephalus	0.54	0.56	0.56	348	86	321	62	82	0.88	64	0.76

The positive and negative rates show the models ability to correctly predict presence and absence data points in the withheld data based on a 0.5 threshold for classification. Overall accuracy was the proportion of true positive and true negative predictions. AUC critical value = 0.70. Dry year predictions were based on models trained without the dry years.

higher detections for that species. Routes were consistently surveyed from stop one, starting around 04.45 AM, to stop 50, ending around 09.00 AM. For all but two species, the first or third section had their highest or second highest number of detections. 'Presence' was defined as ≥1 detection at a minimum of one stop along the route segment. We identified focal species based on their prevalence (section-level occurrence rate) with species detected at fewer than 5% of route sections not included.

Land Cover Data

We extracted land cover variables (Table 1) for North and South Dakota from a GIS raster layer created by the U.S. Fish and Wildlife Service (USFWS; USFWS, Region 6 Habitat and Population Evaluation Team, unpublished data); for Minnesota and Iowa from a GIS raster layer created by the USFWS (USFWS, Region 3 Habitat and Population Evaluation Team, unpublished data); and for uplands in the southern portion of the Iowa Prairie Pothole Region from the 1992 National Land Cover Dataset (NLCD). The USFWS data layers were based on Landsat Thematic Mapper Satellite imagery of scenes from 2000–2003, and the NLCD on scenes from the early to mid-1990s. All raster layers were at a 30-m resolution.

Wetland basins in the land cover layers were areas of contiguous wetland extent. The basins were derived from a GIS wetland polygons layer (USFWS National Wetlands Inventory, NWI) where multiple contiguous polygons of differing wetland regimes were dissolved to a single polygon. The USFWS Habitat and Population Evaluation Team followed the procedures of Cowardin et al. [14] and Johnson and Higgins [15] to describe each wetland basin by its most permanent water regime: temporary, seasonal, semipermanent, lake, and river. Generally, temporary wetlands are flooded in spring for a few weeks after snow-melt, seasonal wetlands hold water until summer, and semipermanent wetlands hold water through the growing season; lake and rivers are permanently flooded wetlands [16]. NWI data are based on aerial photographs taken in the late 1970's and early 1980's. Where water pixels extended beyond NWI polygons, they were labeled as water (wetland regimes, see Cowardin et al. [17]). We characterized wetlands into nine classes: temporary, seasonal, semipermanent, lake, river, forested, shrub, total, and total palustrine (Table 1). Total wetland was the combined composition of temporary, seasonal, semipermanent, lake, river, forested, and shrub; total palustrine wetland was temporary, seasonal, and semipermanent.

We described upland habitat using four land cover classes: cropland, grassland, tree, and developed (Table 1). Cropland included areas planted with crops or fallowed. Grassland included native prairie, planted grasses (i.e. previously cropped but now planted with grasses and forbs such as Conservation Reserve Program land), and hayland. Developed land cover included primarily residential areas. Tree habitat included small sections or rows of trees and occasionally areas of forest. Accuracy of the upland land cover data for North and South Dakota, assessed in 2007, was > 90% (M. Estey, personal communication).

To describe habitat associations for our focal waterbird species, we explored composition-based single scale models. In both single-scale and multi-scale models, composition-based predictors, expressed as the amount of a land cover type within a given area, perform better than their distance-based counterparts, expressed as the distance from a sampling location to a land cover type [18]. We used ArcMap 10.0 to calculate land cover composition for the four upland and nine wetland classes at six spatial scales for the BBS route segments. The scales ranged from 335 ha to 32,200 ha and were based on buffering the segments with radii: 0.2-km, 0.4-

Table 3. Values report projected changes in occurrence in the 2040's, relative to 1981–2000 (baseline).

Species	Change in occurrence (%)		
	CGCM	WRFc	Average
Canada Goose	−76	−66	−71
Wood Duck	−70	−37	−54
Gadwall	49	−87	−19
American Wigeon	−58	−71	−65
Mallard	−30	−23	−27
Blue-winged Teal	−9	−4	−7
Northern Shoveler	−51	−62	−57
Northern Pintail	−45	−37	−41
Green-winged Teal	−46	−18	−32
Redhead	−42	−35	−39
Ruddy Duck	−30	−31	−31
Pied-billed Grebe	−40	−18	−29
Double-crested Cormorant	−11	−20	−16
American Bittern	−42	−42	−42
Great-blue Heron	−72	−82	−77
Sora	−94	−98	−96
American Coot	−38	−38	−38
Killdeer	5	0	3
Upland Sandpiper	8	7	8
Willet	−43	−58	−51
Marbled Godwit	−53	−61	−57
Wilson's Snipe	−99	−100	−100
Wilson's Phalarope	−42	−60	−51
Franklin's Gull	−93	−98	−96
Ring-billed Gull	−39	−83	−61
Black Tern	−67	−64	−66
Sedge Wren	−71	−60	−66
Marsh Wren	−40	−42	−41
Common Yellowthroat	−26	−35	−31
Song Sparrow	−38	−41	−40
Yellow-headed Blackbird	−24	−25	−25

Species distribution models projected species occurrence to 4,957 8-km grid points using climate data for the baseline period and two climate projections (CGCM and WRFc). Negative values indicated the proportion of occupied grid cells for each species, projected to be unoccupied in the future. Positive values indicated the proportion by which occupied cells were projected to increase.

km, 1-km, 2-km, 4-km, and 8-km. BBS surveyors record all birds detected within 0.4-km of the survey point. Thus, assuming no decline in detection probability with increasing distance and no landscape effect, we expected 0.4-km to be the appropriate scale to relate land cover to bird occurrence. However, some waterbird species may decline quickly in detection probability with increasing distance from the survey point—therefore, we also explored a 0.2-km scale. Because other species may respond to land cover heterogeneity at broader extents, we also explored a range (1-km to 8-km) of landscape scales. Land cover data were assumed static across current and future years.

Climate Covariates

We used PRISM (PRISM, Parameter-elevation Regressions on Independent Slopes Model) data for historical climate records. These data are available at a 4-km grid scale as monthly temperature and precipitation and were rescaled to an 8-km grid to match the scale of the projected climate data [19].

Using monthly values of precipitation and temperature, we derived 18 climate variables (Table 1). We calculated mean temperatures for grid points by averaging the minimum and maximum monthly temperatures over different time periods. We delineated seasons as summer (June-August), fall (September-November), winter (December-February), and spring (March-May). We defined year as ending in May to correspond to the June bird surveys. We included seasonal and annual variables because both seasonal and annual climate explain annual variation in the number of prairie pothole wetlands holding water [20]. For semipermanent wetlands and (especially) lakes, wet wetland count is related to long-term climate (at least 3 years) [21]. We included 5-year and 10-year precipitation and mean temperature variables as proxies for long-term climate effects. We also included the

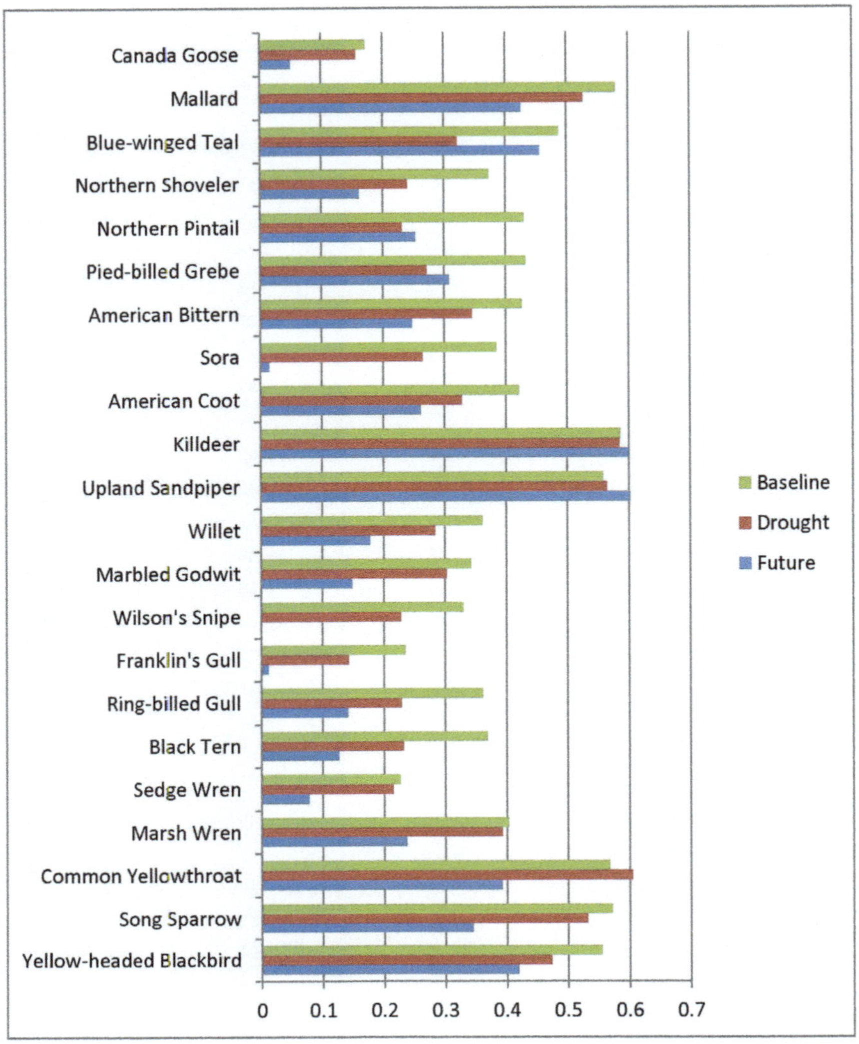

Figure 4. Mean rates of predicted species occurrence at 4,957 8-km grid points. Baseline rate was based on 1981–2000 climate records. Dry years showed predicted occurrence rates for the drought period, 1987–1992. Future rates were based on the average projections of two future climate datasets (CGCM-A2 and WRFc) for 2040–2049.

variances in 5-year and 10-year precipitation and temperature data, because large values of these variables may indicate that wetlands are cycling through wet and dry phases, driving dynamic vegetative conditions [5]. Climate data from 1971–2011 were used to construct the baseline species distribution models. The species distribution models predicted to climate data from 1981–2000 and 2040–2049 to create current and future projections, respectively, of species occurrence.

Future Climate Data

We used statistically downscaled and high resolution climate projections. Statistically downscaled data refine projections from global circulation models (GCMs) using an empirical relationship to local physiography (e.g. topography and water bodies). These projections assume relations will hold into the future and are less computationally intensive than high resolution models. High resolution models nest a dynamical Regional Climate Model within the GCM, re-running the GCM based on mesoscale (a few to a few hundred kilometers) physical relationships with topographical features and surface characteristics [22]. The high resolution projections circumvent the problem associated with lack of "stationarity" when the relationships between GCM output and the fine-scale climate change over time.

The statistically downscaled projections were based on data obtained from output of GCM CGCM3.1MR (Canadian Centre for Climate Modeling and Analysis Third Generation Coupled Global Climate Model Version 3.1, Medium Resolution) [23] and downscaled to an 8-km grid. The high resolution models used the Community Climate System Model (CCSM) to set the boundary condition and a mesoscale model, Weather Research and Forecasting model (WRF) to refine the data to a 36-km regional scale (J. Stamm, personal communication) [24]. Given that we expected high spatial correlation for monthly temperature and precipitation, we interpolated the 36-km data to the 8-km grid [25]. Both climate models were run with a mid-high IPCC emissions scenario, A2 [26]. The high resolution projections were available for 2000–2049, and the statistically downscaled projections were available for 2000–2100. We term the statistically downscaled data "CGCM" after the GCM these data are based on, and we term the dynamically downscaled data "WRFc" after the mesoscale model these data are based on.

Table 4. Variable importance for Random Forest species distribution models.

Rank	Canada Goose	Wood Duck	Gadwall	American Wigeon	Mallard	Blue-winged Teal	Northern Shoveler	Northern Pintail	Green-winged Teal	Redhead	Ruddy Duck
1	W-lake(+)	W-temp(-)	T-5yr_sd(-)	W-lake(+)	W-pal(+)	W-total(+)	W-pal(+)	U-tree(-)	U-tree(-)	W-semi(+)	W-semi(+)
2	P-wint(+)	P-wint(+)	P-10yr_sd(-)	U-tree(-)	W-semi(+)	W-pal(+)	W-total(+)	W-pal(+)	P-wint(+)	W-total(+)	W-total(+)
3	W-river(-)	U-grass(-)	U-tree(-)	W-semi(+)	W-river(-)	W-seas(+)	W-semi(+)	P-10yr(-)	W-lake(+)	W-lake(+)	W-lake(+)
4	T-sum(-)	U-crop(-)	T-spr(-)	W-temp(-)	W-total(+)	W-semi(+)	W-temp(+)	W-seas(+)	W-seas(+)	U-tree(-)	W-pal(+)
5	T-spr(-)	U-tree(+)	U-grass(+)	P-5yr(-)	T-10yr_sd(+)	U-grass(+)	W-seas(+)	W-semi(+)	U-crop(-)	W-temp(-)	U-tree(-)
6	T-10yr_sd(+)	U-devel(-)	P-10yr(m)	P-spr(-)	W-temp(+)	W-temp(+)	U-tree(-)	P-5yr(-)	P-fall(~)	W-pal(+)	W-seas(+)
7	T-10yr(m)	W-semi(+)	U-crop(-)	W-pal(+)	U-tree(-)	U-tree(-)	U-crop(-)	U-grass(+)	W-shrub(-)	W-river(-)	W-river(-)
8	T-5yr(m)	T-10yr_sd(+)	T-10yr_sd(+)	W-total(+)	T-spr(-)	W-river(-)	U-grass(+)	W-river(-)	W-temp(+)	W-seas(+)	W-temp(-)
9	T-1yr(-)	T-sum(~)	W-total(+)	P-10yr(-)	U-grass(+)	U-crop(-)	T-spr(-)	T-spr(-)	P-5yr(-)	U-crop(-)	U-crop(-)
10	W-total(+)	W-total(+)	W-river(-)	P-1yr(-)	T-5yr_sd(+)	P-10yr(-)	P-10yr(-)	W-total(+)	T-5yr(-)	U-grass(+)	U-grass(+)
Rank	Pied-billed Grebe	Double-crested Cormorant	American Bittern	Great-blue Heron	Sora	American Coot	Killdeer	Upland Sandpiper	Willet	Marbled Godwit	
1	W-semi(+)	W-semi(+)	U-crop(-)	T-spring(-)	P-10yr(-)	W-semi(+)	W-semi(+)	W-seas(+)	U-tree(-)	U-tree(-)	
2	W-total(+)	U-tree(~)	W-total(+)	P-10yr(+)	T-sum(-)	W-total(+)	W-pal(+)	U-grass(+)	P-10yr(-)	U-crop(-)	
3	W-pal(+)	W-temp(-)	U-grass(+)	U-tree(+)	T-5yr(-)	W-pal(+)	W-river(-)	U-tree(-)	P-5yr(-)	W-lake(+)	
4	W-seas(+)	T-10yr_sd(+)	W-semi(+)	P-5yr(+)	W-seas(+)	W-seas(+)	U-tree(-)	P-10yr(-)	U-grass(+)	W-total(+)	
5	U-crop(-)	W-lake(+)	W-pal(+)	P-spr(+)	T-1yr(-)	W-river(-)	W-total(+)	W-pal(+)	W-pal(+)	P-10yr(-)	
6	W-lake(+)	T-5yr(~)	W-seas(+)	P-1yr(+)	W-river(-)	P-10yr(-)	W-seas(+)	U-devel(-)	W-semi(+)	U-grass(+)	
7	T-10yr_sd(+)	P-5yr(+)	P-10yr(-)	T-10yr(-)	W-total(+)	W-lake(+)	U-devel(-)	W-river(-)	W-total(+)	T-10yr(-)	
8	U-grass(+)	P-spr(-)	T-5yr(-)	W-shrub(+)	W-pal(+)	U-crop(-)	P-wint(~)	U-crop(-)	U-crop(-)	P-5yr(-)	
9	T-sum(-)	P-fall(~)	T-sum(-)	P-sum(+)	P-5yr(-)	U-grass(+)	W-temp(+)	W-semi(+)	T-10yr(-)	W-semi(+)	
10	U-tree(-)	T-5yr_sd(-)	T-1yr(-)	T-1yr(-)	U-grass(+)	T-5yr(-)	U-grass(+)	P-5yr(-)	W-river(-)	W-river(-)	
Rank	Wilson's Snipe	Wilson's Phalarope	Franklin's Gull	Ring-billed Gull	Black Tern	Sedge Wren	Marsh Wren	Common Yellow-throat	Song Sparrow	Yellow-headed Blackbird	
1	T-5yr(-)	U-tree(-)	U-tree(+)	W-semi(+)	W-seas(+)	P-10yr_sd(-)	W-semi(+)	W-total(+)	U-grass(-)	W-semi(+)	
2	T-10yr(-)	U-grass(+)	W-pal(+)	W-lake(+)	W-pal(+)	U-tree(+)	W-total(+)	W-temp(+)	U-crop(+)	W-pal(+)	
3	W-total(+)	U-crop(-)	T-10yr(-)	W-pal(+)	W-total(+)	U-crop(-)	W-pal(+)	W-seas(+)	U-tree(+)	W-total(+)	
4	U-crop(-)	P-10yr(-)	W-total(+)	W-total(+)	W-semi(+)	P-10yr(+)	U-crop(-)	W-pal(+)	P-10yr(+)	W-seas(+)	
5	W-pal(+)	P-5yr(-)	W-seas(m)	U-tree(-)	W-temp(+)	T-sum(-)	W-seas(+)	U-tree(+)	W-semi(-)	U-grass(+)	
6	W-lake(+)	W-lake(m)	U-grass(~)	U-grass(+)	U-crop(-)	P-5yr(+)	U-grass(+)	W-semi(+)	P-sum(+)	P-10yr(-)	
7	T-sum(-)	W-semi(+)	W-temp(+)	P-5yr(-)	P-10yr(-)	T-10yr_sd(+)	W-lake(+)	U-devel(-)	T-10yr_sd(+)	U-crop(-)	
8	U-tree(~)	W-seas(+)	W-lake(+)	W-seas(+)	T-5yr(-)	W-pal(+)	T-sum(-)	W-lake(+)	T-sum(-)	T-sum(-)	
9	W-river(-)	P-spr(~)	U-crop(-)	W-river(-)	P-wint(+)	U-grass(+)	W-river(-)	T-5yr(-)	T-10yr(-)	U-tree(-)	
10	W-seas(~)	W-total(+)	P-sum(~)	T-5yr(-)	P-5yr(m)	P-1yr(+)	T-10yr(-)	P-5yr_sd(-)	P-5yr(+)	P-5yr(-)	

Top ten variables are shown in descending order of rank. Variable categories were denoted by W (wetland), U (upland), P (precipitation), and T (temperature). Signs indicated the relationship between the predictor and the species response: + (positive), - (negative), m (unimodel), and ~ (equivocal).

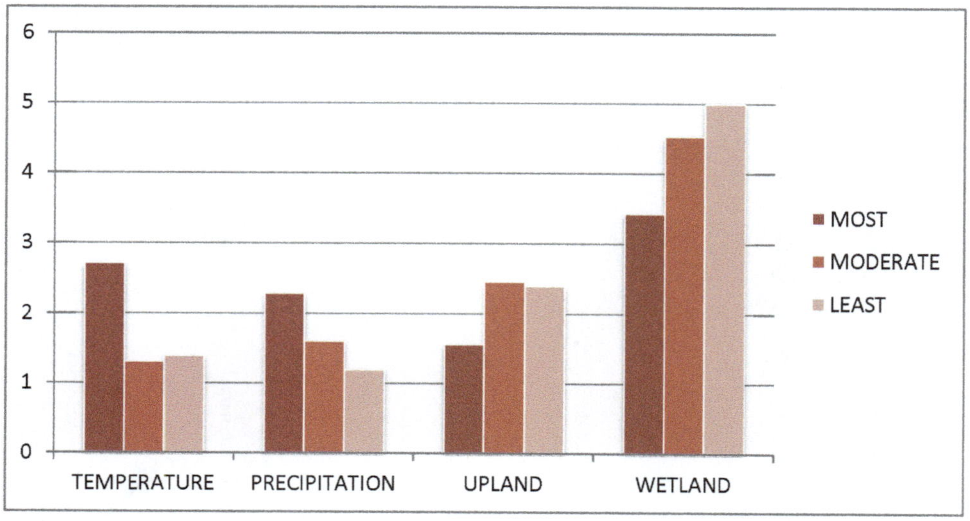

Figure 5. The frequency (y-axis) of variable type (x-axis) in top ten variables for waterbird species distribution models, grouped by species sensitivity to climate change. The most sensitive species were projected to lose ≥66% of their current habitat; moderately sensitive species 33–65%; and least sensitive <33%.

Species Distribution Models

We estimated a species distribution model (SDM) for each waterbird species, relating BBS occurrence records (1971–2011) to climate, and wetland and upland land cover (hereafter grouped as "land cover") predictor variables. We used climate for the same year as the occurrence record from the climate grid point nearest the BBS route segment and land cover surrounding the route segment. We defined occurrence as one or more detection per 10-stop segment by year. The spatial scale used in the final models for land cover calculations was chosen separately for each species based on model performance. We ran six models for each species based on the six different spatial scales of land cover and chose the model with the highest classification accuracy. We used a non-parametric machine learning approach, Random Forests, to create the SDMs [9]. We chose Random Forests because of its high predictive power, ability to model unspecified variable interactions and correlated variables, its ranking of variable importance, and its demonstrated use for bioclimatic species distribution models [27,28]. Random Forests uses an ensemble of classification (categorical response variable) or regression (continuous response variable) trees, each built with a subset of the data, to model the pattern between predictor variables and the response variable. We used permutation procedures to assess variable importance, a method based on reduction in predictive accuracy to internally withheld data when values of a given predictor variable are randomly shuffled. We report the top ten variables for each model. Although the choice of the number of variables to report is arbitrary, we expect the top ten will provide an adequate basis for comparing models.

We used the RandomForests package in R to create our models [29]. We specified 3000 trees which is a sufficiently large number of trees to capture any patterns in the data. Each tree was constructed with a bootstrapped subsample with replacement of the data records (BBS routes). Because the ratio of presence to absence was often skewed, particularly for either very abundant or rare species, we balanced the data by setting Random Forests to randomly use, for each tree, 25 records where the species was present and another 25 where the species was absent [30]. A subsample of five predictor variables was evaluated at each binary split in the tree algorithm.

We partitioned the BBS data in a number of ways to strengthen model evaluation and inference. First, we only excluded consecutive years of surveys to reduce the influence of temporal autocorrelation and maximize information content: the "main training set". The excluded data were used to validate the models created with these data: the "main test set". Second, we separated out six years of data covering a drought period from 1987 through 1992 [21]. We created species distribution models with the drought data to look at variable importance in dry years compared to variable importance for the whole study period (main training set). To assess model transferability, we predicted to the drought data subset using species distribution models created with the remaining wetter years [31,32].

Model Evaluation

To evaluate each model's ability to forecast to the same range of predictor variables, we predicted to the main test set. To evaluate each model's transferability – that is, to project to a new location or time period where predictor variables may be outside the range of the variables used to build the model – we projected to the drought period with models trained with data from the wet years. The transferability assessment should more realistically evaluate how the models extrapolate to a dry future [31].

To assess a model's performance, we report patterns of correct classification in a confusion matrix and the area under the receiver operating characteristic curve (AUC) [33]. From the confusion matrix, we report the counts of true positives, false positives, true negatives, false negatives and overall classification accuracy based on a 0.5 probability of occurrence threshold for concluding presence. Because we set sample sizes of presence and absence points to be equally subsampled in the Random Forests model, we selected a threshold of 0.5 [34]. Overall classification accuracy was calculated by dividing the number of correctly predicted presences and absences by total predictions. AUC is a threshold free assessment of model performance. AUC values range from zero to one and give the probability that a known presence observation has a higher predicted value of presence than an absence observation for a randomly selected pair of presence-absence observations [33]. Models with AUC values of at least 0.7 are

Table 5. Variable importance from 22 Random Forest species distribution models for the particularly dry period, 1987–1992.

Rank	Canada Goose	Mallard	Blue-winged Teal	Northern Shoveler	Northern Pintail	Pied-billed Grebe	American Bittern	Sora	American Coot	Killdeer	Upland Sandpiper
1	P-5yr_sd(+)	W-semi(+)	W-total(+)	W-semi(+)	U-tree(-)	W-lake(+)	U-grass(+)	T-10yr_sd(+)	W-semi(+)	P-10yr_sd(-)	U-grass(+)
2	T-sum(+)	W-total(+)	W-semi(+)	T-1yr(m)	W-semi(+)	U-crop(-)	W-semi(+)	T-sum(-)	W-lake(+)	T-5yr_sd(-)	U-tree(-)
3	T-spr(-)	W-pal(+)	U-crop(-)	T-wint(m)	T-spr(-)	U-grass(+)	W-total(+)	W-river(-)	W-total(+)	U-grass(+)	W-seas(+)
4	T-10yr(-)	W-river(-)	W-lake(+)	W-total(+)	P-10yr(-)	W-semi(+)	U-crop(-)	W-total(+)	W-pal(+)	U-devel(+)	U-crop(-)
5	P-10yr_sd(-)	U-crop(-)	P-fall(+)	W-lake(+)	T-1yr(-)	W-seas(m)	W-lake(+)	P-5yr_sd(-)	U-crop(-)	T-fall(-)	P-10yr(-)
6	T-1yr(-)	P-10yr(-)	W-pal(+)	P-10yr_sd(-)	P-fall(+)	W-total(+)	W-pal(+)	W-pal(+)	T-5yr_sd(+)	W-semi(+)	W-total(+)
7	T-wint(-)	W-temp(+)	U-grass(+)	T-5yr(-)	W-total(+)	T-wint(+)	T-sum(-)	W-seas(+)	T-10yr_sd(+)	T-sum(+)	W-pal(+)
8	T-5yr(-)	T-wint(+)	T-wint(+)	U-crop(-)	T-5yr(-)	P-5yr_sd(-)	U-devel(-)	U-tree(-)	P-spr(-)	U-tree(-)	W-semi(+)
9	W-seas(-)	T-spr(~)	P-sum(+)	T-10yr(-)	T-wint(-)	W-pal(+)	P-fall(+)	W-temp(+)	W-temp(+)	T-5yr(-)	W-temp(+)
10	P-10yr(-)	W-shrub(-)	T-sum(-)	W-seas(-)	W-pal(+)	U-devel(-)	W-seas(+)	W-lake(+)	W-seas(m)	P-1yr(-)	T-wint(+)
Rank	Willet	Marbled Godwit	Wilson's Snipe	Franklin's Gull	Ring-billed Gull	Black Tern	Sedge Wren	Marsh Wren	Common Yellow-throat	Song Sparrow	Yellow-headed Blackbird
1	P-10yr(-)	W-lake(+)	W-lake(+)	W-total(+)	W-lake(+)	U-crop(-)	T-sum(-)	W-semi(+)	W-semi(+)	U-grass(-)	W-semi(+)
2	P-5yr(-)	U-tree(-)	W-total(+)	W-lake(+)	U-crop(-)	W-pal(+)	W-total(+)	W-total(+)	U-crop(-)	U-tree(+)	P-10yr(-)
3	W-total(+)	W-total(+)	T-10yr(-)	T-10yr(-)	P-fall(+)	W-lake(-)	T-1yr(-)	W-lake(+)	W-temp(+)	W-semi(-)	W-pal(+)
4	U-tree(-)	P-10yr(-)	T-5yr(-)	P-fall(~)	P-10yr(-)	W-temp(+)	T-10yr_sd(-)	U-grass(+)	W-pal(+)	U-crop(+)	W-total(+)
5	W-semi(+)	U-crop(-)	P-5yr_sd(-)	P-5yr_sd(+)	P-spr(-)	W-semi(+)	W-lake(+)	W-pal(+)	W-total(+)	P-sum(+)	P-5yr(-)
6	W-pal(+)	W-river(-)	W-temp(+)	T-10yr_sd(-)	U-grass(+)	T-5yr_sd(+)	T-10yr(-)	U-crop(-)	U-tree(+)	P-10yr(+)	W-seas(+)
7	W-seas(+)	T-10yr(-)	T-1yr(-)	T-spr(-)	P-5yr(-)	W-total(+)	U-crop(-)	T-10yr(+)	P-5yr_sd(-)	P-1yr(+)	U-crop(-)
8	P-10yr_sd(-)	P-5yr(-)	U-tree(-)	T-5yr(-)	T-fall(-)	T-spr(-)	P-10yr(+)	T-fall(+)	W-lake(+)	U-devel(+)	U-grass(+)
9	T-sum(m)	T-5yr(-)	W-seas(-)	T-1yr(-)	P-10yr(-)	T-fall(m)	P-5yr(+)	T-5yr(+)	T-5yr(-)	T-10yr(-)	P-10yr_sd(-)
10	U-crop(-)	W-pal(+)	T-spr(-)	U-devel(-)	P-10yr_sd(-)	T-5yr(-)	W-forest(+)	W-seas(-)	W-seas(+)	T-5yr(-)	U-tree(-)

Top ten variables shown in descending order of rank. Variable categories were denoted by W (wetland), U (upland), P (precipitation), and T (temperature). Signs indicated the relationship between the predictor and the species response: + (positive), - (negative), m (unimodel), and ~ (equivocal).

considered acceptable, between 0.8 and 0.9 good, and greater than 0.9 outstanding [35].

Projected Distributional Changes

We created current predictions and future projections of probability of occurrence to each grid cell, for each focal species, by applying the SDMs to the baseline land cover and climate data (the 20-year period for baseline climate data being 1981–2000) and to baseline land cover and future climate data (the 10-year period 2040–2049). Ten to twenty-year time periods were chosen to mitigate the influence of short-term variations in climate.

We created current and future predictive distribution maps for each species in ArcMap 10 based on an assignment of grid point locations as suitable or unsuitable. A grid point was determined suitable if the estimated probability of occurrence (over the time period for the baseline or future data) was greater than 0.5. Three breakpoints within suitable (0.625, 0.75, 0.875) and unsuitable (0.125, 0.25, 0.375) locations showed the degree to which a location was predicted suitable or unsuitable.

We indexed changes between predicted baselines and projected future distributions using change in a species' spatial distribution. To assess change in distribution, we calculated the percent loss (or gain) in the number of grid cells classified as suitable.

Results

Baseline mean temperature (years 1981–2000) was 5.9°C and mean yearly precipitation was 548 mm. By 2040–2049, CGCM projected a 2.9°C increase in mean temperature and a 22 mm (3.9%) increase in annual precipitation while WRFc projected a 3.8°C temperature increase and a 17 mm (3.1%) increase in annual precipitation. Projections of future precipitation fall within the range of historic levels of precipitation, whereas future temperatures projected by both climate models exceed historic temperatures (Figure 2). The climate models differed slightly in the spatial distribution of the precipitation increases, with CGCM projecting greater increase in Iowa and WRFc projecting greater increase in North Dakota than other areas (Figure 3).

The number of data occurrence records in the main training set was 975. The number of years of survey data included in the main training set, for a given route, ranged from one to 21, with a mean of 13. The number of survey routes included for a given year ranged from 15 to 35. Thirty-one waterbird species had prevalence $\geq$ 0.05 and were included in the focal group (Table 2). The number of records in the main test set was 817.The number of data records in the dry years set was 139. We adjusted the prevalence cutoff to 0.07 ($\geq$ 10 detections), at which 22 species qualified.

Model Evaluation

Most models based on known distributional patterns were acceptable to excellent, indicated by AUC values (Table 2). Exceptions were SDMs for the Great Blue Heron and Killdeer. When predicting to dry years only, AUC values indicated the following additional models predicted poorly: Canada Goose, American Wigeon, Mallard, and Green-winged Teal. Overall accuracy of dry year predictions suggested that projected distributional changes for some species should be interpreted with caution, including Blue-winged Teal, Sora, and Common Yellowthroat. For the main datasets, model performance was not related to a species' prevalence according to Spearman's rank correlation (-0.09, p-value 0.62).

Vulnerability

Average projected decline in occurrence rate (spatial distribution) across 31 species under two future climate scenarios was 45%. WRFc models projected slightly more severe distributional changes (-48%) than CGCM (-43%; Table 3). Species expected to experience small to no declines in distribution included Blue-winged Teal, Killdeer, and Upland Sandpiper. Species projected to experience severe declines were Franklin's Gull, Sora, and Wilson's Snipe (Table 3). In general, species maps depicted declines within the baseline range, rather than distributional shifts to new areas (Figures S2-S12).

For most species, future projections of change were consistent with responses of species to historic dry periods (Figure 4). Consistent projections were those that exhibited little to no change between the historic dry period and the future, or those that declined more in the future than in the historic dry period. If a species' habitat was not impacted by dry conditions, the species would be expected to experience little to no impact under future dry conditions. Other species may be impacted by drying conditions, thus responding during the historic dry period, and even more if additional drying occurs in the future. However, inconsistent with expectations, models projected reduced distribution of Blue-winged Teal, Northern Pintail, and Pied-billed Grebe in the dry period relative to future projections. Additionally, several species that remained relatively stable in the historic dry period were projected to decrease in distribution under future scenarios, including Canada Goose, Sedge Wren, Marsh Wren, Common Yellowthroat, and Song Sparrow.

Variable Importance

In general, species distributions were strongly influenced by the distribution of wetland basins and land cover classes and moderately influenced by climate, as evidenced by their influence in the SDMs. Land cover variables, wetland and upland, collectively occurred as 67% of the top ten variables in the SDMs but comprised only 42% of the available predictor variables (Table 4); wetland and upland variables were 1.5 and 1.8 times more likely to appear in lists of top ten predictors than in the list of available predictors, respectively. Species associations with all wetland types, except rivers, were generally positive. All associations with cropland were negative except for the Song Sparrow, whereas associations with grassland were primarily positive except for Wood Duck (Table 4). Climate predictor variables were generally underrepresented in the variables of top importance. Collectively, temperature and precipitation comprised 32% of the top ten variables across the 31 species, although they were 58% of the available predictor variables. Temperature and precipitation variables were similarly influential and were 0.5 and 0.6 times more likely to appear in lists of top ten predictors than in the list of available predictors, respectively. In general, probability of species occurrence was negatively associated with temperature; relationships with temperature variability were often positive (Table 4). Associations with precipitation were often negative, except for Great Blue Heron, Sedge Wren, Song Sparrow, and Wood Duck. Variability in precipitation occurred in the top ten variables for only one species' model (Sedge Wren) and was negatively correlated with probability of occurrence.

Land cover variables were highly influential in observed patterns of species distribution. The importance of these variables is visually apparent when spatial distribution of grasslands and wetlands (Figure S1) and observed climate gradients (Figure 3) were compared to baseline distributions (Figures S2-S12). Many breeding waterbirds have a high probability of occurrence in the western portion of the study area where grasslands and wetlands co-occur.

Temperature and precipitation predictor variables were more often in the top ten variables for the species with the greatest expected declines (Figure 5). Conversely, wetland and upland land cover variables were more often in the top ten variables for the species with smallest expected declines.

Variable Importance: main models versus dry-years models

For dry-years models, climate predictor variables represented 45% of the top ten variables across the 22 species versus 31% for the same 22 species in the main models (Tables 4 and 5). Of land cover predictors, 65% included wetland variables in the top ten variables in the dry years and a similar 67% in the main models. However, representation of different wetland types varied with more seasonal wetlands (positive relationships only) appearing in the non-drought years (30% versus 17%) and more lakes included in the dry years (32% versus 20%).

Discussion

Our projections of large range reductions for waterbirds breeding in the Prairie Pothole Region are not surprising. Globally, freshwater habitats are expected to be particularly vulnerable to climate change [36]. If, as the future climate projections we used indicate, temperatures rise by ~3.0°C and precipitation rises only by 3% by mid-century in the Prairie Pothole Region, many fewer pothole wetlands will exist on the landscape due to an increased deficit in precipitation relative to evapotranspiration. Similarly, other studies of the Prairie Pothole Region have projected a drier future and concomitant reductions in waterbird habitat [5,8,37].

Past studies in the Prairie Pothole Region that extrapolated from relations between climatic factors and wetlands inferred generalized habitat losses for waterfowl [5,8]. Our species-specific approach indicated large variability in the vulnerabilities of waterbird species to climate change. This is expected as patterns of waterbird habitat selection vary among species for wetland attributes such as size, permanence, and vegetative cover [10,38]. Hydrological studies indicate that temperature and precipitation regimes affect not only the number of wetlands and wetland size, but marsh vegetation dynamics and the vegetative coverage patterns at the landscape scale [5]. While reducing the overall number of wetlands, a drier climate will likely lead to more extensive coverage of wetlands by dense vegetation rather than wetland conditions characterized by a mixture of open water and vegetation [5]. Species are expected to respond differentially to these changes in wetland characteristics. Furthermore, individualistic species' responses appear the norm [39,40,41].

Our projections of future change were not always consistent with documented waterbird responses to a historic dry period which represented one possible expression of a drier climate. The dry historic period, a consequence of reduced precipitation, is not a direct analog of future drying which is expected to be driven by increases in evapotranspiration (Figure 2) [42]. Thus, it is unclear to what extent the historic pattern of drought can be used as a benchmark for future climate change. Therefore, the inconsistencies between the dry historic waterbird response relative to projected future responses may indicate our models are under- or over- estimating waterbird response to climate change for some species. It is also possible that changes in temperature versus precipitation may result in divergent, and unprecedented, future wetland habitat conditions. In that case, divergent waterbird responses, relative to the past responses, would not be surprising.

The historic range of temperature variability did not overlap future projections and so, our SDMs were projecting beyond known climatic boundary conditions. Model extrapolation to novel conditions is common when projecting species response to future climate [43]. Our single values for yearly averages (Figure 2), indicated almost no overlap in temperature range between historic and projected time intervals. However, because of spatial variation in temperature regimes (i.e., warmer in the south, as shown in Figure 3), there were likely many individual grid cells in which future temperatures overlapped the historic range even if the study area yearly means do not. SDMs based on the Random Forest algorithm are constrained when extrapolating beyond the observed values of the predictor variables. For example, when projected temperatures are outside of the range of the training set the algorithm holds the prediction constant at the last known value of temperature [44]. Therefore, if future wetland habitat conditions selected by the species become less common with increased temperatures, our estimates of habitat losses for many species may be underestimates.

Ranking predictor variables by their importance provides additional insights into how the 31 waterbird species may respond to changing environmental factors. We included predictors related to suitable waterbird habitat quality, including the amount and type of wetland basins, and temporally scaled temperature and precipitation covariates. Species projected to be most sensitive to anticipated climate change (changes in temperature and precipitation,Table 4) consistently reflected the ecology of the species. For example, the two diving ducks, Ruddy Duck and Redhead, primarily selected large wetlands, such as semipermanent basins, and were less susceptible to total drying [5,45]. As a consequence of their habitat associations, no climate covariates ranked in the top ten for these two waterbird species. In contrast, waterbirds that rely on shallow water habitat, such as Sora or Sedge Wren, or dynamic habitat such as Black Tern or Mallard, showed a much greater projected change in distribution to future climatic conditions [5,38].

The variable importance ranks also suggested that waterbirds may shift their habitat preferences with increased drying. More climate covariates and more permanent wetland regimes appeared in the top variables for dry years. In the Prairie Pothole Region, wetland function can rapidly change with significant changes in the climate. In dry years, for example, semipermanent wetlands may function more like seasonal wetlands, and seasonal wetlands more like temporary wetlands. This differential sensitivity to climate change explains why seasonal wetlands were less important and lakes more important in dry years.

Because bioclimatic SDMs are generally exploratory with many collinear climate predictors, there is concern that these models over-fit the data and thus misrepresent species distributions [46]. However, the inclusion of many collinear climate predictors is often warranted when causal links between specific climate predictors and species' distributions are not established, leading to better model fit and projections [47]. We found that when we reduced our 18 climate and 13 land cover variables to 14 uncorrelated climate and 10 uncorrelated land cover variables model projections were similar: 45% average range reduction for the full model and 48% for the reduced model (results not shown).

Conclusions

Our results indicated, on average, large decreases in suitable habitat by the 2040s for 31 waterbird species breeding in the

Prairie Pothole Region of the U.S.A. Importantly, our results were consistent between two contrasting future climate scenarios. However, there was substantial variability in species specific responses to projected climate change. Therefore, strategic efforts to mitigate climate change effects should preferentially direct management actions to those species expected to be most vulnerable. In continuing research, we are exploring in greater detail various sources of uncertainty in our projections including additional model algorithms, alternative covariates, and other sources of species distribution data [48].

Supporting Information

Figure S1 Distribution of grassland and palustrine wetlands on the U.S. Prairie Pothole Region landscape. Darker shades represent greater coverage of grassland (versus cropland) and greater areal coverage of wetlands (log transformed).

Figure S2 Map of species distributions for baseline and two future climate projections. Brown indicates areas where the species is predicted to occur and green represents areas where the species is not predicted to occur.

Figure S3 Map of species distributions for baseline and two future climate projections. Brown indicates areas where the species is predicted to occur and green represents areas where the species is not predicted to occur.

Figure S4 Map of species distributions for baseline and two future climate projections. Brown indicates areas where the species is predicted to occur and green represents areas where the species is not predicted to occur.

Figure S5 Map of species distributions for baseline and two future climate projections. Brown indicates areas where the species is predicted to occur and green represents areas where the species is not predicted to occur.

Figure S6 Map of species distributions for baseline and two future climate projections. Brown indicates areas where the species is predicted to occur and green represents areas where the species is not predicted to occur.

Figure S7 Map of species distributions for baseline and two future climate projections. Brown indicates areas where the species is predicted to occur and green represents areas where the species is not predicted to occur.

Figure S8 Map of species distributions for baseline and two future climate projections. Brown indicates areas where the species is predicted to occur and green represents areas where the species is not predicted to occur.

Figure S9 Map of species distributions for baseline and two future climate projections. Brown indicates areas where the species is predicted to occur and green represents areas where the species is not predicted to occur.

Figure S10 Map of species distributions for baseline and two future climate projections. Brown indicates areas where the species is predicted to occur and green represents areas where the species is not predicted to occur.

Figure S11 Map of species distributions for baseline and two future climate projections. Brown indicates areas where the species is predicted to occur and green represents areas where the species is not predicted to occur.

Figure S12 Map of species distributions for baseline and two future climate projections. Brown indicates areas where the species is predicted to occur and green represents areas where the species is not predicted to occur.

Acknowledgments

Disclaimer: Any use of trade, firm, or product names is for descriptive purposes only and does not imply endorsement by the U.S. Government.

Climate data were provided by Linda Joyce and David Coulson (U.S. Forest Service, statistically downscaled climate models), and John Stamm, Parker Norton, and Gary Clow (U.S. Geological Survey, high resolution models). Lucy Burris and Diane Granfors provided figures 3 and S1, respectively. Curt Flather, Helen Sofaer, and anonymous reviewers provided comments that improved the manuscript.

Author Contributions

Conceived and designed the experiments: VS SS. Performed the experiments: VS. Analyzed the data: VS. Wrote the paper: VS SS BN.

References

1. Keddy PA (2000) Wetland ecology: principles and conservation. Cambridge, UK; New York, NY: Cambridge University Press. 614 p.
2. Gleason RA, Euliss NH, Tangen BA, Laubhan MK, Browne BA (2011) USDA conservation program and practice effects on wetland ecosystem services in the Prairie Pothole Region. Ecological Applications 21: S65–S81.
3. Beyersbergen GW, Niemuth ND, Norton MR (2004) Northern Prairie and Parkland waterbird conservation plan. A plan associated with the Waterbird Conservation for the Americas initiative Prairie Pothole Joint Venture, Denver, CO, USA.
4. Solomon S, Intergovernmental Panel on Climate Change, Intergovernmental Panel on Climate Change. Working Group I. (2007) Climate change 2007: the physical science basis: contribution of Working Group I to the Fourth Assessment Report of the Intergovernmental Panel on Climate Change. Cambridge; New York: Cambridge University Press. 996 p.
5. Johnson WC, Werner B, Guntenspergen GR, Voldseth RA, Millett B, et al. (2010) Prairie Wetland Complexes as Landscape Functional Units in a Changing Climate. Bioscience 60: 128–140.
6. Niemuth ND, Solberg JW (2003) Response of waterbirds to number of wetlands in the Prairie Pothole Region of North Dakota, USA. Waterbirds 26: 233–238.
7. Johnson DH, Grier JW (1988) Determinants of Breeding Distributions of Ducks. Wildlife Monographs: 1–37.
8. Sorenson LG, Goldberg R, Root TL, Anderson MG (1998) Potential effects of global warming on waterfowl populations breeding in the Northern Great Plains. Climatic Change 40: 343–369.
9. Breiman L (2001) Random forests. Machine Learning 45: 5–32.
10. Kantrud HA, Krapu GL, Swanson GA (1989) Prairie basin wetlands of the Dakotas: a community profile. Washington, DC: U.S. Dept. of the Interior, Fish and Wildlife Service, Research and Development. 111 p.
11. Johnson RR, Oslund FT, Hertel DR (2008) The past, present, and future of prairie potholes in the United States. Journal of Soil and Water Conservation 63: 84a–87a.
12. Dahl TE (1990) Wetlands losses in the United States 1780's to 1980's. Washington, D.C.: U.S. Department of the Interior, Fish and Wildlife Service.
13. Sauer JR, Hines JE, Fallon J (2007) The North American Breeding Bird Survey, Results and Analysis 1966 – 2006. Version 10.13.2007. Laurel, MD, USGS Patuxent Wildlife Research Center.
14. Cowardin LM, Shaffer TL, Arnold PM, United States National Biological Service (1995) Evaluations of duck habitat and estimation of duck population

sizes with a remote-sensing-based system. Washington, D.C.: U.S. Dept. of the Interior, National Biological Service. 26 p.
15. Johnson R, Higgins K (1997) Wetland resources of eastern South Dakota. Brookings, SD, USA, South Dakota State University.
16. Stewart RE, Kantrud HA (1971) Classification of natural ponds and lakes in the glaciated prairie region.
17. Cowardin LM, Carter V, Golet FC, LaRoe ET (1979) Classification of wetlands and deepwater habitats of the United States. Washington, D.C.: US Department of the Interior/Fish and Wildlife Service.
18. Martin AE, Fahrig L (2012) Measuring and selecting scales of effect for landscape predictors in species-habitat models. Ecological Applications 22: 2277–2292.
19. Coulson DP, Joyce LA (2010) Historical climate (1940–2006) for the conterminous United States at the 5 arc minute grid spatial scale based on PRISM climatology. Fort Collins, Colorado: U.S. Department of Agriculture, Forest Service, Rocky Mountain Research Station.
20. Larson DL (1995) Effects of Climate on Numbers of Northern Prairie Wetlands. Climatic Change 30: 169–180.
21. Winter TC, Rosenberry DO (1998) Hydrology of prairie pothole wetlands during drought and deluge: a 17-year study of the Cottonwood Lake wetland complex in North Dakota in the perspective of longer term measured and proxy hydrological records. Climatic Change 40: 189–209.
22. Giorgi F, Mearns LO (1991) Approaches to the simulation of regional climate change: A review. Reviews of Geophysics 29: 191–216.
23. Coulson DP, Joyce LA, Price DT, McKenney DW, Siltanen RM, et al. (2009) Climate Scenarios for the conterminous United States at the 5 arc minute grid spatial scale using SRES scenarios A1B and A2 and PRISM climatology. Fort Collins, Colorado: U.S. Department of Agriculture, Forest Service, Rocky Mountain Research Station.
24. Skamarock W, Klemp J, Dudhia J, Gill D, Barker D, et al. (2008) A description of the advanced research WRF version 3, NCAR technical note. NCAR/TN–475+ STR.
25. National Center Atmospheric Research Staff (2014) The Climate Data Guide: Regridding Overview.
26. Nakicenovic N, Intergovernmental Panel on Climate Change. Working Group III (2000) Special report on emissions scenarios: a special report of Working Group III of the Intergovernmental Panel on Climate Change. Cambridge; New York: Cambridge University Press. 599 p.
27. Prasad AM, Iverson LR, Liaw A (2006) Newer classification and regression tree techniques: Bagging and random forests for ecological prediction. Ecosystems 9: 181–199.
28. Lawler JJ, White D, Neilson RP, Blaustein AR (2006) Predicting climate-induced range shifts: model differences and model reliability. Global Change Biology 12: 1568–1584.
29. R Development Core Team (2012) R: A language and environment for statistical computing. Vienna, Austria: R Foundation for Statistical Computer.
30. Chen C, Liaw A, Breiman L (2004) Using Random Forest to Learn Imbalanced Data. Berkeley, University of California.
31. Schröder B, Richter O (2000) Are habitat models transferable in space and time? Zeitschrift für Ökologie und Naturschutz 8: 195–205.
32. Guisan A, Thuiller W (2005) Predicting species distribution: offering more than simple habitat models. Ecology Letters 8: 993–1009.
33. Hastie T, Tibshirani R, Friedman JH (2009) The elements of statistical learning: data mining, inference, and prediction.
34. Liu CR, Berry PM, Dawson TP, Pearson RG (2005) Selecting thresholds of occurrence in the prediction of species distributions. Ecography 28: 385–393.
35. Hosmer DW, Lemeshow S (2000) Applied logistic regression: Wiley-Interscience.
36. Kundzewicz ZW, Mata LJ, Arnell N, Doll P, Kabat P, et al. (2007) Freshwater resources and their management.
37. Poiani KA, Johnson WC (1991) Global Warming and Prairie Wetlands - Potential Consequences for Waterfowl Habitat. Bioscience 41: 611–618.
38. Weller MW, Spatcher CS (1965) Role of habitat in the distribution and abundance of marsh birds. Iowa, Ames.
39. Matthews SN, Iverson LR, Prasad AM, Peters MP (2011) Changes in potential habitat of 147 North American breeding bird species in response to redistribution of trees and climate following predicted climate change. Ecography 34: 933–945.
40. Peterson AT (2003) Projected climate change effects on Rocky Mountain and Great Plains birds: generalities of biodiversity consequences. Global Change Biology 9: 647–655.
41. Tingley MW, Koo MS, Moritz C, Rush AC, Beissinger SR (2012) The push and pull of climate change causes heterogeneous shifts in avian elevational ranges. Global Change Biology 18: 3279–3290.
42. Cook B, Smerdon J, Seager R, Coats S (2014) Global warming and 21st century drying. Climate Dynamics: 1–21.
43. Elith J, Leathwick JR (2009) Species Distribution Models: Ecological Explanation and Prediction Across Space and Time. Annual Review of Ecology Evolution and Systematics 40: 677–697.
44. Elith J, Graham CH (2009) Do they? How do they? WHY do they differ? On finding reasons for differing performances of species distribution models. Ecography 32: 66–77.
45. Kantrud HA, Stewart RE (1977) Use of Natural Basin Wetlands by Breeding Waterfowl in North Dakota. The Journal of Wildlife Management 41: 243–253.
46. Beaumont LJ, Hughes L, Poulsen M (2005) Predicting species distributions: use of climatic parameters in BIOCLIM and its impact on predictions of species' current and future distributions. Ecological Modelling 186: 250–269.
47. Braunisch V, Coppes J, Arlettaz R, Suchant R, Schmid H, et al. (2013) Selecting from correlated climate variables: a major source of uncertainty for predicting species distributions under climate change. Ecography 36: 971–983.
48. Beale CM, Lennon JJ (2012) Incorporating uncertainty in predictive species distribution modelling. Philosophical Transactions of the Royal Society B-Biological Sciences 367: 247–258.

3

Microrefugia and Shifts of *Hippophae tibetana* (Elaeagnaceae) on the North Side of Mt. Qomolangma (Mt. Everest) during the Last 25000 Years

Lu Xu[1,2]⁹, Hao Wang[1]⁹, Qiong La[1,3]⁹, Fan Lu[1], Kun Sun[2], Yang Fang[1], Mei Yang[1], Yang Zhong[1,3], Qianhong Wu[1], Jiakuan Chen[1], H. John B. Birks[4,5,6], Wenju Zhang[1]*

1 Institute of Biodiversity Science, School of Life Sciences, Fudan University, Shanghai, China, **2** College of Life Sciences, Northwest Normal University, Lanzhou, China, **3** Department of Biology, Tibet University, Lhasa, China, **4** Department of Biology, University of Bergen, Bergen, Norway, **5** Environmental Change Research Centre, University College London, London, United Kingdom, **6** School of Geography and the Environment, University of Oxford, Oxford, United Kingdom

Abstract

Microrefugia at high altitudes or high latitudes are thought to play an important role in the post-glacial colonization of species. However, how populations in such microrefugia have responded to climate changes in alternating cold glacial and warm interglacial stages remain unclear. Here we present evidence to indicate the Rongbuk Valley of the Mt. Qomolangma (Mt. Everest) area, the highest region on earth, had microrefugia for *Hippophae tibetana* and discuss how this low shrub was adapted to the extreme climate fluctuations of the last 25,000 years by shifts. By integrating geological, glaciological, meteorological, and genetic information, we found that the Rongbuk Valley was not only a glacial microrefugium but also an interglacial microrefugium for *H. tibetana*: the former was located on the riverbank below 4800 m above sea level (asl) or lower area and the latter at ~5000 m asl. Our results show that after the Last Glacial Maximum (LGM), *H. tibetana* in the valley has undergone upward and downward migrations around ~5000 m driven by climate fluctuations and the population in the glacial microrefugium has suffered extinction or extreme contraction. Moreover, with the rise of temperature in the last four decades, the upper limit of *H. tibetana* has shifted at least 30 m upward. Combining population history and recent range shift of this species is important in predicting the fate of this endemic species to future climate changes.

Editor: Navnith K. P. Kumaran, Agharkar Research Institute, India

Funding: This research was funded by National Basic Research Program of China (Grant No. 2014CB954103; http://www.973.gov.cn/AreaMana.aspx) and the National Natural Science Foundation of China (Grant No. 41061007 and Grant No. 91131901; http://isisn.nsfc.gov.cn). The funders had no role in study design, data collection and analysis, decision to publish, or preparation of the manuscript.

Competing Interests: The authors have declared that no competing interests exist.

* E-mail: wjzhang@fudan.edu.cn

⁹ These authors contributed equally to this work.

Introduction

The alternation between glacial and interglacial stages during the Quaternary has greatly affected species distributions [1], [2], [3]. Many temperate species retreated to refugia at lower latitudes or altitudes during glacials and expanded to higher latitudes or altitudes during interglacials [1], [4], [5], [6]. Apart from these broad-scale refugia, more and more scattered small refugia have been found at high latitudes or in alpine areas in recent decades, with local favourable environmental features outside the species' main distribution area, so-called microrefugia (or cryptic refugia) [3], [4], [5], [7], [8], [9], [10], [11], [12]. Understanding microrefugia is of critical importance in interpreting species genetic diversity and evolutionary processes such as adaptation, speciation, and extinction [1], [2], [3], [9], [13], [14], [15], [16]. For example, the incongruence between estimated post-glacial migration rates and tree dispersal capacity ('Reid's Paradox') can be explained, in part, by the widespread existence of microrefugia [5], [14], [15], [17], [18]. The concept of microrefugia is now well recognized through empirical studies which have mainly focused on identifying microrefugia during the Last Glacial Maximum (LGM) by palaeoecological or phylogeographical methods [4], [5], [10], [17]. However, as Rull (2009) points out, the concept of microrefugia lacks appropriate biogeographical and ecological characterization [9]. Although the importance of landscape physiography and microclimate for plant microrefugia to occur has been highlighted [15], and some potential microrefugia may be deduced according to topoclimate, climate stability, and isolation from the matrix [19], little is known about actual microrefugial situations, their time span, and their exact elevation or location, as well as how species (especially plants) have managed to track a favourable microclimate for their survival [8], [9], [15], [16], [20].

At present, most studies on microrefugia have paid close attention to glacier refugia, especially to the LGM microrefugia, in which some genotypes have been supposed to survive this cold period [4], [5], [7], [8], [9], [10], [14], [17], [21], [22], [23]. But as Hampe et al. (2013) pointed out that, with wide-ranging implications, glacial refugia no longer exist and can hence only be inferred by indirect means [16]. Populations in microrefugia, defined as climate relicts by Hampe et al. (2011) [24], should be the key to understanding the effects of microrefugia. An important fact often has been ignored by many researchers: although populations in the LGM microrefugia were able to persist through this cold period, these populations may have been dramatically changed in interglacial stages because of warm climate (e.g. American pika) [25]. Thus,when we infer glacial refugia by indirect means, especially by genetic information, the dynamics of these climate relics after the LGM are needed to survey at the smaller scale. A meta-population model has been considered to characterize these climate relicts [26]; however, few "climate relict" has been studied carefully in detail.

On the other hand, investigating species' responses to past climate changes is also important in understanding how species might respond to recent and future climate changes [4], [5], [16], [18], [20], [27]. Global temperature has increased 0.6°C in the past three decades and 0.8°C in the past century [28], which has led to both latitudinal and altitudinal shifts in species ranges [29], [30], [31], [32], [33], [34] and even caused some species to be possibly on the brink of extinction [35], [36]. However, present studies on the responses of species to past climate changes and recent warming are very disconnected. The former often focuses on tracking species population history through phylogeographical or paleoecological methods whereas the latter usually involves comparison of well-documented historical records with present distributional data [29], [36], [37]. According to oxygen isotope ($\delta^{18}O$) data from ice cores and sea sediments, past climate oscillations are often more pronounced than recent warming [38], [39], [40], [41], but almost all the species existing today have survived these past oscillations [1], [5]. Thus, we could better predict a species response in the future by combining the present and past responses of the species to climate changes over a range of time scales.

In the present study, we investigate *Hippophae tibetana* Schlecht. (Elaeagnaceae) as a means of characterizing the biogeographical and ecological features of a particular microrefugium and the responses of this species to climate changes since the LGM and to recent warming. *H. tibetana* is a small dioecious shrub propagated by seeds or by horizontal roots [42]. It is endemic to the Qinghai-Tibet Plateau (QTP) and ranges from the west Himalaya to the east-north QTP [43]. In the eastern plateau, *H.tibetana* occurs in the lowlands and in alpine meadows at an altitude below 4000 m, but in the central plateau and the Himalayas, it has a fragmented distribution along several valleys. It is one of the shrubs to occur at the highest altitudes [44], growing up to ~5200 m asl. Our previous study [43] investigated the phylogeography of *H. tibetana* and found that three main lineages (A, B, and C) of the present populations of this species occupy the middle, the western, and the eastern parts of its geographical range, respectively. Based on the distribution of a large number of private haplotypes, we concluded that *H. tibetana* had multiple LGM microrefugia on the Plateau and inferred that the Rongbuk Valley, north of Mt. Qomolangma (Mt. Everest) is a possible microrefugium for *H. tibetana* even though it is the highest region within the geographical range of *H. tibetana*. However, in this valley, where and how *H. tibetana* has survived since the LGM remains unclear. Also, we chose this area for our current study because the area is highly sensitive to global climate change, and the geological and meteorological characteristics of the valley have already been studied well [45], [46], [47], [48]. This area provides an excellent opportunity to consider the favourable local environment required for microrefugia to occur and to study the responses of *H. tibetana* to climate change since the LGM [4], [8] by integrating the available geological, meteorological, and genetic information. In the present study, we hope to clarify two questions: 1) Do microrefugia of the LGM for *H. tibetana* really exist in the Rongbuk valley, one of the highest areas on earth? 2) On the local scale, how has the population of *H. tibetana* in the microrefugia responded to climate changes since the LGM?

Materials and Methods

Ethics Statement

In this study, all field works were carried out in the Rongbuk Valley of Mountain Qomolangma National Nature Preserve (QNNP) and were permitted by QNNP. There is no endangered or protected species involved in this study. The plant species studied in this work, *Hippophae tibetana* (Elaeagnaceae), has a large distribution on the QTP and has not been listed in any protection lists. No animals were used in this study. Coordinate data of sample locations of this study were shown in Table 1.

Study Area

The Rongbuk Valley is located on the northern slopes of Mt. Qomolangma (27.98°N, 86.92°E; elevation 8844 m, the highest mountain on Earth), which lies toward the eastern end of the Himalaya. Three large glaciers occur at the upper end of the valley: West Rongbuk Glacier, Rongbuk Glacier, and East Rongbuk Glacier. These glaciers flow for 13~18 km down to the proglacial plain at 5200 m asl near Everest Base Camp used by many mountaineering expeditions [49], [50]. Glacial meltwater runs through the Rongbuk Valley, which is 0.4−1 km wide, ~91 km long, and ~1500 m fall at altitude (5200 m to 3700 m), and empties into the Pengqu River at 3700 m asl.

The modern Equilibrium Line Altitude (ELA) of these glaciers is above 6000 m asl [51]. In the past, their extents shifted in response to climate change, leaving terminal moraines of different ages in the valley [47], [52], [53], [54], [55] (Fig. 1a). The Far East Rongbuk ice-core only records climate changes for the last 200 years, but these can be matched with the longer Guliya ice-core records [38], [45]. We can thus infer past temperature change over a long time span in the Rongbuk Valley from the Guliya ice-core. An increase (or decrease) of 1‰ in mean annual $\delta^{18}O$ in the Guliya ice-core corresponds to an increase (or decrease) of ~1.5°C in mean annual air temperature [56] as shown in Fig. 1b. There are also detailed studies on the glacial moraines of different ages. From 25 ka B.P. until now, there have been four main glaciations in the valley, and their terminal moraines are the Jilong moraine (~4800 m), Rongbuk moraine (~5000 m), Rongbude moraine (~5100 m), and the Little Ice Age moraine (~5150 m) from the oldest to the youngest (M4, M3, M2, and M1 in Fig. 1a, respectively) [47], [52], [53], [54]. The ages of the moraines have been estimated using optically stimulated luminescence (OSL) and terrestrial cosmogenic nuclide (TCN) dating by Owen *et al.* (2009) [47] and the OSL dates are indicated along the time axis in Fig. 1b.

Main vegetation in study area is alpine steppe vegetation with small shrubs in the bottom of the valley at an altitude between 4400 and 5000 m asl, dominated by *Artemisia wellbyi*, *Stipa purpurea*, *Orinus thoroldii*, *Carex montis-everestii*, *Potentilla parvifolia*. At the higher level, between 5000–5600 m, some alpine sparse and cushion vegetation occurs locally on the upper part of area, such as *Potentilla fruticosa* var. *pumila*, *Androsace tapete*, *Arenaria polytrichoides*, *Kobresia pygmaea*, *Kobresia prainii*, *Saussurea gnaphalodes* and other *Saussurea* sp. In addition, some lichen common species occurs on the leeward side of rocks and gravels, for example *Lecidea auriculata* and *Cloplaca elegans* etc.

Table 1. List of populations (POP) and patches (PAT) analysed in the present study with their sampling localities, number of specimens, coordinates, genetic diversity parameters, and chlorotype composition of each population.

Population No.	N	Altitude (m)	Latitude (N)	Longitude (E)	Nuclear microsatellites				Chloroplast Haplotypes		
					F_{IS}	A	H_O	H_E	Haplotypes (Frequencies, %)	D	π
POP 1	39	4200	28°24′23″	86°59′14″	−0.061	2.10	0.419	0.390	R1(5), R2(85), R3(10)	0.2780	0.00053
POP 2	23	4400	28°18′49″	86°55′53″	−0.041	2.18	0.430	0.403	R1(44), R2(39), R3(17)	0.6561	0.00148
POP 3	51	4465	28°18′46″	86°53′39″	0.084	2.15	0.328	0.354	R1(39), R2(49), R3(12)	0.6039	0.00127
POP 4	25	4690	28°16′38″	86°48′26″	−0.048	2.28	0.410	0.384	R1(52), R2(36), R3(12)	0.6100	0.00135
POP 5	31	4805	28°14′38″	86°49′05″	0.031	2.25	0.422	0.429	R1(36), R2(61), R4(3)	0.5140	0.00098
POP 6	28	4946	28°12′25″	86°49′19″	−0.850	1.69	0.564	0.304	R1(100)	0.0000	0.00000
POP 7	33	5000	28°10′03″	86°50′23″	0.011	2.15	0.405	0.403*	R1(30), R2(3), R3(18), R4(12), R5(6), R6(31)	0.7879	0.00343
POP 8	35	5035	28°09′42″	86°50′38″	−0.051	2.19	0.429	0.402	R1(57), R2(5), R3(26), R4(11)	0.6084	0.00201
PAT 1	25	>5035	28°09′08″	86°50′52″	−0.778	1.20	0.176	0.099*	R3(100)	0.0000	0.00000
PAT 2	22	5047	28°09′34″	86°50′42″	−0.707	1.40	0.305	0.177*	R3(100)	0.0000	0.00000
PAT 3	15	5046	28°09′35″	86°50′42″	−0.774	1.49	0.373	0.209*	R2(100)	0.0000	0.00000
PAT 4	24	5047	28°09′37″	86°50′42″	−0.378	1.80	0.426	0.305*	R1(100)	0.0000	0.00000
PAT 5	32	5066	28°09′24″	86°50′49″	−1.000	1.20	0.200	0.100*	R2(100)	0.0000	0.00000
PAT 6	16	>5035	28°09′21″	86°50′56″	−0.408	1.78	0.390	0.267*	R2(100)	0.0000	0.00000
PAT 7	15	5058	28°09′32″	86°50′52″	−0.647	1.60	0.373	0.224*	R1(100)	0.0000	0.00000

The exact elevations of PAT 1 and PAT 6 are not given because the GPS data of the two spots are not very accurate. N, number of individuals analysed; Nuclear microsatellites data, including F_{IS}, fixation index; A, allelic richness; H_O, observed heterozygosity; H_E, expected heterozygosity;
*significant Hardy–Weinberg disequilibrium. cpDNA haplotypes and their frequencies, as well as estimates of gene diversity (D) and nucleotide diversity averaged across loci (π) of the populations and patches studied.

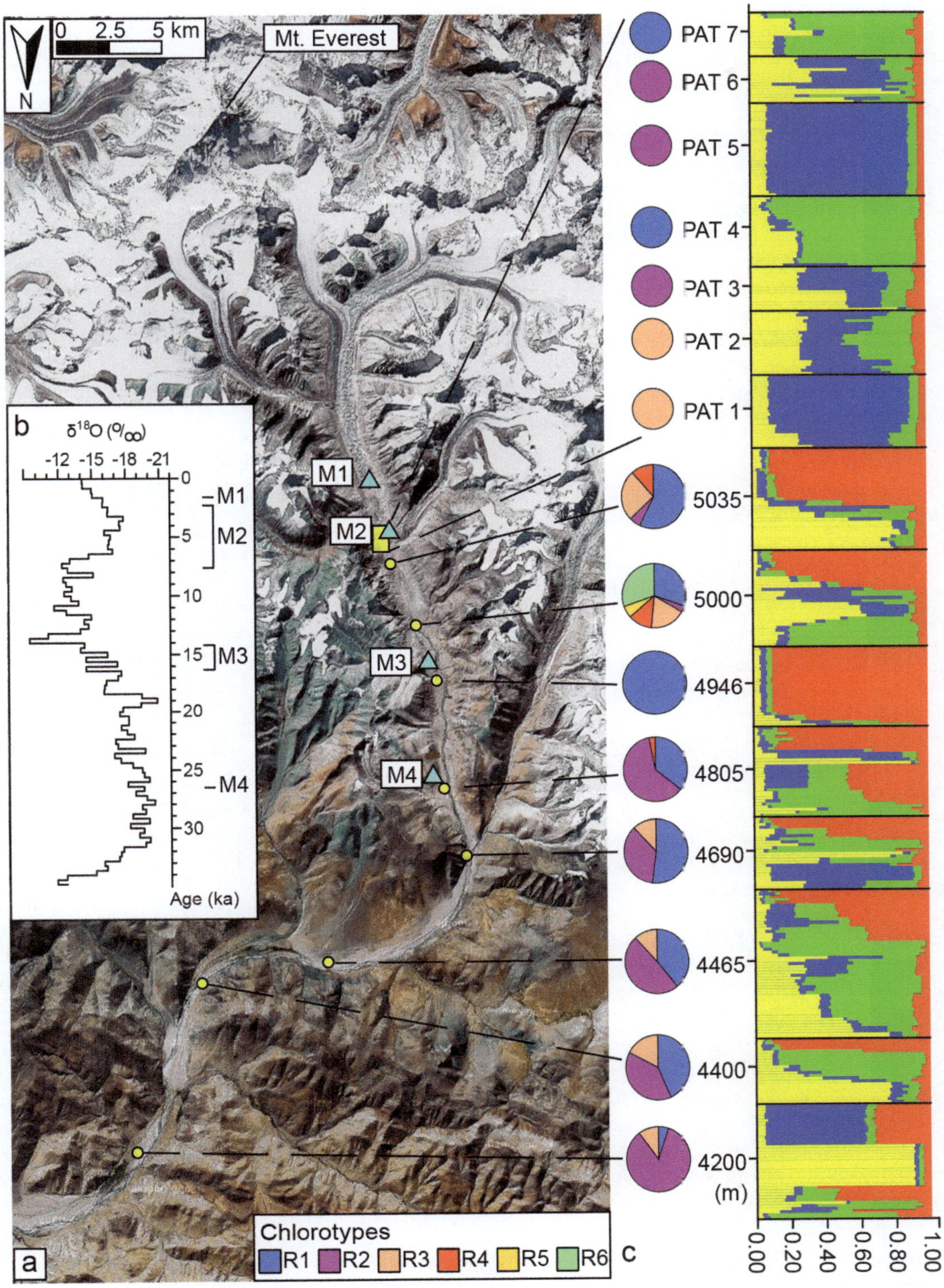

Figure 1. Sampling locations of *Hippophae tibetana* and the genetic composition of each population. (*a*) Map showing the sampling locations along the Rongbuk Valley, north of Mt. Qomolangma. Populations are represented by yellow circles and the patches are in the yellow rectangular area. Blue triangles represent the end-moraine of each glaciation [47], [52], [54] (M1, Little Ice Age moraine; M2, Rongbude moraine; M3 Rongbuk moraine; M4, Jilong moraine). (*b*) Temperature change revealed in the Guliya ice-core $\delta^{18}O$ record over the past 35 ka [38]. OSL ages of the moraines [47] are indicated along the time axis. (*c*) Elevation of each population, chlorotype composition, and proportion shown as pie charts (left column) and the population assignment test results with the software STRUCTURE (right column).

Climate Conditions

In terms of climate conditions, the Himalaya and the QTP are located in the inter-tropical convergence zone. Mt. Qomolangma is influenced by the mid-latitude westerlies and the south Asian monsoon and the climate in the Rongbuk valley is semi-arid and cold [57], [58]. According to data obtained from 2008 to 2011, annual mean air temperature at Qomolangma Station (4300 m alt.) is 4.3°C and average annual precipitation is 213.4 mm [58]. The rainy season is from June to August in summer, when moisture comes with the monsoon from the Indian Ocean and the Bay of Bengal. Although it is cold in the valley, air temperature has increased in the last four decades at a rate of 0.302°C per decade (from 1971 to 2004), reflecting a high sensitivity to recent global warming [46]. This warming trend has contributed to the retreat of glaciers in recent decades [55]. In this valley, mean air temperature drops 0.7–0.8°C with a 100 m increase in altitude as a result of changes in atmospheric pressure and surface conditions [48].

Sample Collection and Field Investigation

In the Rongbuk valley, *H. tibetana* extends from ~4200 m to ~5200 m asl along the banks of the river and is one of dominant species in the valley. There are some large colonies (area>1000 m^2) between 4200~5035 m asl, but only a few small colonies of 20–250 m^2 areas are scattered above 5035 m (>5035 m but < 5100 m asl). We are not sure that all the individuals of *H. tibetana* in this valley consist of one or some populations, so when we collected samples, we treated large colonies at different altitudes as different populations (POP) and a small colony above 5035 m as a patch (PAT). Leaves of *H. tibetana* were collected in the summers of 2007−2010. Population samples (POP) were collected along the elevational gradient from 4200 m to 5035 m every 200 m at most, and individuals are at least 10 m apart (Fig. 1a). We also surveyed the area from 5035 m to 5200 m in a ~150-metre-wide belt transect, and collected leaf samples from all the patches of the shrub that we found (Table 2; Fig. 1a). In the patches, the spacing of individuals is not always over 10 m because of the small area of some patches. Leaves were dried in silica gel. In total, we analyzed 414 individuals from 8 populations and 7 patches, and their GPS data are shown in Table 1. Field measurements were conducted in January 2011 to determine the colonization times of the patches. Areas, crown diameters, and colonization times of the patches were estimated. The five biggest individuals in each patch and the highest large colony (POP 8 at 5035 m) were measured. The basal stems of the biggest individual in each patch and POP8 at 5035 m were cut (~5 cm length) and conserved in plastic bags. Transverse section toward root was smoothed in the laboratory and annual rings were identified by using a dissecting microscope.

Genetic Analysis

Total genomic DNA was extracted using the modified CTAB method [43]. Both the chloroplast trnT-trnF sequences and nuclear microsatellites are used in the present study.

Based on the sequences from our previous phylogeographic study [43], we selected partial *trn*T-*trn*F sequences of chloroplast DNA to identify chlorotypes. Amplification and sequencing were carried out using primers INP1 5′ TAGATCGTTCAAGTATT-CAAAATA 3′ and INP2 5′ CAGGTCGTCATTAATCATTTT-CAGA 3′, following the procedures of Wang *et al.* (2010). The DNA sequences of *H. tibetana* individuals were aligned using the program CLUSTAL X with subsequent manual adjustments, and assigned to different haplotypes using DnaSP 5.10 [59]. Our previous study has shown that all the *trn*T-*trn*F haplotypes of populations in the Mt. Qomolangma area and adjoining region belong to the B-lineage [43] To evaluate the evolutionary relationships among chlorotypes, the chlorotypes obtained in the present study and all the B-lineage chlorotypes from Wang *et al.* (2010) were used to construct a network by the program NETWORK version 4.5.1.0 (http://www.fluxus-engineering.com) which uses the median joining approach to combine a minimum spanning tree within a single network and then, by the criterion of parsimony, median vectors are added to the network (Fig. 2). The indices of unbiased genetic diversity (D) and nucleotide diversity (π) [60] were calculated for each population using the program ARLEQUIN version 3.1 [61]. Using part of the *trn*T-*trn*F sequences that we had obtained for all individuals, pairwise mismatch distributions were carried out with ARLEQUIN version 3.1 for populations containing different chlorotypes to infer their demographic history. The sum of the squared differences (SSD) was used as a statistical test to accept or reject the hypothesis of sudden population expansion. The raggedness index and its significance were calculated to quantify the smoothness of the observed mismatch distribution [62].

Microsatellite loci were developed using 5′-anchored PCR [63], [64] and five microsatellite loci exhibiting polymorphism among individuals from the Rongbuk Valley (Table S1 in File S1) were amplified following the procedures of Song *et al.* (2003) to analyze the genetic variation [65]. All the primers had reliable scoring and the interpretation of electropherograms was performed by the same person in the same laboratory for all samples. For each population, the intra-population genetic diversity was evaluated (Table 2). Fixation index (F_{IS}) and allelic richness (A, mean number of alleles per locus based on the minimal sample size) were calculated with FSTAT version 2.9.3.2 [66] using five microsatellite loci, and H_E and H_O were calculated with GENETIX version 4.03 (www.genetix.univ-montp2.fr). For tests of deviation from the Hardy-Weinberg equilibrium and genotypic disequilibrium, the p-values obtained (with a 0.05 significance threshold) were adjusted in FSTAT version 2.9.3.2 by applying sequential Bonferroni corrections to avoid false-positives [66]. Pairwise F-statistics (F_{ST}) among populations and patches were calculated using five microsatellite loci with GENETIX version 4.03, the significance of which was tested by comparison of the 95 and 99% confidence intervals derived from 1000 bootstrap permutations

Table 2. Area, crown diameter, and annual ring of 5 patches and the highest population.

Patch No.	Altitude (m)	Area (m^2)	Crown diameter (cm)[a]	Annual rings[b]
POP 8	5035	>5000	56.6±2.70	40
PAT 4	5047	~250	26.6±3.21	37
PAT 3	5046	~50	17.0±2.34	30
PAT 2	5047	~30	14.4±6.95	15
PAT 7	5058	~30	8.8±1.30	21
PAT 5	5066	~20	8.4±1.14	15

[a]the mean value ± S.D. of crown diameters of five individuals with the biggest crow; [b]the annual rings of the individual with the largest basal stem.

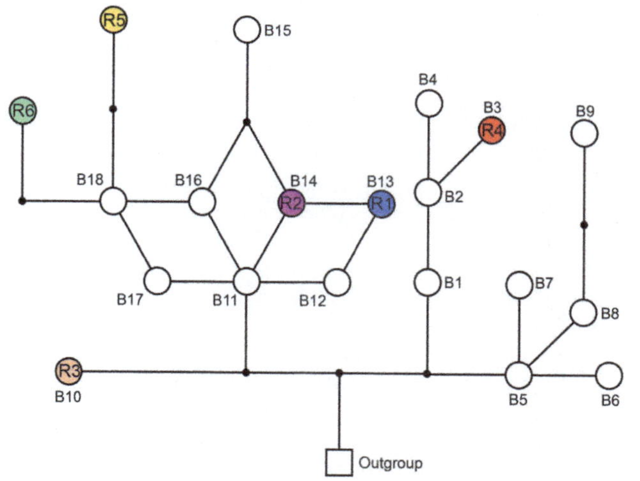

Figure 2. Chlorotype network based on the *trn*T-*trn*F sequences of *H. tibetana*. Circles with letters inside indicate chlorotypes found in the Rongbuk Valley. The other circles represent the remaining B-lineage chlorotypes found in the western part of the species' geographical range and the outgroup consists of the A-lineage chlorotypes that occupy the central part of the range [43]. Genbank accession numbers of these haptotypes were listed in Table S2 in File S1.

individuals across sites. To illustrate the shortest dispersal route between populations, we connected populations on a non-metric multidimensional scaling (NMDS) of F_{ST} using a minimum-spanning tree (MST) [67]. The bottleneck test was performed using BOTTLENECK version 1.2.02 [68] with five microsatellite loci. We employed the Wilcoxon sign-rank test because of the small number of loci [69], [70]. The estimates are based on 1000 simulations performed under both the strict one-step stepwise mutation model (SMM) and the two-phase model (TPM: 70% single-step mutation and 30% multi-step mutation) assumptions. A Bayesian clustering approach was used to infer population structure with STRUCTURE version 2.2 [71]. We used the admixture model and correlated allele models without any prior information. Twenty independent runs with a burn-in of 10,000 and 100,000 Markov chain Monte Carlo simulations for each value of K (from 1 to 15) were used to evaluate the genomic composition of the populations. We then used the ΔK statistics to evaluate the change in likelihood [72].

Results

Size of Patch and Annual Rings of the Biggest Individual

The sizes of the patches above 5035 m asl and the annual rings of the biggest individual in each patch are shown in Table 2. Five out of seven patches were found in the field investigation of 2011 according to the previous GPS data, but PAT 1 and PAT 6 could not be re-found. The topography of the sample region is rugged and some patches are small, thus we guess that the inaccurate GPS data of some patches and the harsh weather in January 2011 led us not to re-find these two patches.

The basal stems of all the cut individuals have clear annual rings when viewed under a dissecting microscope. The age (annual ring) of the biggest individual of the large colony (POP8) at 5035 m is 40 years and that of the largest patch at 5047 m is 37 years. The patch with the smallest size and crown diameter, PAT 5, grows at the highest altitude, 5066 m, and has the youngest age (15 years). The crown diameter of patch is related negatively to its altitude ($R^2 = 0.69$, $F = 8.83$, $p = 0.041$), and the same relationship exists between the annual ring of the biggest individual and its altitude ($R^2 = 0.56$, $F = 5.14$, $p = 0.086$).

Chlorotypes and Distribution

Chloroplast DNA *trn*T-*trn*F sequences of 414 individuals from 8 populations (large colonies) and 7 patches (small colonies over 5035 m asl) at different elevations in the Rongbuk Valley were sequenced in this study and six chlorotypes were identified (Fig. 1c and Table 1). Genebank accession numbers of the chlorotypes are R1 - GU561447, R2 - GU561448, R3 - GU561444, R4 - GU561441, R5 - JF268789, and R6 - JF268790. Among these six chlorotypes, R5 and R6 were first found in this study and the others (R1–R4) were reported in our previous study [43]. Comparing with all the chlorotypes found in *H.tibetana* [43], four of these six haplotypes, R1, R4, R5 and R6, are endemic to this valley, i.e. private haplotypes. The chlorotype compositions of the populations at different elevations are different (Fig. 1c). All the populations below 4700 m contain the same chlorotype component: three chlorotypes (R1, R2, R3), while the population at 5000 m has all six chlorotypes (Fig. 1c); the patches over 5035 m also have three chlorotypes (R1, R2, R3), but each patch consists of only one (Fig. 1c). Unbiased genetic diversity (D) within the populations in the Rongbuk Valley ranges from 0 to 0.7879, and nucleotide diversity (π) from 0 to 0.00343 (Table 1). A network of the six chlorotypes and other chlorotypes occurred in the west part of the species' geographical range (the B-lineage chlorotype, sequences from Wang et al. 2010) shows that chlorotypes found in the Rongbuk Valley are scattered within the network (Fig. 2).

Genetic Structure

Five microsatellite loci exhibiting polymorphism among individuals from the Rongbuk Valley were obtained. Their primer sequences are listed in Table S1 in File S1. Bayesian clustering of the information from the five microsatellite loci demonstrates that the model with $K = 4$ explains the data satisfactorily based on the ΔK statistics. Each patch contains genetically similar individuals revealed by the STRUCTURE analysis, whereas populations usually have more complex components except for the population at 4946 m (Fig. 1c). All the patches and the 4946 m-population also exhibit large negative F_{IS} values and a significant deviation from the Hardy-Weinberg equilibrium (Table 1). Pairwise F_{ST} shows little differentiation between populations except for the 4946 m-population (Table 3). The NMDS projection accounts for 88.5% (stress = 0.115) of the variance in the F_{ST} matrix (Table 3, Fig. 3). The 4690 m-population and the 5000 m-population are at the centre of the MST; populations higher than 4690 m are all linked to the 5000 m-population while others are linked to the 4690 m-population. All the patches are connected to populations below 4690 m in the MST (Fig. 3).

Population Expansion and Population Bottleneck

Results of the Wilcoxon sign-rank test and the mismatch distribution analysis are given in Table 4. The Wilcoxon test values under both the strict one-step stepwise mutation model (SMM) and the two-phase model (TMP) are significant in POP 1, 2, 5, 7, and 8, indicating that these five populations have been through recent bottlenecks. Furthermore, when performing mismatch analysis, the SSD p-value and raggedness index for POP 1, 2, 4, 5, and 7 are not significant at the 95% significance level, suggesting that these five populations have experienced demographic expansion (Table 4).

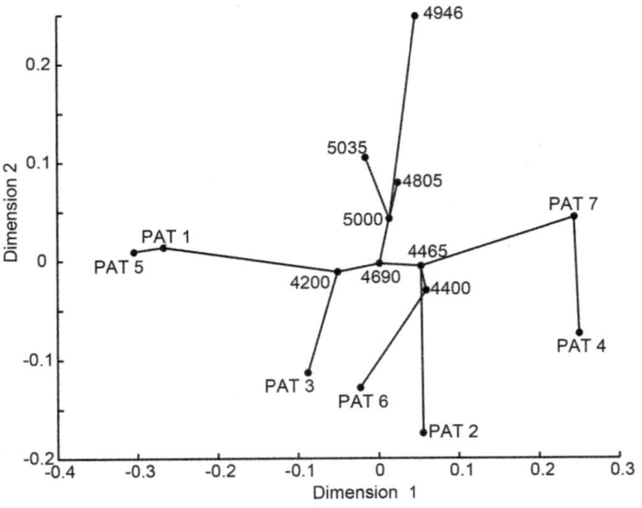

Figure 3. Non-metric multidimensional scaling of pairwise F_{ST} and a minimum-spanning tree linking populations and patches. Populations are indicated by their altitudes (m).

Discussion

The Microrefugia of *H. tibetana* in the Rongbuk Valley

In previous study, we inferred that the Rongbuk Valley may have been a LGM microrefugium for *H.tibetana* based on private chlorotypes [43]. In the present study, we analysed 414 individuals from this valley using chloroplast DNA *trn*T-*trn*F sequences and five nuclear microsatellite loci. More private chlorotypes were found and their distribution in the valley was revealed. By comparing chlorotypes here to those of *H. tibetana* elsewhere, including the two known nearest populations in the Mt. Qomolangma area (Dingri population and Nielamu population, see reference 43), we found that four (R1, R4, R5, R6) of six haplotypes found in the Rongbuk Valley are endemic to this valley, i.e. private haplotypes. In addition, these six chlorotypes are scattered in the chlorotype network (Fig. 2), and their divergence time could date back to 1.52±0.814 million years ago [43]. Considering the Rongbuk Valley is so high and had dramatically influenced by glaciations during the LGM [47], [49], the fact that so many private chlorotypes occur here is surprising. These private haplotypes only have three potential origins: origination from common haplotypes (R2 and R3) after the LGM, coming from other areas after the LGM, and survival through the LGM in the valley. As Opgenoorth et al. discussed in the previous study on *Juniperus tibetica* complex [22], these private chlorotypes are very unlikely to have originated from common haplotypes after the LGM according to the substitution rate of cpDNA *trn*T-*trn*F regions (from the fastest, 8.24×10^{-9} substitutions per site per year, to the lowest, 4.87×10^{-10}) [73]. Also, they are very unlikely to have come recently from other areas because in these other areas, including regions very near this site, the above private chlorotypes have not found according to our previous study [43]. Thus, the third origin, survival in the valley through the LGM is the only acceptable hypothesis. These results provide strong evidence that the Rongbuk valley was a refuge for *H.tibetana* during the LGM and possibly earlier glaciations. According to the definition of microrefugium [9], it is a microrefugium for *H.tibetana* because the Rongbuk valley is far away from its main range and a very small part of its whole range.

Shifts of *H. tibetana* in the Last 25,000 Years

As the 5000 m-population has all six chlorotypes and the highest genetic diversity (D and π in Table 1), including two endemic types (R5 and R6) in this population, it seems that in the valley, the microrefugium for *H. tibetana* in the past was located in the area at 5000 m. However, the integration of geological [47], [52], [54], glaciological [38], [45], [55], meteorological [46], [48], and genetic information reveals that the survival process may have been complex. From 25 ka B.P. until today, there have been four major periods of glacier advance that formed four terminal moraines in the Rongbuk valley (M1-M4, Fig. 1a) [52]. A terminal- or end-moraine, forms at the end of the glacier tongue and marks the maximum advance of the glacier. During the LGM, the valley area over 4800 m was covered by the glacier tongue, which consists of ice and gravel and would prevent *H. tibetana* from colonizing. The valley, except riversides and riverbed, is very arid today because of its special topography (46) and *H. tibetana* is not able to survive on the xeric side slopes in the valley. During the glacial stages, the QTP became more arid [74], [75], and the valley would be more arid. Thus, the xeric side slopes in the valley were not suitable habitats for *H. tibetana*. In fact, all populations of *H. tibetana* found in our field investigation and reported by other collectors did not occur over glacier tongues. These results support that the upper limit of *H. tibetana* in the valley must have been below the glacier tongue.

According to data from the Guliya ice-core (Fig. 1b), temperature during the LGM was ~9°C lower than present [38], [56], and the glacier tongue dropped to ~4800 m [47], [54] (M4 in Fig. 1a, b), so the upper limit of *H. tibetana* must then have been below 4800 m. According to the altitudinal difference between the recent glacier tongue and the recent upper limit of *H. tibetana* in this valley (~200 m), we infer that the upper limit of *H. tibetana* during the LGM might have been about 4600 m. It means that the microrefugium for *H. tibetana* during the LGM was located below 4800 m, and all six chlorotypes of *H. tibetana* should occur in the area below 4800 m. But now, though we checked all the populations below 4800 m, three private chlorotypes (R4, R5, and R6) could not be found in this area. This fact indicates that after the LGM, populations of *H. tibetana* in the LGM microrefugium, i.e. climate relict according to Hampe & Jump (2011) [24], have already undergone dramatic contraction or extinction. From the LGM to the Younger Dryas (YD, a cold episode around 12 ka ago), temperature fluctuated several times [38]. Annual average temperature at ~14.2 ka BP may have been up to 7°C higher than present, but it decreased dramatically by ~10°C around 12 ka BP, and suddenly increased by 4.5°C about 11 ka BP according to data from the Guliya ice-core (Fig. 1b). The glacier formed the conspicuous Rongbuk end-moraine complex (5000 m, M3 in Fig. 1a). After the YD, there was a ~4 ka warm period (~11 to 7 ka ago, Fig. 1b) and temperature in this period was ~3°C higher than present (Fig. 1b). Although these rapid and dramatic climate fluctuations were very likely to drive *H. tibetana* to shift up and down in the valley, the history of this species from the LGM to the YD in this valley is difficult to be proved. But at least during the warm period after the YD (~11 to 7 ka ago, Fig. 1b), *H. tibetana* had migrated upward to 5000 m. The reasons are as follows: 1) only the 5000 m-population now has the highest genetic divergence with all six chlorotypes, 2) two of the six chlorotypes, R5 and R6, are endemic to this 5000 m-population, strongly indicating that not less than these two endemic chlorotypes had colonized at this altitude during this period, and 3) these two endemic chlorotypes are unlikely to originate here from other chlorotypes in such a short period (Fig. 2). It means that from the LGM to the warm time after the YD, the upper limit of *H. tibetana*

Table 3. Genetic differences (*F*st) among populations and patches using data from five microsatellite loci (Table S1).

FST	POP2	POP3	POP4	POP5	POP6	POP7	POP8	PAT1	PAT2	PAT3	PAT4	PAT5	PAT6	PAT7
POP1	0.0603^{ns}	0.0928**	0.0303^{ns}	0.0884^{ns}	0.1950**	0.0488^{ns}	0.0631^{ns}	0.2147**	0.1671**	0.0403^{ns}	0.2473**	0.2336**	0.0552^{ns}	0.2564**
POP2		0.0130^{ns}	0.0158^{ns}	0.0632^{ns}	0.2196**	0.0377^{ns}	0.0860^{ns}	0.2948**	0.0863**	0.1308**	0.1425**	0.3274**	0.0436^{ns}	0.1157**
POP3			0.0119^{ns}	0.0525**	0.1609**	0.0200^{ns}	0.0834**	0.2351**	0.0715**	0.1225**	0.1542**	0.2664**	0.0897**	0.0751^{ns}
POP4				0.0274^{ns}	0.1638**	−0.0006^{ns}	0.0548^{ns}	0.1820**	0.0876**	0.0718^{ns}	0.1757**	0.2096**	0.0512^{ns}	0.1592**
POP5					0.1503**	0.0068^{ns}	0.0345^{ns}	0.2583**	0.1864**	0.1882**	0.1994**	0.2866**	0.1632**	0.1923**
POP6						0.1244**	0.1254**	0.4405**	0.3579**	0.2944**	0.3667**	0.4813**	0.3480**	0.3050**
POP7							0.0253^{ns}	0.2024**	0.1357**	0.1138**	0.1767**	0.2330**	0.1144**	0.1450**
POP8								0.2835**	0.2057**	0.1794**	0.2997**	0.3171**	0.1759**	0.2464**
PAT1									0.3075**	0.2777**	0.4712**	0.0054^{ns}	0.2863**	0.5185**
PAT2										0.2148**	0.3252**	0.3430**	0.0785^{ns}	0.2545**
PAT3											0.3123**	0.2995**	0.0710^{ns}	0.3420**
PAT4												0.5007**	0.2666**	0.1225**
PAT5													0.3066**	0.5612**
PAT6														0.2920**

The *F*-statistics (F_{ST}) values were calculated with GENETIX version 4.03, the significance of which was tested by comparison of the 95 and 99% confidence intervals derived from 1,000 bootstrap permutations individuals across sites.
$^{ns}p>0.05$,
$^{*}p\leq0.05$,
$^{**}p<0.001$.

Table 4. Bottleneck tests and mismatch distribution analysis.

Population No.	Altitude (m)	Wilcoxon Test		Mismatch distribution analysis	
		SMM	TPM	SSD *p*-value	Raggedness index
POP 1	4200	0.0469*	0.0312*	ns	0.273^{ns}
POP 2	4400	0.0312*	0.0156*	ns	0.164^{ns}
POP 3	4465	0.3125^{ns}	0.3125^{ns}	0.028	0.195*
POP 4	4690	0.5000^{ns}	0.3125^{ns}	ns	0.148^{ns}
POP 5	4805	0.0469*	0.0312*	ns	0.218^{ns}
POP 6	4946	0.0625^{ns}	0.0625^{ns}	–	–
POP 7	5000	0.0312*	0.0312*	ns	0.067^{ns}
POP 8	5035	0.0312*	0.0312*	ns	0.476*
PAT 1	>5035	0.2500^{ns}	0.2500^{ns}	–	–
PAT 2	5047	0.1250^{ns}	0.1250^{ns}	–	–
PAT 3	5046	0.1250^{ns}	0.1875^{ns}	–	–
PAT 4	5047	0.0625^{ns}	0.0625^{ns}	–	–
PAT 5	5066	0.2500^{ns}	0.2500^{ns}	–	–
PAT 6	>5035	0.1250^{ns}	0.1250^{ns}	–	–
PAT 7	5058	0.1250^{ns}	0.1250^{ns}	–	–

($^{ns}p>0.05$, $*p\leq0.05$, $**p<0.001$).

shifted at least 200 m upward, and most probably over 400 m (from ~4600 m to 5000 m asl or higher). Meanwhile, the lower limit of this species here might upward shift to 4800 m or higher in this warm period, and at least populations in the LGM microrefugium have undergone extreme contraction, which resulted in the loss of three chlorotypes (R4, R5, R6) in the LGM microrefugium. These results demonstrate that the area about 5000 m plays an interglacial refuge and that some area below 4800 m was merely a glacial refuge.

When populations shift up and down repeatedly, populations will undergo genetic bottlenecks and expansions [76]. These genetic events have been observed in most populations based on microsatellite and chlorotype data, respectively (Table 4). Analysis of microsatellites provides us with more information. Non-metric multidimensional scaling of pairwise *F*-statistics (Table 3) and a minimum-spanning tree show that the 4690 m-population and the 5000 m-population are at the centre of the net (Fig. 3), and that populations higher than 4690 m are all linked to the 5000 m-population while others are linked to the 4690 m-population. These results imply that three present populations below 4690 m are the result of the expansion of populations at higher altitude, and that the lower limit of *H. tibetana* moved upward to at least 4690 m when climate became warmer after the LGM, especially after the YD.

Information from both chlorotypes and microsatellites strongly demonstrate that the genetic structure of populations of *H. tibetana* in the LGM microrefugium has been changed greatly after the LGM. If we infer glacial microrefugia only based on genetic information, we are likely to be misled, because present populations in glacial microrefugia may not be true climate relicts of glacial stages. Similar patterns have been found at a broader scale. The study on chloroplast DNA variation in 22 widespread European trees and shrubs sampled in the same forests has shown that the genetically most diverse populations were not located in glacial refugia in southern Europe but at intermediate latitudes [77]. The authors proposed that it was a likely consequence of the admixture of divergent lineages colonizing the continent from separate refugia. We suspect that present populations in glacial refugia in southern Europe have not been true glacial relics.

Recent Shifts of *H. tibetana*

In the last 100 years, altitudinal ranges of many plants have shifted upward, presumably in response to global temperature increases [29], [32], [33], [37], [78]. The Mt. Qomolangma area is one of the most vulnerable areas to global change [79], [80], [81], and in the last four decades (from 1971 to 2004), annual mean temperature has increased here at a rate of 0.302°C per decade and from 1976 to 2006, the ends of Central Rongbu glacier has retreated about 500 m in length [55], while precipitation has stayed at a constant level [46]. In our field investigation, we only found seven scattered patches in the region over 5035 m (Table 1). These patches are mostly located on the Rongbude terminal moraine (M2, Fig. 1a) and have not been covered by landslides. No dead patches could be found in this region and few or no dead individuals could be found in these patches. These results show that these patches are not the relict of an old large colony but new small colonies, and that the annual ring of the largest individual of a patch can indicate the successful colonization age of this patch. These patches have a colonization age of 15~37 years and show a significant negative relationship between colonization time and altitude, i.e. the higher the altitude of the patch, the younger the patch (Table 2), demonstrating that these patches must have colonized here in the last 40 years and that these patches shifted step by step from lower to higher sites. The highest large colony (POP 8) is located at 5035 m, and the highest patch (PAT 5) at 5066 m. In fact, the annual ring of PAT 5 at 5066 m is 15 years, indicating that *H. tibetana* had arrived at 5066 m 15 years ago and that the shift of *H. tibetana* from 5035 m to 5066 m only spent about 25 years. This rate is slower than the mean rate of shift of alpine plants above tree-line in the European Alps and of forest plants in western Europe (27.8 m per decade; 29.4 m per decade) [33], [82]. The shift of *H. tibetana* in the valley during the past 40 years is also smaller than the inferred value (~100 m) based on an average decrease of 0.7~0.8°C per 100 m increase in altitude and an average increase of 0.302°C per decade in the Qomolangma area [46], [48]. Lenoir *et al.* (2008) showed that species with different ecological properties displayed different rates of shift and that small grasses moved faster than large woody plants [33]. We suspect that the high altitude (over 5000 m) of our study area may influence the upward rate of shift of plants. Grabherr *et al.* (1994) found that mountain plants that grew above 3000 m in the central part of the European Alps have slower rates of shift (<4 m per decade) than those that grow below 3000 m [29]. Besides climate warming and glacier shrinkage, precipitation and other ecological factors have not obviously changed during the past 40 years in the Rongbuk Valley, and human activity was very infrequent in this valley 20 years ago [52], [46]. Thus, global warming is likely to be the main reason for the recent upward shift of *H. tibetana*.

This study also highlights the colonization pattern of primary habitats. Each patch contains only one chlorotype and consists of highly genetically similar individuals inferred from the microsatellite data (Fig. 1c), indicating that all the patches expanded mainly by clonal reproduction, which then resulted in a deviation from the Hardy-Weinberg equilibrium and high F_{ST} values (Table 1). Although the 4946 m-population area is much larger than the patches, it is on a landslide and almost entirely consists of small and genetically identical individuals (Fig. 1c, Table 1), implying it is also a new clonal population after the landslide formed.

Factors Determining the Survival of *H. tibetana* in the Rongbuk Valley

Plant survival depends on a favourable local environment [9], [15], and extreme conditions, stable climates, and distinct differences from the surrounding matrix are the basis for microrefgia [19]. According to detailed topoclimate, Ashcroft et al. (2012) deduced that a mountain area may have not only warm refugia but cool refugia for striking differences in climate among different locations [19]. In the Rongbuk Valley, mountains play an important role in the ability of *H. tibetana* to withstand climate changes. Here, the elevational range of the valley is nearly 1500 m (from 3700 m to 5200 m asl) and mean temperature drops 0.7–0.8°C with each 100 m increase in altitude in this area [48], implying that there is nearly a ~11°C temperature gradient, which not only provides *H. tibetana* with a wide range within which to shift its occurrence in response to climate change, but also provides a series of fine-scale ecological niches with different microclimates that can buffer against temperature fluctuations [15], [83], [84]. In fact, with the retreat of glaciers in warmer periods, the area over 5200 m in the valley is likely to become a suitable habitat for *H. tibetana*.

In addition, glaciers are critical for *H. tibetana*. Historically when it became cold on the QTP, widespread aridity also occurred [74], and the special topography of the Mt. Qomolangma area caused the Rongbuk Valley to receive lower rainfall and experience more droughts. Glacial meltwater have thus provided almost all the moisture for *H. tibetana*. Our field investigation also confirms the importance of glaciers. In the semi-arid and arid areas in the QTP, all the eleven populations of *H. tibetana* were located in the valley below large glaciers except for the Ritu population [43], which was once near glacier [85] but the population is now shrinking as the local glacier disappears. Besides *H. tibetana*, there are many other plants which mainly depend on glacial meltwater for their moisture sources in the Rongbuk Valley [52]. We thus think that this valley may have been a favourable microhabitat for some of these plants in both cold and/or warm stages. Furthermore, in arid areas on the QTP, water is a limiting factor for the growth of many plants [86]. Similar to the Rongbuk Valley, water supply depends largely on glaciers in many places on the Plateau [87]. In these areas, glaciers become a crucial resource for the growth of many plants not only in warm periods but also in cold glacial stages.

Taken together, we present evidence to indicate the Rongbuk Valley, Mt. Qomolangma is a microrefugium of the LGM for *H. tibetana*, and how this plant here has responded to climate changes in the last 25 ka. By combining historical and recent range shifts, we have derived a better understanding of the rates of range shift, the dynamics of populations in the microrefugium after the LGM, and the mode of colonization of *H. tibetana*, all of which are important in inferring the potential locations of microrefugia for *H. tibetana* and predicting the response of this endemic species in relation to future climate warming.

Supporting Information

File S1 Tables S1–S2. Table S1. Characterization of polymorphic microsatellite loci in this study. PCR products were electrophoresed on 7.5 m urea 6% polyacrylamide gel, sized with the DNA ladder pUC19/MspI (Fermentas Life Sciences) and visualized by silver stain as in ref 36. All the primers had reliable scoring. Ta, annealing temperature; NA, number of alleles; HO, observed heterozygosity; HE, expected heterozygosity. HO and HE were calculated with Genetix version 4.03 (www.genetix.univ-montp2.fr). When POP 7 and the seven patches were not counted because they are most probably new clone populations, no significant Hardy–Weinberg disequilibrium was detected for each locus (for the seven remaining populations each or as a whole). Table S2. Genbank accession numbers of the haptotypes in Fig. 2. Haptotype B1–B18 and the haptotypes of outgroup were found in our previous work (reference 43). The outgroup consists of Haptotype A1–A6. R5 and R6 were found in the present study.

Acknowledgments

We thank Ou La and Zhiping Song for their help in sample collection. We are indebted to Cathy Jenks, Wen Fan and Xinji Li for editorial help.

Author Contributions

Conceived and designed the experiments: WZ. Performed the experiments: WZ QL FL HW YZ YF LX. Analyzed the data: MY LX HW WZ. Contributed reagents/materials/analysis tools: LX HW MY YZ KS JC QW. Wrote the paper: WZ LX HW HJBB. Obtained materials: WZ KS HW FL.

References

1. Hewitt GM (2000) The genetic legacy of the Quaternary ice ages. Nature 405:907–913.
2. Shafer A, Cullingham CI, Cote SD, Coltman DW (2010) Of glaciers and refugia: a decade of study sheds new light on the phylogeography of northwestern North America. Mol Ecol 19: 4589–4621.
3. Stewart JR, Lister AM, Barnes I, Dalen L (2010) Refugia revisited: individualistic responses of species in space and time. Philos T R Soc B 277: 661–671.
4. Bennett KD, Provan J (2008) What do we mean by 'refugia'? Quaternary Sci Rev 27: 2449–2455.
5. Birks HJB, Willis KJ (2008) Alpines, trees, and refugia in Europe. Plant Ecol Divers 1: 147–160.
6. Keppel G, Van Niel KP, Wardell-Johnson GW, Yates CJ, Byrne M, et al. (2012) Refugia: identifying and understanding safe havens for biodiversity under climate change. Global Ecol Biogeogr 21: 393–404.

7. Abbott RJ, Brochmann C (2003) History and evolution of the arctic flora: in the footsteps of Eric Hultén. Mol Ecol 12: 299–313.
8. Holderegger R, Thiel-Egenter C (2009) A discussion of different types of glacial refugia used in mountain biogeography and phylogeography. J Biogeogr 36: 476–480.
9. Rull V (2009) Microrefugia. J Biogeogr 36: 481–484.
10. Binney HA, Willis KJ, Edwards ME, Bhagwat SA, Anderson PM, et al. (2009) The distribution of late-Quaternary woody taxa in northern Eurasia: evidence from a new macrofossil database. Quaternary Sci Rev 28: 2445–2464.
11. Rull V (2010) On microrefugia and cryptic refugia. J Biogeogr 37: 1623–1625.
12. Provan J, Bennett KD (2008) Phylogeographic insights into cryptic glacial refugia. Trends Ecol Evol 23: 564–571.
13. Stewart IT, Cayan DR, Dettinger MD (2005) Changes toward earlier streamflow timing across western North America. J Climate 18: 1136–1155.
14. Anderson LL, Hu FS, Nelson DM, Petit RJ, Paige KN (2006) Ice-age endurance: DNA evidence of a white spruce refugium in Alaska. P Natl Acad Sci U S A 103: 12447–12450.
15. Dobrowski SZ (2011) A climatic basis for microrefugia: the influence of terrain on climate. Global Change Biol 17: 1022–1035.
16. Hampe A, Rodriguez-Sanchez F, Dobrowski S, Hu FS, Gavin DG (2013) Climate refugia: from the Last Glacial Maximum to the twenty-first century. New Phytol 197: 16–18.
17. McLachlan JS, Clark JS, Manos PS (2005) Molecular indicators of tree migration capacity under rapid climate change. Ecology 86: 2088–2098.
18. Pearson RG (2006) Climate change and the migration capacity of species. Trends Ecol Evol 21: 111–113.
19. Ashcroft MB, Gollan JR, Warton DI, Ramp D (2012) A novel approach to quantify and locate potential microrefugia using topoclimate, climate stability, and isolation from the matrix. Global Change Biol 18: 1866–1879.
20. Graham CH, VanDerWal J, Phillips SJ, Moritz C, Williams SE (2010) Dynamic refugia and species persistence: tracking spatial shifts in habitat through time. Ecography 33: 1062–1069.
21. Previsic A, Walton C, Kucinic M, Mitrikeski PT, Kerovec M (2009) Pleistocene divergence of Dinaric *Drusus* endemics (Trichoptera, Limnephilidae) in multiple microrefugia within the Balkan Peninsula. Mol Ecol 18: 634–647.
22. Opgenoorth L, Vendramin GG, Mao KS, Miehe G, Miehe S, et al. (2010) Tree endurance on the Tibetan Plateau marks the world's highest known tree line of the Last Glacial Maximum. New Phytol 185: 332–342.
23. Moeller DA, Geber MA, Tiffin P (2011) Population genetics and the evolution of geographic range limits in an annual plant. Am Nat 178: S44–S61.
24. Hampe A, Jump AS, Futuyma DJ, Shaffer HB, Simberloff D (2011) Climate relicts: past, present, future. Annu Rev Ecol Evol S 42: 313–333.
25. Galbreath KE, Hafner DJ, Zamudio KR (2009) When cold is better: climate-driven elevation shifts yield complex patterns of diversification and demography in an alpine specialist (American pika, *Ochotona princeps*). Evolution 63: 2848–2863.
26. Mosblech N, Bush MB, van Woesik R (2011) On metapopulations and microrefugia: palaeoecological insights. J Biogeogr 38: 419–429.
27. McLaughlin BC, Zavaleta ES (2012) Predicting species responses to climate change: demography and climate microrefugia in California valley oak (*Quercus lobata*). Global Change Biol 18: 2301–2312.
28. Hansen J, Sato M, Ruedy R, Lo K, Lea DW, et al. (2006) Global temperature change. P Natl Acad Sci U S A 103: 14288–14293.
29. Grabherr G, Gottfried M, Pauli H (1994) Climate effects on mountain plants. Nature 369: 448.
30. Walther GR, Post E, Convey P, Menzel A, Parmesan C, et al. (2002) Ecological responses to recent climate change. Nature 416: 389–395.
31. Parmesan C (2006) Ecological and evolutionary responses to recent climate change. Annu Rev Ecol Evol S 37: 637–669.
32. Colwell RK, Brehm G, Cardelus CL, Gilman AC, Longino JT (2008) Global warming, elevational range shifts, and lowland biotic attrition in the wet tropics. Science 322: 258–261.
33. Lenoir J, Gegout JC, Marquet PA, de Ruffray P, Brisse H (2008) A significant upward shift in plant species optimum elevation during the 20th century. Science 320: 1768–1771.
34. Lenoir J, Svenning JC (2013) Latitudinal and elevational range shifts under contemporary climate change. Encyclopedia of Biodiversity Vol. 4. San Diego, CA: Academic Press. 599–611p.
35. Thomas CD, Cameron A, Green RE, Bakkenes M, Beaumont LJ, et al. (2004) Extinction risk from climate change. Nature 427: 145–148.
36. Parmesan C, Yohe G (2003) A globally coherent fingerprint of climate change impacts across natural systems. Nature 421: 37–42.
37. Pauli H, Gottfried M, Reiter K, Klettner C, Grabherr G (2007) Signals of range expansions and contractions of vascular plants in the high Alps: observations (1994–2004) at the GLORIA master site Schrankogel, Tyrol, Austria. Global Change Biol 13: 147–156.
38. Thompson LG, Davis ME, Mosley-Thompson E, Sowers TA, Henderson KA, et al. (1998) A 25,000-year tropical climate history from Bolivian ice cores. Science 282: 1858–1864.
39. Thompson LG, Yao T, Davis ME, Henderson KA, Mosley-Thompson E, et al. (1997) Tropical climate instability: The last glacial cycle from a Qinghai-Tibetan ice core. Science 276: 1821–1825.
40. Stott L, Poulsen C, Lund S, Thunell R (2002) Super ENSO and global climate oscillations at millennial time scales. Science 297: 222–226.
41. Bush MB, Silman MR, Urrego DH (2004) 48,000 years of climate and forest change in a biodiversity hot spot. Science 303: 827–829.
42. Zhang ZR (1983) Elaeagnaceae, genus *Hippophae*. Delectis Florae Reipublicae Popularis Sinicae. Beijing: Science Press. 61–64p.
43. Wang H, Qiong L, Sun K, Lu F, Wang YG, et al. (2010) Phylogeographic structure of *Hippophae tibetana* (Elaeagnaceae) highlights the highest microrefugia and the rapid uplift of the Qinghai-Tibetan Plateau. Mol Ecol 19: 2964–2979.
44. Chen XL, Ma RJ, Sun K, Lin YS (2003) Germplasm resource and habitat types of Seabuckthorn in China. Acta Botanica Boreali-Occidentalia Sinica 23: 5.
45. Kang SC, Qin DH, Mayewski PA, Wake CP, Ren JW (2001) Climatic and environmental records from the Far East Rongbuk ice core, Mt. Qomolangma (Mt. Everest). Episodes 24: 176–181.
46. Yang XC, Zhang YL, Zhang W, Yan YP, Wang ZF, et al. (2006) Climate change in Mt. Qomolangma region since 1971. J Geogr Sci 16: 11.
47. Owen LA, Robinson R, Benn DI, Finkel RC, Davis NK, et al. (2009) Quaternary glaciation of Mount Everest. Quaternary Sci Rev 28: 1412–1433.
48. Yang XG, Zhang TJ, Qin DH, Kang SC, Qin XA (2011) Characteristics and changes in air temperature and glacier's response on the north slope of Mt. Qomolangma (Mt. Everest). Arct Antarct Alp Res 43: 147–160.
49. Xie ZC, Su Z (1975) Developing conditions, amounts and distributions of glaciers in the Mt. Qomolangma area. Dissertation of the Scientific Expedition to Mt. Qomolangma (1966–1968), Modern Glacier and Topography. Beijing: Science Press. 1–14p.
50. Pecci M, Pignotti S, Smiraglia C, Mortara G (2010) Geomorphology of the central and frontal Rongbuck glacier area (Mt. Everest, Tibet). Geogr Fis Din Quat 33: 79–92.
51. Burbank DW, Kang JC (1991) Relative dating of Quaternary Moraines, Rongbuk Valley, Mount Everest, Tibet - Implications for an ice-sheet on the Tibetan Plateau. Quaternary Res 36: 1–18.
52. Team of Scientific Expedition to Mt. Qomolangma C (1962) Report of the Scientific Expedition to Mt. Qomolangma. Science Press: Beijing.
53. Zheng BS, Shi YF (1976) A study of Quaternary glaciations of the Mount Qomolangma Region. Report of the Scientific Expedition to Mt. Qomolangma (1966–1968), Quaternary Geology. Beijing: Science Press. 29–62p.
54. Hu E, Yi CI, Li YJ (2010) Observation and evolution investigation of the moraine geomorphology in the Rongbuk Valley of Mount Qomolangma. J Glaciol Geocryol, 32, 316–324.
55. Nie Y, Zhang YL, Liu LS, Zhang JP (2010) Glacial change in the vicinity of Mt. Qomolangma (Everest), central high Himalayas since 1976. J Geogr Sci 20: 667–686.
56. Yao TD, Thompson LG, Qin DH, Tian LD, Jiao KQ, et al. (1996) Variations in temperature and precipitation in the past 2000a on the Xizang (Tibet) Plateau - Guliya ice core record. Sci China Ser D 39: 425–433.
57. Yang XG, Qin DH, Zhang TJ, Kang SC, Qin X (2012) Characteristics of the air temperature and humidity on the north slope of Mt. Qomolangma. Acta Meteorol Sin 70: 12.
58. Wang ZY, Ma YM, Liu JS, Han CB (2013) Characteristic analyses on hydrological and related meteorological factors on the north slope of Mt. Qomolangma. Plateau Meteorol 32: 7.
59. Rozas J, Sanchez-DelBarrio JC, Messeguer X, Rozas R (2003) DnaSP, DNA polymorphism analyses by the coalescent and other methods. Bioinformatics 19: 2496–2497.
60. Nei M (1987) Molecular Evolutionary Genetics. New York: Columbia University Press.
61. Excoffier L, Laval G, Schneider S (2005) Arlequin (version 3.0): An integrated software package for population genetics data analysis. Evol Bioinform 1: 47–50.
62. Harpending HC (1994) Signature of ancient population-growth in a low-resolution mitochondrial-DNA mismatch distribution. Hum Biol 66: 591–600.
63. Fisher PJ, Gardner RC, Richardson TE (1996) Single locus microsatellites isolated using 5′ anchored PCR. Nucleic Acids Res 24: 4369–4371.
64. Zhou YB, Wang H, Yang M, Chen JK, Zhang WJ (2009) Development of microsatellites for *Scirpus mariqueter* Wang et Tang (Cyperaceae) and cross-species amplification in *Scirpus planiculmis* F. Schmidt. Mol Ecol Resour 9: 370–372.
65. Song ZP, Xu X, Wang B, Chen JK, Lu BR (2003) Genetic diversity in the northernmost *Oryza rufipogon* populations estimated by SSR markers. Theor Appl Genet 107: 1492–1499.
66. Goudet J (1995) FSTAT (Version 1.2): A computer program to calculate F-statistics. J Hered 86: 485–486.
67. Geffen E, Waidyaratne S, Dalen L, Angerbjorn A, Vila C, et al. (2007) Sea ice occurrence predicts genetic isolation in the Arctic fox. Mol Ecol 16: 4241–4255.
68. Cornuet JM, Luikart G (1996) Description and power analysis of two tests for detecting recent population bottlenecks from allele frequency data. Genetics 144: 2001–2014.
69. Pope LC, Estoup A, Moritz C (2000) Phylogeography and population structure of an ecotonal marsupial, *Bettongia tropica*, determined using mtDNA and microsatellites. Mol Ecol 9: 2041–2053.
70. Hampton JO, Spencer P, Alpers DL, Twigg LE, Woolnough AP, et al. (2004) Molecular techniques, wildlife management and the importance of genetic population structure and dispersal: a case study with feral pigs. J Appl Ecol 41: 735–743.
71. Pritchard JK, Stephens M, Donnelly P (2000) Inference of population structure using multilocus genotype data. Genetics 155: 945–959.

72. Evanno G, Regnaut S, Goudet J (2005) Detecting the number of clusters of individuals using the software STRUCTURE: a simulation study. Mol Ecol 14: 2611–2620.
73. Richardson JE, Pennington RT, Pennington TD, Hollingsworth PM (2001) Rapid diversification of a species-rich genus of neotropical rain forest trees. Science 293: 2242–2245.
74. Gasse F, Arnold M, Fontes JC, Fort M, Gibert E, et al. (1991) A 13,000-year climate record from western Tibet. Nature 353: 742–745.
75. Shi YF, Liu SY (2000) Estimation on the response of glaciers in China to the global warming in the 21st century. Chinese Sci Bull 45: 668–672.
76. Ruedi M, Castella V (2003) Genetic consequences of the ice ages on nurseries of the bat *Myotis myotis*: a mitochondrial and nuclear survey. Mol Ecol 12: 1527–1540.
77. Petit RJ, Aguinagalde I, de Beaulieu JL, Bittkau C, Brewer S, et al. (2003) Glacial refugia: Hotspots but not melting pots of genetic diversity. Science 300: 1563–1565.
78. Harsch MA, Hulme PE, McGlone AS, Duncan RP (2009) Are treelines advancing? A global meta-analysis of treeline response to climate warming. Ecol Lett 12: 1040–1049.
79. Hou SG, Qin DH, Wake CP, Mayewski PA, Ren JW, et al. (2000) Climatological significance of an ice core net-accumulation record at Mt. Qomolangma (Everest). Chinese Sci Bull 45: 259–264.
80. Ren JW, Qin DH, Kang SC, Hou SG, Pu JC, et al. (2004) Glacier variations and climate warming and drying in the central Himalayas. Chinese Sci Bull 49: 65–69.
81. Liu XD, Cheng ZG, Yan LB, Yin ZY (2009) Elevation dependency of recent and future minimum surface air temperature trends in the Tibetan Plateau and its surroundings. Global Planet Change 68: 164–174.
82. Walther GR, Beissner S, Burga CA (2005) Trends in the upward shift of alpine plants. J Veg Sci 16: 541–548.
83. Körner C (2000) Why are there global gradients in species richness? Mountains might hold the answer. Trends Ecol Evol 15: 513–514.
84. Scherrer D, Körner C (2011) Topographically controlled thermal-habitat differentiation buffers alpine plant diversity against climate warming. J Biogeogr 38: 406–416.
85. Li B, Li J, Cui Z (1991) Quaternary Glacial Distribution Map of Qinghai-Xizang (Tibet) Plateau. Beijing: Science Press (Map).
86. Tilman D, Lehman C (2001) Human-caused environmental change: Impacts on plant diversity and evolution. P Natl Acad Sci U S A 98: 5433–5440.
87. Yao TD, Pu J, Lu A, Wang Y, Yu W (2007) Recent glacial retreat and its impact on hydrological processes on the Tibetan plateau, China, and surrounding regions. Arct Antarct Alp Res 39: 642–650.

4

Estimating Soil Organic Carbon Stocks and Spatial Patterns with Statistical and GIS-Based Methods

Junjun Zhi[1], Changwei Jing[2], Shengpan Lin[1], Cao Zhang[1], Qiankun Liu[1], Stephen D. DeGloria[3], Jiaping Wu[2]*

1 College of Environmental and Resource Sciences, Zhejiang University, Hangzhou, China, **2** Ocean College, Zhejiang University, Hangzhou, China, **3** Department of Crop and Soil Sciences, Cornell University, Ithaca, New York, United States of America

Abstract

Accurately quantifying soil organic carbon (SOC) is considered fundamental to studying soil quality, modeling the global carbon cycle, and assessing global climate change. This study evaluated the uncertainties caused by up-scaling of soil properties from the county scale to the provincial scale and from lower-level classification of Soil Species to Soil Group, using four methods: the mean, median, Soil Profile Statistics (SPS), and pedological professional knowledge based (PKB) methods. For the SPS method, SOC stock is calculated at the county scale by multiplying the mean SOC density value of each soil type in a county by its corresponding area. For the mean or median method, SOC density value of each soil type is calculated using provincial arithmetic mean or median. For the PKB method, SOC density value of each soil type is calculated at the county scale considering soil parent materials and spatial locations of all soil profiles. A newly constructed 1:50,000 soil survey geographic database of Zhejiang Province, China, was used for evaluation. Results indicated that with soil classification levels up-scaling from Soil Species to Soil Group, the variation of estimated SOC stocks among different soil classification levels was obviously lower than that among different methods. The difference in the estimated SOC stocks among the four methods was lowest at the Soil Species level. The differences in SOC stocks among the mean, median, and PKB methods for different Soil Groups resulted from the differences in the procedure of aggregating soil profile properties to represent the attributes of one soil type. Compared with the other three estimation methods (i.e., the SPS, mean and median methods), the PKB method holds significant promise for characterizing spatial differences in SOC distribution because spatial locations of all soil profiles are considered during the aggregation procedure.

Editor: Ben Bond-Lamberty, DOE Pacific Northwest National Laboratory, United States of America

Funding: This study was supported by the National Natural Science Foundation of China (No. 30771253) and the Key Project of the Science and Technology Department of Zhejiang Province (No. 2006C22026). The funders had no role in study design, data collection and analysis, decision to publish, or preparation of the manuscript.

Competing Interests: The authors have declared that no competing interests exist.

* E-mail: jw67@zju.edu.cn

Introduction

Soil organic carbon (SOC), which plays a critical role in the global carbon cycle, comprises a major part of the terrestrial carbon reservoir [1–3]. In terrestrial ecosystems, SOC stock is almost three times the size of carbon storage in the vegetation of terrestrial ecosystems [4], and approximately twice as large as carbon storage in the atmosphere [5]. Because of the important role of SOC and its large quantity stored in terrestrial ecosystems, a slight change in SOC stock may influence global climate [6–8]. Accurately quantifying SOC stock has become a focus of present research on global climate change, and is considered essential for studying soil quality, modeling the global carbon cycle, and assessing global climate change [8–12].

Different SOC stock estimates, however, can vary greatly at both global and regional scales [13–17]. For global scales, Bohn [18] estimated the total SOC stock was 3,000 Pg (1 Pg = 10^{15} g), whereas Bolin [19] estimated only 710 Pg, over a four-fold difference. In China, estimated SOC stocks for terrestrial ecosystems range from 50 Pg [20] to 185.7 Pg [21], also approximately a four-fold difference. The methodology for estimating SOC stocks, such as soil profile-based [22–24] and model-based (e.g., the CENTURY [25] and DeNitrification-DeComposition models [26]), is one of the main factors attributed to the wide range of differences in SOC stock estimation from different studies. The soil profile-based methodology calculates SOC stock using soil profiles and their corresponding areas obtained from soil survey products (the soil type method), vegetation type maps (the vegetation type method), or life zone maps (the life zone method), among which the soil type method is the most widely used [9,18–19,27–28]. According to the sources of soil area, the soil type method contains the Soil Profile Statistics (SPS) method and the GIS-based Soil Type (GST) method [2]. The SPS method calculates SOC stock by multiplying the SOC density value of a soil type by its corresponding field survey area recorded in soil survey reports (e.g., Soil Species of China [29]). The GIS-based Soil Type method calculates areas of various soil types accurately based on digital soil map and can provide information on the spatial distribution of SOC stocks.

In China, different scale (i.e., county, provincial, and national) soil survey products (e.g., soil survey reports, soil maps) of the Second National Soil Survey of China are the most important data sources for SOC stock estimations [15,20–21,30–33]. For the

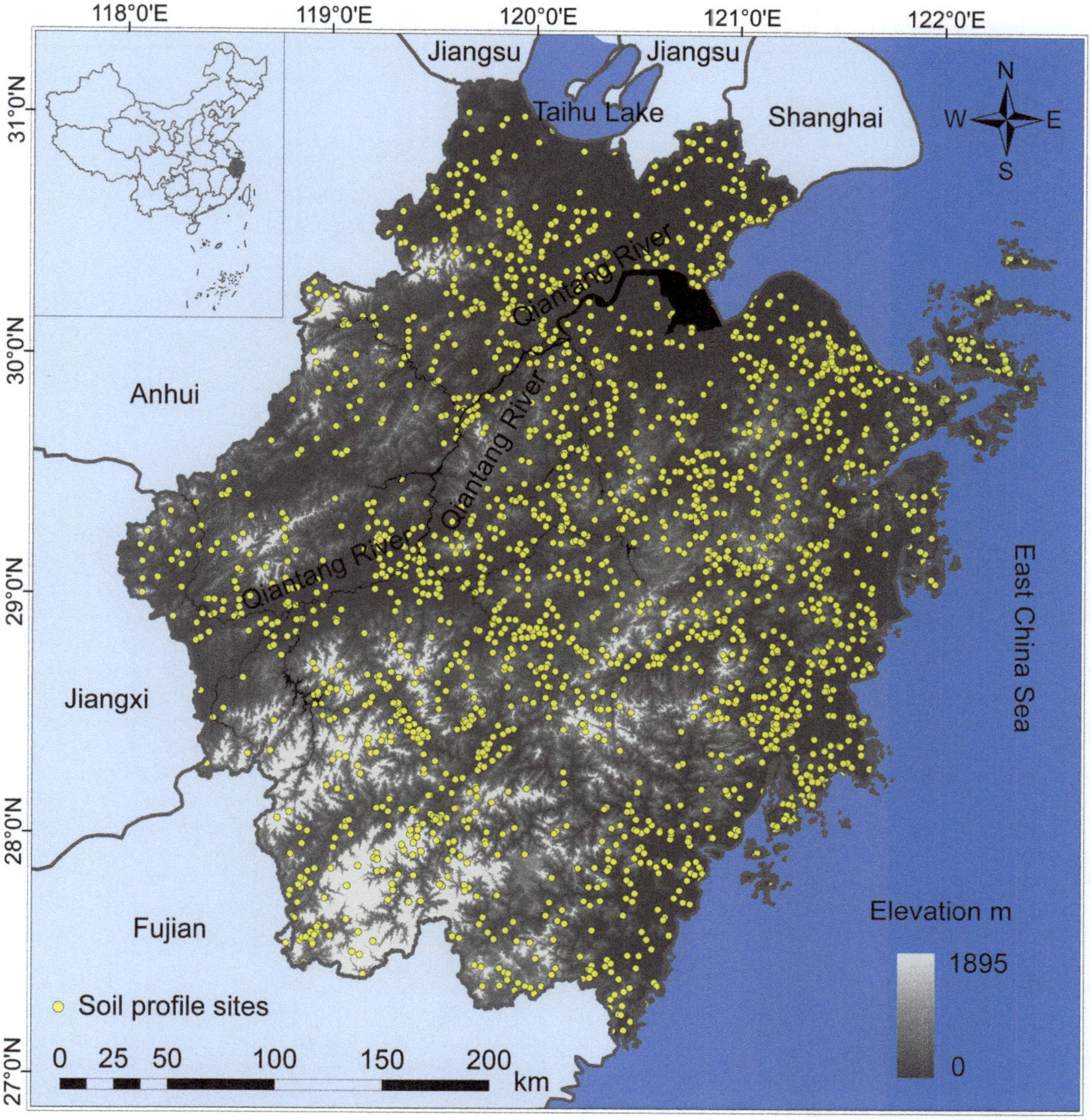

Figure 1. Study area and distribution of soil profile sites.

Second National Soil Survey of China, soil profile properties were aggregated sequentially in the county, provincial, and national scales. Different number of sampling profiles can be used to represent the attributes of one soil type at different scales and can cause uncertainties in SOC stock estimation. At the provincial scale, calculating the mean and median of SOC density values of multiple soil profiles of the same soil type name to represent the SOC density value of that soil type are the two most commonly used GIS-based Soil Type methods [32–34]. At the county scale, the SPS method [20,30–32] and the pedological professional knowledge based (PKB) method [2,15,34–36] are the two most commonly used methods. Compared with the median or mean method, the PKB method aggregates soil profile properties downscaling from the provincial scale to more detail soil map units, which links a SOC density value of each soil profile to a digital soil map according to the identity or similarity in soil parent materials and spatial locations of all soil profiles at the county scale. However, the uncertainties of soil profile properties caused by up-scaling from the county scale to the provincial scale among the four methods remain unknown.

Soil map up-scaling includes soil classification level up-scaling and resolution up-scaling. The former aggregates Soil Species to higher soil classification levels (e.g., Soil Group, Subgroup, Soil Family); the latter aggregates higher resolution soil maps to lower resolution ones, such as from 1:50,000-county-scale map to 1:1,000,000-country-scale map. Both the soil classification level up-scaling and resolution up-scaling can cause uncertainties in SOC stock estimation. Zhao et al. [34] tested the effects of soil map resolution up-scaling from 1:500,000 to 1:10,000,000 using the three GIS-based Soil Type methods (i.e., the mean, median, and PKB methods) on SOC stock estimation for Hebei Province, China. However, the effects of these three GIS-based Soil Type

Table 1. Methods used to estimate SOC (soil organic carbon) stocks in Zhejiang Province, China.

Method	SOC density value	Scale	Note
SPS[a]	Mean	County; soil species	(1) One soil species has multiple areas, which were surveyed county by county. (2) Mean SOC density value was calculated from one or multiple profiles within the county. (3) 2154 soil profiles were used.
Mean or median	Mean or median	Province; soil species	(1) One soil species has one area calculated from the digital soil map. (2) Mean or median SOC density value was calculated from one or multiple profiles within the Province. (3) 2154 soil profiles were used.
PKB[a]	Mean	County; soil map unit	(1) Soil map units were derived from the digital soil map county by county. (2) One soil map unit in one county may have one or multiple areas calculated from the digital soil map. (3) Mean SOC density value was calculated from one or multiple profiles located within one polygon; polygons belong to one soil map unit in one county may assigned different SOC density values. (4) 2154 soil profiles were used.

[a]SPS, Soil Profile Statistics; PKB, pedological professional knowledge based.

methods on SOC stock estimation at soil map resolution larger than 1:500,000 remain unknown. Moreover, the uncertainties in SOC stock estimation caused by soil classification level up-scaling using the SPS, mean, median, and PKB methods remain unknown.

The goal of this study is to evaluate the uncertainties of changes in scale among the four estimation methods (i.e., the SPS, mean, median, and PKB methods) by using a newly completed 1:50,000 soil survey geographic database of Zhejiang Province, China. Specifically, the main objectives of this study were to: (1) evaluate the uncertainties caused by up-scaling of soil profile properties from the county scale to the provincial scale and soil classification levels from Soil Species to Soil Group among the four methods; and (2) quantify spatial differences in SOC stock estimation among the three GIS-based Soil Type methods.

Materials and Methods

Study Area

Zhejiang Province is located between 27°06′ and 31°03′ N latitude and 118°01′ and 123°10′ E longitude in southeastern China (Figure 1). The Province covers a land area of $1.02*10^5$ km^2. Hills and mountains are dominant terrain, occupying 70.4% of the total land area. The plains and basins account for 23.2%, and the remaining 6.4% is comprised of lakes and rivers. Provincial topography is characterized by high mountains in the southwest and low plains in the northeast. The elevation ranges from 0 to 1895 m with an average of 296 m above sea level.

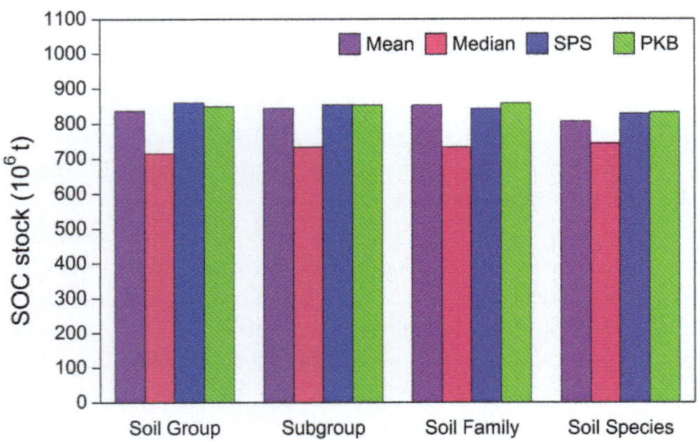

Figure 2. The estimates of SOC (soil organic carbon) stocks for soil classification levels up-scaling from Soil Species to Soil Group using the mean, median, SPS (soil profile statistic), and PKB (pedological professional knowledge based) methods.

With a subtropical monsoon climate, Zhejiang Province has an annual average temperature from 15 to 18°C and an annual precipitation of 1200 to 1800 mm. As substantial differences exist in climate, geomorphology, geohydrology, land use, and parent material throughout the Province, its vegetation and soils and their spatial distribution patterns vary greatly. According to Soils of Zhejiang Province [37], Red soils are the dominant Soil Group, accounting for approximately 39% of the total area of Zhejiang soils.

Data Sources

A recently completed 1:50,000 soil survey geographic database of Zhejiang Province [38] was used in the study, which is the most detailed soil survey data at the provincial scale in China to date. The database includes Soil Spatial Data and Soil Attributes Data. The Soil Spatial Data include a 1:50,000 digital soil map of the Province and a digital map of 2154 geo-referenced soil sampling profile sites, both of which were derived by digitizing and re-compiling field soil survey maps at 1:50,000 from 76 counties in Zhejiang. The Soil Attributes Data contain the properties of the 2154 geo-referenced soil sampling profiles, which were taken from original county soil survey reports. For each profile, there are one to seven soil layers; for each layer, there are up to 104 soil physical and chemical properties, including geographic location, depth, bulk density, organic matter content, and gravel content, etc.

The soil survey geographic database was used to estimate SOC stocks with the mean, median, and PKB methods. This information from the Second National Soil Survey of China conducted in early 1980's was the most comprehensive and detailed inventory of soil characteristics in Zhejiang Province. The Genetic Soil Classification System of Zhejiang Province was used during the field soil survey. The soils were classified using a six-level hierarchical scheme: Order, Suborder, Group, Subgroup, Family, and Species. The Soil Species is the basic classification level, and the Soil Group is the most commonly used classification level in China [2]. According to the Genetic Soil Classification

Table 2. Descriptive statistics of profile soil organic carbon densities by Soil Group.

Soil Group	Soil Order of U.S. Taxonomy[a]	N[b]	Range	Min[b]	Max[b]	Mean	Median	SD[b]	CV/%[b]	Skew[b]	Kurt[b]
Red soils	Alfisols, ultisols, inceptisols	372	32.98	0.87	33.85	6.97	6.15	3.93	15.5	2.29	9.48
Yellow soils	Alfisols, inceptisols	126	91.26	1.93	93.19	16.56	13.39	14.61	213.3	3.25	13.16
Purple soils	Inceptisols, entisols	84	12.10	0.29	12.38	5.19	4.84	2.36	5.6	0.70	0.28
Limestone soils	Mollisols, inceptisols	22	20.52	1.95	22.47	9.14	9.68	5.04	25.4	0.51	0.92
Skel soils	Inceptisols, entisols	114	36.71	0.10	36.81	4.89	3.30	5.27	27.8	3.56	16.24
Red clay soils	Inceptisols, alfisols	4	2.22	4.80	7.02	5.48	5.04	1.04	1.1	1.92	3.75
Mountain meadow soils	Histosols, inceptisols	4	241.54	37.98	279.52	104.32	49.89	116.98	13680.0	1.98	3.94
Fluvio-aquic soils	Inceptisols, entisols	189	20.34	0.94	21.28	6.38	5.78	3.48	12.1	0.96	1.66
Coastal saline soils	Inceptisols	64	14.25	0.92	15.17	7.37	7.50	3.19	10.2	−0.05	−0.55
Paddy soils	Anthrosols	1175	143.90	1.97	145.87	9.82	8.54	7.13	50.9	8.10	124.50
All profiles		2154	279.42	0.10	279.52	9.06	7.62	9.41	88.6	14.17	346.41

[a]Reference conversion between Soil Group of the Genetic Soil Classification System of Zhejiang Province and Soil Order of the U.S. Taxonomy.
[b]N, Min, Max, SD, CV, Skew, Kurt are the abbreviations of the number of soil profiles occurring in a Soil Group, minimum, maximum, standard deviation, coefficient of variation, skewness and kurtosis, respectively.

System of Zhejiang Province, the collected 2154 soil profiles originated from 10 Soil Groups, 21 Subgroups, 99 Soil Families and 277 Soil Species.

Estimation of Organic Carbon Density for Soil Profiles

For each of the 2 154 soil profiles, SOC density was calculated with the formula [34,39]:

$$\mathrm{SOCD}_D = \sum_{i=1}^{n} [(1-\theta_i\%) \times \rho_i \times C_i \times T_i/100] \tag{1}$$

where SOCD_D (kg m^{-2}) is the SOC density of a soil profile within a depth of D (cm), n is the number of soil layers in the soil survey, $\theta_i\%$ represents the volumetric percentage of gravel (>2 mm) content, ρ_i is the soil bulk density (g/cm^3), C_i is the organic carbon content (C g/kg), and T_i represents the thickness (cm) of the layer i. Density of SOC was estimated to a maximum depth of 100 cm to facilitate comparison among data sets. For profiles with actual depths greater than or equal to 100 cm, but less than 100 cm was observed, data for the unobserved profile sections were derived from the mean values of all the corresponding soil profiles of the same Soil Species or Soil Family [15,33,40]. Organic carbon content is calculated by multiplying soil organic matter content by 0.58 (the Bemmelen index), which is based on the assumption that soil organic matter contains approximately 58% organic carbon [41].

Estimation of Regional Soil Organic Carbon Stocks

The SOC stocks of the Province were estimated by using the SPS, mean, median, and PKB methods (Table 1).

For the SPS method, the mean SOC density value of each soil type in one county was calculated from one or multiple profiles belonging to that soil type; then the SOC stock value for that soil type was calculated by multiplying its SOC density value by its corresponding area; finally, the SOC stock values of all soil types were summed up as the total SOC stock for the Province. The area calculated from the 1:50,000 digital soil map was used in this study to facilitate comparison among different methods.

For the mean or median method, the arithmetic mean or median of SOC density values of all soil profiles of the same soil type name in the Province was calculated first as the SOC density value of that soil type; secondly, SOC stocks for polygons on the 1:50,000 digital soil map were calculated by multiplying the mean or median values by corresponding areas of the soil types and then were summed up as the total stock for the Province.

For the PKB method, the SOC density value of each of the 2154 soil profiles was linked to corresponding polygons on the 1:50,000 digital soil map according to the same soil type name, the soil parent materials, and the spatial locations by county. When two or more soil profiles of the same soil type name were located in one polygon, the mean SOC density value of these soil profiles was calculated and used for the linkage. The SOC stock values for all polygons were calculated and then were summed up as the total stock for the Province.

The estimation of SOC stocks using the three GIS-based Soil Type methods were performed using ESRI software ArcGIS 10.0 (Redlands, CA) with water and urban areas excluded from the calculation.

Table 3. Estimates of SOC (soil organic carbon) stocks and SOC density values of various Soil Groups using the mean, median, SPS (soil profile statistic), and PKB (pedological professional knowledge based) methods.

Soil Group	Area[a]	PKB		Mean			Median			SPS		
	(km^2)	SOC density ($kg\ m^{-2}$)	SOC stock (10^6 t)	SOC density ($kg\ m^{-2}$)	SOC stock (10^6 t)	%[b]	SOC density ($kg\ m^{-2}$)	SOC stock (10^6 t)	%[b]	SOC density ($kg\ m^{-2}$)	SOC stock (10^6 t)	%[b]
Red soils	39681.3	6.53	259.10	6.73	266.96	3.0	6.24	247.71	−4.4	6.50	258.11	−0.4
Yellow soils	10013.5	16.92	169.45	14.00	140.15	−17.3	12.24	122.54	−27.7	16.72	167.44	−1.2
Purple soils	3597.7	5.73	20.60	5.00	18.01	−12.6	4.77	17.15	−16.7	5.65	20.34	−1.2
Limestone soils	1571.0	10.18	16.00	8.56	13.45	−15.9	9.43	14.82	−7.4	10.47	16.45	2.8
Skel soils	13736.8	5.11	70.20	5.26	72.30	3.0	4.29	58.99	−16.0	5.09	69.95	−0.4
Red clay soils	29.1	5.45	0.16	5.45	0.16	0.0	5.45	0.16	0.0	5.45	0.16	0.0
Mountain meadow soils	3.3	45.30	0.15	104.32	0.34	130.3	49.89	0.16	10.1	45.30	0.15	0.0
Fluvio-aquic soils	4318.3	7.43	32.10	7.33	31.66	−1.4	7.17	30.98	−3.5	7.44	32.13	0.1
Coastal saline soils	2793.3	7.79	21.75	7.58	21.19	−2.6	7.50	20.95	−3.7	7.80	21.78	0.1
Paddy soils	24995.8	9.68	241.98	9.67	241.66	−0.1	9.20	229.86	−5.0	9.64	241.06	−0.4
Total soils	100740.1	8.25	831.49	8.00	805.88	−3.1	7.38	743.32	−10.6	8.21	827.57	−0.5

[a]Area used for the four methods in this study was from the 1:50,000 digital soil map.
[b]Percentage of difference of estimated SOC stocks between the mean, median, or SPS method and the PKB method.

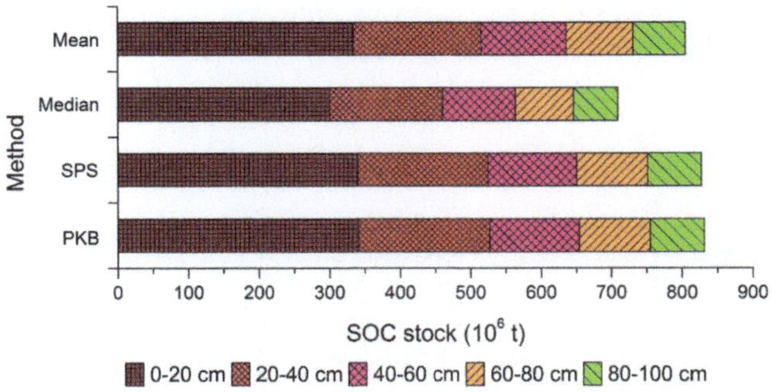

Figure 3. Estimated SOC stocks at different soil depths using the mean, median, SPS (soil profile statistic), and PKB (pedological professional knowledge based) methods.

Results

Effect of Soil Classification Levels Up-Scaling from Soil Species to Soil Group on SOC Estimation Using Different Methods

The SOC stocks for soil classification levels up-scaling from Soil Species to Soil Group using the mean, median, SPS, and PKB methods were calculated (Figure 2). Among the four methods, the estimated SOC stocks for the study area using the median method were always the lowest regardless of soil classification level. Because the SOC densities of the 2154 soil profiles showed a positively skewed distribution with skewness of 14.17 (Table 2), the estimated SOC stocks using the mean method were much higher than those using the median method at all four soil classification levels.

With soil classification levels up-scaling from Soil Species to Soil Group, the estimated SOC stocks presented small variations with coefficient of variations of 2.1%, 1.4%, 1.4%, and 1.2% for the mean, median, SPS, and PKB methods, respectively. The coefficient of variations of estimated SOC stocks among different methods presented an increasing trend, with 4.4%, 6.2%, 6.3%, and 7.1% for soil classification levels of Soil Species, Soil Family, Subgroup, and Soil Group, respectively. The estimated SOC stocks using the mean, median, and SPS methods at the Soil Species level were 3.1%, 10.6%, and 0.5% lower than that obtained by the PKB method, respectively. The variation of estimated SOC stocks among different methods was obviously larger than that among different soil classification levels.

Effect of Different Methods on the Estimates of SOC Stocks for Various Soil Groups

Soil Group is the most stable and consistent soil classification level commonly used in China [2]. In this study, the SOC stocks and SOC densities for various Soil Groups estimated by the mean, median, SPS, and PKB methods are presented in Table 3. For the Soil Groups with SOC density values of soil profiles presented positively skewed distribution, the estimated SOC stocks using the median method were usually lower than that using the mean and PKB methods (Table 2, Table 3). The estimated SOC stocks for various Soil Groups using the SPS method were most similar to that obtained by the PKB method because both methods aggregate soil profile properties at the county scale. The estimated SOC stocks for various Soil Groups using the mean and median methods showed large differences compared with the PKB method because both the mean and median methods aggregating soil profile properties at the provincial scale that the SOC density values of all soil profiles belonging to a soil type in the Province were aggregated to one value.

The most pronounced differences of the estimated SOC stocks for the ten Soil Groups (i.e., Red soils, Yellow soils, Purple soils, Limestone soils, Skel soils, Red clay soils, Mountain meadow soils, Fluvio-aquic soils, Coastal saline soils and Paddy soils) among the three GIS-based Soil Type methods occurred on Yellow soils and Mountain meadow soils, which presented high SOC density values with high coefficient of variances. Thus, the process of linking the

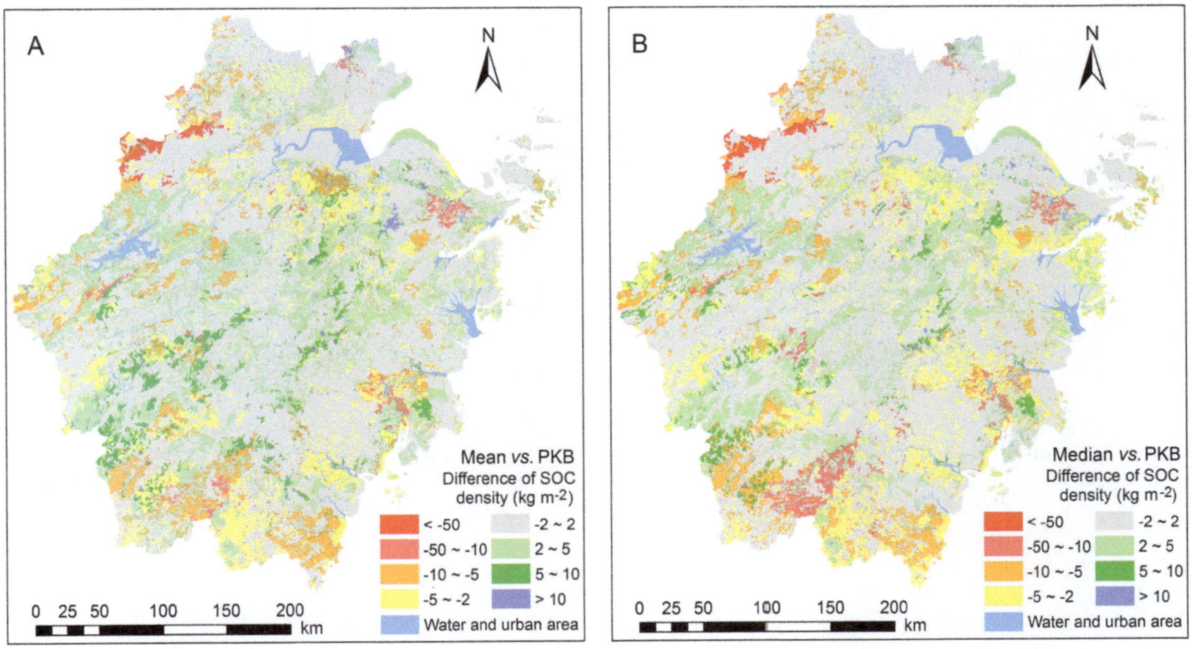

Figure 4. Spatial differences of estimated SOC (soil organic carbon) densities between the mean and PKB (pedological professional knowledge based) methods (A), and between the median and PKB methods (B).

aggregated SOC density values of soil profiles belonging to these two Soil Groups to corresponding polygons on the 1:50,000 digital soil map would significantly influence the estimates of SOC stocks. One important factor leading to the significantly high variation of SOC in Mountain meadow soils was the limited number of soil profiles (n = 4) in the soil survey area.

Effect of Different Methods on the SOC Stock Estimation at Different Soil Depths

For the four methods, the trend of estimated SOC stocks at each of the five soil depths (0–20, 20–40, 40–60, 60–80, and 80–100 cm) is consistent with the soil depth of 0–100 cm, followed the order: PKB>SPS>mean>median (Figure 3). At each of the five soil depths, the estimated SOC stock using the SPS method presented the smallest difference while the median method presented the largest difference compared with the PKB method. Decreasing SOC stock with increasing soil depth was evident independent of estimation method, and these decreasing trends are consistent among all four methods. At the depth of 0–20 cm, SOC stocks estimated by the mean, median, SPS, and PKB methods account for 41.5%, 42.3%, 41.1%, and 41.1%, of total SOC stocks, respectively, while the SOC stocks at the depth of 80–100 cm estimated by the four methods account for only 9.2%, 8.9%, 9.1%, and 9.1%, of the total SOC stocks, respectively.

Comparison of the Differences in the Spatial Patterns of SOC Distribution among the Three GIS-based Soil Type Methods

Estimated SOC densities using three GIS-based Soil Type methods showed clearly spatial differences (Figure 4). The largest differences were as high as -229.63 kg m^{-2} between the median and PKB methods and -175.20 kg m^{-2} between the mean and PKB methods. Polygons mostly belonging to the Soil Groups of Mountain meadow soils and Yellow soils showed substantial differences (<-50 kg m^{-2}) in SOC densities among the three GIS-based Soil Type methods. For the polygons with large spatial differences in SOC densities ranging from -50 kg m^{-2} to -5 kg m^{-2} and >5 kg m^{-2}, the locations were usually in mountain and hilly areas with dominate Soil Groups of Yellow soils, Red soils, and Skew soils, with the remainder in plain areas with dominate Soil Group of Paddy soils. The total areas for spatial differences in SOC densities with absolute value larger than 5 kg m^{-2} between the mean and PKB methods and the median and PKB methods were estimated to be 12883.4 km^2 and 11428.0 km^2, accounting for 12.8% and 11.3% of the total area of Zhejiang soils, respectively. The total areas for spatial differences in SOC densities with absolute value lower than 2 kg m^{-2} between the mean and PKB methods and the median and PKB methods were estimated to be 55316.1 km^2 and 59147.7 km^2, accounting for 54.9% and 58.7% of the total area of Zhejiang soils, respectively.

Discussion

With soil classification levels up-scaling from Soil Species to Soil Group, the estimated SOC stock values using the SPS method were most similar to that obtained by the PKB method. The difference of the estimated SOC stocks between the SPS method and the PKB method occurred in the cases where two or more soil profiles belonging to one soil species existed in one county. However, for most cases, there was only one soil profile belonging to one soil species in one county, the SOC density value of that soil profile was used to represent the SOC density value of that soil species in that county, which resulted in the same estimated SOC stock value of that soil species in that county using the SPS and PKB methods. Therefore, more number of soil profiles is needed to further compare the difference between these two methods.

The estimated SOC stock values using the mean method were similar to that using the PKB method. When two or more soil profiles located in the same polygon within a county, the mean SOC density value of these soil profiles was calculated and used to link with that polygon, so the PKB method approaches the mean method in linking SOC density values with polygons to some extent [34], especially for higher soil classification levels (e.g., Soil Group, Subgroup, Soil Family).

With soil profile properties up-scaling from the county scale (the SPS and PKB methods) to the provincial scale (the mean and median methods), the SOC density values of all soil profiles belonging to a soil type in the Province were aggregated to one value. The aggregating procedure for the mean or median method could cause large uncertainties because information about how much area one profile can represent was missing, thus different polygons belonging to the same map unit on the digital soil map were linked with the same SOC density value. For the PKB method, different polygons belonging to the same map unit may be assigned different SOC density values depending on their locations, thus has an obvious advantage in the demonstration of spatial differences in SOC distribution. The difference in aggregating procedure led to Soil Groups (e.g., Mountain meadow soils, Yellow soils) with soil profiles of high variations in SOC density values presented significant spatial differences among the three GIS-based Soil Type methods.

Conclusions

The up-scaling of soil classification levels from Soil Species to Soil Group has small effect on SOC stock estimation. However, obvious differences occurred with different estimation methods, especially for Soil Groups with high variations in SOC densities and spatial distribution. The PKB method, which links soil profile properties to spatial databases by county, produced more stable results for soil types, soil parent materials, and spatial locations of all soil profiles under consideration. Thus this method has an obvious advantage in the demonstration of differences in spatial patterns of SOC distribution than both the mean and median methods. We recommend the PKB method as a prior option rather than the mean, median, and SPS methods for SOC stock estimation in China, especially when 1:50,000 soil survey geographic database (soil map) is available. It potentially reduces uncertainties related to up-scaling soil profile properties.

Acknowledgments

We are very grateful to the Academic Editor and two reviewers for their useful comments and suggestions on the original manuscript.

Author Contributions

Conceived and designed the experiments: JJZ JPW. Performed the experiments: JJZ CWJ CZ QKL. Analyzed the data: JJZ CZ QKL. Contributed reagents/materials/analysis tools: JJZ. Wrote the paper: JJZ. Revised the manuscript: SPL JPW SDD.

References

1. Davidson EA, Janssens IA (2006) Temperature sensitivity of soil carbon decomposition and feedbacks to climate change. Nature 440: 165–173.
2. Zhang Y, Zhao YC, Shi XZ, Lu XX, Yu DS, et al. (2008) Variation of soil organic carbon estimates in mountain regions: a case study from Southwest China. Geoderma 146: 449–456.
3. Zhou Y, Pei Z, Su J, Zhang J, Zheng Y, et al. (2012) Comparing Soil Organic Carbon Dynamics in Perennial Grasses and Shrubs in a Saline-Alkaline Arid Region, Northwestern China. PLoS ONE 7 (8): e42927. doi:10.1371/journal.-pone.0042927
4. Schlesinger WH (1990) Evidence from chronosequence studies for a low carbon-storage potential of soils. Nature 348: 232–234.
5. Eswaran H, Van Den Berg E, Reich PF (1993) Organic carbon in soils of the world. Soil Science Society of America Journal 57: 192–194.
6. Lal R, Kimble JM, Follett RF, Cole CV (1998) The Potential of U.S. Cropland to Sequester Carbon and Mitigate the Greenhouse Effect. Chelsea: Ann Arbor Press.
7. Uri ND (2000) Conservation practices in U.S. agriculture and their implication for global climate change. Science of the Total Environment 256: 23–38.
8. Li MM, Zhang XC, Pang GW, Han FF (2013) The estimation of soil organic carbon distribution and storage in a small catchment area of the Loess Plateau. Catena 101: 11–16.
9. Batjes NH (1996) Total carbon and nitrogen in the soils of the world. European Journal of Soil Science 47: 151–163.
10. Tan G, Shibasaki R (2003) Global estimation of crop productivity and the impacts of global warming by GIS and EPIC integration. Ecological Modelling 168: 357–370.
11. Janzen HH (2004) Carbon cycling in earth systems-a soil science perspective. Agriculture, Ecosystems & Environment 104: 399–417.
12. Morisada K, Ono K, Kanomata H (2004) Organic carbon stock in forest soils in Japan. Geoderma 119: 21–32.
13. Wang SQ, Huang M, Shao XM, Robert AM, Li KR, et al. (2004) Vertical distribution of soil organic carbon in China. Environmental Management 33: 200–209.
14. Leifeld J, Bassin S, Fuhrer J (2005) Carbon stocks in Swiss agricultural soils predicted by land-use, soil characteristics, and altitude. Agriculture, Ecosystems & Environment 105: 255–266.
15. Yu DS, Shi XZ, Wang HJ, Sun WX, Chen JM, et al. (2007) Regional patterns of soil organic carbon stocks in China. Journal of Environmental Management 85: 680–689.
16. Arnold RW (1995) Role of soil survey in obtaining a global carbon budget. In: Lal R, Kimble J, Levine E, Stewart BA, editors. Soils and global change. Boca Raton: CRC Press. 257–263.
17. Rapalee G, Trumbore SE, Davidson EA, Harden JW, Veldhuis H (1998) Soil carbon stocks and their rates of accumulation and loss in a boreal forest landscape. Global Biogeochemical Cycles 12: 687–701.
18. Bohn HL (1976) Estimate of organic carbon in world soils. Soil Science Society of America Journal 40: 468–470.
19. Bolin B (1977) Change of land biota and their importance for the carbon cycle. Science 196: 613–615.
20. Pan GX (1999) Study on carbon reservoir in soils of China. Bulletin of Science and Technology 15: 330–332.
21. Fang JY, Liu GH, Xu SL (1996) Carbon sinks of terrestrial ecosystems in China. In: Wang RS, editor. Study on Key Issues of Modern Ecology. Beijing: Press of Chinese Sciences and Technologies, 251–267 p.
22. Homann PS, Sollins P, Fiorella M, Thorson M, Kern JS (1998) Regional soil organic carbon storage estimates for western Oregon. Soil Science Society of America Journal 62: 789–796.
23. Batjes NH (2000) Effects of mapped variation in soil conditions on estimates of soil carbon and nitrogen stocks for South America. Geoderma 97: 135–144.
24. Wu J, Ransom MD, Kluitenberg GJ, Nellis MD, Seyler HL (2001) Land-use management using a soil survey geographic database for Finney County, Kansas. Soil Science Society of America Journal 65: 169–177.
25. Alvaro-Fuentes J, Plaza-Bonilla D, Arrue JL, Lampurlanes J, Cantero-Martinez C (2014) Soil organic carbon storage in a no-tillage chronosequence under Mediterranean conditions. Plant and Soil 376: 31–41.
26. Xu SX, Zhao YC, Shi XZ, Yu DS, Li CS, et al. (2013) Map scale effects of soil databases on modeling organic carbon dynamics for paddy soils of China. Catena 104: 67–76.
27. Batjes NH (2006) Soil carbon stocks of Jordan and projected changes upon improved management of croplands. Geoderma 132: 361–371.
28. Liebens J, VanMolle M (2003) Influence of estimation procedure on soil organic carbon stock assessment in Flanders, Belgium. Soil Use and Management 19: 364–371.
29. National Soil Survey Office (1993) Soil Species of China, Volumes 1–6. Beijing: China Agriculture Press.
30. Wang SQ, Zhou CH (1999) Estimating soil carbon reservoir of terrestrial ecosystem in China. Geographical Research 18: 349–356.
31. Jin F, Yang H, Cai ZC, Zhao QG (2001) Calculation of density and reserve of organic carbon in soils. Acta Pedologica Sinica 38: 522–528.
32. Yu JJ, Yang F, Wu KN, Li L, Lü QL (2008) Soil organic carbon storage and its spatial distribution in Henan Province. Chinese Journal of Applied Ecology 19: 1058–1063.
33. Zhao YC, Shi XZ, Yu DS, Pagella TF, Sun WX, et al. (2005) Soil organic carbon density in Hebei Province, China: estimates and uncertainty. Pedosphere 15: 293–300.
34. Zhao YC, Shi XZ, Yu DS, Sun WX (2006) Map scale effect on soil organic carbon stock estimation in north China. Soil Science Society of America Journal 70: 1377–1386.
35. Shi XZ, Yu DS, Warner ED, Pan XZ, Petersen GW, et al. (2004) Soil database of 1:1000000 digital soil survey and reference system of the Chinese Genetic Soil Classification System. Soil Survey Horizons 45: 111–148.
36. Liu QH, Shi XZ, Weindorf DC, Yu DS, Zhao YC, et al (2006) Soil organic carbon storage of paddy soils in China using the 1:1000000 soil database and their implications for C sequestration. Global Biogeochemical Cycles 20: GB3024.
37. Zhejiang Soil Survey Office (1994) Soils of Zhejiang Province. Hangzhou: Zhejiang Science and Technology Press.
38. Wu JP, Hu YL, Zhi JJ, Jing CW, Chen HJ, et al. (2013) A 1:50000 soil database of Zhejiang Province, China. Acta Pedologica Sinica 50: 30–40.
39. Kazuhito M, Ono K, Kanomata H (2004) Organic carbon stock in forest soils in Japan. Geoderma 119: 21–32.
40. Sun WX, Shi XZ, Yu DS (2003) Distribution pattern and density calculation of soil organic carbon in profile. Soils 35: 236–241.
41. Nelson DW, Sommers LE (1982) Total carbon, organic carbon and organic matter. In: Page AL, Miller RH, Keeney DR, editors. Methods of Soil Analysis. Part 2: Chemical and Microbiological Properties. Wisconsin: American Society of Agronomy. 539–579.

Spatio-Temporal Dynamics of Maize Yield Water Constraints under Climate Change in Spain

Rosana Ferrero[1,4]*, Mauricio Lima[2,3,4], Jose Luis Gonzalez-Andujar[1,3]

1 Departamento Protección de Cultivos, Instituto de Agricultura Sostenible, Consejo Superior de Investigaciones Científicos (CSIC), Córdoba, Spain, 2 Departamento de Ecología, Pontificia Universidad Católica de Chile, Santiago, Chile, 3 Laboratorio Internacional de Cambio Global, LINCG (CSIC-PUC), Santiago, Chile, 4 Center of Applied Ecology and Sustainability (CAPES), Santiago, Chile

Abstract

Many studies have analyzed the impact of climate change on crop productivity, but comparing the performance of water management systems has rarely been explored. Because water supply and crop demand in agro-systems may be affected by global climate change in shaping the spatial patterns of agricultural production, we should evaluate how and where irrigation practices are effective in mitigating climate change effects. Here we have constructed simple, general models, based on biological mechanisms and a theoretical framework, which could be useful in explaining and predicting crop productivity dynamics. We have studied maize in irrigated and rain-fed systems at a provincial scale, from 1996 to 2009 in Spain, one of the most prominent "hot-spots" in future climate change projections. Our new approach allowed us to: (1) evaluate new structural properties such as the stability of crop yield dynamics, (2) detect nonlinear responses to climate change (thresholds and discontinuities), challenging the usual linear way of thinking, and (3) examine spatial patterns of yield losses due to water constraints and identify clusters of provinces that have been negatively affected by warming. We have reduced the uncertainty associated with climate change impacts on maize productivity by improving the understanding of the relative contributions of individual factors and providing a better spatial comprehension of the key processes. We have identified water stress and water management systems as being key causes of the yield gap, and detected vulnerable regions where efforts in research and policy should be prioritized in order to increase maize productivity.

Editor: Pilar Hernandez, Institute for Sustainable Agriculture (IAS-CSIC), Spain

Funding: R. Ferrero gratefully acknowledges receipt of a grant from the Fundación Carolina. J. L. Gonzalez-Andujar and R. Ferrero were supported by FEDER (Fondo Europeo de Desarrollo Regional) and the Spanish Ministry of Economy and Competitiveness funds (AGL2012-33736). R. Ferrero and M. Lima acknowledge financial support from Fondo Basal-CONICYT grant FB-0002. We are grateful to LINCGlobal (Laboratorio Internacional en Cambio Global) for their support. The funders had no role in study design, data collection and analysis, decision to publish, or preparation of the manuscript.

Competing Interests: The authors have declared that no competing interests exist.

* E-mail: rferrero@ias.csic.es

Introduction

Spatio-temporal patterns of agricultural production are clearly influenced both by climate change and agricultural management practices. Recently, many studies have analyzed the impact of climate change on crop productivity [1], but comparing the performance of different crop management systems has rarely been explored (exc. [2]). To be specific, we need to evaluate how and where irrigation practices (e.g. rain-fed versus irrigated) are effective in mitigating the effects of climate change, because water constraints and crop demand in agro-systems could be increased due to climate change [3–7]. Identifying whether there are any differences in the principal bio-physical factors and mechanisms that explain both systems will enable us to improve crop productivity without expanding the cropland area and to diminish the adverse impacts of agriculture for social and ecological systems [8].

We do not know much about crop response to climate change yet, and still less about the differential response between irrigated and rain-fed systems [4]. Increases in agriculture production could potentially come from increases in irrigated crops, because higher yields could be attained with reduced production variability [9]. However, this also depends on soil and management factors that result in spatial patterns of yields [10]. Secondly, irrigation can influence local climate by inducing cooling, but this may depend on the extent of the irrigated area, the level of soil moisture alteration and cloud response to irrigation [11]. Third, average yields in rain-fed systems are commonly 50% or less of yield potential (high yield gap), suggesting ample room for improvement [12] but, again a great spatial variability has been found [13]. Yield gaps could be bigger in cropping systems that experience wider ranges of variation under climate conditions [10]. Fourth, plant population (or density) is known to affect the yield potential at a given location [12] and grain yield stability [14]. However, to our knowledge, there are no previous studies explicitly comparing endogenous processes under different water management systems. Finally, simulation at a broad scale level cannot fully explain the above process, and process-based crop models do not always relate to observed yields [15]. Finer spatial scales and historical data of irrigated versus rain-fed systems could help to compare modelled or simulated yield potentials [12].

Analyzing the sensitivity of irrigated and non-irrigated (rain-fed) crops to past climate changes is crucial to an understanding of the vulnerability of agriculture to climate change in the future, particularly in regions that already suffer from this under present conditions. This paper explores biophysical factors and water management practice constraints to maize (*Zea mays* L.) in Spain. Spatial shifts northwards have been projected for maize, due to the extremely hot, dry summers in south-central Europe [16,17], particularly in Spain [18]. The expected effects of climate change on Spain's agriculture would not be uniform. Mediterranean (arid and semiarid) regions may be particularly sensitive, where a decrease in the general availability of hydric resources and an increase in evaporative demand, especially during summer, will affect irrigation requirements [19]. Namely, it is one of the most prominent "hot-spots" in future climate change projections [20], where a mean reduction of 17% in water resources [21,22] has been predicted. For this drought-prone zone, all climate change scenarios imply the need to significantly increase the contribution of irrigation water. Therefore, identifying and quantifying the links between water management practices and food production is crucial in addressing the intensified conflicts between water scarcity and food safety.

The objective of this paper is to determine how climate variability affects maize production in Spain under irrigated and rain-fed conditions. First, we have analyzed the regulatory structure of maize production dynamics under both water management systems. Second, we have evaluated the mechanisms (in ecological parameters) underlying climate perturbations on maize yields. Third, we have assessed whether the importance of maize production structures (i.e. intrinsic regulation) and climate change perturbations (i.e. exogenous factors) could change according to the type of management (i.e. rain-fed and irrigated) and the geographical location. Fourth, we have estimated the potential yield of each region and water management using the previous models and analyzing the spatial variability of yield losses due to water stress [23]. We have combined information on spatial autocorrelation water stress patterns for maize yields to identify the importance of climate constraints at a regional scale.

Methods and Materials

Database

Provincial maize yield levels (*Zea mays*; production per hectare, *kg/ha*) for 1996–2009 were obtained from statistical yearbooks [24]. We studied selected provinces that had both rain-fed and irrigated systems (Fig. 1), and displayed trends in yield fluctuation in Fig. S1. We used Global Historical Climatology Network (GHCND) data on monthly temperature and rainfall (mean, minimum, maximum and extreme; [25]). Various summary statistics of the growing season (July to October) weather were then computed: *EMNT* extreme minimum temperature (°C), *EMXT* extreme maximum temperature (°C), *MMNT* mean minimum temperature (°C), *MMXT* mean maximum temperature (°C), *MNTM* mean temperature (°C), *EMXP* extreme maximum daily precipitation total (l/m2) and *TPCP* total precipitation (l/m2). We also examined carbon dioxide emission (CO_2), an important atmospheric gas that contributes to global warming. The annual country-level emissions of CO_2 (*kt*) were taken from the World Bank's World Development Indicators (WDI; [26]).

Diagnosis and statistical models of yields dynamics

We have analyzed and predicted maize yield responses to the impact of climate change in Spain through the use of models based on the population dynamics theory. Of course this is not a true population in the reproductive sense, but crop systems obey the same rules as all other dynamic systems, both natural and engineered.

First, where necessary, we used sequencing (i.e., splitting the series into two stationary segments) and detrending (i.e., rotating the series around the linear or quadratic trend) to generate a stationary time series. Second, we estimated the logarithmic rate of change of the yield as $R_t = Y_t - Y_{t-1}$ (the same response variable as [1,27]), where Y_t represents the provincial yield in a year t(the logarithm of the detrended yield) and Y_{t-1} is the same series with one year of delay (lag 1).

We were able to detect and analyze non-trivial feedback processes by examining their relationship $R_t = f(Y_{t-d})$, where the function f described how the crop yield change rate varied with yield level, and this has been called the *R*-function. We used the partial rate correlation function (or *PRCF*) to estimate the order of the dynamical process and determine how many time lags (*d*) should be included in the model for representing the feedback structure. This function detects the feedback order removing the confounding effect by calculating the partial correlation between R_t and Y_{t-d} with the effects of lower lags removed [28].

We then used the generalized version of the exponential form of the discrete time logistic model [29,30] in terms of the *R*-function to represent pure endogenous models in the function f:

$$R_t = r_{\max} - \exp(aY_{t-d} + c) \tag{1}$$

where Y_{t-d} represents the yield data at time $t-d$ (where d was obtained from *PRCF* function), $r_{\max}$ is a positive constant representing the maximum finite rate of change (and is estimated as the maximum rate of change from the observed data), c is a measure of the ratio between demand and offer of limiting resources and a is the nonlinearity of the curve. The nonlinearity of this model includes a biological realistic property: its net reproductive rate is bounded [29], that is, the performance of any crop must have an upper bound simply because no crop can produce an infinite number of grains that subsequently contribute to the crop yield.

Finally, we used the Royama classification of exogenous effects as a framework to deduce causal mechanisms of the climate change impact on these crop yields in a spatial-temporal study [29]. To include exogenous perturbations, we modelled $r_{\max}$ and c of (1) as linear functions of climate conditions, each of which has an explicit biological interpretation. In this way, we set up mechanistic hypotheses about the exogenous effects of climate on these yields data [29].

If an exogenous factor (i.e. climate or gas emissions) changes $r_{\max}$ and has an additive or independent perturbation effect on crop yield levels, it shifts the *R*-function curve along the *y*-axis ("*vertical*" perturbations):

$$\begin{aligned} r^*_{\max} &= r_{\max} + bZ_{t-d'} \\ R_t &= r_{\max} - \exp(aY_{t-d} + c) + bZ_{t-d'} \end{aligned} \tag{2}$$

where $Z_{t-d'}$ is the exogenous factor (for lags or d 0 and 1; in logarithm scale). This model produces alterations to both $r_{\max}$ and the carrying capacity (equilibrium point of the population, $R_t = 0$), changing the level of equilibrium and its stability.

If an exogenous factor (i.e. climate or gas emissions) changes c and has a non-additive perturbation effect on crop yield levels, and influences the equilibrium point of the population shifting the *R*-function curve along the *x*-axis ("*lateral*" perturbations):

$$\begin{aligned} c* &= c + bZ_{t-d'} \\ R_t &= r_{\max} - \exp(aY_{t-d} + c + bZ_{t-d'}) \end{aligned} \tag{3}$$

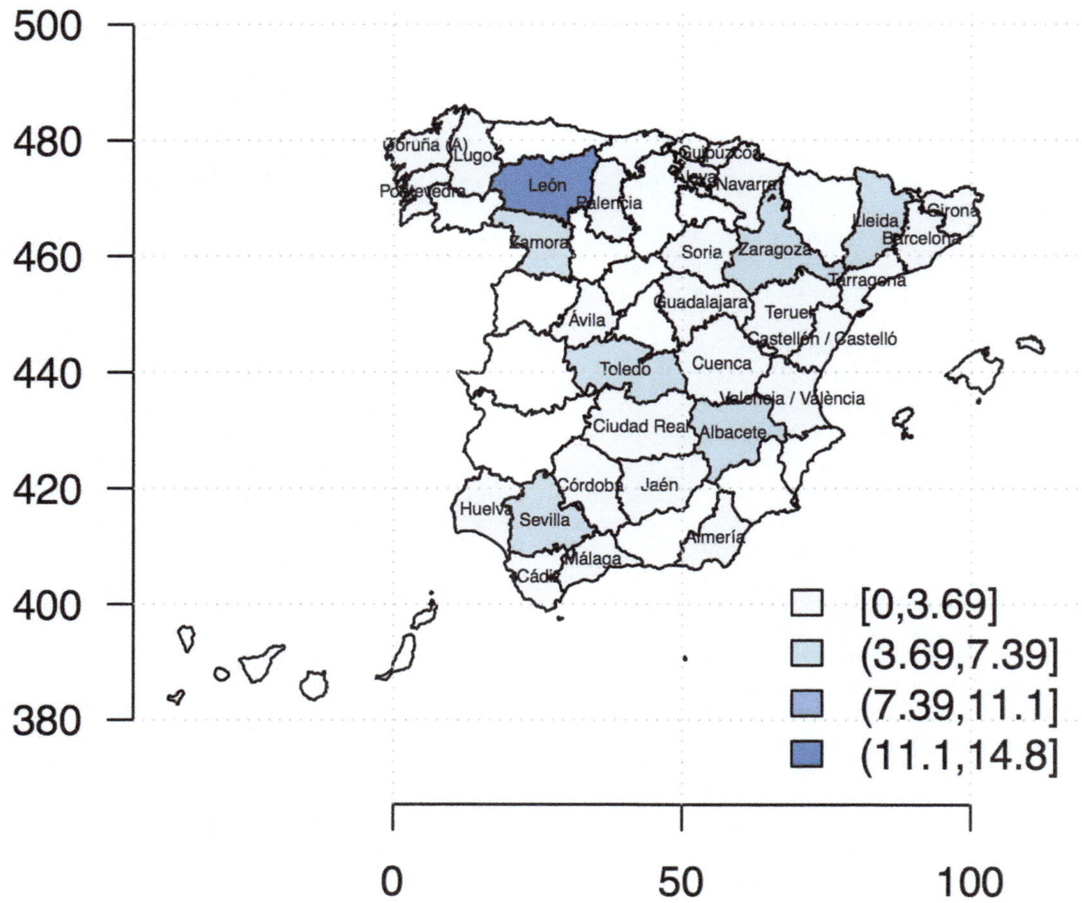

Figure 1. Definition of study regions (provinces) with percentage of total maize production for 1996–2009. Only provinces with both irrigated and rain-fed systems were analyzed.

Lateral perturbations do not change the pattern of dynamics around equilibrium because they do not change the slope at the equilibrium.

We fitted Eqs. 1–3 using nonlinear least squares regressions with the *nls* library in the software *R* [31,32]. In particular, the models were fitted by minimizing the Akaike criterion with a correction for finite sample sized (AIC_c):

$$AIC_c = (2k - 2\ln L) + \frac{2k(k+1)}{n-k-1}$$

where k is the number of parameters and L is the maximized value of the likelihood function for the model, and n denotes the sample size. Also, we maximized the pseudo R^2 measures based on the deviance residual [33]. Models were chosen on the basis of their goodness-of-fit (assessed using root mean square error RMSE and the log-likelihood values), their ability to describe the correct feedback structure, and their appropriateness.

Yield losses due to suboptimal water availability (YGRw)

We propose a new estimation of the potential yield or equilibrium productivity [34] at the provincial level as the equilibrium value of the models. By solving Eqn. (1–3) for the equilibrium dynamics $Y_t = Y_{t-1} = K$ (when $R_t = 0$), we calculated the maize yield level at equilibrium, sometimes called the carrying capacity (*Mg/ha*). For non-pure endogenous models we made potential yield estimations for each year as the exogenous factor changed. Then we calculated the percentage of yield losses due to suboptimal water availability (*YGRw*; Eqn. 4; view [35]), which indicated how close the rain-fed yield potential is to the irrigated value for a given site (%).

$$YGR_w = \frac{YP_{IR} - YP_{RF}}{YP_{IR}} \tag{4}$$

We obtained some time-invariant YGR_w values when, in the same province, irrigated and rain-fed *YP* were estimated from pure endogenous models, so that we calculated the averaged YGR_w for each province, and studied its spatial variability without taking into account the temporal dimension of the data.

We determined whether there was any spatial autocorrelation in YGR_w with the global Moran's *I* (spatial correlation on average, of an entire map). At this stage, we were not yet trying to determine the causes, although the results could have motivated a hypothesis. We assumed: 1) that there was no spatial patterning due to some underlying but unmodelled factor, and 2) that the assigned spatial weights were those that generated the autocorrelation. Then we tested whether YGR_w was more spatially clustered than by chance. The matrix that represents spatial dependence (*W*) uses a binary indicator of neighbourhood (i.e. the spatial weights, w_{ij}, are defined as $w_{ij} = 1$ if the *i* and *j* provinces are contiguous neighbours, $w_{ij} = 0$ otherwise, based on rook contiguity; [36]). We used row-standardisation (style *W*) that favours observations with few neighbours. We calculated a non-parametric approach to infer-

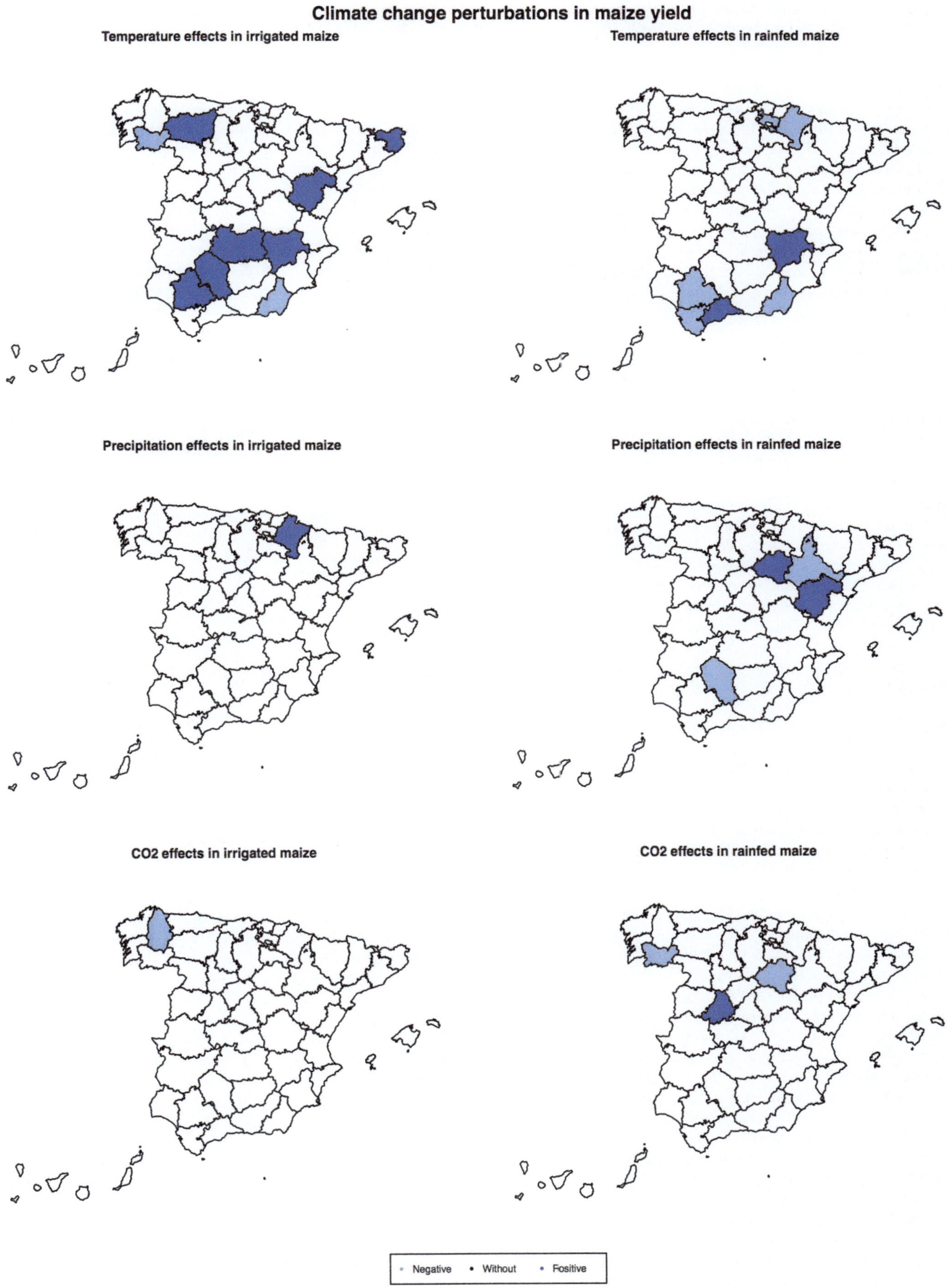
Climate change perturbations in maize yield
Temperature effects in irrigated maize
Temperature effects in rainfed maize
Precipitation effects in irrigated maize
Precipitation effects in rainfed maize
CO2 effects in irrigated maize
CO2 effects in rainfed maize
Negative
Without
Positive

Figure 2. Effects of temperature, precipitation and CO2 emission, on maize productivity for rain-fed and irrigated crops. Provinces for both water management systems were selected for the analysis. All models are from Table S1.

ence on Moran's *I* using 999 simulations (Monte Carlo permutation test). Also, local indicators of spatial association (or LISA) were calculated to detect "hot spots" where there was a strong autocorrelation, and "cold spots", where there were none. The results were plotted on a Moran scatterplot: the target variable on the *x*-axis, and the (spatially-weighted) sum of neighbouring values on the *y*-axis; these are called spatially lagged values. We identified the high-influence areas.

We analyzed the environmental spatially distributed causes of averaged YGR_w through a Simultaneous Autoregressive Model (SAR; [36]) that considers spatial autocorrelation of residuals:

$$YGR_w = Z^T\beta + e \qquad (5)$$

where, for each province, YGR_w is the percentage of yield losses due to suboptimal water availability, *Z*is a matrix of averaged climate variables (see Database section; except country-level CO_2 emissions), $e = B(Y - Z^T\beta) + \epsilon$ is the error term, and ϵ represents residual errors (assumed to be independently distributed according to a Normal distribution with zero mean and diagonal covariance matrix σ_e^2). The error terms are modelled so that they depend on each of the other areas to account for their spatial dependence (*B* is a matrix that contains the dependence parameters; $B = \lambda W$, where λ is a spatial autocorrelation parameter and *W* is a matrix that represents the spatial dependence explained above). Global Moran's *I* was computed for the residuals to test if the SAR model accounts for all the spatial autocorrelations in YGR_w. For the spatial analysis we used *spdep* library in the software *R* [37].

Results

Regulatory structure and exogenous perturbation models

After sequencing and detrending, all the sites exhibited first-order negative feedback (*PRCF(1)*) as being the most important component of yield growth rate (Figure S2; except for irrigated maize in Vizcaya and rain-fed systems in Tarragona). Major sites showed the highly significant ($p<0.05$) effect of endogenous processes as determinants of the structure of crop productivity regulation (Table S1).

We evaluated gas emission (CO_2) and climate factors (temperature and precipitation; see Database section), as exogenous perturbations of the production curve (*R*-function). Table S1 shows several models that were selected as climate change impacts on maize production for each Spanish region and management system. The stochastic versions of the step-ahead predictions of the models are shown in Fig. S3. As expected, the effects of climate on maize production were not uniform, and depended on the irrigation management system (Figure 2). Maize yields were significantly related to minimum temperatures (possibly night ones) in 11 sites and by maximum temperature in another 5 sites. Generally, there were positive effects of temperature for irrigated systems, except for Almería (for minimum temperature *–EMNT-*) and Ourense (mean temperature *–MNTM-*). However, for rain-fed systems, we detected negative effects of warming on major sites, with the exception of Málaga and Albacete (both for *EMNT*). As expected, precipitation was not important for irrigated systems (except for maximum precipitation *–EMXP-* in Navarra), but it was an important factor in some rain-fed managements. There were positive effects of precipitation on Teruel and Soria (for a total *–TPCP-* and maximum rainfall), and negative ones in Córdoba (*TPCP* and *EMXP*) and Zaragoza (*TPCP*) on rain-fed crops. Finally, CO_2 emissions negatively affected maize in Lugo (irrigated), Ourense and Soria (both rain-fed), and positively only in Ávila (rain-fed).

Temperature acted mainly as having non-additive (lateral) effects on maize yield dynamics, whereas CO_2 emission acted as additive (vertical) effects (Table S1; Figure 3 and S4). Finally, rainfall exerted non-additive effects when it had a negative impact on maize, but when it obtained positive responses the effects were of both types (additive or non-additive; Table S1; Figure S4). For example, Figure 3 shows positive and non-additive (lateral) effects of temperature on rain-fed maize in Albacete and irrigated maize in Sevilla. That is, the increase in temperature had a positive effect on both maize systems, and more so at high yield levels. Figure 3 also indicates a negative and additive (vertical) effect of CO_2 emission on Ourense (same strength for all yield levels), and a positive and non-additive (lateral) rainfall effect on rain-fed maize in Soria (more important for high yield levels).

Relative yield losses due to suboptimal water availability (YGR_W)

We first visualized the spatial relation of YGR_w (Figure 4), where several high YGR_w values were shown in central and southern provinces of Spain. The global Moran's *I* value ($I = 0.39$) was of an opposite sign and much larger in absolute value than the expectation ($E[I] = -0.034$); this was quite unlikely to be equal to the expectation of no spatial association. The probability of incorrectly rejecting the null hypothesis of no association (type I error) was 0.0021. The Monte Carlo approach also rejects the null hypothesis (the true value for Moran's *I* is zero; $I_{mc} = 0.406$, $p = 0.005$; Figure S5). The Moran scatterplot (Figure 5; the vector of values and the neighbour list with weights) showed points with a great influence, which are identified by a special symbol and their name. The highest-leverage area is marked on Almería; it has the highest YGR_w (84.56) and a zero weighted spatially-lagged proportion, because it did not have any adjacent areas in the study. Soria and Palencia had low YGR_w, and a low spatially-lagged proportion; these are the low-YGR_w neighbourhoods adjacent to low-YGR_w neighbourhoods. They have a great influence on the slope (global Moran's *I*). From Figure 5 it is clear that most of the global Moran's *I* significance comes from the local Moran's *I* from high YGR_w in Almería, and low YGR_w associated with low YGR_w, in the Soria and Palencia area in the north.

There was clear evidence of local clustering, 6 areas (Ciudad Real, Cuenca, Albacete, Valencia, A Coruña and Pontevedra) showed sufficiently high local Moran's *I* to reject the null hypothesis with less than a 5% chance of Type I error. These areas were not highlighted in the Moran scatterplot, as they did not greatly influence the global Moran's *I* but were locally-clustered.

There was a significant spatial correlation in the residuals, because the estimated value of lambda was 0.141 and the *p*-value of the likelihood ratio test 0.0354. Only averaged temperature (*MNTMt*) was significant for the SAR model, suggesting that provinces with higher temperature have larger YGR_w percentages. The model found was: $YGR_w = -64.42 + 5.69 * MNTM_t$ the SAR model, which accounted for the whole spatial autocorrelation

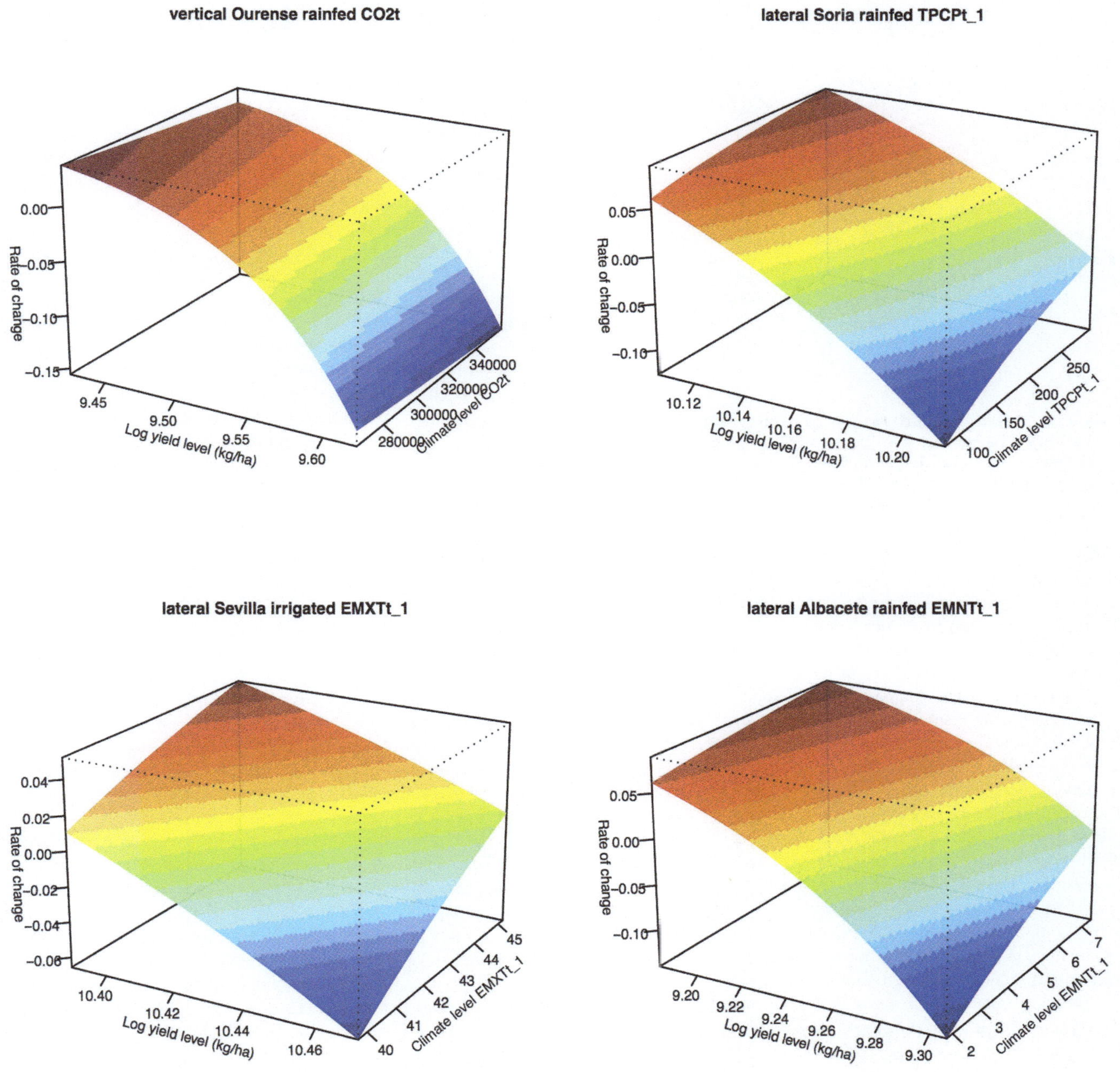

Figure 3. Yield rate of change against the log observed yield level (with one year of delay) and the exogenous factor that perturbs the productivity function (*R*-function). Exogenous factors include carbon emissions ($CO2_t$), precipitation ($TPCP_{t\_1}$), and maximum and minimum temperature ($EMXT_{t\_1}$ and $EMNT_{t\_1}$). Additive (vertical) and non-additive (lateral) perturbation effects were detected. Colours indicate the *R*-function value. See Table S1 for description of models and Figure S4 for their graphs.

in YGR_w (global Moran's test for residuals was $I = -0.00811$, $p = 0.422$). Thus, the autocorrelation in the linear model residuals was explained.

Discussion

In the present study, the impact of climate variability on maize yields in Spanish rain-fed and irrigated systems was investigated for the period 1996–2009. We explored the endogenous structure (regulation) and the exogenous perturbations effects on maize production at a regional scale.

Regulatory structure: endogenous feedback

We found that maize productivity had a persistently negative effect on crop yields for a one year time delay (first order negative feedback, *PRCF(1)*). Maize productivity was characterized by negative first-order feedback structure in major sites and in both irrigation systems. Namely, there were biomass or density-induced feedback loops in the growth, survival rates, seed germination or grain production rates of individual plants, tending to stabilize their dynamics [38]. In Spain, the seeds produced are used for the next year and, therefore, a year's crop performance could change seed viability and vigour, which also affects the performance of the following crops (changing the demand for resources). Also, a crop

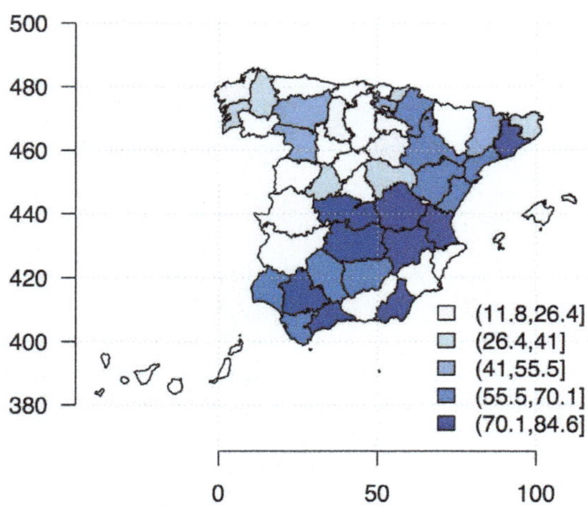

Figure 4. Relative yield losses due to suboptimal water availability (*YGRw; %*). The percentage of yield losses due to suboptimal water availability indicates how close rain-fed yield potential are to the irrigated value for a given site.

system could alter habitat conditions; in fact, the frequent practice of crop rotation is a testimony to the importance of negative feedbacks in agricultural systems (i.e. it modifies resource supplies). This produces high-frequency dynamics due to year-to-year endogenous variability in maize yields. Our logistic models appear to capture the essential features of the fluctuations observed, and suggest a mechanistic explanation for the latter. This implies that, to understand the response of maize productivity to climate, we must also know the endogenous feedback structure of the system.

Our models are important to conceptualizing the problem of regulated versus unregulated systems. If a system were to be controlled entirely by an exogenous process (unregulated systems), then the series would perform a random walk and we saw no sign of the generated series becoming stabilized, but it drifted increasingly away from the origin with the passing of time [29]. However, persistence implies regulation (but not necessarily vice versa) and, therefore, the rate of change in a persistent crop productivity system is not statistically independent of the yield level and should be bounded (i.e. regulated systems).

Climate change effects: exogenous perturbations

In line with previous studies, temperature during the growing season was the most important weather variable influencing maize yields [39]. However, we deciphered the effects of climate on maize productivity providing new interpretations. First, diagnostic analysis suggested that temperature acts mainly as a non-additive (lateral) perturbation in maize productivity. Therefore, the relationship between temperature and maize yields was nonlinear and could not be captured adequately by a linear or quadratic functional relation as in previous studies [40]. Our analysis suggests a biological reason for the nonlinear interaction between climate and maize yield level. Temperature had no direct impacts on yield rate of change (affecting r_{max}; additive or vertical effects), but influenced the availability or requirements of some limiting factor or resource (changing c; non-additive or lateral effects). There is probably a relationship between extreme heat and plant water stress, increasing water demand and/or soil water content in rain-fed systems, in agreement with the recent results of Lobell *et al.* [41]. This is because, the effects of high temperature are experienced only when the maize yield level is close to equilibrium [29]. This kind of perturbation exerts strong effects on the average level of yield but few on the intrinsic periodicity induced by endogenous feedback.

Secondly, rain-fed maize yields are negatively affected by temperature increases, but irrigated systems may gain from warming in some regions. As expected, rain-fed crop damage may result from greater water and heat stress during hot growing seasons. However, unexpected positive effects of temperature in irrigated systems are possibly a consequence of heat tolerance, which is consistent with other studies on local adaptation to hot temperatures being able to minimize stress effects [40] or the cooling effect of irrigation [42]. Therefore, we detected some adaptation to heat stress that could mitigate the projected heat-related losses, at least in a few regions with irrigated systems.

Thirdly, climate variability and extreme events are more important than averages. Thus, we detected that minimum temperature was the dominant factor in maize production, in agreement with other recent studies for maize [43–45] and rice [46]. Currently, a new paradigm has been originated: crop yields have declined with a higher minimum or night temperature [46,47] or when there was a marked asymmetry between maxima and minima [48]. One possible explanation includes the facts that the grain-growth rate has increased and that the duration of grain-filling has been shortened as the temperature increased, producing lower crop production (yield levels) [49]. Mohammed & Tarpley [47] proposes a list of the effects of high night temperatures on crop production. Also, our findings are in line with the results of recent research which argue that global minimum temperatures are increasing faster than maximum temperatures, and the need to explore the ecological consequences of this phenomenon [41,50,51]. Therefore, we wish to highlight the importance of considering extreme climate variables in crop production studies, and limiting the use of averages or accumulative climate data which ignore inter-annual variability of climate and extreme events. Our results differ from those of most studies which do not take into account food production structure regulation, and those which use degree-days [40] concepts which assume a cumulative or additive effect of temperature on crop yield and do not adequately account for the effects of extreme temperatures (high or low) either.

In the study period, precipitation was not a major abiotic factor limiting maize yield of cultivated rain-fed crops in Spain. We only detected positive effects of precipitation for irrigated maize in Navarra, Teruel and Soria. Also, growing season rainfall negatively affected rain-fed maize yield in Córdoba and Zaragoza, possibly due to flood and waterlogging problems causing production losses. Again, we agree with Lobell *et al.* [41], who argue that the apparent paradox of the scant effect of precipitation on rain-fed maize yield whereas, on the contrary, there is a water stress effect of temperature, can be solved with the following reflection "*large precipitation changes are required to rival the effect of temperature on water stress, because high temperature affects both water demand and supply*".

As in the study of Long *et al.* 2006 [52], ours study indicates that there was a smaller CO_2 effect on maize yield than previously presumed. Impacts of higher CO_2 on maize yield were reduced probably because it is a C4 plant, and also because of the national scale of the variable in our study.

Spatial variability of yield losses due to water stress

We found that the global spatial pattern of yield losses due to water stress is not a random one (Figure 4); there was a high influence in Palencia, Soria (lowest) and Almería (highest). We detected clusters of "cold spots" in northern Spain (A Coruña and Pontevedra) and "hot spots" in central provinces (Ciudad Real,

Cuenca, Albacete, Valencia; Figure 5). Neither cluster greatly influenced the global Moran's *I* but they were locally-clustered. Moreover, we modelled spatial YGR_w values with climate variables and found that the mean temperature was the highest constraint of maize productivity due to water stress. In conclusion, policy action to decrease the relative yield gap due to water stress on maize productivity has the potential to geographically target high YGR_w areas. Future work will help determine other non-climatic causal relationships between YGR_w and an array of factors that could influence water management practices in maize (e.g. access to water, management technology, soil conditions, etc.).

A recent comparison of simulated and observed yield patterns highlights the value of data in the spatial distribution of yields for understanding the causes of landscape yield variability [10].

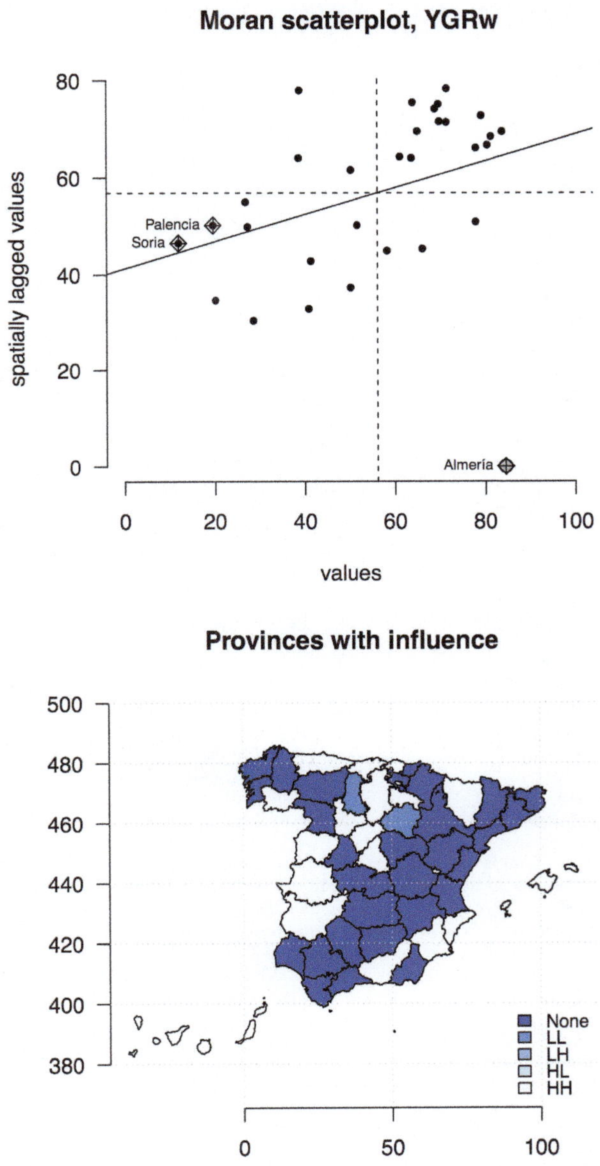

Figure 5. Spatial autocorrelation analysis of the relative yield losses due to suboptimal water availability (*YGRw*). Top: Moran scatterplot; bottom: high-influence areas neighbours: no influence (*None*), high proportion with low proportion neighbours (*HL*), the reverse (*LH*), and both high (*HH*). We define the break between "low" and "high" as the third quartile.

However, to our knowledge, this is the first study explicitly evaluating the spatial pattern of real relative yield gaps due to a water management system and its sensitivity to underlying climate factors. The results demonstrate that spatial patterns of yield loss due to water stress possess substantial information on the relative importance of water management factors for maize productivity.

The need for an analysis to identify and implement adaptation options in agriculture emphasizes the importance of regional scales (federal, provincial, and territorial governments). Global and non-spatial studies can provide only a very partial and potentially misleading insight into the true impact of climate change, where aggregation can indeed conceal vulnerability and climate change costs [53]. However, individual regions (provinces) allow a better analysis of uncertainty and risks, thus providing practical recommendations to farmers.

Conclusions

We identified the same regulation structure for both management systems, i.e. a negative first-order feedback process that tends to stabilize the crop's dynamics. We analyzed the underlying mechanisms of the interaction between climate variation and regulatory structure on maize production. Different climate variables appear to operate differently on maize productivity. We found that the effect of temperature (mainly extreme values) cannot be evaluated independently of crop productivity as in previous studies, because its consequences are experienced only when maize yield level is close to equilibrium (lateral perturbation). We suggest that high maize yield crops are especially vulnerable to weather-related yield variations. These data support the belief that lower yields are more suitable for low-input conditions, because climate might be more severe in crops that interact strongly with productivity [14].

Our results also indicate that it may be important to consider explicitly the irrigation system and spatial variability. Rain-fed agriculture may be at risk as heat waves will be more intense, more frequent and longer (particularly in Seville, Cádiz, Almería, Navarra and Ávila; see Fig. 2). Irrigation seems to allow some tolerance to warming but future levels of water availability would be compromised if water restrictions and irrigation costs increased, as climate change projections indicate. We propose a new framework to estimate yield potential as the equilibrium yield or yield carrying capacity. Climate change is not uniform over Spain and the effectiveness of irrigated and rain-fed management varies with the location, producing different regional vulnerabilities and potential yields. Accordingly, the general strategies for adapting maize productivity to climate change will vary between different zones in Spain.

Supporting Information

Figure S1 Time series of maize yield level for rain-fed (red) and irrigated (blue) systems. Each provinces of Spain were analyzed for 1996–2009.

Figure S2 Partial rate correlation function (*PRCF*).

Figure S3 Comparison of observed crop yield levels (points, *obs*) for the period 1997–2009 with stochastic predictions from models fitted to the data until the year 1996 (broken line, *sim*) and 95% confidence intervals for forecasts (shaded area, *95PPU*). P-factor is the percent of observations that are within the given uncertainty bounds and R-factor represents the average width of the given uncertainty bounds divided by the standard deviation of the observations. See Table S1 for description of models and variables.

Figure S4 *R*-functions: yield rate of change against the log observed yield level (with one year of delay). Climate factors had vertical (additive) and lateral (non-additive) perturbations on the *R*-function. Colors indicate the value of the *R*-function. See Table S1 for description of models and variables.

Figure S5 A non-parametric approach to inference on Moran's *I* using 999 simulations (Monte Carlo permutation test).

Table S1 Summary statistics of nonlinear logistic models, 1996–2009. We evaluated pure Endogenous models (E), and additive (or Lateral, L) and non-additive (or Vertical, V) models that also represent the effect of exogenous perturbations. Different crop management systems were analyzed (*IR* = irrigated and *RF* = rainfed). *%Total* percentage of total crop production in Spain, *K* carrying capacity or potential yield, *rmax* maximum finite reproductive rate, *a* non-linearity coefficient, *c* the ratio between demand and offer of limiting resources, *b* coefficients for different exogenous effects, R^2 pseudo-coefficient of determination, *logLIK* log-likelihood, *RMSE* root-mean-square error and *AICc* corrected Akaike information criterion. NOTE: *$p<0.05$, **$p<0.01$, Number of not avaiable data (NA) were indicated by I. *CO2* carbon dioxide emission (*kt*, country-level emissions), and summary statistics of the growing season weather: *EMNT* extreme minimum temperature (°C), *EMXT* extreme maximum temperature (°C), *MMNT* mean minimum temperature (°C), *MMXT* mean maximum temperature (°C), *MNTM* mean temperature (°C), *EMXP* extreme maximum daily precipitation total (l/m2), *TPCP* total precipitation (l/m2).

Acknowledgments

We thank the editor and two anonymous reviewers for their constructive comments, which helped us to improve the manuscript.

Author Contributions

Conceived and designed the experiments: RF ML JLG-A. Performed the experiments: RF. Analyzed the data: RF. Contributed reagents/materials/analysis tools: RF. Wrote the paper: RF ML JLG-A.

References

1. Lobell DB, Schlenker W, Costa-Roberts J (2011) Climate trends and global crop production since 1980. Science 333: 616–620.
2. Licker R, Johnston M, Foley JA, Barford C, Kucharik CJ, et al. (2010) Mind the gap: how do climate and agricultural management explain the "yield gap" of croplands around the world? Glob Ecol Biogeogr 19: 769–782.
3. Döll P (2002) Impact of climate change and variability on irrigation requirements: a global perspective. Clim Change 54: 269–293.
4. Turral H, Burke J, Faurès JM, Faurés JM (2011) Climate change, water and food security. Rome Food Agric Organ United Nations: 204.
5. Easterling WE, Aggarwal PK, Batima P, Brander KM, Erda L, et al. (2007) Food, fibre and forest products. In: Parry ML, Canziani OF, Palutikof JP, Linden PJ van der, Hanson CE, editors. Climate Change 2007: Impacts, Adaptation and Vulnerability. Contribution of Working Group II to the Fourth Assessment Report of the Intergovernmental Panel on Climate Change. Cambridge University Press, Cambridge, UK. pp. 273–313.
6. Smith P, Martino D, Cai Z, Gwary D, Janzen H, et al. (2007) Agriculture. In: Metz B, Davidson OR, Bosch PR, Dave R, Meyer LA, editors. Climate Change 2007: Mitigation. Contribution of Working Group III to the Fourth Assessment Report of the Intergovernmental Panel on Climate Change. Cambridge University Press, Cambridge, United Kingdom and New York, NY, USA. pp. 497–540.
7. Medici LO, Reinert F, Carvalho DF, Kozak M, Azevedo RA (2014) What about keeping plants well watered? Environ Exp Bot 99: 38–42.
8. Zou X, Li Y, Gao Q, Wan Y (2011) How water saving irrigation contributes to climate change resilience—a case study of practices in China. Mitig Adapt Strateg Glob Chang 17: 111–132.
9. Mueller ND, Gerber JS, Johnston M, Ray DK, Ramankutty N, et al. (2012) Closing yield gaps through nutrient and water management. Nature 490: 254–257.
10. Lobell DB, Ortiz-Monasterio JI (2006) Regional importance of crop yield constraints: Linking simulation models and geostatistics to interpret spatial patterns. Ecol Modell 196: 173–182.
11. Lobell D, Bala G, Mirin A, Phillips T, Maxwell R, et al. (2009) Regional Differences in the Influence of Irrigation on Climate. J Clim 22: 2248–2255.
12. Lobell DB, Cassman KG, Field CB (2009) Crop Yield Gaps: Their Importance, Magnitudes, and Causes. Annu Rev Environ Resour 34: 179–204.
13. Meng Q, Hou P, Wu L, Chen X, Cui Z, et al. (2013) Understanding production potentials and yield gaps in intensive maize production in China. F Crop Res 143: 91–97.
14. Tokatlidis IS (2013) Adapting maize crop to climate change. Agron Sustain Dev 33: 63–79.
15. Reidsma P, Ewert F, Boogaard H, van Diepen K (2009) Regional crop modelling in Europe: The impact of climatic conditions and farm characteristics on maize yields. Agric Syst 100: 51–60.
16. Olesen JE, Carter TR, Díaz-Ambrona CH, Fronzek S, Heidmann T, et al. (2007) Uncertainties in projected impacts of climate change on European agriculture and terrestrial ecosystems based on scenarios from regional climate models. Clim Change 81: 123–143.
17. Wolf J, Diepen CA (1995) Effects of climate change on grain maize yield potential in the european community. Clim Change 29: 299–331.
18. Iglesias A, Mínguez MI (1995) Prospects for maize production in Spain under climate change. In: Rosenzweig C, editor. Climate Change and Agriculture: Analysis of Potential International Impacts. American Society of Agronomy, Vol. asaspecial. pp. 259–273.
19. Moreno JM, Aguiló E, Alonso S, Cobelas MÁ, Anadón R, et al. (2005) A Preliminary General Assessment of the Impacts in Spain Due to the Effects of Climate Change A Preliminary Assessment of the Impacts in Spain due to the Effects of Climate Change. Available: http://www.magrama.gob.es/en/cambio-climatico/temas/impactos-vulnerabilidad-y-adaptacion/plan-nacional-adaptacion-cambio-climatico/eval_impactos_ing.aspx. Accessed 2014 May 8.
20. Rodríguez-Puebla C, Ayuso S, Frías M, García-Casado L (2007) Effects of climate variation on winter cereal production in Spain. Clim Res 34: 223–232.
21. Iglesias A, Garrote L, Quiroga S, Moneo M (2011) A regional comparison of the effects of climate change on agricultural crops in Europe. Clim Change 112: 29–46.
22. Iglesias A, Mougou R, Moneo M, Quiroga S (2010) Towards adaptation of agriculture to climate change in the Mediterranean. Reg Environ Chang 11: 159–166.
23. Neumann K, Verburg PH, Stehfest E, Müller C (2010) The yield gap of global grain production: A spatial analysis. Agric Syst 103: 316–326.
24. MAGRAMA (2012) Ministerio de Agricultura, Alimentación y Medio Ambiente (MAGRAMA). MAGRAMA Stat Databases. Available: http://www.magrama.gob.es/es/estadistica/temas/publicaciones/anuario-de-estadistica/. Accessed 1 April 2012
25. Lawrimore JH, Menne MJ, Gleason BE, Williams CN, Wuertz DB, et al. (2011) An overview of the Global Historical Climatology Network monthly mean temperature data set, version 3. J Geophys Res 116: D19121. Available: http://doi.wiley.com/10.1029/2011JD016187. Accessed 1 April 2012.
26. World Bank, World Development Indicators (WDI) (2012). ESDS Int Univ Manchester. Available: http://data.worldbank.org/data-catalog/world-development-indicators/wdi-2012. Accessed 1 April 2012.
27. Lobell DB, Bänziger M, Magorokosho C, Vivek B (2011) Nonlinear heat effects on African maize as evidenced by historical yield trials. Nat Clim Chang 1: 42–45.
28. Berryman A, Turchin P (2001) Identifying the density-dependent structure underlying ecological time series. Oikos 92: 265–270.
29. Royama T (1992) Analytical population dynamics. London, New York: Chapman & Hall. Springer.
30. Ricker WE (1954) Stock and recruitment. J Fish Res Board Canada 11: 559–623.
31. R Development Core Team (2011) R: A language and environment for statistical computing. R Found Stat Comput Vienna. Available: http://www.r-project.org/. Accessed 2014 May 8.
32. Bates D, Chambers JM (1991) Nonlinear Models. In: Chambers JM, Hastie TJ, editors. Statistical Models in S. Wadsworth & Brooks/Cole, Pacific Grove, California.
33. Cameron AC, Windmeijer FAG (1996) R-Squared Measures for Count Data Regression Models With Applications to Health-Care Utilization. J Bus Econ Stat 14: 209–220.

34. De la Maza M, Lima M, Meserve PL, Gutierrez JR, Jaksic FM (2009) Primary production dynamics and climate variability: ecological consequences in semiarid Chile. Glob Chang Biol: 1116–1126.
35. Liu Z, Yang X, Hubbard K, Lin X (2012) Maize potential yields and yield gaps in the changing climate of Northeast China. Glob Chang Biol 86.
36. Bivand RS, Pebesma E, Gómez-Rubio V (2008) Applied Spatial Data Analysis with R. New York, NY: Springer New York.
37. Bivand R, Altman M, Anselin L, Assunçao R, Berke O, et al. (2013) spdep: Spatial dependence: weighting schemes, statistics and models. R package version 0.5-56.
38. Berryman AA (1999) Principles of population dynamics and their application. Stanley Thrones (Publishers) Limited.
39. Sun BJ, Van Kooten GC (2013) Weather effects on maize yields in northern China. J Agric Sci: 1–11.
40. Butler EE, Huybers P (2012) Adaptation of US maize to temperature variations. Nat Clim Chang 3: 68–72.
41. Lobell DB, Hammer GL, McLean G, Messina C, Roberts MJ, et al. (2013) The critical role of extreme heat for maize production in the United States. Nat Clim Chang 3: 497–501.
42. Lobell DB, Bonfils CJ, Kueppers LM, Snyder MA (2008) Irrigation cooling effect on temperature and heat index extremes. Geophys Res Lett 35: L09705.
43. Muchow RC, Sinclair TR, Bennett JM (1990) Temperature and Solar Radiation Effects on Potential Maize Yield across Locations. 82: 338–343.
44. Chen C, Lei C, Deng A, Qian C, Hoogmoed W, et al. (2011) Will higher minimum temperatures increase corn production in Northeast China? An analysis of historical data over 1965–2008. Agric For Meteorol 151: 1580–1588.
45. Lobell DB (2007) Changes in diurnal temperature range and national cereal yields. Agric For Meteorol 145: 229–238.
46. Peng S, Huang J, Sheehy JE, Laza RC, Visperas RM, et al. (2004) Rice yields decline with higher night temperature from global warming. Proc Natl Acad Sci U S A 101: 9971–9975.
47. Mohammed A, Tarpley L (2007) Effects of High Night Temperature on Crop Physiology and Productivity: Plant Growth Regulators Provide a Management Option. Global Warming Impacts – Case Studies on the Economy, Human Health, and on Urban and Natural Environments. doi: 10.5772/24537.
48. Rosenzweig C, Tubiello FN (1996) Effects of changes in minimum and maximum temperature on wheat yields in the central US A simulation study. Agric For Meteorol 80: 215–230.
49. Muchow RC (1990) Effect of high temperature on grain-growth in field-grown maize. F Crop Res 23: 145–158.
50. Alward RD, Detling J, Milchunas D (1999) Grassland Vegetation Changes and Nocturnal Global Warming. Science (80-) 283: 229–231.
51. Katz RW, Brown BG (1992) Extreme events in a changing climate: Variability is more important than averages. Clim Change 21(3):289–302.
52. Long SP, Ainsworth EA, Leakey ADB, Nösberger J, Ort DR (2006) Food for thought lower-than-expected crop yield stimulation with rising CO2 concentrations. Science 312: 1918–1921.
53. Bosello F, Carraro C, De Cian E (2009) An Analysis of Adaptation as a Response to Climate Change. SSRN Electron J.

6

Range-Wide Latitudinal and Elevational Temperature Gradients for the World's Terrestrial Birds: Implications under Global Climate Change

Frank A. La Sorte[1]*, Stuart H. M. Butchart[2], Walter Jetz[3], Katrin Böhning-Gaese[4,5]

1 Cornell Laboratory of Ornithology, Cornell University, Ithaca, New York, United States of America, **2** BirdLife International, Cambridge, United Kingdom, **3** Department of Ecology and Evolutionary Biology, Yale University, New Haven, Connecticut, United States of America, **4** Biodiversity and Climate Research Centre (BiK-F), Senckenberg Gesellschaft für Naturforschung, Frankfurt (Main), Germany, **5** Department of Biological Sciences, Goethe Universität, Frankfurt (Main), Germany

Abstract

Species' geographical distributions are tracking latitudinal and elevational surface temperature gradients under global climate change. To evaluate the opportunities to track these gradients across space, we provide a first baseline assessment of the steepness of these gradients for the world's terrestrial birds. Within the breeding ranges of 9,014 bird species, we characterized the spatial gradients in temperature along latitude and elevation for all and a subset of bird species, respectively. We summarized these temperature gradients globally for threatened and non-threatened species and determined how their steepness varied based on species' geography (range size, shape, and orientation) and projected changes in temperature under climate change. Elevational temperature gradients were steepest for species in Africa, western North and South America, and central Asia and shallowest in Australasia, insular IndoMalaya, and the Neotropical lowlands. Latitudinal temperature gradients were steepest for extratropical species, especially in the Northern Hemisphere. Threatened species had shallower elevational gradients whereas latitudinal gradients differed little between threatened and non-threatened species. The strength of elevational gradients was positively correlated with projected changes in temperature. For latitudinal gradients, this relationship only held for extratropical species. The strength of latitudinal gradients was better predicted by species' geography, but primarily for extratropical species. Our findings suggest threatened species are associated with shallower elevational temperature gradients, whereas steep latitudinal gradients are most prevalent outside the tropics where fewer bird species occur year-round. Future modeling and mitigation efforts would benefit from the development of finer grain distributional data to ascertain how these gradients are structured within species' ranges, how and why these gradients vary among species, and the capacity of species to utilize these gradients under climate change.

Editor: David L. Roberts, University of Kent, United Kingdom

Funding: This work was funded through a postdoctoral fellowship at the Cornell Lab of Ornithology. The funders had no role in study design, data collection and analysis, decision to publish, or preparation of the manuscript.

Competing Interests: The authors have declared that no competing interests exist.

* E-mail: fal42@cornell.edu

Introduction

A variety of responses are available for species as global climate change progresses [1,2] and changing climatic conditions increasingly impact species' performance and fitness. The two primary options, which are not necessarily independent, are for species' populations to shift their geographic ranges to regions containing suitable climatic conditions, or to remain and adapt phenotypically or genetically [3,4]. One of the most important immediate responses is for species to track their geographic climatic associations across space. This may be possible provided that opportunities are available for geographic range-shifts, species have the physiological and behavioral capacity to take advantage of them, and they can keep pace with the velocity of climate change [5]. Geographic niche tracking has been observed under climate change over geological time scales [6] and more recently under modern climate change [7,8].

Two environmental gradients appear particularly relevant when considering geographic niche tracking under past and current climate change: surface latitudinal and elevational temperature gradients. Both gradients represent natural environmental features whose location and general form remain consistent across ecological time-scales, and there is empirical evidence that many taxa including birds are currently tracking these gradients under climate change [7,9]. Latitudinal temperature gradients are strongest outside the tropics, especially in the Northern Hemisphere where the most extensive land masses are located [10]. Elevational temperature gradients are stronger and less variable than latitudinal temperature gradients but are geographically more restricted [11]. Latitudinal and elevational temperature gradients are not equally available for all species nor are they uniformly distributed within species' geographic ranges. Under current climate change projections, there is no evidence to suggest the strength of elevational temperature gradients are expected to change; whereas there is evidence that the strength of latitudinal temperatures gradients may change over broad geographic regions [2]. Due to latitudinal asymmetry in surface warming, latitudinal temperature gradients are projected to weaken in the Northern

Hemisphere [2,12], which is expected to result in diminished seasonality and increased vegetation growth at the high northern latitudes [13,14].

Life history, morphological, and physiological traits are known to vary among species' populations at geographic scales [15] and, in some cases, this variation is correlated with latitudinal or elevational gradients [16,17] and associated temperature gradients [18]. This variation in ecological traits across latitudinal or elevational gradients can in turn affect reproductive potential and rates of population growth, especially at range limits [19,20]. Under climate change, the structure of latitudinal and elevational temperature gradients can be modified, affecting population growth rates, especially at the leading and trailing edge of the gradient within species' ranges [21,22]. Specifically, climate change can result in population growth increasing at the leading edge and declining at the trailing edge of the gradient, with the increased likelihood of range shifts at the leading edge and extirpation of populations at the trailing edge.

The strength of latitudinal and elevational temperature gradients within species current geographic ranges relative to projected changes in temperature under climate change can determine the potential for range-shifts to occur under climate change. For species whose geographic ranges have similar spatial dimensions in relationship to latitudinal or elevational temperature gradients, the steeper the gradient, the broader the species' geographic climatic niche (i.e., species' association with climatic conditions across space) and the greater the likelihood that a larger component of current climatic associations will be retained within the geographic range as temperatures increase (Figure 1A,B,C). In contrast, the shallower the gradient, the narrower the species' geographic climatic niche and the smaller the likelihood that a significant geographic component of current climatic conditions will be retained within the geographic range as temperatures increase (Figure 1D,E,F). One potential consequence of shallow latitudinal or elevational temperature gradients is the formation of range-shift gaps [10]. Here, increasing temperatures can result in the complete loss of current climatic associations within species' current geographic ranges, an outcome that may substantially hinder successful range-shifts. Range-shift gap width is also likely to be negatively correlated with the strength of the temperature gradients (Figure 1), suggesting species' distributions containing very weak gradients, as found with many tropical species, are particularly susceptible to the consequences of climate change [10]. In total, climate change can modify latitudinal or elevational temperature gradients within species' geographic ranges, which in turn can alter population dynamics at range boundaries and range-shift potential based on the strength of those gradients relative to the magnitude of climate change.

Here, we estimate latitudinal and elevational temperature gradients within the geographic ranges of *ca.* 90% of the world's avifauna, numbering 9,014 species. To evaluate whether bird species that are currently threatened with extinction also face particularly high risks under climate change, we structure our analysis to compare species classified as threatened ($n = 878$) vs. non-threatened with extinction ($n = 8,136$) on the IUCN Red List [23]. We additionally examine the ability of species' geography (range size, shape, and orientation relative to the equator) to explain variation in elevational and latitudinal temperature gradients among species. These characteristics of species' ranges tend to be geographically similar across taxa [24] and are determined in part by spatial variation in climate and topography [25]. Lastly, to assess the degree of correspondence between projected warming and the availability of elevational and latitudinal temperature gradients within species' ranges, we consider how gradient strength is related to variation in projected changes in temperature under climate change.

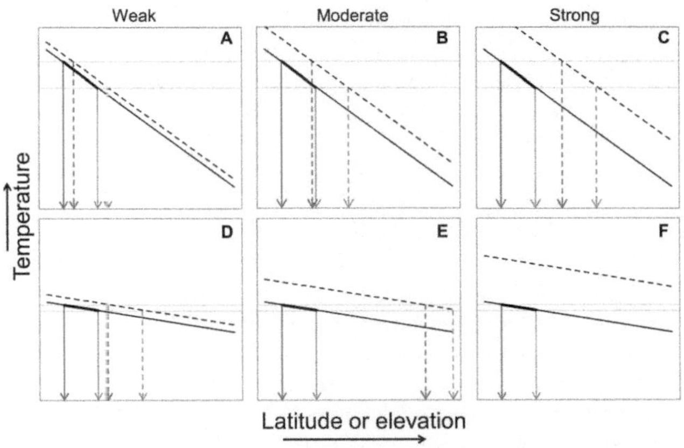

Figure 1. Conceptual figures of six geographic range-shift scenarios for species containing steep (A,B,C) or shallow (D,E,F) latitudinal or elevational temperature gradients within their geographic ranges. The spatial breadth of the latitudinal or elevational gradient within the geographic range is identical in each case. The current temperature gradient (solid line) and the projected temperature gradient under climate change (dashed line) are shown for each scenario. The location of the geographic range of the species along the temperature gradient is depicted with a bold line between the solid blue and red arrows. The species' leading (blue arrow) and trailing (red arrow) range limits are shown based on the current gradient (solid arrows) and the projected gradient under climate change (dashed arrows). Three gradations in projected climate change are shown weak (A, D), moderate (B, E), and strong climate change (C, F). The horizontal dotted lines identify the boundaries of each species' geographic climatic niche as defined along the temperature gradient. Everything else being equal, species with steeper environmental temperature gradients have broader geographic climatic niches and need to shift their geographic range less in order to track climate change.

Materials and Methods

Data sources and preparation

We acquired range maps for the world's birds from BirdLife International and NatureServe [26] whose data sources are described in Buchanan, Donald and Butchart [27]. We considered breeding/resident ranges only and excluded species that were not well suited for our analysis, specifically those associated with marine environments. This resulted in a total of 9,014 extant species for analysis (see Table S1). These species were classified as either threatened ($n = 878$) or non-threatened with extinction ($n = 8,136$) using the IUCN Red List [23]. Here threatened refers to species identified as vulnerable, endangered, or critically endangered under the IUCN Red List.

Species range data was analyzed using a gridded system having a cylindrical equal-area projection and a cell area of 3,091 km^2. We defined species' geographic breeding ranges as the spatial combination of equal area cells with ≥50% terrestrial surface area (based on the proportion of the cell that contained non-marine terrain) that intersected each species' geographic range polygon. We placed each species into one of six biogeographical realms [28] based on the realm that contained the greatest proportion of each species' gridded range (Nearctic, Palaearctic, IndoMalaya, Neotropics, Afrotropics, and Australasia). Species whose ranges occurred primarily in Antarctica or Oceania were excluded from analysis. From the 9,014 species, we acquired minimum and maximum elevation associations for 4,978 species from BirdLife International [23]. The distribution of these species across the six realms did not show geographic biases, with the proportion of species in each realm not diverging significantly from expectation ($\chi^2 = 0.029$, df = 5, $P = 1.0$).

Temperatures within species' geographic ranges were estimated using the annual average of 1950–2000 average-monthly temperatures from WorldClim [29] gridded and analyzed at a 30-arcsec resolution (*ca.* 1 km at the equator). Elevations within species' ranges were estimated using the USGS global digital elevation model (GTOPO30) gridded and analyzed at a 30-arcsec resolution. Tropospheric lapse rates, the association between temperature and elevation within the troposphere, were estimated globally using a gridded model at a resolution of 2.5° (*ca.* 278 km at the equator) for the period 1948–2001 [11].

Broad-scale, expert-based range maps are only coarsely defined estimates of species occurrence and reliably characterize presences only above 100 km grain [30,31]. Here, we are interested in the form of latitudinal and elevational temperature gradients (i.e., shallow or steep) as defined across the geographic breadth of species' distributions independent of patterns of occurrence. The use of relatively high resolution global estimates of temperature and elevation provided the detail of data necessary to estimate the full structure of these gradients for individual species, especially in situations where geographic ranges were small or temperature gradients were non-linear. For the purpose of this baseline study we assume that low probability of occurrence naturally incurred at the finer spatial grain are not biasing the geographic and species comparisons. We acknowledge that this assumption is not necessarily valid and will require careful follow-up assessments at local scales or at broader scales once more detailed distribution data becomes available.

Projected changes in temperature (temperature anomalies) were estimated using projections averaged over four Atmosphere Ocean General Circulation Models (AOGCMs). The four AOGCMs were CNRMCM3, CSIRO-Mk3.0, ECHam5, and MIROC 3.2. The gridded projections from these models were downscaled to 30 arc-minute resolution using pattern scaling with the MarkSim weather generator [32]. The projections were based on the A2 scenario, which is currently considered to be the most relevant under current greenhouse gas emission rates [33,34]. The anomalies represent the change in temperature between the twentieth century control 30-year normal (1961–1990) and the 30-year period 2081–2100. The temperature anomalies were bilinearly interpolated to match the resolution of the 3,091 km^2 equal-area grid.

Elevational temperature gradients

We estimated latitudinal and elevational temperature gradients separately in order to better represent each gradient's unique characteristics. The presence of elevational temperature gradients is dictated by the presence of mountain systems or orographic features, which show substantial geographic variation (see Figure S1). Mountain systems often contain strong topographic heterogeneity resulting in a mosaic of climatic conditions, which are poorly estimated using interpolated weather station data [35,36]. Here, we use elevation data, which does not rely on similar interpolation methods, to estimate elevational temperature gradients within species' ranges.

Unlike latitudinal temperature gradients, the strength of elevational temperature gradients are broadly consistent from the equator to the poles. Tropospheric lapse-rate averages 6.2°C km^{-1} over the continents, with a range of 4.5 to 6.5°C km^{-1} from the polar latitudes to the equator, respectively [11]. To account for this geographic variation, we calculated the average tropospheric lapse-rate within each species' geographic range. To account for the high variability in the extent and spatial distribution of orographic features and the decline in terrestrial area with increasing elevation resulting in strongly skewed distributions of elevation within species' ranges, we estimated two components of the elevational temperature gradient for each species: (1) the overall magnitude and (2) the evenness of the gradient. We then combined tropospheric lapse-rate with magnitude and evenness to generate one metric designed to represent the overall strength of elevational temperature gradients within each species' range.

More specifically, for species whose minimum and maximum elevation associations had both been estimated ($n = 4{,}978$), we excluded elevations within the geographic range below and above these associations. We then identified the actual minimum and maximum elevations within each species' range based on our gridded elevation data, which in combination with species' documented elevation associations were used to calculate each species' realized vertical range extent. Vertical range extent was used to estimate the magnitude of each species' elevational temperature gradient. To assess how evenly the vertical range extent was distributed across each species' range, we first summed the number of 30 arc-second elevation cells in all 100-m elevational bands found within each species range. We then used the Lorenz curve and associated Gini coefficient [37–39], explained below, to estimate how evenly the cells were distributed across elevation bands. Here, the 100-m bands were first ranked from lowest to highest elevation. The frequency of cells within these bands was then used to plot the Lorenz curve, i.e. the cumulative proportion of the number of cells in each band (x-axis) against the corresponding cumulative proportion of their elevations (y-axis). Gini coefficients (or Gini ratio) were then calculated, which is defined as twice the area contained between the Lorenz curve and the line of perfect equality (or perfect evenness). The Gini coefficient has a range from 0 to 1, high to low evenness, respectively. To ease interpretation, for each species we took the product of the vertical range extent, average tropospheric lapse rate, and the inverse of evenness (1 – Gini coefficient) to generate a single index for the elevational temperature gradient. We interpret this index as the overall strength of the elevational temperature gradient across each species' range: larger values indicate the presence of a strong and more evenly distributed gradient; lower values indicate the presence of a weak gradient due to either a small vertical range extent or an unevenly distributed vertical range extent.

Latitudinal temperature gradients

Outside the tropics, average annual temperature declines on average 0.7°C for each degree of latitude in the Northern Hemisphere and on average 0.5°C for each degree of latitude in the Southern Hemisphere (Figure S2). With one degree of latitude equal to approximately 111 km, this translates to a decline of 1°C for every 150 km in the Northern Hemisphere and a decline of 1°C for every 197 km in the Southern Hemisphere.

We estimated the latitudinal temperature gradient within the breeding ranges of 9,014 species while controlling for variation in elevation. For our approach, we first found the average tropospheric lapse rate within each species' range. We then compiled gridded 30 arc-second average annual temperatures and elevations within each species' range. We chose annual average temperature because matching each species' breeding season distribution with their unique breeding season temperatures, especially for migratory species that only spend a few weeks to months on the breeding grounds, was not feasible. A likely consequence of using annual average temperature is that estimates of latitudinal temperature gradients for species that breed outside the tropics are likely to be steeper than they would be if estimates were based on the breeding season alone. Beyond these qualitative differences, using annual average temperature is unlikely to alter our qualitative conclusions. To control for the effect of elevation on temperature, we applied quantile regression to temperature as a function of elevation for 11 quantiles (0.0, 0.1,...0.9, 1.0). The residuals for the quantile whose slope most closely matched the tropospheric lapse rate estimated for each species was retained for further analysis. This approach allowed us to interpret latitudinal temperature gradients as the gradient along a fixed elevation within each species' range.

The relationship between these residuals and latitude was then assessed for each species using segmented linear regression [40]. If species' ranges occurred in both hemispheres, the range was first split in half at the equator and each portion was analyzed separately. We tested if the relationship between temperature and latitude contained one or two segments using Davie's test [41], which tests for a non-zero difference in the slope parameters of the segmented relationship. Relationships with two segments are more likely to occur if species' distributions are intersected by the Tropics of Cancer or Capricorn (Figure S2). When there was statistical evidence for a segmented relationship, we estimated a break point between segments and slope coefficients for each segment. When there was no statistical evidence for multiple segments, a single slope coefficient was estimated. Coefficients estimated south of the equator were multiplied by -1 to allow for comparison between the two hemispheres (i.e., latitudinal temperature gradients north and south of the equator are both negative; Figure S2). For species whose ranges were estimated by one slope coefficient, this coefficient was used to represent the overall gradient. For species whose ranges contained multiple slope coefficients, we used a weighted average to summarize the gradient. Weights were based on the latitudinal extent of the region where each slope coefficient was estimated. Based on this procedure, larger negative values indicate steeper latitudinal temperature gradients and values approaching zero indicate shallower latitudinal temperature gradients.

Analysis

To summarize how steep elevational and latitudinal temperature gradients are on average within species' ranges globally, we first calculated the average elevational temperature gradient and median latitudinal temperature gradient within each equal-area cell for all species whose geographic ranges intersected that cell. To summarize how gradients differed for threatened and non-threatened species, we quantified how the gradients varied by species' median latitude. This analysis was implemented separately for species in each of the six biogeographical realms. For elevational temperature gradients, we used generalized additive models (GAM; [42]) with a Gaussian error distribution. For latitudinal temperature gradients, we used robust MM-type polynomial regression [43,44]. GAM was selected because it adjusts automatically to the nonlinear associations observed between elevational temperature gradients and species' median latitude. This feature also supported the use of GAM to examine how species richness, range size, elevational extent, and Gini coefficient varied by species' median latitude for threatened and non-threatened species. Robust polynomial regression was selected for latitudinal temperature gradients because of the presence of many extreme outliers that were poorly modeled with GAM. In addition, the general curvilinear form of the relationship was well represented with a polynomial function. The order of the polynomial function was determined using robust Wald-type or deviance-type tests of nested model pairs. Median latitude was defined as the median of the full latitudinal extent of each species' range.

We used robust multiple regression to examine the ability of geographic and climatic predictors to explain the variability observed in species latitudinal and elevational temperature gradients. The four predictors were estimated within each species' geographic range and included average projected temperature anomaly, and geographic range size, shape, and orientation. Range size was $\log_{10}$ transformed before analysis. Range shape and orientation were estimated based on a principle component analysis (PCA) of the latitude and longitude of the grid cells contained within each species geographic range. PCA was conducted using a singular value decomposition of the centered data matrix. The square roots of the two eigenvalues were used to estimate range shape, where the minimum eigenvalue was divided by the maximum eigenvalue. Here, values approaching zero indicating elongated ranges and values approaching one indicating circular ranges. Range orientation was estimated based on the angle of the major axis of the PCA eigenvector from the equator. Values for range orientation were defined from 0° (parallel with the equator) to 90° (perpendicular to the equator). We evaluated our four predictors for multi-collinearity and singularity using variance inflation factors (VIFs) where predictors with VIF>5 indicate cause for concern and VIF>10 indicate the presence of significant collinearity [45]. We retained all four predictors after they were deemed to be statistically independent (VIF≤1.6).

To evaluate differences between threatened and non-threatened species, we included in addition to the four continuous predictors, a species' IUCN Red List status as threatened or non-threatened with extinction. For the analysis of the latitudinal temperature gradient, we also considered if the median latitude of a species' range was located within or outside the tropics. Median latitudes between 23.5°S and 23.5°N latitude were considered tropical. The tropical/extratropical classification was included because it defined a major feature of the gradient (Figure S2). In total, we examined eight robust multiple regression models that consisted of the two gradients, the four continuous predictors, associated categorical predictors, and all pairwise interactions.

All analyses were conducted in R version 3.0.1 [46]. Segmented regression was conducted using the 'segmented' library, robust regression using the 'robust' and 'robustbase' libraries, and GAM using the 'mgcv' library [42]. We used the default optimization procedure to estimate the degree of smoothing in the GAMs [42]. The Lorenz curves were estimated using the 'ineq' library and the Gini coefficients were estimated using the 'reldist' library.

Results

Threatened and non-threatened bird species presented similar latitudinal trends in species richness with peaks for both groups occurring in the tropics (Figure 2A and Figure S3). Range sizes for threatened species were significantly smaller on average except for species located at the most extreme northern and southern latitudes (Figure 2B and Figure S3). For the 4,978 species with documented elevational associations, threatened species had smaller vertical range extents within their geographic ranges (Figure 2C and Figure S3) and threatened species had higher elevational evenness within their geographic ranges in the central tropics and more equivalent elevational evenness elsewhere (Figure 2D and Figure S3). Projected increases in temperature within species' ranges were highest for species located at higher

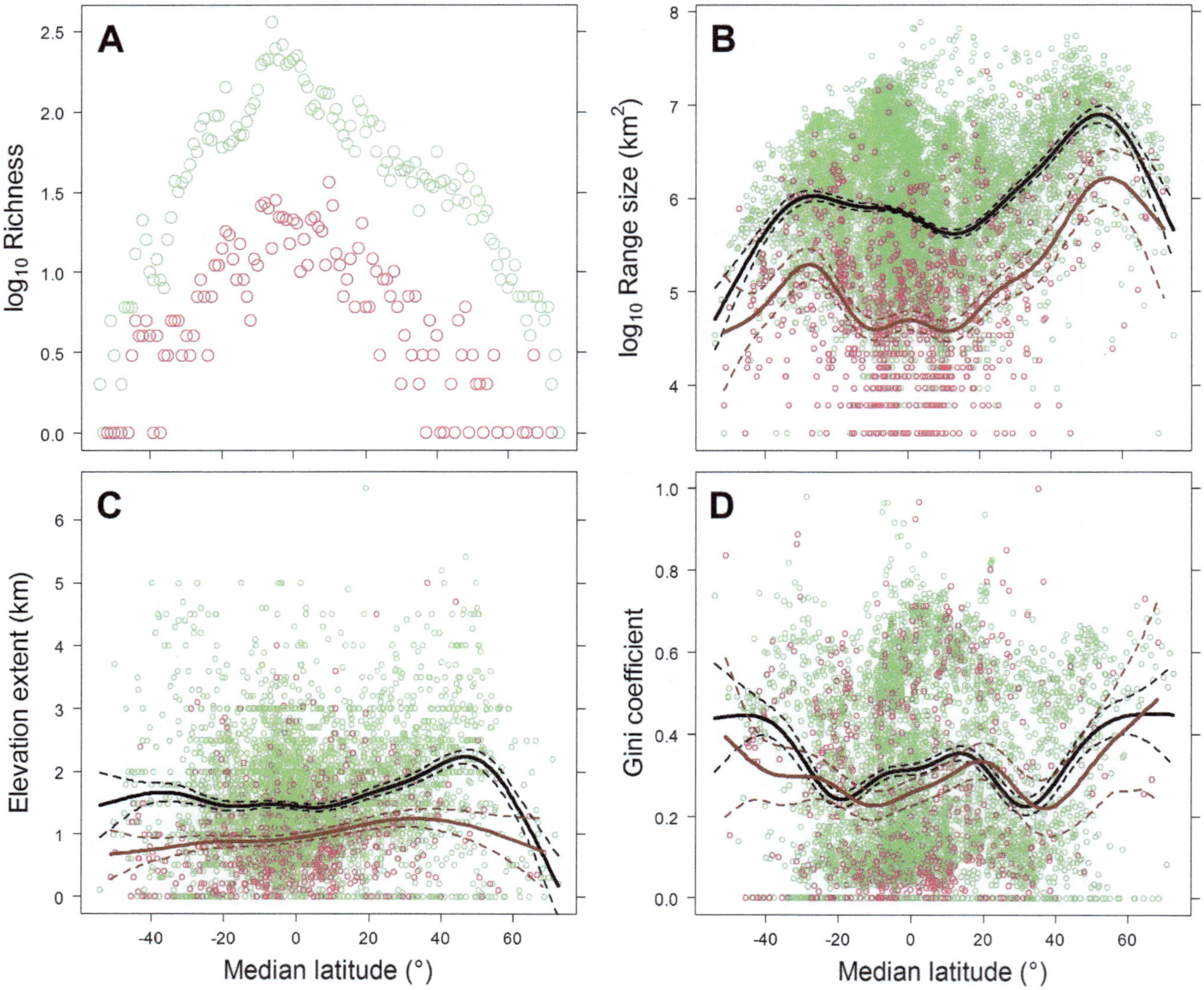

Figure 2. Plots summarizing geographic patterns for species' currently identified as threatened (red points; $n = 878$) and non-threatened (green points; $n = 8,136$) with extinction by the median latitude of each species' range. (A) The number of species' whose median latitudes occur within one degree latitudinal bands; (B) species' range size; (C) the elevational extent within species' ranges for 4,978 species with minimum and maximum elevation associations; and (D) the evenness of the distribution of 100 m elevations bands within species' ranges based on the Gini coefficient (0 = high evenness; 1 = low evenness; see Materials and Methods for details) for 4,978 species with minimum and maximum elevation associations. The trend lines are the fits of generalized additive models for species threatened (red) and non-threatened (black) with extinction.

latitudes in the Northern Hemisphere and lowest for species located at lower latitudes in the Southern Hemisphere (Figure S4).

There was strong geographic variation in the steepness of elevational temperature gradients (Figure 3). When considering the combined effects of elevational extent and elevational evenness (Figure S3) with tropospheric lapse rate, the strongest gradients were found for species located in Africa, the western part of North and South America, and the Tibetan Plateau region. In contrast, species in Australasia, the islands of IndoMalaya and the lowlands of the Neotropics had the weakest elevational temperature gradients (Figure 3). Threatened species were associated with weaker elevational temperature gradients across the six biogeographical realms, with the most significant differences occurring in the Neotropics and Afrotropics (Figure 3).

After controlling for elevation, latitudinal temperature gradients were on average shallowest for species in the tropics and steepest for species outside the tropics (Figure 4). The steepest latitudinal temperature gradients occurred in the Northern Hemisphere. Australasia had the strongest gradients in the Southern Hemisphere, which did not match the strength observed in the Northern Hemisphere. There was no evidence that the latitudinal temperature gradients differed between threatened and non-threatened species across the six biogeographical realms (Figure 4).

Variation in elevational temperature gradients among species was poorly explained by our four predictors (Table 1 and Figure S5). Species with higher projected temperature increases and larger geographic ranges had steeper elevational temperature gradients, and this was the case for both threatened and non-threatened species. Species with geographic ranges that were elongated in shape and oriented perpendicular to the equator had stronger gradients, and this was more pronounced for non-threatened species.

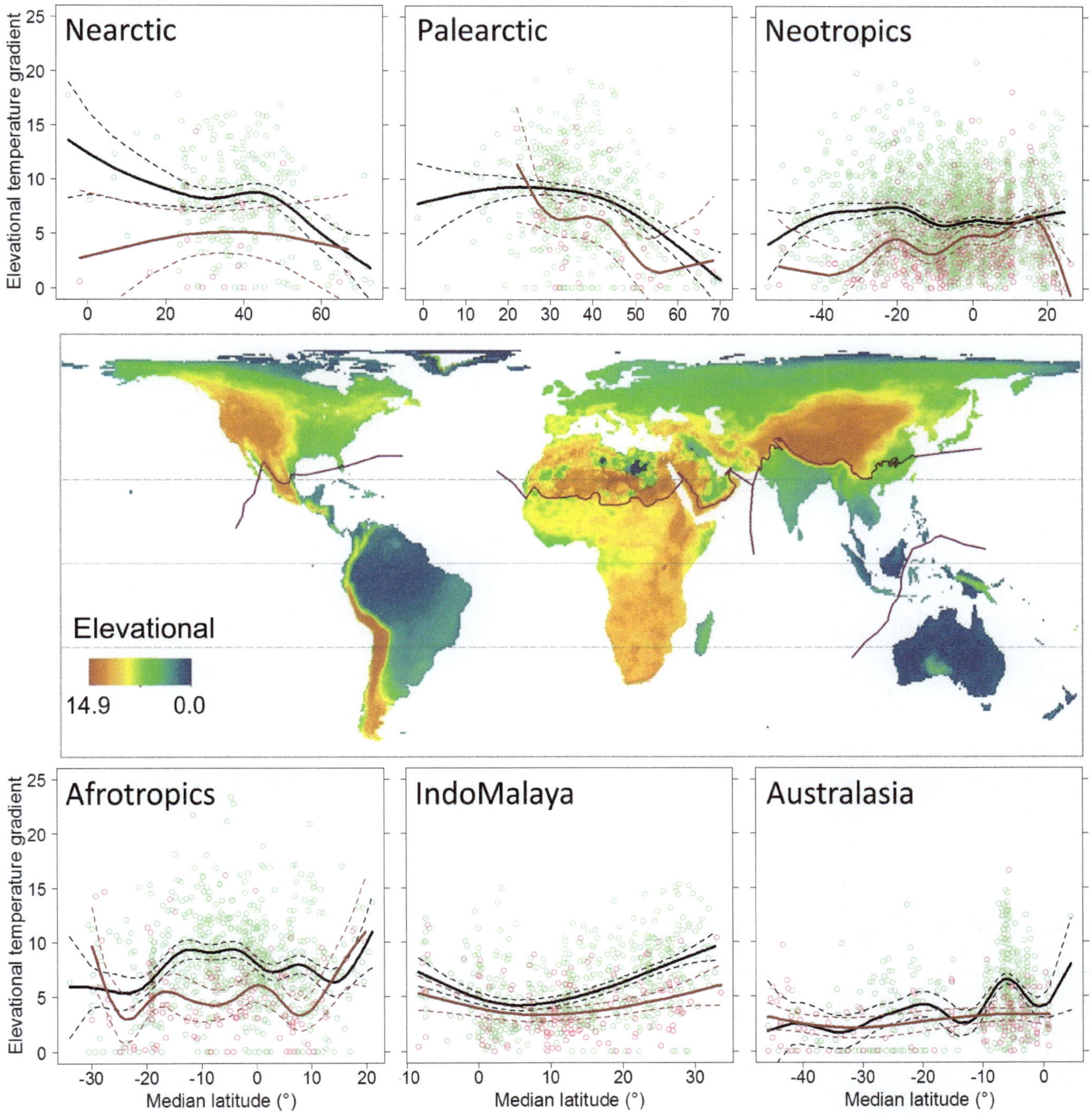

Figure 3. The elevation temperature gradient index summarized across species' geographic ranges (map) and summarized within six biogeographical realms (plots) as a function of the median latitude of each species' range. The temperature gradient was estimated spatially within species' current ranges (see Materials and Methods for details). In the plots, red points are threatened species ($n = 766$) and green points are non-threated species ($n = 4,212$). Trend lines are the fits of generalized additive models for threatened (red) and non-threatened species (black). The solid line is the equator and the dashed lines are the Tropics of Cancer and Capricorn (23.5°N and 23.5°S latitude, respectively).

Variation in latitudinal temperature gradients was better explained by our four predictors, with evidence for significant and strong differences between tropical and non-tropical species (Table 2 and Figure S6). Tropical species had weaker relationships for all four predictors with little evidence for differences between threatened and non-threatened species. For threatened and non-threatened extratropical species, species with higher projected temperature increases, larger ranges and ranges that were elongated in shape and oriented parallel with the equator had steeper latitudinal temperature gradients. For threatened extratropical species, gradients were even steeper for species with larger ranges.

Discussion

Species face unique challenges under climate change, with the availability and strength of range-shift opportunities along latitudinal or elevational temperature gradients playing a signif-

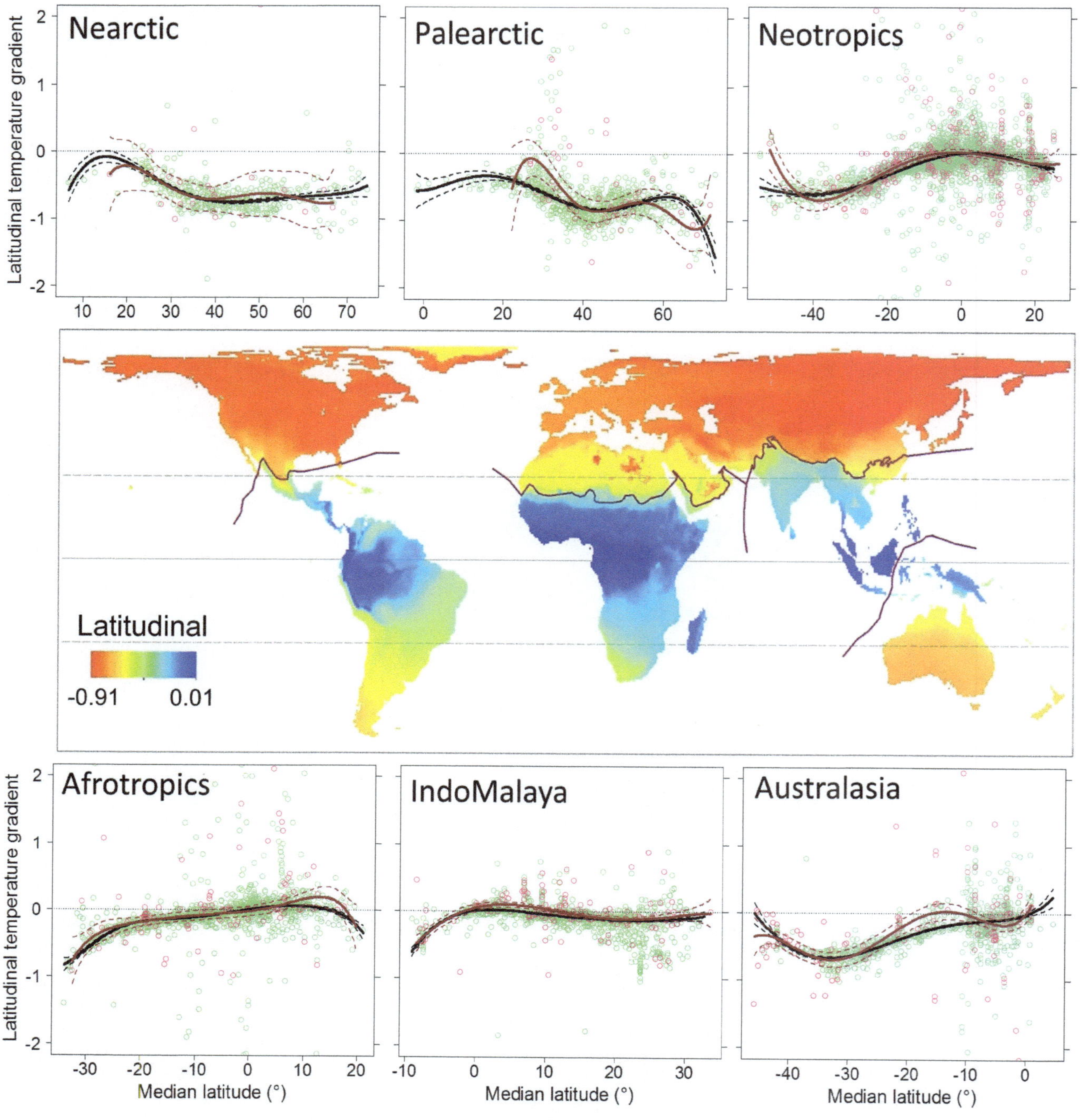

Figure 4. The latitudinal temperature gradient index summarized across species' geographic ranges (map) and summarized within six biogeographical realms (plots) as a function of the median latitude of each species' range. The temperature gradient was estimated spatially within species' current ranges (see Materials and Methods for details). In the plots, red points are threatened species ($n = 878$) and green points species non-threated species ($n = 8{,}136$). Trend lines are the fits of generalized additive models for threatened (red) and non-threatened species (black). The solid line is the equator and the dashed lines are the Tropics of Cancer and Capricorn (23.5°N and 23.5°S latitude, respectively).

icant role in determining species' likelihood of persistence. In this study we provide the first estimation of these gradients for the world's avifauna relative to species' IUCN Red List category, geography, and projected exposure to climate change. Our approach expands upon current methods for estimating species' geographic climatic associations by measuring these two gradients individually using fine-grained environmental data, which more accurately captures their unique spatial structures. The application of this approach has the potential to increase biological realism and predictive quality in current large-scale modeling efforts by removing problematic assumptions and minimizing potential biases. Our findings suggest threatened species are at a distinct disadvantage globally based on their association with shallower elevational temperature gradients. Latitudinal temperature gradients are primarily relevant for species located outside the tropics, a region where few bird species occur year-round, including threatened species whose associations with latitudinal temperature gradients did not differ from non-threatened species. Our results

Table 1. Coefficients and test statistics from robust linear models examining predictors of elevational temperature gradients as estimated within the geographic ranges of 4,978 bird species.

Model	Coef.	SE	t	P
Intercept	3.149	0.380	8.29	<0.001
Anomaly	0.976	0.108	9.05	<0.001
Threat	−3.117	0.895	−3.48	<0.001
Anomaly × Threat	0.344	0.273	1.26	0.2090
(R^2 = 0.07)				
Intercept	3.600	0.434	8.30	<0.001
Size	0.511	0.074	6.92	<0.001
Threat	−1.872	1.075	−1.74	0.082
Size × Threat	0.002	0.213	0.01	0.993
(R^2 = 0.06)				
Intercept	6.988	0.134	52.01	<0.001
Shape	−1.155	0.310	−3.73	<0.001
Threat	−2.940	0.295	−9.96	<0.001
Shape × Threat	1.580	0.707	2.23	0.026
(R^2 = 0.06)				
Intercept	5.845	0.105	55.85	<0.001
Orientation	0.018	0.002	8.18	<0.001
Threat	−1.824	0.280	−6.52	<0.001
Orientation × Threat	−0.014	0.005	−2.68	0.008
(R^2 = 0.06)				

Predictors include projected temperature Anomaly, geographic range Size, Shape, and Orientation and if the species is threatened with extinction (Threat).

also indicate a degree of correspondence between range-shift opportunities and climate change, in that species in regions that are projected to experience greater warming are more likely to be associated with steeper temperature gradients.

The strength of latitudinal temperature gradients is determined in part by the location, size, shape and orientation of species' geographic ranges. Not surprisingly, the strongest latitudinal temperature gradients were associated with species whose ranges occurred outside the tropics, especially at the higher latitudes in the North Hemisphere where the largest land masses are located. Species in these regions tend to have larger geographic ranges (see Figure 2B) and ranges that tend to be elongated parallel to the equator. As our results indicate, the strength of the latitudinal temperature gradient is of sufficient size within this region that geographic ranges with narrow latitudinal extents are still able to capture a significant gradient. In contrast, the spatial distribution of elevation temperature gradients reflects the form and extent of orographic features and how species' distributions intersect these features. We found that the strength of these gradients varied substantially across the globe and, unlike latitudinal temperature gradients, differed significantly between threatened and non-threatened species. In agreement with other studies [47,48], species in Australia and the Islands of IndoMalaya stood out as particularly vulnerable, with some of the shallowest elevational gradients, followed by species located in the lowlands of the Neotropics. In contrast to latitudinal gradients, our ability to predict species' elevational temperature gradients was limited, reflecting the greater geographic heterogeneity in the location and structure of these gradients.

The strength of both temperature gradients was positively correlated with projected changes in temperature, but only for species outside the tropics with latitudinal temperature gradients. The greatest increases in temperature under climate change are projected to occur at the high northern latitudes (see Figure S4). Our findings therefore suggest a geographic correspondence with the greatest projected warming occurring in regions with the strongest latitudinal temperature gradients. However, few bird species occur at the high northern latitudes year round. The proportion of migratory species (*ca.* 19% of extant bird species [49]) within breeding communities increasing linearly as you travel north from the equator, approaching 100% at the highest latitudes [50]. Migratory species spend the bulk of their annual cycle in migration or on the non-breeding ranges at lower latitudes. Thus, very few bird species are likely to benefit substantially from this correspondence.

IUCN Red List categories of extinction risk are assigned using the Red List criteria, which have quantitative thresholds relating to population and range size, structure and trends [23]. Only 3.5% of threatened or near threatened bird species (77/2193) are assessed as threatened by climate change so severely that the global population may be declining rapidly or very rapidly over three generations, and hence potentially contributing to their Red List category through the A criterion. Of these, only 26 (1.2%) actually qualify under the A criterion. Our findings suggest this proportion is likely to increase, especially for threatened species associated with shallow elevational temperature gradients. As confirmed in our analysis, threatened species occur primarily in the tropics and their distributions are defined by smaller geographic ranges and elevational extents. Hence, threatened species may be at a geographic disadvantage under climate change with limited associations with elevational or latitudinal temperature gradients, although we found no evidence for the latter. Our findings suggest

Table 2. Coefficients and test statistics from robust linear models examining predictors of latitudinal temperature gradients as estimated within the geographic ranges of 9,014 bird species.

Model	Coef.	SE	*t*	*P*
Intercept	−0.288	0.021	−13.61	<0.001
Anomaly	−0.092	0.006	−16.52	<0.001
Tropics	0.199	0.028	7.14	<0.001
Threat	0.027	0.047	0.58	0.564
Tropics × Threat	0.028	0.019	1.48	0.138
Anomaly × Tropics	0.094	0.008	12.20	<0.001
Anomaly × Threat	−0.016	0.013	−1.21	0.225
(R^2 = 0.34)				
Intercept	0.148	0.045	3.32	0.001
Size	−0.120	0.007	−17.09	<0.001
Tropics	0.014	0.048	0.28	0.776
Threat	0.130	0.061	2.12	0.034
Tropics × Threat	−0.034	0.020	−1.66	0.098
Size × Tropics	0.080	0.008	10.38	<0.001
Size × Threat	−0.026	0.011	−2.47	0.014
(R^2 = 0.34)				
Intercept	−0.655	0.011	−61.97	<0.001
Shape	0.096	0.025	3.82	<0.001
Tropics	0.591	0.012	47.91	<0.001
Threat	−0.005	0.335	−0.02	0.988
Tropics × Threat	−0.032	0.061	−0.52	0.606
Shape × Tropics	−0.134	0.028	−4.70	<0.001
Shape × Threat	0.129	1.018	0.13	0.900
(R^2 = 0.35)				
Intercept	−0.687	0.006	−107.55	<0.001
Orientation	0.002	0.000	13.52	<0.001
Tropics	0.613	0.008	77.18	<0.001
Threat	−0.027	0.019	−1.42	0.155
Tropics × Threat	−0.026	0.019	−1.34	0.180
Orientation × Tropics	−0.002	0.000	−12.49	<0.001
Orientation × Threat	0.001	0.000	4.49	<0.001
(R^2 = 0.34)				

Predictors included projected temperature Anomaly and geographic range Size, Shape, and Orientation and if the species is tropical (Tropics) or threatened with extinction (Threat).

threatened species that occur in the central tropics have higher elevational evenness across their range, which is a likely consequence of their distributions occurring on steep montane slopes outside of lowland areas, which should result in stronger elevational temperature gradients for these species. However, our analysis suggests that the differences in elevational evenness between threatened and non-threatened species was not substantial enough to overcome the smaller elevational extents contained within threatened species' geographic ranges. Thus, our findings suggest the weaker elevational temperature gradients identified for threatened species is due in large part to a consistent association with smaller elevational extents within their ranges.

When considering both gradients in combination, two regions stand out as having both strong elevational and latitudinal temperature gradients: western North America and Central Asia (see Figures 3–4). These results suggest species in these regions will have greater opportunities to track their climatic niches geographically. Depending on how these temperature gradients intersect spatially within species' ranges, some combination of these gradients could be used. Similarly, two regions stand out as having both shallow elevational and latitudinal temperature gradients: the islands of IndoMalaya and the Amazon Basin in South America (see Figures 3–4). Species in these regions are likely to develop substantial range-shift gaps under climate change [10], resulting in limited opportunities to track their climatic niche geographically. Bird species occurring in the Amazon Basin are also considered to have low adaptive capacity, which may increase their vulnerability to climate change [51].

Species in tropical montane regions are more likely to have restricted lateral and vertical distributions [48,52] and are also more likely to be threatened with extinction (see Figure S3). Even though in many cases these species occur in some of the most extensive montane systems in the world, our findings suggest that the overall steepness of elevational gradients is constrained by their limited representation within species' geographic ranges. Local climatic heterogeneity, however, may allow for microclimatic adjustments that could compensate for broader climatic trends [53–56]. The level of climatic heterogeneity is determined by local topographic variation and also by species' habitat associations [57]. For example, species in forested habitats may have greater opportunities for microclimatic adjustments relative to species in open habitats [58]. In general, however, behavior resulting in microclimatic adjustments might not be sufficient to address the full consequences of climate change [59].

Even if latitudinal and elevational temperature gradients are readily available, dispersal along these gradients is not guaranteed. Dispersal abilities are determined in large part by species' morphological, physiological and behavioral characteristics and adaptations [60,61], all of which are likely to be altered under rapid climate change [62]. Dispersal under climate change may be constrained by unsuitable habitat, population or habitat fragmentation [58,63], or interspecific competition [64]. In addition, gene flow that occurs in association with dispersal can both support and hinder range-shift potential [65]. Gene flow among populations at the leading edge of the range has been found to benefit fitness, whereas gene flow from the interior to the leading edge has been identified as maladaptive [66,67]. Dispersal abilities also vary tremendously among species and between taxa [60]. Even highly vagile taxa such as birds do not necessarily have the full complement of behavioral or physiological characteristics needed to support successful dispersal, which is considered particularly relevant for tropical species [68,69]. Although there is evidence tropical montane birds have moved upslope under climate change [70], the low annual variation in temperatures within the tropics relative to temperate regions may result in stronger physiological barriers to dispersal for tropical species [52,71,72]. When tropical habitats are fragmented through human activities – a process that is occurring at an increasing rate – dispersal can become even more constrained [73,74]. If species' responses are hindered by dispersal limitations or interspecific competition, as is expected for many tropical montane bird communities, substantial ecological disruptions are likely to occur [75].

In contrast, species that breed outside the tropics that are also migratory [76] or have larger geographic ranges tend to have stronger natal dispersal abilities [77–79]. Thus migratory species that are broadly distributed and breed outside the tropics might be in a better position to track elevational or latitudinal temperature gradients. However, exceptions exist [80] and factors that are strongly correlated with dispersal abilities might not be similarly correlated with broad-scale distributional responses under climate change [81]. Nevertheless, current evidence indicates distributional responses for birds along latitudinal [7] and elevational temperature gradients [82] are lagging behind warming trends, suggesting stronger responses could develop as climate change progresses and time lags are overcome [7].

We estimated temperature gradients in this study using relatively high resolution environmental data that was finer than the resolution of the species' range data [30]. This grain mismatch resulted in temperature gradients likely being estimated over fine-grain sections of species' ranges where probability of occurrence is low. Our main results focus on the geographic form of the gradient across the spatial breadth of species' distributions, but future work will benefit from further examination of potential biases this grain mismatch may have on comparisons. How patterns of occurrence are structured within species' ranges, based on proportion of occupancy and level of aggregation [83], may affect the chances of successful dispersal along these gradients. For example, sparse or fragmented patterns of occurrence may diminish the chances of successful dispersal, while more continuous patterns of occurrence may improve them. Alternatively, large contiguous regions with low probabilities of occurrence might contain the steepest gradients with very little dispersal value, as when lowland species' distributions border mountain systems. These remaining uncertainties highlight the need for advancing the spatial grain of our global biodiversity knowledge and for the necessary collaborative data mobilization, integration and quality control to realize this [31].

In addition to temperature, other climate factors and conditions are likely to be important in determining the strength and importance of latitudinal and elevation temperature gradients in defining range-shift potential. Two that have received particular attention are changes in precipitation regimes and changes in climatic variability. In contrast to temperature, precipitation has more complex spatial and temporal variability and more complex associations with latitudinal and elevational gradients. Projected changes in precipitation are also more ambiguous [2], which adds uncertainty to current estimates on how species are likely to respond to climate change [84]. A second factor is climatic variability, which is projected to increase through an increased frequency of extreme climate events [2]. Temporal climatic variability is considered an important factor defining species' distributions limits [85] and increasing climatic variability can have varied effects on species' populations [86] including the potential to hinder range shifts [87] and, especially under rapid climate change, increase extinction risk [88].

Additional research is needed to improve our understanding of how latitudinal and especially elevational temperature gradients are structured within species' ranges and the ability of species, especially tropical, to take advantage of these gradients. In particular, studies using higher resolution climate data (e.g., [22]) for species with varying spatial patterns of occurrence and forms of elevational or latitudinal gradients would be informative. Birds are often used as a model biological system due to high data availability, but less vagile taxa also need to be considered such as plants or vertebrate ectotherms that have alternative modes of dispersal and different physiological and behavior capacities. With birds, examining how these gradients are structured within non-breeding ranges, where migratory birds tend to occur during the majority of the annual cycle and are typically less well studied, would be beneficial [9]. Efforts to better understand the role of precipitation and climate variability in defining range-shift potential along latitudinal and elevational temperature gradients would support more comprehensive inferences.

Current efforts to project species' distributions or to estimate movement corridors under climate change often assume perfect and complete dispersal along temperature gradients (e.g., [89]). Temperature gradients are typically estimated using coarse-grained interpolated weather-station data based on current climatic conditions where the contrasting spatial forms of latitudinal and especially elevational temperature gradients are often poorly summarized [35,36]. These methods may exaggerate estimated dispersal distances [90], especially at high northern latitudes where latitudinal temperature gradients are projected to weaken under climate change [12]. These considerations in addition to the correlative nature of these models have promoted the development of more mechanistic or process-based methods intended to improve biological realism and minimize bias in current projections [91]. Our findings suggest the quality of model predictions may be improved by considering the spatial form and strength of species' demographic responses [92] to projected latitudinal and elevational temperature gradients within species' ranges and along possible movement corridors

Lastly, in order to maximize range-shift potential under climate change, conservation efforts are needed to maximize the size of areas of contiguous habitat, minimize the extent of habitat fragmentation and promote the permeability of the matrix of habitats around key sites. In particular, efforts to assess and promote connectivity at broad geographic extents are likely to be the most beneficial (e.g., [93,94]) where existing gradients are represented as continuously as possible across space [95]. The more thoroughly this can be achieved, the greater the potential for species to track changing climatic conditions based on their unique dispersal abilities and ecological and environmental requirements.

Conclusions

The availability of opportunities for geographic responses under climate change reflects the interplay between each species' geographic distribution and their associated latitudinal and elevational temperature gradients (with the extent to which these opportunities will be utilized depending on each species' dispersal abilities and the consequences of changing gene flow). The distribution of these gradients across the world follows clearly defined geographic patterns, but their availability depends on how they intersect with species' distributions. The location, size, and shape of species' geographic range and projected changes in temperature all contain a degree of relevance in determining the strength of these temperature gradients. However, the strength of latitudinal and elevational temperature gradients varied considerably among species, constraining our ability to make robust predictions for all regions globally. Nevertheless, we found that species that are currently at risk of extinction are consistently at the greatest disadvantage, especially based on the availability of elevational temperature gradients.

Supporting Information

Figure S1 Global patterns of elevation based on the USGS global digital elevation model (GTOPO30) gridded at a 30 arc-second resolution (*ca.* 1 km at the equator). The solid grey line is the equator and the dashed grey lines are the Tropics of Cancer and Capricorn (23.5°N and 23.5°S latitude, respectively).

Figure S2 Global terrestrial annual mean temperature for 1950–2000 (WorldClim) averaged within 0.1 degree latitudinal bands. The dotted lines indicate the Tropics of Cancer and Capricorn (23.5°N and 23.5°S latitude, respectively).

Figure S3 Maps summarizing features of the global distribution of terrestrial avifauna across an equal area grid (cell area: 3,091 km^2). (A) The richness of all species (n = 9,014) and (B) the richness of species threatened with extinction (n = 878). (C) Geographic range size for all species (n = 9,014) and (D) elevational extent for 4,978 species with minimum and maximum elevation associations. (E) The evenness of the distribution of elevation within each species' range based on the Gini coefficient and Lorenz curve (0 = high evenness; 1 = low evenness) for 4,978 species with minimum and maximum elevation associations. Maps are displayed using the Behrmann equal-area cylindrical projection and the color ramps use Jenk's natural break classification. The solid grey line is the equator and the dashed grey lines are the Tropics of Cancer and Capricorn (23.5°N and 23.5°S latitude, respectively). Red lines distinguish boundaries between six biogeographical realms.

Figure S4 Projected temperature anomalies summarized across species' geographic ranges (map) and summarized within six biogeographical realms (plots) as a function of the median latitude of each species' range. In the plots, red points are threatened species (n = 878) and green points species non-threatened species (n = 8,136). Trend lines are the fits of generalized additive models for threatened (red) and non-threatened species (black). The solid line is the equator and the dashed lines are the Tropics of Cancer and Capricorn (23.5°N and 23.5°S latitude, respectively).

Figure S5 Fit of robust linear regression models to four predictors of elevational temperature gradients estimated within the geographic ranges of 4,978 bird species. Red points and lines are threatened species (n = 766) and green points and black line are non-threatened species (n = 4,212).

Figure S6 Fit of robust linear regression models to four predictors of latitudinal temperature gradients estimated within the geographic ranges of 9,014 bird species. Green points and lines are for tropical species (n = 6,720), brown points and lines are for extratropical species (n = 2,294). The solid lines are the fits for non-threatened species (n = 4,214) and the dashed lines are the fits for threatened species (n = 766).

Table S1 The 9,014 species considered in the analysis and their estimated latitudinal and elevational temperature gradients, designated biogeographical realm and threat status (1 = threatened and 0 = non-threatened with extinction), projected temperature anomaly, and geographic range size, shape and orientation.

Acknowledgments

We thank I. I. Mokhov for providing global lapse-rate values and three anonymous reviewers for their constructive suggestions. We thank the large number of experts and organizations who contribute information incorporated into BirdLife International's extinction risk assessments for all species on the IUCN Red List, including spatial information incorporated into the distribution maps.

Author Contributions

Conceived and designed the experiments: FALS KBG WJ. Analyzed the data: FALS. Wrote the paper: FALS KBG WJ SHMB. Collected the data: SHMB.

References

1. Karl TR, Trenberth KE (2003) Modern global climate change. Science 302: 1719–1723.
2. IPCC (2013) Climate change 2013: The physical science basis. In: T. Stocker, Q. Dahe, G.-K. Plattner, editors. Fifth assessment report of the intergovernmental panel on climate change. Cambridge and New York: Intergovernmental Panel on Climate Change. 127 p.
3. Skelly DK, Joseph LN, Possingham HP, Freidenburg LK, Farrugia TJ, et al. (2007) Evolutionary responses to climate change. Conserv Biol 21: 1353–1355.

4. Parmesan C (2006) Ecological and evolutionary responses to recent climate change. Annu Rev Ecol Evol Syst 37: 637–669.
5. Loarie SR, Duffy PB, Hamilton H, Asner GP, Field CB, et al. (2009) The velocity of climate change. Nature 462: 1052–1055.
6. Martínez-Meyer E, Peterson AT, Hargrove WW (2004) Ecological niches as stable distributional constraints on mammal species, with implications for Pleistocene extinctions and climate change projections for biodiversity. Glob Ecol Biogeogr 13: 305–314.
7. La Sorte FA, Jetz W (2012) Tracking of climatic niche boundaries under recent climate change. J Anim Ecol 81: 914–925.
8. Tingley MW, Monahan WB, Beissinger SR, Moritz C (2009) Birds track their Grinnellian niche through a century of climate change. Proc Natl Acad Sci U S A 106: 19637–19643.
9. La Sorte FA, Jetz W (2010) Avian distributions under climate change: Towards improved projections. J Exp Biol 213: 862–869.
10. Colwell RK, Brehm G, Cardelús CL, Gilman AC, Longino JT (2008) Global warming, elevational range shifts, and lowland biotic attrition in the wet tropics. Science 322: 258–261.
11. Mokhov II, Akperov MG (2006) Tropospheric lapse rate and its relation to surface temperature from reanalysis data. Izvestiya, Atmospheric and Oceanic Physics 42: 430–438.
12. Xu YY, Ramanathan V (2012) Latitudinally asymmetric response of global surface temperature: Implications for regional climate change. Geophys Res Lett 39: L13706.
13. Xu L, Myneni RB, Chapin III FS, Callaghan TV, Pinzon JE, et al. (2013) Temperature and vegetation seasonality diminishment over northern lands. Nature Clim Change 3: 581–586.
14. Mao J, Shi X, Thornton P, Hoffman F, Zhu Z, et al. (2013) Global latitudinal-asymmetric vegetation growth trends and their driving mechanisms: 1982–2009. Remote Sens 5: 1484–1497.
15. Jetz W, Ashton KG, La Sorte FA (2009) Phenotypic population divergence in terrestrial vertebrates at macro scales. Ecol Lett 12: 1137–1146.
16. Jonas CS, Geber MA (1999) Variation among populations of *Clarkia unguiculata* (Onagraceae) along altitudinal and latitudinal gradients. Am J Bot 86: 333–343.
17. Naya DE, Spangenberg L, Naya H, Bozinovic F (2012) Latitudinal patterns in rodent metabolic flexibility. Am Nat 179: E172–E179.
18. Munch SB, Salinas S (2009) Latitudinal variation in lifespan within species is explained by the metabolic theory of ecology. Proc Natl Acad Sci U S A 106: 13860–13864.
19. Stanton-Geddes J, Tiffin P, Shaw RG (2012) Role of climate and competitors in limiting fitness across range edges of an annual plant. Ecology 93: 1604–1613.
20. Hargrove L, Rotenberry JT (2011) Breeding success at the range margin of a desert species: Implications for a climate-induced elevational shift. Oikos 120: 1568–1576.
21. Webb T, Bartlein PJ (1992) Global change during the last 3 million years - climatic controls and biotic responses. Annu Rev Ecol Syst 23: 141–173.
22. Bennie J, Hodgson JA, Lawson CR, Holloway CTR, Roy DB, et al. (2013) Range expansion through fragmented landscapes under a variable climaBirdLife Internationalte. Ecol Lett 16: 921–929.
23. BirdLife International (2012) The BirdLife checklist of the birds of the world, with conservation status and taxonomic sources. BirdLife International, Cambridge.
24. Baselga A, Lobo JM, Svenning J-C, Araújo MB (2012) Global patterns in the shape of species geographical ranges reveal range determinants. J Biogeogr 39: 760–771.
25. Pigot AL, Owens IPF, Orme CDL (2010) The environmental limits to geographic range expansion in birds. Ecol Lett 13: 705–715.
26. BirdLife International and NatureServe (2012) Bird species distribution maps of the world. BirdLife International, Cambridge, and NatureServe, Arlington.
27. Buchanan GM, Donald PF, Butchart SHM (2011) Identifying priority areas for conservation: A global assessment for forest-dependent birds. PLoS ONE 6: e29080.
28. Udvardy MDF (1975) A classification of the biogeographical provinces of the world. IUCN Occasional Paper. Morgues: IUCN. 48 p.
29. Hijmans RJ, Cameron SE, Parra JL, Jones PG, Jarvis A (2005) Very high resolution interpolated climate surfaces for global land areas. Int J Climatol 25: 1965–1978.
30. Hurlbert AH, Jetz W (2007) Species richness, hotspots, and the scale dependence of range maps in ecology and conservation. Proc Natl Acad Sci U S A 104: 13384–13389.
31. Jetz W, McPherson JM, Guralnick RP (2012) Integrating biodiversity distribution knowledge: Toward a global map of life. Trends Ecol Evol 27: 151–159.
32. Jones PG, Thornton PK, Heinke J (2009) Generating characteristic daily weather data using downscaled climate model data from the IPCC's fourth assessment. CGIAR Research Program on Climate Change, Agriculture and Food Security. 23 p.
33. Raupach MR, Marland G, Ciais P, Le Quéré C, Canadell JG, et al. (2007) Global and regional drivers of accelerating CO_2 emissions. Proc Natl Acad Sci U S A 104: 10288–10293.
34. Beaumont LJ, Hughes L, Pitman AJ (2008) Why is the choice of future climate scenarios for species distribution modelling important? Ecol Lett 11: 1135–1146.
35. Scherrer D, Schmid S, Körner C (2011) Elevational species shifts in a warmer climate are overestimated when based on weather station data. Int J Biometeorol 55: 645–654.
36. Graae BJ, De Frenne P, Kolb A, Brunet J, Chabrerie O, et al. (2012) On the use of weather data in ecological studies along altitudinal and latitudinal gradients. Oikos 121: 3–19.
37. Nijssen D, Rousseau R, Cleland EE (1998) The Lorenz curve: A graphical representation of evenness. Coenoses 13: 33–38.
38. Lorenz MO (1905) Methods of measuring concentrations of wealth. J Am Statist Assoc 9: 209–219.
39. Dagum C (1980) The generation and distribution of income, the Lorenz curve and the Gini ratio. Econ Appl 33: 327–367.
40. Muggeo VMR (2003) Estimating regression models with unkown break-points. Stat Med 22: 3055–3071.
41. Davies RB (1987) Hypothesis testing when a nuisance parameter is present only under the alternative. Biometrika 74: 33–43.
42. Wood SN (2006) Generalized additive models: An introduction with R. Boca Raton: Chapman & Hall/CRC. 391 p.
43. Yohai VJ (1987) High breakdown-point and high efficiency estimates for regression. Ann Stat 15: 642–665.
44. Koller M, Stahel WA (2011) Sharpening wald-type inferences in robust regression for small samples. Comput Stat Data Anal 55: 2504–2515.
45. Menard S (1995) Applied logistic regression analysis. Thousand Oaks: Sage Publications. 98 p.
46. R Development Core Team (2014) R: A language and environment for statistical computing. Vienna: R Foundation for Statistical Computing.
47. Anderson AS, Reside AE, Vanderwal JJ, Shoo LP, Pearson RG, et al. (2012) Immigrants and refugees: The importance of dispersal in mediating biotic attrition under climate change. Glob Change Biol 18: 2126–2134.
48. La Sorte FA, Jetz W (2010) Projected range contractions of montane biodiversity under global warming. Proc R Soc Lond B Biol Sci 277: 3401–3410.
49. Kirby JS, Stattersfield AJ, Butchart SHM, Evans MI, Grimmett RFA, et al. (2008) Key conservation issues for migratory land- and waterbird species on the world's major flyways. Bird Conserv Int 18: S49–S73.
50. Somveille M, Manica A, Butchart SHM, Rodrigues ASL (2013) Mapping global diversity patterns for migratory birds. PLoS ONE 8: e70907.
51. Foden WB, Butchart SHM, Stuart SN, Vié J-C, Akçakaya HR, et al. (2013) Identifying the world's most climate change vulnerable species: A systematic trait-based assessment of all birds, amphibians and corals. PLoS ONE 8: e65427.
52. McCain CM (2009) Vertebrate range sizes indicate that mountains may be 'higher' in the tropics. Ecol Lett 12: 550–560.
53. Luoto M, Heikkinen RK (2008) Disregarding topographic heterogeneity biases species turnover assessments based on bioclimatic models. Glob Change Biol 14: 1–12.
54. Randin CF, Engler R, Normand S, Zappa M, Zimmermann NE, et al. (2009) Climate change and plant distribution: Local models predict high-elevation persistence. Glob Change Biol 15: 1557–1569.
55. Engler R, Randin CF, Thuiller W, Dullinger S, Zimmermann NE, et al. (2011) 21st century climate change threatens mountain flora unequally across Europe. Glob Change Biol 17: 2330–2341.
56. Gillingham PK, Huntley B, Kunin WE, Thomas CD (2012) The effect of spatial resolution on projected responses to climate warming. Divers Distrib 18: 990–1000.
57. Suggitt AJ, Gillingham PK, Hill JK, Huntley B, Kunin WE, et al. (2011) Habitat microclimates drive fine-scale variation in extreme temperatures. Oikos 120: 1–8.
58. Reif J, Flousek J (2012) The role of species' ecological traits in climatically driven altitudinal range shifts of central European birds. Oikos 121: 1053–1060.
59. Suggitt AJ, Stefanescu C, Paramo F, Oliver T, Anderson BJ, et al. (2012) Habitat associations of species show consistent but weak responses to climate. Biol Lett 8: 590–593.
60. Clobert J, Danchin E, Dhondt AA, Nichols JD (2001) Dispersal. Oxford: Oxford University Press. 452 p.
61. Watkinson AR, Gill JA (2001) Climate change and dispersal. In: J. M. . Bullock, R. E. . Kenward, R. S. . Hails, editors. Dispersal ecology. Oxford: Blackwell Publishing. pp. 410–428.
62. Travis JMJ, Delgado M, Bocedi G, Baguette M, Bartoń K, et al. (2013) Dispersal and species' responses to climate change. Oikos 122: 1532–1540.
63. Forero-Medina G, Joppa L, Pimm SL (2010) Constraints to species' elevational range shifts as climate changes. Conserv Biol 25: 163–171.
64. Jankowski JE, Robinson SK, Levey DJ (2010) Squeezed at the top: Interspecific aggression may constrain elevational ranges in tropical birds. Ecology 91: 1877–1884.
65. Møller AP, Flensted-Jensen E, Mardal W (2006) Dispersal and climate change: A case study of the arctic tern *Sterna paradisaea*. Glob Change Biol 12: 2005–2013.
66. Kremer A, Ronce O, Robledo-Arnuncio JJ, Guillaume F, Bohrer G, et al. (2012) Long-distance gene flow and adaptation of forest trees to rapid climate change. Ecol Lett 15: 378–392.
67. Sexton JP, Strauss SY, Rice KJ (2011) Gene flow increases fitness at the warm edge of a species' range. Proc Natl Acad Sci U S A 108: 11704–11709.
68. Moore RP, Robinson WD, Lovette IJ, Robinson TR (2008) Experimental evidence for extreme dispersal limitation in tropical forest birds. Ecol Lett 11: 960–968.

69. Dexter KG (2010) The influence of dispersal on macroecological patterns of lesser Antillean birds. J Biogeogr 37: 2137–2147.
70. Freeman BG, Class Freeman AM (in press) Rapid upslope shifts in new guinean birds illustrate strong distributional responses of tropical montane species to global warming. Proc Natl Acad Sci U S A.
71. Janzen DH (1967) Why mountain passes are higher in the tropics. Am Nat 101: 233–249.
72. Ghalambor CK, Huey RB, Martin PR, Tewksbury JJ, Wang G (2006) Are mountain passes higher in the tropics? Janzen's hypothesis revisited. Integr Comp Biol 46: 5–17.
73. Gillies CS, Beyer HL, St Clair CC (2011) Fine-scale movement decisions of tropical forest birds in a fragmented landscape. Ecol Appl 21: 944–954.
74. Ibarra-Macias A, Robinson WD, Gaines MS (2011) Experimental evaluation of bird movements in a fragmented neotropical landscape. Biol Conserv 144: 703–712.
75. Urban MC, Tewksbury JJ, Sheldon KS (2012) On a collision course: Competition and dispersal differences create no-analogue communities and cause extinctions during climate change. Proc R Soc Lond B Biol Sci 279: 2072–2080.
76. Dawideit BA, Phillimore AB, Laube I, Leisler B, Böhning-Gaese K (2009) Ecomorphological predictors of natal dispersal distances in birds. J Anim Ecol 78: 388–395.
77. Böhning-Gaese K, Caprano T, van Ewijk K, Veith M (2006) Range size: Disentangling current traits and phylogenetic and biogeographic factors. Am Nat 167: 555–567.
78. Laube I, Graham CH, Böhning-Gaese K (2013) Intra-generic species richness and dispersal ability interact to determine geographic ranges of birds. Glob Ecol Biogeogr 22: 223–232.
79. Laube I, Korntheuer H, Schwager M, Trautmann S, Rahbek C, et al. (2013) Towards a more mechanistic understanding of traits and range sizes. Glob Ecol Biogeogr 22: 233–241.
80. Winkler DW, Wrege PH, Allen PE, Kast TL, Senesac P, et al. (2005) The natal dispersal of tree swallows in a continuous mainland environment. J Anim Ecol 74: 1080–1090.
81. Angert AL, Crozier LG, Rissler LJ, Gilman SE, Tewksbury JJ, et al. (2011) Do species' traits predict recent shifts at expanding range edges? Ecol Lett 14: 677–689.
82. Forero-Medina G, Terborgh J, Socolar SJ, Pimm SL (2011) Elevational ranges of birds on a tropical montane gradient lag behind warming temperatures. PLoS ONE 6: e28535.
83. La Sorte FA, Hawkins BA (2007) Range maps and species richness patterns: Errors of commission and estimates of uncertainty. Ecography 30: 649–662.
84. McCain CM, Colwell RK (2011) Assessing the threat to montane biodiversity from discordant shifts in temperature and precipitation in a changing climate. Ecol Lett 14: 1236–1245.
85. Mac Arthur RH (1972) Geographical ecology. New York: Harper & Row. 269 p.
86. Drake JM (2005) Population effects of increased climate variation. Proc R Soc Lond B Biol Sci 272: 1823–1827.
87. Early R, Sax DF (2011) Analysis of climate paths reveals potential limitations on species range shifts. Ecol Lett 14: 1125–1133.
88. Mustin K, Dytham C, Benton TG, Travis JMJ (2013) Red noise increases extinction risk during rapid climate change. Divers Distrib 19: 815–824.
89. Lawler JJ, Ruesch AS, Olden JD, McRae BH (2013) Projected climate-driven faunal movement routes. Ecol Lett 16: 1014–1022.
90. Bocedi G, Pe'er G, Heikkinen RK, Matsinos Y, Travis JMJ (2012) Projecting species' range expansion dynamics: Sources of systematic biases when scaling up patterns and processes. Methods Ecol Evol 3: 1008–1018.
91. Dormann CF, Schymanski SJ, Cabral J, Chuine I, Graham C, et al. (2012) Correlation and process in species distribution models: Bridging a dichotomy. J Biogeogr 39: 2119–2131.
92. Schurr FM, Pagel J, Cabral JS, Groeneveld J, Bykova O, et al. (2012) How to understand species' niches and range dynamics: A demographic research agenda for biogeography. J Biogeogr 39: 2146–2162.
93. Hole DG, Willis SG, Pain DJ, Fishpool LD, Butchart SHM, et al. (2009) Projected impacts of climate change on a continent-wide protected area network. Ecol Lett 12: 420–431.
94. Hole DG, Huntley B, Arinaitwe J, Butchart SHM, Collingham YC, et al. (2011) Toward a management framework for networks of protected areas in the face of climate change. Conserv Biol 25: 305–315.
95. Kubisch A, Degen T, Hovestadt T, Poethke HJ (2013) Predicting range shifts under global change: The balance between local adaptation and dispersal. Ecography 36: 873–882.

7

Uncertainty in Model Predictions of *Vibrio vulnificus* Response to Climate Variability and Change: A Chesapeake Bay Case Study

Erin A. Urquhart[1]*, Benjamin F. Zaitchik[1], Darryn W. Waugh[1], Seth D. Guikema[2], Carlos E. Del Castillo[3]

1 Department of Earth and Planetary Sciences, Johns Hopkins University, Baltimore, Maryland, United States of America, **2** Department of Geography and Environmental Engineering, Johns Hopkins University, Baltimore, Maryland, United States of America, **3** Ocean Ecology Laboratory, NASA Goddard Space Flight Center, Greenbelt, Maryland, United States of America

Abstract

The effect that climate change and variability will have on waterborne bacteria is a topic of increasing concern for coastal ecosystems, including the Chesapeake Bay. Surface water temperature trends in the Bay indicate a warming pattern of roughly 0.3–0.4°C per decade over the past 30 years. It is unclear what impact future warming will have on pathogens currently found in the Bay, including *Vibrio* spp. Using historical environmental data, combined with three different statistical models of *Vibrio vulnificus* probability, we explore the relationship between environmental change and predicted *Vibrio vulnificus* presence in the upper Chesapeake Bay. We find that the predicted response of *V. vulnificus* probability to high temperatures in the Bay differs systematically between models of differing structure. As existing publicly available datasets are inadequate to determine which model structure is most appropriate, the impact of climatic change on the probability of *V. vulnificus* presence in the Chesapeake Bay remains uncertain. This result points to the challenge of characterizing climate sensitivity of ecological systems in which data are sparse and only statistical models of ecological sensitivity exist.

Editor: Ben Bond-Lamberty, DOE Pacific Northwest National Laboratory, United States of America

Funding: This work was financially supported by the Johns Hopkins University Department of Earth and Planetary Sciences. The funders had no role in study design, data collection and analysis, decision to publish, or preparation of the manuscript.

Competing Interests: The authors have declared that no competing interests exist.

* E-mail: erinu@jhu.edu

Introduction

Vibrio spp. bacteria are a threat in many coastal aquatic ecosystems around the world [1–4]. In the Chesapeake Bay, the number of annual human *Vibrio* cases of infection has nearly doubled in the past decade [5,6]. Furthermore, *Vibrio* spp. is frequently detected in shellfish harvested for human consumption during the warm summer months [7]. In general, this seasonality correlates with peak incidence of *Vibrio* disease caused by *Vibrio* spp. bacteria in many coastal regions [8–10]. The probability of finding various *Vibrio* spp. in the Bay varies spatially and seasonally, and researchers have modeled these probability patterns as a statistical function of surface water temperature and salinity [11–16]. These temperature and salinity-based *Vibrio* models have demonstrated skill for available datasets in the Bay and structurally similar statistical models have been applied to predictions of *V. cholerae*, *V. vulnificus*, and *V. parahaemolyticus* in other regions [1,4,17,18]. The environmental range of *V. vulnificus* can vary by region, but in general the bacteria are found in waters with salinity between 5 and 25 (practical salinity units) and temperature above 15°C [12,19–21].

Recent studies show that surface water temperatures in the Chesapeake Bay have warmed by 0.3–0.4°C per decade over the past 30 years [22,23]. This trend has resulted in an expansion of the warm season period during which water temperatures are high enough to support *V. vulnificus* growth: the onset of spring time temperatures (>15°C) has advanced by nearly three weeks [22]. Salinity patterns are also sensitive to climate change, as changes in springtime flow of the Susquehanna River - the primary freshwater input to the Bay - can influence salinity throughout the Bay over the *V. vulnificus* growth season. The consensus of climate models is that there will be a rise in winter and spring precipitation in the northern portion of the watershed [24,25] implying an increase in January to May Susquehanna River steam flow. A study by Gibson and Najjar (2000; [26]) showed that an increase in the January-May Susquehanna stream flow could potentially decrease winter and springtime salinity values by 7% in the upper Chesapeake Bay.

Even though there is considerable uncertainty in the magnitude of projected warming and freshening of the Chesapeake Bay [27], it is valuable to understand how a temperature and salinity sensitive pathogen like *V. vulnificus* might respond to observed and projected trends in these environmental parameters. Here we examine three statistical models of *V. vulnificus* probability of presence that demonstrate skill in predicting *V. vulnificus* probability of presence in Chesapeake Bay. All three models use water surface temperature and salinity as the only predictors, but they differ in their structure and/or in the data used for training and evaluation. One model is the generalized linear model (GLM) of Jacobs et al. (2010; [12]) trained on data collected in the Chesapeake Bay in 2007 and 2008, the second model is also a

GLM but trained on a 2011 and 2012 data set, while the third model is generalized additive model (GAM) also trained on the 2011–2012 data. The latter two models are from Urquhart et al. (2014; [15]), where surface temperature and salinity were used to model both the probability of presence and concentration *Vibrio* spp. in the upper Chesapeake Bay. Here we focus just on probability models to enable comparison with Jacobs et al. (2010; [12]), who considered only a probability model. Furthermore, we can evaluate the effect that these differences in structure and training data have on modeled estimates of *V. vulnificus* probability under current climate conditions, which is relevant for pathogen risk assessment and early warning, and consider the implications of these differences for projected *V. vulnificus* risk under climate change.

Methods

The Chesapeake Bay Estuary, adjacent to the Maryland, Delaware, and Virginia coastline, covers an area of approximately 11,500 km^2 and is characterized by a sharp north-to-south salinity gradient. Salinity ranges from 0–6 in the northern Bay to 18–30 near the mouth of the Bay. Surface water temperatures follow a seasonal cycle, ranging from local wintertime temperatures of −0.5°C to summertime temperatures of 31°C [28]. The Susquehanna River, the largest and northernmost tributary, accounts for roughly 45% of the yearly freshwater inflow into the Bay. This paper focuses on the upper portion of Chesapeake Bay (Fig. 1). The upper region of the Bay was selected to avoid model predictions outside of the original training data salinity range (salinity >14).

The climatological analysis presented here used historical environmental data collected by the Chesapeake Bay Data Program [29]. Bi-monthly surface water temperature, salinity, and chlorophyll a data were obtained for 16 main stem and tributary monitoring stations (Fig. 1) collected from 1985 through 2013. For salinity, the absolute difference between observed salinity and the *V. vulnificus* optimal salinity value of 11.5 [12] was calculated, and use of deviation from this was used as an explanatory covariate. The 16 monitoring stations were selected based on their geographic location serving as a representation of the upper Chesapeake Bay. In situ data were used to delineate three different salinity zones: upper-upper Bay (hereafter: "upper region"), middle-upper Bay (hereafter: "mid region"), and lower-upper Bay (hereafter: "lower region"). These stations cover the upper main-stem Bay as well as tributary locations, with six stations in the upper region, five stations in the mid region, and five stations in the lower region. Observational data were averaged at monthly intervals for each zone resulting in 337 data records for the upper region and 342 data records for both the mid and lower regions.

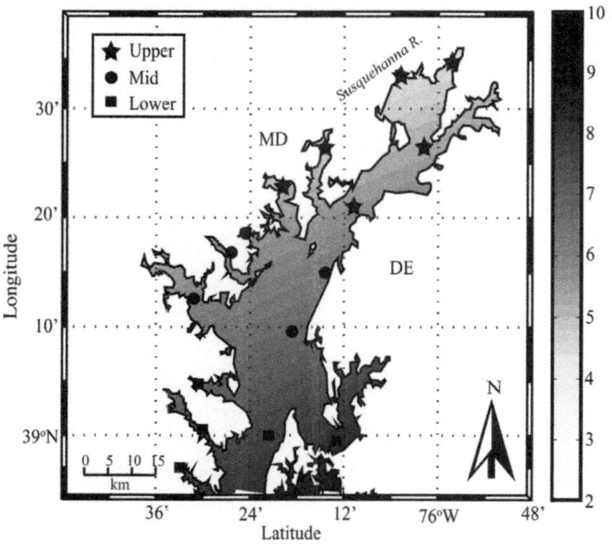

Figure 1. Map of the study area, showing contours of average surface water salinity. Dark markers represent in situ monitoring stations used for each of the subregions in this study: upper (star), mid (circle), and lower (square).

These salinity and temperature data were applied to the three statistical *V. vulnificus* probability models available for Chesapeake Bay:

1. NOAA_GLM: The generalized linear model (GLM) of Jacobs et al. (2010; [12]): [$z(V.v) = \beta_0 + \beta_1 Temp + \beta_3 |SalnOpt|$, where β_0 is the intercept, β_n is the regression coefficient for the independent covariates, *Temp* is daily surface temperature, and $|SalnOpt|$ is the absolute distance from optimal salinity of 11.5], which was trained using 235 *V. vulnificus* samples collected during the months of July and October of 2007, and April, July, and October of 2008 and were analyzed by the NOAA Chesapeake Bay Office.
2. JHU_GLM: The GLM of Urquhart et al., (2014; [15]) of the same structural form as the NOAA_GLM [$z(V.v) = \beta_0 + \beta_1 Temp + \beta_3 |SalnOpt|$] trained using 148 *V. vulnificus*, surface temperature, and surface salinity samples collected in the upper Chesapeake Bay during the months of July and September of 2011 and March through June of 2012 (Table S1; [15]). Samples were collected by The Johns Hopkins University (JHU) in collaboration with the Maryland Department of the Environment and NASA and were processed at the University of Maryland College Park.
3. JHU_GAM: A generalized additive model (GAM; [30]) trained and evaluated using the same data that were used for JHU_GLM: [$z(V.v) = \beta_0 + s_1(Temp) + s_2(Saln)$, where $s_i(x_i)$ is a parameter of the smoothing function, and *Saln* is the salinity value].

We included a GAM in addition to the two structurally identical GLMs because GAM models allow a more flexible regression modeling of the transformed response that combine the predictor variables in a nonparametric manner [31]. The NOAA_GLM and JHU_GLM models have the same structure but vary in the training data, whereas the JHU_GLM and JHU_GAM models use the same training data but vary in the structure of model. All models were implemented in logistic form using a "logit" link function for an optimal prediction point and were trained using observational bacteria data transformed to binary presence/absence. Probability of *V. vulnificus* presence was calculated using $p = e^z/(1+e^z)$. Diagnostics for each model were performed using Akaike's Information Criterion (AIC) and accuracy (ACC) in an out-of-bag (OOB) cross validation [32]. ACC is defined as ACC = (TP+TN)/(P+N) where TP is true positive, TN is true negative, P is the number of presence instances, and N is the number of absence instances.

To explore sensitivity of the *V. vulnificus* models to temperature and salinity, we used a range of surface water temperature (0–40°C) and surface salinity (0–13) values as independent model input. Here the range of model input extends past the range of the in situ temperature (8–31°C) observations, and is constrained to the range of the surface salinity (0–14) observations that were used

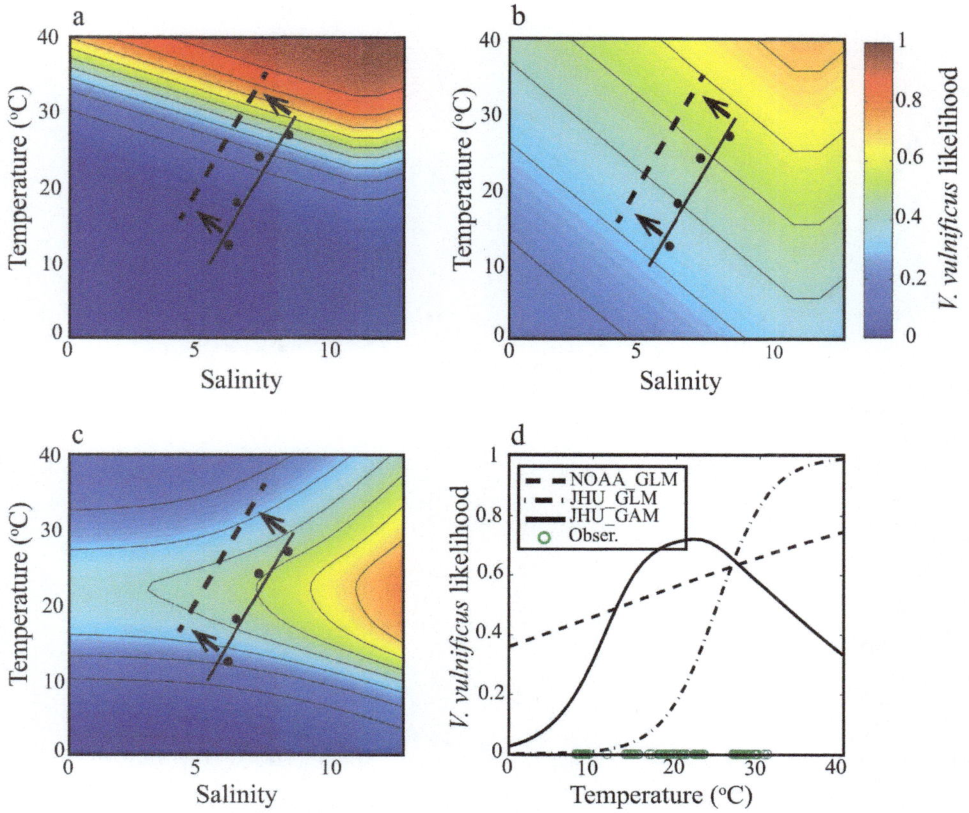

Figure 2. Contour plots of *V. vulnificus* probability with temperature and salinity for (a) NOAA GLM, (b) JHU GLM, and (c) JHU GAM. Black dots represent monthly average (April-July) of in situ conditions; black lines represents in situ trend line, and dashed line represents shift in present day temperature and salinity, (d) Plot of temperature regressed against V.vulnificus probability at 11.5 salinity for each empirical method. Green circles represent the range of temperature observations during bacterium sampling.

in the training of the two JHU models. Additionally, historical temperature and salinity data were tested as model input, enabling identification of *V. vulnificus* climatology and seasonal trends. To further assess the geographic distribution of the predicted *V. vulnificus* probability for each method, geospatially-interpolated satellite-derived surface temperature and surface salinity [33,34] were used to map spatially complete estimates of probability throughout the upper Bay. Interpolated satellite estimates were developed using monthly, level-2 Moderate Resolution Imaging Spectroradiometer (MODIS) surface water temperature (MOD 28) and ocean color (Rrs 412–678) products.

All statistical computations were carried out in the R Statistical Environment version 2.14, using the 'mgcv' and 'stats' packages, on an Intel Xeon W3580 Processor, 3.33 GHz machine with 12 GB RAM. Computation time for all statistical models was less than one minute.

Results and Discussion

For model evaluation, goodness of fit and predictive skill for the JHU models were determined using AIC and ACC indices. AIC results indicated that the JHU GAM (145.9) offered better model fit than the JHU GLM (160.4), but performance differences between models were small relative to measurement uncertainty. NOAA GLM model fit using the NOAA training dataset yielded an AIC of 164.3 [12]). A direct comparison of model fit of could not be calculated due to lack of access to NOAA GLM training data. We stress that the difference in training data between the NOAA and JHU models is the primary reason for differences between NOAA_GLM and JHU_GLM, as the models are structurally identical. To predict bacterial presence, selection of an optimal prediction point was required. Rather than setting a prediction point at 0.5 arbitrarily, the prediction point was based on three performance indices: true positive rate, true negative rate, and ACC, yielding an optimal threshold of 0.4 for *V. vulnificus*. To determine the prediction skill of each model, ACC was calculated using the JHU validation dataset (ACC: 0.47, for NOAA GLM, 0.59 for JHU GLM, and 0.60 for JHU GAM). The AIC and ACC values indicated that the JHU models performed significantly better than a null model that only included seasonality as a predictor.

Figure 2 shows the relationship between temperature, salinity, and the mean estimate of predicted *V. vulnificus* probability for each of the tested models, with likelihood levels plotted as contour curves. NOAA GLM (Fig. 2a) exhibits a sharp increase in *V. vulnificus* probability with increasing temperatures along the axis of optimal salinity (11.5). Similarly, JHU GLM (Fig. 2b) exhibits a steady increase in probability with higher temperatures, though the rate of change with temperature is less steep than NOAA GLM. In contrast to the GLMs, JHU GAM (Fig. 2c) shows a probability maximum dependent on temperature, indicating a temperature optimum *V. vulnificus* growth above which probability gradually declines. Figure 2d offers an alternative view of predicted *V. vulnificus* probability with temperature, at optimal salinity, including temperature observations during in situ bacteria collection. Furthermore, the wide range of observed temperatures

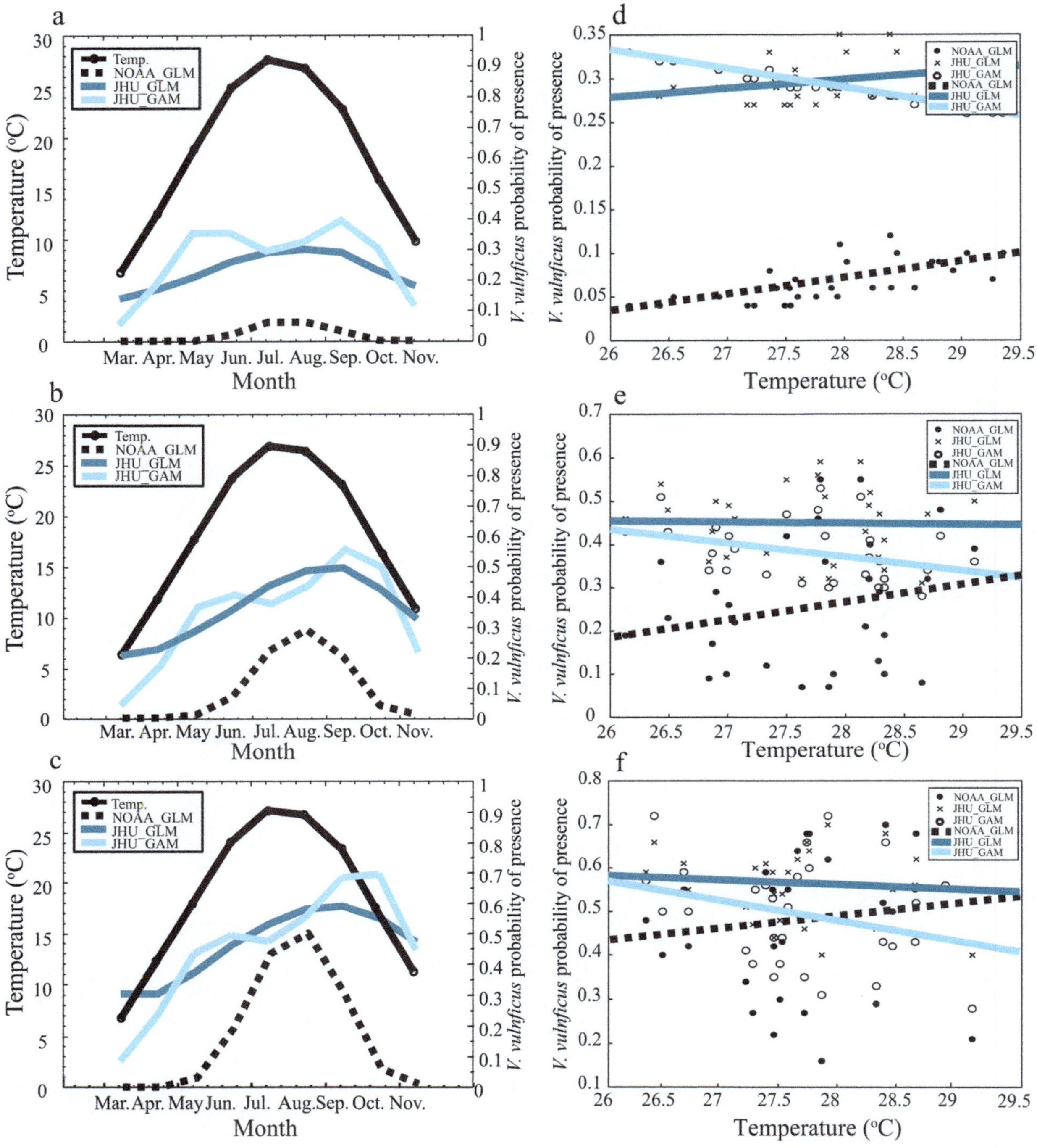

Figure 3. Monthly climatology of temperature and *V. vulnificus* probability for each method in the upper (a), mid (b), and lower (c) regions of the Chesapeake Bay. Peak temperature observations by year versus *V. vulnificus* probability for each method in the upper (d), mid (e), and lower (f) regions of the Chesapeake Bay. Trend lines are included for each method's observations.

confirms that the declining GAM probability above optimal temperature is a valid model response and not an issue of limited observations at high temperature.

These differences in model response also have implications for retrospective or near real-time estimation of risk of *V. vulnificus* presence. Using a 27-year in situ record of temperature and salinity in the upper Chesapeake Bay, we estimated *V. vulnificus* monthly probability of presence according to each statistical model. Fig. 3 shows the climatology of surface water temperature and mean estimate model predictions in each region of the upper Bay for March through November. A southward increase in predicted probabilities for all statistical methods during summer months suggests that distance from optimal salinity plays a role in the spatial distribution of *V. vulnificus* presence. Predicted probabilities are likely lower in the upper region due to decreased salinity and larger deviation from optimal salinity. Seasonal patterns in all regions indicate that NOAA_GLM and JHU_GLM predict highest probabilities during the warmest summertime months. JHU_GAM exhibits a bimodal seasonal pattern with peaks in early and late summer across all regions. These

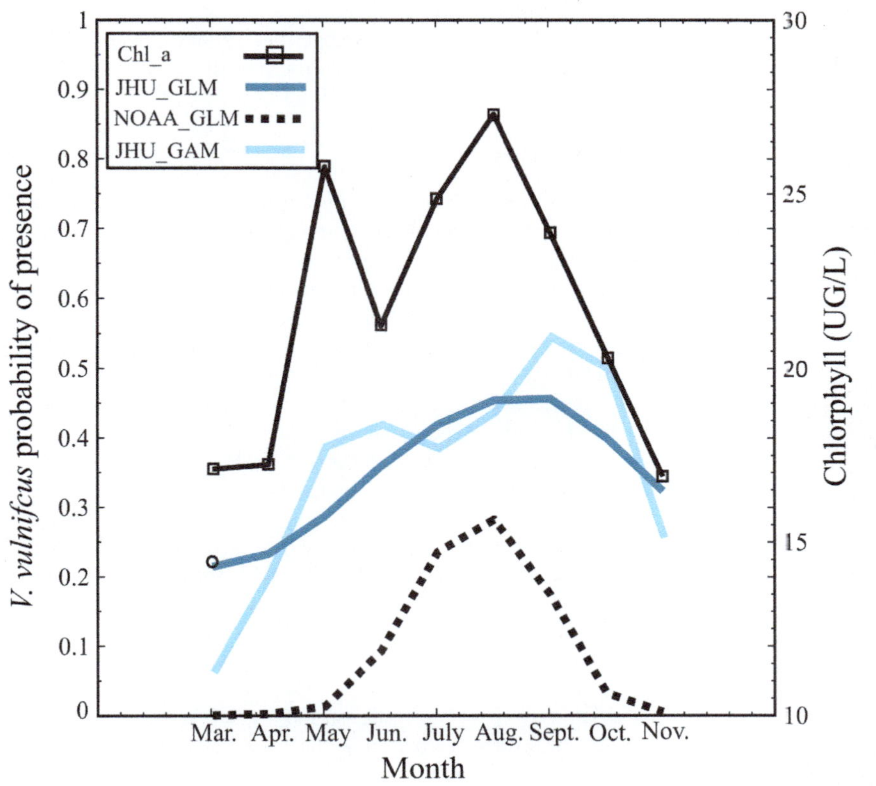

Figure 4. Monthly climatology of Chlorophyll a and *V. vulnificus* probability for each method averaged over the entire upper Chesapeake Bay.

JHU_GAM results are consistent with the temperature dependency shown in Fig. 2c.

The difference in model sensitivity to temperature has implications for characterizing interannual variability in risk. Fig. 3d–f show mean predicted *V. vulnificus* probability for the upper, middle, and lower portions of the study area plotted against annual peak monthly SST for the available historical record. In all three subregions, NOAA_GLM predicts that peak probabilities were highest in warmer years, while JHU_GAM predicts the opposite and JHU_GLM falls in between. We emphasize that these are the mean probability estimates for each model, and that there may not be statistically significant differences between model predictions in any given year. Nevertheless, mean estimates are commonly used to communicate risk and to project trends, so the fact that two comparably high performing models – NOAA_GLM and JHU_GAM – yield opposite mean estimates of the relationship between warm summers and *V. vulnificus* probability is relevant.

The differences in these model response surfaces also have clear implications for projections of *V. vulnificus* probability under climate change. As a simple demonstration, we consider the consensus prediction of warming and freshening of the Bay (dashed line in Figure 2 a–c). NOAA_GLM projects steady or increasing probabilities: freshening moves conditions away from the salinity optimum but this effect is offset by increases in predicted probability with rising water temperatures. The JHU_GLM shows a similar pattern but with lower sensitivity to changing environmental conditions. In contrast, warming only increases predicted probability of *V. vulnificus* presence in JHU_GAM for relatively cool temperatures, representative of spring or fall conditions. Peak summertime temperatures are already above the temperature optimum in this model, so further warming results in a predicted decline in peak summertime *V. vulnificus* probability.

While we cannot presently determine which sensitivity pattern is correct—the JHU_GLM and NOAA_GLM increase with higher temperatures or the JHU_GAM decline under warmest conditions—the JHU_GAM behavior might indicate that present-day summertime water temperatures are already above the optimal temperature for *V. vulnificus* growth in Chesapeake Bay. Alternatively, the result might be understood in the context of previous studies that have shown *Vibrio* dependence on zooplankton due to attachment and/or *Vibrio's* chitinoclastic activity [35,36]. Unfortunately we do not have adequate co-located measurements of zooplankton and *V. vulnificus* to include zooplankton in a predictive model. However, we do find that the climatology of Chesapeake Bay Program in situ chlorophyll a concentrations, which generally correlate with zooplankton presence, exhibits a bimodal seasonal pattern with a slight lead over the JHU GAM predicted *V. vulnificus* peaks (Fig. 4).

To examine the geographic extent of each methods' predicted *V. vulnificus* probability, monthly interpolated satellite surface water temperature and surface salinity products were used to create spatially complete probability hind-casts for 2012 in the upper Bay (Fig. 5). Consistent with results shown in Fig. 3, these maps show highest predicted probability towards the south of the analysis region, where salinity values are closest to optimum. NOAA_GLM and JHU_GLM both show the most widespread zones of high probability in the warmest summer months, while JHU_GAM predicts higher probabilities at the beginning and end of the warm season. Interesting spatial structures are also apparent in these maps. For example, NOAA_GLM predicts slightly

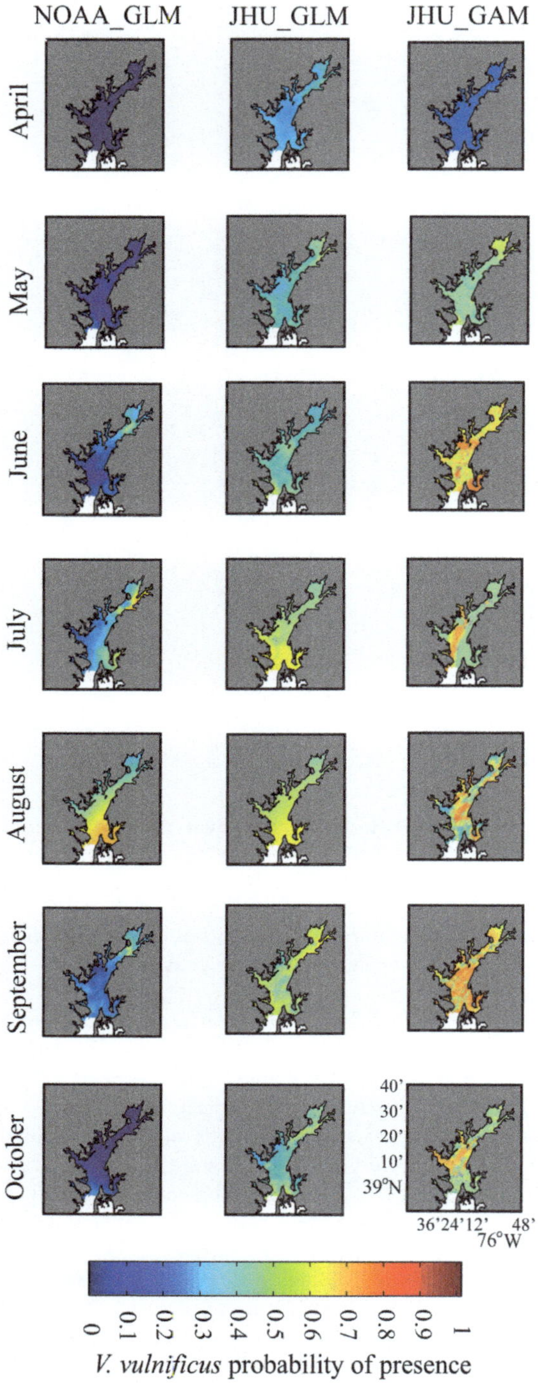

Figure 5. Upper Chesapeake Bay monthly *V. vulnificus* probability hind-casts for April through October 2012, for NOAA GLM, JHU GLM, and JHU GAM methods.

elevated *V. vulnificus* probabilities in the eastern waters of the Chesapeake Bay during warmer months, while JHU_GAM predicts high probability zones in the western Bay during months with lower overall probability (Fig. 5). These patterns likely reflect the Bay's two-layer physical circulation scheme in which we see fresher surface waters along the western shore and saltier waters along the eastern shore of the Bay. The predictions of statistical *V. vulnificus* probability models compared in this study clearly differ in the implied relationships between the structure of this circulation and the location of high *V. vulnificus* risk areas.

Conclusions

In summary, this study presents a comparison of three statistical ecological habit models for estimating the probability of *V. vulnificus* presence in the upper Chesapeake Bay. We examined individual model sensitivity to climatic variability and change within the upper Bay by assessing model response to a range of temperature and salinity values. We find that the three models differ systematically in the predicted response of *V. vulnificus* probability to high temperatures in the upper Chesapeake Bay.

These results indicate that more data are required to constrain estimates of climate sensitivity of *V. vulnificus* in Chesapeake Bay: statistical models are limited by the paucity of publicly available data from *V. vulnificus* collections and co-located measurements of ecologically relevant variables, and process-based models would require further research on the *V. vulnificus* life cycle in the Bay. Our results also caution against predicting or projecting climate-based changes in *V. vulnificus* exposure risk on the basis of the mean predictions of existing statistical models, as skillful and statistically indistinguishable models differ systematically in predicted *V. vulnificus* sensitivity to rising surface water temperature, even within the range of environmental conditions under which the models were trained.

The challenges facing *V. vulnificus* modeling in Chesapeake Bay are not unique. Indeed, predictive capabilities for climate sensitivity of many pathogens are limited to statistical models based on scarce data. Other recent studies [37–38] emphasize that the inadequacy of available data hamper climate change projections for a diversity of waterborne pathogen systems in many regions. In the case of *V. vulnificus* in Chesapeake Bay we have a specific example of closely related modeling efforts that suggest systematically different impacts of climate change due to differences in model structure—i.e., the difference between JHU_GLM and JHU_GAM—and training data—i.e., the difference between JHU_GLM and NOAA_GLM. These kinds of structural comparisons of statistical models, however, are not always performed in studies of climate sensitivity in ecological systems. The results of this study suggest that such model comparisons can be quite important when evaluating uncertainty in climate-based predictions and projections.

Supporting Information

Table S1 In situ training dataset used for JHU_GLM and JHU_GAM likelihood models.

Acknowledgments

The authors would like to acknowledge Anand Gnanadesikan for providing insightful comments on this manuscript.

Author Contributions

Conceived and designed the experiments: EU BZ DW SG CDC. Performed the experiments: EU. Analyzed the data: EU. Contributed reagents/materials/analysis tools: SG. Wrote the paper: EU BZ.

References

1. Baker-Austin C, Trinanes JA, Taylor NG, Hartnell R, Siitonen A, et al. (2012) Emerging *Vibrio* risk at high latitudes in response to ocean warming. Nature Climate Change 3: 73–77.
2. Deepanjali A, Kumar HS, Karunasagar I (2005) Seasonal variation in abundance of total and pathogenic *Vibrio parahaemolyticus* bacteria in oysters along the southwest coast of India. Applied and Environmental Microbiology 71: 3575–3580.
3. Cantet F, Hervio-Heath D, Caro A, Le Mennec C, Monteil C, et al. (2013) Quantification of *Vibrio parahaemolyticus, V. vulnificus* and *V. cholerae* in French Mediterranean coastal lagoons. Research in Microbiology.
4. Oberbeckmann S, Fuchs BM, Meiners M, Wichels A, Wiltshire KH, et al. (2012) Seasonal dynamics and modeling of a *Vibrio* community in coastal waters of the North Sea. Microbial Ecology 63: 543–551.
5. Maryland Department of Health (2013) Cases of selected notifiable conditions reported in Maryland. URL http://phpa.dhmh.maryland.gov/SitePages/disease-conditions-count rates.aspx.
6. Virginia Department of Health (2013) Virginia reportable disease surveillance data. URL http://www.vdh.virginia.gov/Epidemiology/Surveillance/SurveillanceData.
7. de Magny GC, Murtugudde R, Sapiano MR, Nizam A, Brown CW, et al. (2008) Environmental signatures associated with cholera epidemics. Proceedings of the National Academy of Sciences 105: 17676–17681.
8. Klontz KC, Lieb S, Schreiber M, Janowski HT, Baldy LM, et al. (1988) Syndromes of *Vibrio vulnificus* infections. Clinical and epidemiologic features in Florida cases, 1981–1987. Ann Intern Med. 109: 318–23
9. Shapiro RL, Altekruse S, Hutwagner L, Bishop R, Hammond R, et al. (1998) The role of Gulf Coast oysters harvested in warmer months in *Vibrio vulnificus* infections in the United States, 1988–1996. *Vibrio* Working Group. The Journal of Infectious Diseases 178: 752–759.
10. Lipp EK, Huq A, Colwell RR (2002) Effects of global climate on infectious disease: the cholera model. Clinical Microbiology Reviews 15: 757–770.
11. Heidelberg J, Heidelberg K, Colwell R (2002) Seasonality of Chesapeake Bay bacterioplankton species. Applied and Environmental Microbiology 68: 5488–5497.
12. Jacobs J, Rhodes M, Brown C, Hood R, Leigh A, et al. (2010) Predicting the distribution of *Vibrio vulnificus* in Chesapeake Bay. NOAA Technical Memorandum NOS NCCOS 112: 1–12.
13. Wright AC, Hill RT, Johnson JA, Roghman MC, Colwell RR, et al. (1996) Distribution of *Vibrio vulnificus* in the Chesapeake Bay. Applied and Environmental Microbiology 62: 717–724.
14. Louis VR, Russek-Cohen E, Choopun N, Rivera IN, Gangle B, et al. (2003) Predictability of *Vibrio cholerae* in Chesapeake Bay. Applied and Environmental Microbiology 69: 2773–2785.
15. Urquhart EA, Zaitchik BF, Guikema SD, Haley BJ, Taviani E, et al. (2014) Use of environmental parameters to model pathogenic *Vibrios* in Chesapeake Bay. Journal of Environmental Informatics (Accepted).
16. de Magny GC, Long W, Brown CW, Hood RR, Huq A, et al. (2009) Predicting the distribution of *Vibrio* spp. in the Chesapeake Bay: a *Vibrio cholerae* case study. EcoHealth 6: 378–389.
17. Eiler A, Johansson M, Bertilsson S (2006) Environmental influences on *Vibrio* populations in north- ern temperate and boreal coastal waters (Baltic and Skagerrak Seas). Applied and Environmental Microbiology 72: 6004–6011.
18. Johnson CN, Bowers JC, Griffitt KJ, Molina V, Clostio RW, et al. (2012) Ecology of *Vibrio parahaemolyticus* and *Vibrio vulnificus* in the coastal and estuarine waters of Louisiana, Maryland, Mississippi, and Washington (United States). Applied and Environmental Microbiology 78: 7249–7257.
19. Colwell R, Kaper J, Joseph S (1977) *Vibrio cholerae, Vibrio parahaemolyticus*, and other *vibrios*: occurrence and distribution in Chesapeake Bay. Science 198: 394–396.
20. Kaper J, Remmers E, Lockman H, Colwell R (1981) Distribution of *Vibrio parahaemolyticus* in Chesapeake Bay during the summer season. Estuaries 4: 321–327.
21. Lipp EK, Rodriguez-Palacios C, Rose JB (2001) Occurrence and distribution of the human pathogen *Vibrio vulnificus* in a subtropical Gulf of Mexico estuary. In: The Ecology and Eti- ology of Newly Emerging Marine Diseases,Springer. pp. 165–173.
22. Austin HM (2002) Decadal oscillations and regime shifts, a characterization of the Chesapeake Bay marine climate. In: American Fisheries Society Symposium. volume 32, pp. 155–170.
23. Secor D, Wingate R (2008). A 69-year record of warming in the Chesapeake Bay.
24. Najjar R, Patterson L, Graham S (2009) Climate simulations of major estuarine watersheds in the Mid-Atlantic region of the US. Climatic Change 95: 139–168.
25. Hayhoe K, Wake CP, Huntington TG, Luo L, Schwartz MD, et al. (2007) Past and future changes in climate and hydrological indicators in the US Northeast. Climate Dynamics 28: 381–407.
26. Gibson JR, Najjar RG (2000) The response of Chesapeake Bay salinity to climate-induced changes in streamflow. Limnology and Oceanography 45: 1764–1772.
27. Najjar RG, Pyke CR, Adams MB, Breitburg D, Hershner C, et al. (2010) Potential climate-change impacts on the Chesapeake Bay. Estuarine, Coastal and Shelf Science 86: 1–20.
28. Baird D, Ulanowicz RE (1989) The seasonal dynamics of the Chesapeake Bay ecosystem. Ecological Monographs 59: 329–364.
29. Chesapeake Bay Program (2013) CBP Water Quality Database (1984-present). URL http://www.chesapeakebay.net/data waterquality.aspx.
30. Hastie T, Tibshirani R (1986) Generalized additive models. Statistical Science: 297–310.
31. Faraway JJ (2004) Extending the linear model with R: generalized linear, mixed effects and non- parametric regression models. CRC press.
32. Breiman L (1996) Out-of-bag estimation. Technical report, Citeseer.
33. Urquhart EA, Zaitchik BF, Hoffman MJ, Guikema SD, Geiger EF (2012) Remotely sensed estimates of surface salinity in the Chesapeake Bay: A statistical approach. Remote Sensing of Environment 123: 522–531.
34. Urquhart EA, Hoffman MJ, Murphy RR, Zaitchik BF (2013) Geospatial interpolation of MODIS- derived salinity and temperature in the Chesapeake Bay. Remote Sensing of Environment 135: 167–177.
35. Kaneko T, Colwell RR (1973) Ecology of *Vibrio parahaemolyticus* in Chesapeake Bay. Journal of Bacteriology 113: 24–32.
36. Ebi KL (2008) Healthy people 2100: modeling population health impacts of climate change. Climatic Change 88: 5–19.
37. Hofstra N (2011) Quantifying the impact of climate change on enteric waterborne pathogen concentrations in surface water. Current Opinion in Environmental Sustainability 3: 471–479.
38. Schets F, Engels G, Evers E (2004) *Cryptosporidium* and *Giardia* in swimming pools in the Netherlands. Journal of Water Health 2: 191–200.

8

Predicting Potential Global Distributions of Two *Miscanthus* Grasses: Implications for Horticulture, Biofuel Production, and Biological Invasions

Heather A. Hager[1]*, Sarah E. Sinasac[1], Ze'ev Gedalof[2], Jonathan A. Newman[1]

1 School of Environmental Sciences, University of Guelph, Guelph, Ontario, Canada, 2 Department of Geography, University of Guelph, Guelph, Ontario, Canada

Abstract

In many regions, large proportions of the naturalized and invasive non-native floras were originally introduced deliberately by humans. Pest risk assessments are now used in many jurisdictions to regulate the importation of species and usually include an estimation of the potential distribution in the import area. Two species of Asian grass (*Miscanthus sacchariflorus* and *M. sinensis*) that were originally introduced to North America as ornamental plants have since escaped cultivation. These species and their hybrid offspring are now receiving attention for large-scale production as biofuel crops in North America and elsewhere. We evaluated their potential global climate suitability for cultivation and potential invasion using the niche model CLIMEX and evaluated the models' sensitivity to the parameter values. We then compared the sensitivity of projections of future climatically suitable area under two climate models and two emissions scenarios. The models indicate that the species have been introduced to most of the potential global climatically suitable areas in the northern but not the southern hemisphere. The more narrowly distributed species (*M. sacchariflorus*) is more sensitive to changes in model parameters, which could have implications for modelling species of conservation concern. Climate projections indicate likely contractions in potential range in the south, but expansions in the north, particularly in introduced areas where biomass production trials are under way. Climate sensitivity analysis shows that projections differ more between the selected climate change models than between the selected emissions scenarios. Local-scale assessments are required to overlay suitable habitat with climate projections to estimate areas of cultivation potential and invasion risk.

Editor: Jose L. Gonzalez-Andujar, Instituto de Agricultura Sostenible (CSIC), Spain

Funding: This research was funded by the Ontario Ministry of Agriculture and Food and Ministry of Rural Affairs (project 027167, www.omafra.gov.on.ca), a Mitacs Elevate post-doctoral fellowship (www.mitacs.ca), and the Natural Sciences and Engineering Research Council of Canada (http://www.nserc-crsng.gc.ca). The funders had no role in the study design, data collection and analysis, decision to publish, or preparation of the manuscript.

Competing Interests: The authors have declared that no competing interests exist.

* E-mail: hhager@uoguelph.ca

Introduction

Plant species are often introduced to new regions through human intervention. Plants that were introduced historically for medicinal, agricultural, or horticultural uses compose a large proportion (>60%) of the currently naturalized angiosperms in the United States and elsewhere [1]. Once established, these species have the potential to become invasive, with subsequent negative ecological and economic effects [2]. Many jurisdictions have introduced weed risk assessment methods to evaluate the risk that deliberately introduced plant species will become invasive in the future (e.g., [3], [4], [5], [6]). However, for species that were introduced prior to widespread use of such methods, their proliferation through increasing horticultural sales or industrial cultivation could increase their risk of escape. Assessing the risks and potential invasive outcomes of such species is therefore important in developing best management practices [7], [8], [9].

Several species of *Miscanthus* have been introduced to novel ranges in North America, Europe, and Scandinavia for both horticultural and agricultural purposes. These are tall, perennial, rhizomatous, C4 grasses native to temperate, humid subtropical, and tropical savannah climates of Asia [10]. *Miscanthus sinensis* Andersson was first introduced to North America as an ornamental plant in the 1890 s; it has since escaped cultivation in the northeastern United States and is considered invasive in some states [11]. Less is known about the first introduction of *M. sacchariflorus* (Maxim.) Franch. as an ornamental plant in North America, but escaped specimens were noted in the mid-western United States by 1950 [12], and the first escaped specimens in Ontario, Canada, were collected in 1952 [13]. These two species produce a sterile hybrid *M. x giganteus* J.M. Greef & Deuter ex Hodkinson & Renvoize that has been encountered infrequently in the wild [14]. *M. sinensis* has become a very popular ornamental plant in areas of the United States and Canada [15], and all three species are of interest as potential biofuel crops (e.g., [16], [17], [18]). Breeding programs for horticultural and agricultural improvement could enhance the potential of these species to be invasive [19], [11].

The role of weed risk assessment is to evaluate the risk of escape from cultivation and the extent of possible economic and environmental damage [4]. Thus, many assessments include estimating the potential climate suitability of the risk assessment area for candidate species for import (e.g., [3], [4], [5], [6]) because climate is a good predictor of plant distributions [20], [21]. Climate suitability provides a coarse-scale estimate of a species' distribution while ignoring factors such as substrate geology, biotic interactions, and infrastructure, which affect regional habitat suitability.

Various methods have been used to estimate potential species distributions, for example, plant hardiness zones [22], [6], climate regions occupied [23], and a wide range of bioclimatic and niche models (e.g., [24], [25], [9]). The latter use detailed information on temperature and moisture in the species' native range to determine areas with potentially suitable climate for population persistence. Estimating species preferences for certain growing conditions can be difficult, leading to uncertainty in model parameterization. However, model sensitivity to the choice of parameter values is seldom evaluated (but see [26], [27]), but is important in understanding which parameters have the greatest effect on model output and therefore should be estimated most carefully or where results should be treated most cautiously. Such information can be used in interpreting model output and in directing future research to improve model reliability.

Estimating species' potential ranges under projected climate change is also a priority. This work is of particular concern for species of agronomic and horticultural importance, as well as for invasive and potential pest species [28]. Shifts in distributions of agronomic and horticultural species are likely to keep pace with climate change if people continue to plant them where conditions are suitable [29]. Distribution shifts of pests and invaders could also be favoured under climate change because these species tend to have wide physiological tolerances and traits that allow them to take advantage of long-distance (human-assisted) dispersal vectors [30]. Previous assessments of climatic suitability indicate that the modelled distributions differ depending on the choice of climate change model and scenario (e.g., [31], [32]). Comparing multiple models and climate change scenarios therefore allows assessment of the sensitivity of results to these choices.

The purpose of this analysis was threefold. First, we estimated the current potential global distributions of *M. sacchariflorus* and *M. sinensis* using a commonly employed niche model. Second, we evaluated the model sensitivity to estimate which parameters are most critical and whether this differs between the species. Third, we examined how the potentially suitable area is projected to change under future climates, as well as how sensitive these results are to the selection of climate model and emissions scenario. The resulting potential distributions indicate areas that might be suitable for horticultural or agricultural cultivation of these species, but also susceptible to their invasion, should they escape.

Methods

We estimated the current global native and introduced distributions of *M. sacchariflorus* and *M. sinensis* using species occurrence data from the Global Biodiversity Information Facility (GBIF; www.gbif.org), botanical garden records, and a literature search of agronomic trials. All data were sorted to remove duplicate records and were separated based on whether geolocation information (latitude and longitude) was provided or whether we could geocode the record using information such as street name, city, county, or region ("inferred" location). Records that did not provide location information below the country level were omitted.

We used the native distribution data to model the potential climatic range of each species using the CLIMEX Compare Locations model [33], [34]. We used 0.5° world grid climate data provided with the software (from the Climate Research Unit at Norwich, UK [35]). CLIMEX assumes that the geographical distribution of the species is limited by climate; it does not generally account for biotic interactions [36] or substrate type. It calculates an annual growth index based on the species' fitted temperature and moisture response functions, as well as four stress indices (hot, cold, dry, wet, and their combinations) to calculate an ecoclimatic index (EI). The EI is an estimate of a location's climatic suitability to support a persistent population of the species being modelled [34]. EI ranges from 0 to 100, with 0 meaning the area is not suitable for species persistence, and 100 meaning the climate is optimal for the species year-round. In practice, EI of 100 is only attained for species in stable and ideal climate [34].

For each species, we began with parameter values from the default temperate template and adjusted them iteratively based on the species' biology until the modelled distributions approximated the native distributions [34]; these included humid subtropical (*M. sacchariflorus, M. sinensis*) and tropical savannah (*M. sinensis*) climates of Asia. The model was then validated by comparison with the observed distribution records in the species' introduced ranges in Europe and North America. Given the satisfactory fit, no further adjustments were required at this stage for either model.

Parameter Fitting

Temperature and cold/hot stress. Based on their native distributions, both *M. sacchariflorus* and *M. sinensis* are well suited for cold-temperate regions (e.g., [37], [38]). Their percentage shoot emergence at experimental, low temperatures ranges from 10 to 100% at 7 to 15°C [39]. Their *M.* x *giganteus* hybrid shows only minor reduction in leaf photosynthetic capacity when grown at 10°C or 14°C compared to 25°C [40], [41]. Therefore, the limiting low temperature (DV0) and lower optimal temperature (DV1) were set at 5°C and 15°C, respectively, for both species (Table 1). This and the cold stress (below) accounted for the northernmost occurrence of both species in far northeastern China and the southeastern Primorsky Krai region of Russia.

The upper optimal temperature (DV2) and limiting high temperature (DV3) were adjusted based on maximum temperatures occurring in the native region [42] and observations that the experimental optimal temperature for photosynthesis of their hybrid is between 30°C and 35°C [43]. The native distribution of *M. sinensis* extends much further south into tropical regions than does that of *M. sacchariflorus* [37]. Therefore, both DV2 and DV3 for *M. sinensis* were greater than those for *M. sacchariflorus* (Table 1).

For the cold stress index, we used parameters related to overwinter survival (lethal temperatures: cold stress temperature threshold, TTCS, and cold stress temperature rate, THCS) and cold stress affecting metabolism (cold stress degree day threshold, DTCS, and cold stress degree day rate, DHCS). Both species have strong cold tolerance and overwinter underground as rhizomes. The lethal temperature at which 50% of rhizomes were killed in a freezing experiment ranged from −3.4°C to −6.3°C for both species [44]. Therefore, TTCS was set to −5°C and THCS to a very low accumulation rate below this threshold air temperature because of the expectation that rhizomes would be insulated by plant litter and snow pack in colder areas. DTCS was adjusted downward slightly from the temperate template (15°C) to 12°C for *M. sacchariflorus* and 14°C for *M. sinensis,* with a very low accumulation rate (DHCS; Table 1) because both species have high cold tolerance. DTCS was lower for *M. sacchariflorus* than for *M. sinensis* because native distribution records show *M. sacchariflorus* persisting at slightly higher latitudes.

There has been little investigation of heat stress effects on the growth of *Miscanthus*. Therefore, heat stress parameters (heat stress threshold, TTHS, and heat stress rate, THHS) were adjusted based on native distribution records for the species. TTHS was set to begin accumulating at or above the upper optimal growth temperature for *M. sacchariflorus* and *M. sinensis*, respectively, with a

Table 1. Fitted parameter values used to generate bioclimatic envelope models of *Miscanthus sacchariflorus* and *M. sinensis* distributions using CLIMEX.

Parameter description	Parameter	*Miscanthus sacchariflorus*	*Miscanthus sinensis*
Moisture			
Limiting low moisture[a]	SM0	0.25	0.25
Lower optimal moisture[a]	SM1	0.8	0.8
Upper optimal moisture[a]	SM2	1.2	1
Limiting high moisture[a]	SM3	1.8	2.5
Temperature			
Limiting low temperature[b]	DV0	5	5
Lower optimal temperature[b]	DV1	15	15
Upper optimal temperature[b]	DV2	28	30
Limiting high temperature[b]	DV3	32	35
Cold Stress			
Cold stress temperature threshold[b]	TTCS	−5	−5
Cold stress temperature rate[c]	THCS	−0.0002	−0.0002
Cold stress degree-day temperature threshold[d]	DTCS	12	14
Cold stress degree-day rate[c]	DHCS	−0.00005	−0.00005
Heat Stress			
Heat stress temperature threshold[b]	TTHS	32	36
Heat stress temperature rate[c]	THHS	0.06	0.05
Dry Stress			
Dry stress moisture threshold[a]	SMDS	0.1	0.1
Dry stress rate[c]	HDS	−0.02	−0.02
Wet Stress			
Wet stress moisture threshold[a]	SMWS	1.8	3
Wet stress rate[c]	HWS	0.02	0.05
Hot-wet Stress			
Hot-wet temperature threshold[b]	TTHW	31	-
Hot-wet moisture threshold[a]	MTHW	1	-
Hot-wet stress rate[c]	PHW	0.01	-
Degree-days			
Minimum degree days above DV0 to complete one generation[d]	PDD	600	600

[a]Proportion soil capacity.
[b]°C.
[c]$Week^{-1}$.
[d]°D.

high accumulation rate (THHS). TTHS was higher and THHS was slightly lower for *M. sinensis* than for *M. sacchariflorus* to account for the former species' more tropical native distribution.

Soil moisture and dry/wet stress: Limiting low soil moisture (SMO) and lower optimal soil moisture (SM1) were set according to the temperate template. Upper optimal soil moisture (SM2) and limiting high soil moisture (SM3) were set according to the wet-tropical template for *M. sinensis*. In comparison, SM2 and SM3 were reduced slightly for *M. sacchariflorus* in conjunction with the wet stress parameters to limit its potential distribution from occurring widely in wet tropical areas.

The dry stress threshold (SMDS) and dry stress rate (HDS) were set to accommodate the high drought tolerance of both species [45]. Both species were assigned an SMDS of 0.1, which is close to the minimum moisture content at which plants can extract water from the soil [46], [47]. HDS was set at a moderate rate, given that the hybrid can recover from short-term (30 days) but not long-term (60 days) drought [48].

Both *M. sacchariflorus* and *M. sinensis* grow well in saturated areas such as drainage ditches (HAH, personal observation), but not when submerged such as in streams [49], [50]. Therefore, the wet stress threshold (SMWS) and wet stress rate (HWS) were set so that the species would grow in wet areas but would experience a high rate of stress accumulation. Parameters for *M. sacchariflorus* were adjusted to restrict its southerly distribution.

Hot-wet stress: A hot-wet stress was added to the model for *M. sacchariflorus* to exclude its distribution from tropical equatorial areas of southeastern Asia-Pacific [37]. No hot-wet stress was used for *M. sinensis*.

Sensitivity to Model Parameters

Once the models were validated, we performed a sensitivity analysis for each species to determine the response of the modelled distributions to small changes in parameter values. Each parameter was adjusted upward or downward while holding all other parameters constant, and the resulting EI values for the global distribution were generated in CLIMEX. Rate and soil moisture variables were adjusted by 10%; temperature variables were adjusted by 1°C (D. Kriticos, *personal communication*). EI was divided into five classes of suitability for species growth for further analysis: 0, unsuitable; 1–10, marginal; 10–20, suitable; 20–30, favourable; and >30, highly favourable [51], [34].

EI values from each iteration of the sensitivity analysis were compared with those of the original model to determine the change in area for each EI class. To do this, EI data generated by CLIMEX were reprojected to a cylindrical equal area projection in ArcGIS 10.1, masked to land area, and a nearest neighbour, inverse distance weighting interpolation was performed so that each grid cell represented an equal amount of area (50×50 km). The resulting grid cells were then reclassified corresponding to the five EI classes and the number of cells counted to determine the total area in each class. We then computed the proportional change in area from the original model for each EI class. Sensitivity was evaluated as a greater proportional change in area than in parameter value.

Climate Change Scenarios

To compare the species' potential distributions under future climate change predictions, and to assess the sensitivity of these potential distributions to model selection and emissions assumptions, we used two general circulation models with two emissions scenarios (A2 and B1). We used the Bergen Climate Model 2.0 (BCM) from the Bjerknes Center for Climate Research, and the Coupled Global Climate Model 3 (CGCM3-T63) from the Canadian Center for Climate Modeling and Analysis [52]. The selected emissions scenarios represent comparatively low (B1) and high (A2) future greenhouse gas concentrations, thus spanning the likely range of probable future conditions [53].

Climate projection data for each model–scenario combination were input into CLIMEX as the 30-yr mean for three standard time periods: baseline (1971–2000), 2050s (2041–2070), and 2080s (2071–2100). First, monthly means of maximum temperature, minimum temperature, and relative humidity, and daily means of precipitation for the world (excluding Greenland, Antarctica, and the Arctic) were downloaded from the Canadian Climate Change Scenarios Network (CCCSN: http://www.cccsn.ec.gc.ca/?page=dd-gcm) for each model–scenario combination. The 30-yr means were then calculated to obtain a single value for each climate variable for each 2.75°×2.75° grid cell. Because CLIMEX requires monthly climate data, the average daily precipitation data were multiplied by the number of days in the month to produce monthly means. CLIMEX also requires two relative humidity values, RH% at 0900 h and at 1500 h. Although only mean monthly values are available from the CCCSN, there is a strong diurnal cycle in RH%, with maximum values at night and minimum values in mid-afternoon. Daily mean values are therefore reasonably close to observed values at 0900 h and were considered RH% at 0900 h. RH% at 1500 h was estimated using the CLIMEX method of multiplying the 0900 h value by 0.85 [34], [31].

Using these model–scenario data, the CLIMEX bioclimatic model was run for each species. EI values were obtained for a total of 20 model runs ([circulation model baseline + two scenarios × two future time periods] × two models × two species). The EI data were reprojected into a cylindrical equal area projection, masked to land area, and interpolated using inverse distance weighting of the three nearest points in ArcGIS 9.3. The areas of suitable climate under baseline and future conditions were calculated by reclassifying the raw EI data into the five EI classes. Projected changes in the potential distributions of each species were determined by overlaying future and baseline projections.

For each species and time, areas of agreement between models or scenarios were determined by overlaying projected distributions based on the two models for a given scenario or the two scenarios for a given model, respectively. To quantify the similarity between pairs of projected potential distributions, we calculated a simple index of agreement by dividing the area where the two projections agree about the potential presence of *Miscanthus* (all area classified as suitable, favourable, and highly favourable, or EI >10) by the area where the two projections disagree. The higher the value, the more similar the two projections are. Values >1.0 indicate that the models agree more than they differ; values <1.0 indicate more disagreement than agreement.

Results

Occurrence records indicate that both the native and introduced distributions of *M. sacchariflorus* are less widespread than those of *M. sinensis* (Fig. 1). In the native range, both species occur throughout Japan, Korea, and south-central and eastern China, and in parts of northern and northeastern China. However, *M. sinensis* extends further south and east into Taiwan, the Philippines, and Vanuatu [37] (Fig. 1). In the introduced range, *M. sacchariflorus* occurs mainly in northwestern Europe, Denmark, Sweden, northeastern United States, and southeastern Canada. *M. sinensis* has been introduced to these areas as well as southeastern and parts of western United States, Mexico, Puerto Rico, Colombia, Chile, Argentina, Uruguay, southern Australia, Tasmania, and New Zealand.

For the *M. sacchariflorus* model, 100% of given native, 90.1% of inferred native, 89.1% of given introduced and 97.7% of inferred introduced occurrence records are in areas deemed suitable, favourable, or highly favourable (EI>10). Zero and 9.4% of given and inferred native, and 10.9% and zero of given and inferred introduced occurrences are in marginal areas. One inferred native record (0.6%) and one inferred introduced record (2.3%) are in areas deemed unsuitable. For the *M. sinensis* model, 100% of given native, 94.2% of inferred native, 99.7% of given introduced, and 97.5% of inferred introduced occurrences are in areas deemed suitable, favourable, or highly favourable. Zero and 5.8% of given and inferred native, and 2.0% and 2.5% of given and inferred introduced occurrences are in marginal areas. No observed occurrences are in areas deemed unsuitable. The models indicate that *M. sinensis* has a wider potential global distribution than does *M. sacchariflorus* (Fig. 1).

Sensitivity to Model Parameters

The modelled distributions differed in their sensitivity to a ±10% or 1°C change in parameter values (Table 2). Eight of 22 (36%) parameters for *M. sacchariflorus* and 5 of 19 (26%) parameters for *M. sinensis* showed sensitivity in at least one EI class. Models for both species were very sensitive to changes in heat stress temperature threshold (TTHS) and upper optimal and limiting temperatures (DV2, DV3) and were mildly sensitive to changes in upper optimal and limiting moistures (SM2, SM3). In addition, the *M. sacchariflorus* model was highly sensitive to an increase in dry stress rate (HDS) and a decrease in hot-wet temperature threshold (TTHW), and mildly sensitive to a decrease in wet stress moisture threshold (SMWS).

Sensitive temperature parameters all related to upper temperatures and heat tolerance thresholds (Table 2). A decrease in the

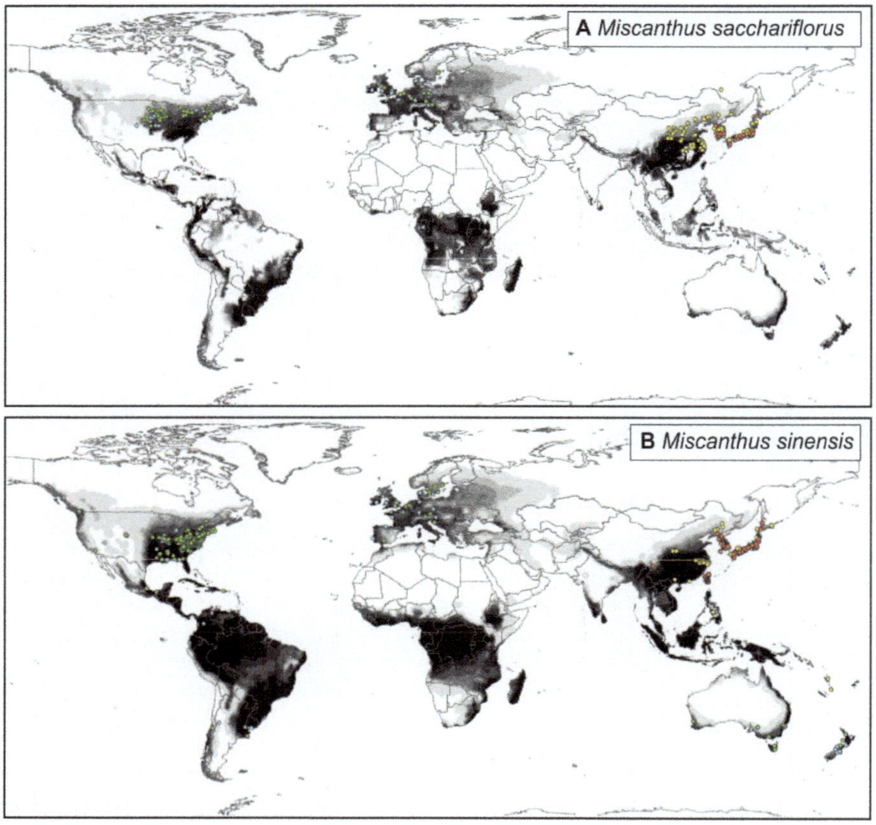

Figure 1. Observed (circles) and modelled (shaded) native and introduced plant distributions. Circles indicate native (red), native inferred (yellow), introduced (green), and introduced inferred (blue) geolocations. The density of shading represents the bioclimatic suitability of a region for plant population persistence, with white areas indicating no suitability, lighter grey areas indicating low suitability, and darker grey areas indicating high suitability. (a) *Miscanthus sacchariflorus*. Geolocations: native (n = 94), native inferred (n = 171), introduced (n = 119), and introduced inferred (n = 43). (b) *Miscanthus sinensis*. Geolocations: native (n = 335), native inferred (n = 52), introduced (n = 297), and introduced inferred (n = 81).

upper temperature parameters tended to decrease EI towards less favourable values, whereas an increase tended to increase EI towards more favourable values for both species. The heat stress temperature threshold followed a similar pattern. The *M. sacchariflorus* model was more sensitive than the *M. sinensis* model to changes in these parameters. For *M. sacchariflorus*, a decrease in the hot-wet temperature threshold also decreased EI towards less favourable values, whereas an increase had only a small effect.

Sensitive soil moisture parameters all related to upper soil moistures and the wet stress threshold. Changes in upper soil moisture parameters had weaker effects than for temperature. A decrease in the upper soil moisture parameters tended to decrease the most favourable EI values but had little effect on the unsuitable climate area. An increase in these parameters had little effect, except for slight increases in EI towards more favourable values for *M. sacchariflorus*. For *M. sacchariflorus*, a decrease in the wet stress moisture threshold tended to decrease the suitable climate area.

The only rate parameter that exhibited sensitivity was the dry stress rate for the *M. sacchariflorus* model. An increase in this parameter increased EI values from unsuitable towards more favourable values, with the largest change in marginal climate area (Table 2).

Range Shifts Under Climate Change Projections

Projected changes in the potential area occupied by *M. sacchariflorus* and *M. sinensis* are generally large, ranging from global decreases in potential area by 2080 of 4 to 6%, depending on the species, model, and scenario chosen (Table 3, World; Fig. 2). In all cases, the area of climatically suitable locations (EI >10) continues to decrease over time.

Limiting this analysis to North America reveals some important regional differences (Table 3, North America). Projected percent changes in climatically suitable area are smaller for North America than for the world and are often different in sign. The direction of change also differs between the two models: For both species by 2080, regardless of scenario, the BCM projects reductions in the total suitable area and the CGCM projects increases in suitable area. There are also large changes in the relative proportions of habitat categories. In all projections, including those in which the area of climatically suitable habitat increases, highly favourable area (EI >30) decreases and favourable area (EI 20–30) increases.

Comparing the future potential distributions of *Miscanthus* species to their baseline potential distributions provides an indication of how rapidly shifts in suitable range might occur. This analysis indicates that projected suitability shifts are moderately large for both species and under both emissions scenarios (Table 4). The smallest projected shifts in suitable range occur under the B1 emissions scenario, with *M. sacchariflorus* projected to have larger shifts than *M. sinensis*. The mean overlap between the baseline and 2080s potentially suitable areas is only 61% for *M. sacchariflorus* and 78% for *M. sinensis*. Very little range contraction is projected to occur in the native range, except for some projections for *M. sacchariflorus*. In contrast, range expansion is projected in northern parts of the native range for both species.

Table 2. Sensitivity analysis results for *Miscanthus sacchariflorus* and *M. sinensis* indicating the proportional change in area from that of the baseline model global distribution for each category of ecoclimatic index (EI) modelled using CLIMEX software.

	Miscanthus sacchariflorus						*Miscanthus sinensis*					
Parameter	**Value**[a]	**Proportional change in area from baseline (%)**					**Value**[a]	**Proportional change in area from baseline (%)**				
		EI = 0	**EI = 1–10**	**EI = 10–20**	**EI = 20–30**	**EI = 30–100**		**EI = 0**	**EI = 1–10**	**EI = 10–20**	**EI = 20–30**	**EI = 30–100**
DV0	4 (20)	−0.61	0.98	1.17	−1.08	2.50	4 (20)	−0.71	0.36	1.81	−0.76	1.09
	6 (20)	0.69	−1.39	−1.28	2.13	−3.01	6 (20)	0.85	−1.16	−1.75	1.88	−1.30
DV1	14 (6.7)	−0.14	−2.13	0.19	−1.66	3.68	14 (6.7)	−0.19	−2.45	0.79	−0.88	1.55
	16 (6.7)	0.12	2.15	0.11	2.67	−4.32	16 (6.7)	0.19	2.12	−0.79	2.61	−1.83
DV2	27 (3.6)	0.06	**5.31**	**9.94**	−0.27	**−11.68**	29 (3.3)	0.11	1.77	**8.97**	**16.08**	**−7.08**
	29 (3.6)	−0.07	**−5.62**	**−7.63**	−2.98	**12.25**	31 (3.3)	−0.13	−1.01	**−3.92**	**−13.90**	**5.10**
DV3	31 (3.1)	0.54	**10.64**	−0.84	**−7.20**	**−9.72**	34 (2.9)	0.26	0.46	**5.41**	**5.22**	**−3.28**
	33 (3.1)	−0.15	**−5.80**	**−4.26**	3.08	**7.80**	36 (2.9)	−0.12	−0.46	−1.62	**−5.13**	2.11
SM0	0.225	−0.77	1.62	1.62	−0.10	2.03	0.225	−1.46	2.97	3.59	−1.03	1.11
	0.275	0.71	−1.43	−1.67	0.57	−2.10	0.275	1.41	−2.80	−3.76	1.52	−1.17
SM1	0.72	−0.28	−3.06	−0.42	−2.27	6.18	0.72	−0.54	−3.30	1.42	−4.28	3.36
	0.88	0.24	4.14	−1.59	2.13	−5.68	0.88	0.38	4.48	−3.69	5.13	−3.26
SM2	1.08	0.03	4.32	7.60	7.37	**−12.94**	0.9	0.00	1.13	2.74	**10.08**	−3.65
	1.32	−0.01	−3.89	−6.71	−4.50	**10.39**	1.1	−0.01	−0.91	−3.23	−8.07	3.18
SM3	1.62	0.17	7.86	7.60	1.22	**−14.21**	2.25	0.00	1.01	4.39	**13.26**	−4.76
	1.98	−0.05	−5.12	−5.40	1.93	7.81	2.75	−0.01	−0.50	−2.77	−7.37	2.71
TTCS	−6 (20)	0.76	−1.86	−2.26	−0.91	−0.83	−4 (20)	−1.10	2.60	2.67	0.88	0.26
	−4 (20)	−0.78	2.32	2.42	0.37	0.67	−6 (20)	1.00	−1.66	−4.25	−0.36	−0.25
THCS	−0.00022	0.66	−2.56	−1.45	−0.44	−0.31	−0.00022	0.81	−1.96	−2.80	−0.24	−0.08
	−0.00018	−0.83	3.34	2.03	0.44	0.19	−0.00018	−1.07	3.18	2.61	0.24	0.11
DTCS	11 (8.3)	−0.36	0.95	1.20	0.10	0.40	13 (7.1)	−0.42	0.61	1.48	−0.21	0.31
	13 (8.3)	0.38	−1.15	−0.53	−0.41	−0.57	15 (7.1)	0.44	−0.63	−1.78	0.03	−0.23
DHCS	−0.000055	0.26	−0.69	−0.39	−0.17	−0.55	−0.000055	0.33	−0.23	−1.62	−0.12	−0.17
	−0.000045	−0.27	0.71	0.95	0.07	0.31	−0.000045	−0.42	0.68	1.32	−0.12	0.28
TTHS	31 (3.1)	**8.79**	**−13.96**	**−22.28**	**−21.78**	**−13.97**	35 (2.8)	2.29	**−3.12**	−2.01	**−7.37**	−1.03
	33 (3.1)	**−8.15**	**26.66**	**16.04**	**8.12**	**8.21**	37 (2.8)	−1.82	**3.30**	**3.99**	**5.01**	0.11
THHS	0.054	−0.21	0.46	0.72	0.30	0.21	0.045	−0.05	0.10	0.16	0.09	0.00
	0.066	0.18	−0.46	−0.47	−0.41	−0.14	0.055	0.08	−0.10	−0.43	−0.06	0.00
SMDS	0.09	−0.12	−0.15	−1.64	0.98	1.39	0.09	−0.42	0.28	−2.84	0.55	1.25
	0.11	−0.36	−1.91	−6.35	−0.91	8.48	0.11	0.43	−0.02	1.39	0.21	−1.25
HDS	−0.022	−0.36	−1.91	−6.38	−0.88	8.48	−0.022	0.15	0.45	0.53	0.09	−0.66
	−0.018	**−17.00**	**88.48**	6.10	7.51	5.97	−0.018	−0.25	0.05	−1.65	0.76	0.68

Table 2. Cont.

	Miscanthus sacchariflorus						*Miscanthus sinensis*					
Parameter	**Value**[a]	**Proportional change in area from baseline (%)**					**Value**[a]	**Proportional change in area from baseline (%)**				
		EI = 0	**EI = 1–10**	**EI = 10–20**	**EI = 20–30**	**EI = 30–100**		**EI = 0**	**EI = 1–10**	**EI = 10–20**	**EI = 20–30**	**EI = 30–100**
SMWS	1.62	1.03	−2.96	**−10.05**	−2.71	4.37	2.7	0.01	0.23	−0.20	0.00	−0.08
	1.98	−0.63	−2.77	−5.35	0.57	9.55	3.3	−0.01	0.02	−0.03	0.03	0.02
HWS	0.018	−0.39	−2.08	−6.04	−0.78	8.57	0.045	0.00	0.00	−0.03	0.00	0.01
	0.022	−0.32	−1.94	−6.52	−1.05	8.45	0.055	0.00	0.00	0.00	0.00	0.00
TTHW	30 (3.2)	**3.67**	**−5.43**	**−6.91**	**−9.77**	**−7.37**						
	32 (3.2)	−1.15	2.87	**3.93**	0.95	1.12						
MTHW	0.9	0.14	−0.05	−0.58	−0.07	−0.40						
	1.1	−0.19	0.52	0.11	0.51	0.28						
PHW	0.009	−0.09	0.21	0.28	0.10	0.14						
	0.011	0.05	−0.02	−0.22	0.00	−0.15						
PDD	540	−0.60	−0.53	−6.35	−0.95	8.52	540	−0.27	1.21	0.03	0.00	0.00
	660	−0.13	−3.11	−6.74	−0.95	8.52	660	0.27	−1.14	−0.16	0.00	0.00

[a]Parameter value used in the sensitivity analysis. The percent change from the original model parameter value was 10%, except for temperature variables, which were changed by 1°C and for which percent change is indicated in parentheses. Changing temperature variables by 10% gave qualitatively similar results (data not shown).

Parameter values were each decreased and increased by 10% from the baseline value (Table 1) unless indicated otherwise[a]. Changes in area that exceed the percent change in parameter value are shown in boldface font; positive and negative proportions respectively indicate increases and decreases in area.

Table 3. Percent change in climatically suitable area for the World and for North America projected using the BCM and CGCM models and the A2 and B1 emissions scenarios relative to the baseline projections.

		World						North America					
Model	**Year, scenario**	**EI 0–1**	**EI >1–10**	**EI >10–20**	**EI >20–30**	**EI >30**	**EI >10**	**EI 0–1**	**EI >1–10**	**EI >10–20**	**EI >20–30**	**EI >30**	**EI >10**
Miscanthus sacchariflorus								*Miscanthus sacchariflorus*					
BCM	2050 (A2)	+4.1	+6.9	+15.3	−1.7	−20.6	**−10.4**	−1.6	+1.5	+15.4	+74.5	−32.4	**+3.9**
	2080 (A2)	+12.4	+14.7	+8.1	−31.8	−38.6	**−28.9**	−4.4	+18.5	−2.7	+16.7	−18.8	**−6.7**
	2050 (B1)	+3.2	+2.1	+12.6	−0.2	−14.6	**−6.9**	−1.1	−0.3	+6.6	+58.6	−18.4	**+4.4**
	2080 (B1)	+6.5	+5.4	+9.3	−4.6	−24.3	**−14.3**	−1.1	+4.3	+4.8	+49.8	−26.2	**−1.4**
CGCM	2050 (A2)	+8.7	+7.4	−6.3	−12.0	−58.2	**−31.1**	−7.2	+32.8	+27.0	+5.0	−51.5	**+0.6**
	2080 (A2)	+13.0	+10.0	−11.8	−41.7	−71.9	**−46.1**	−13.9	+55.9	+47.7	+21.0	−70.1	**+9.9**
	2050 (B1)	+7.0	+6.4	−9.7	−9.9	−44.3	**−25.4**	−6.5	+26.0	+23.0	+18.2	−42.3	**+5.0**
	2080 (B1)	+8.6	+6.5	−3.8	−13.9	−57.5	**−30.5**	−6.7	+27.2	+27.6	+23.8	−59.9	**+4.3**
Miscanthus sinensis								*Miscanthus sinensis*					
BCM	2050 (A2)	+4.0	+8.3	−7.4	−14.4	−5.0	**−6.9**	+1.9	−2.8	−6.5	+23.1	−7.1	**−1.9**
	2080 (A2)	+9.7	+19.1	−8.5	−32.4	−14.9	**−16.5**	−5.4	+28.9	−22.0	+10.3	−9.3	**−11.4**
	2050 (B1)	+1.8	+4.9	+0.7	−11.4	−2.7	**−3.5**	+1.6	+1.8	−11.4	+15.7	−6.8	**−5.0**
	2080 (B1)	+5.1	+12.8	−12.9	−16.1	−6.9	**−9.4**	+0.0	+17.9	−34.6	+17.4	−6.3	**−14.3**
CGCM	2050 (A2)	+8.9	−5.6	+8.3	−0.4	−24.5	**−13.4**	−8.2	+24.3	+13.0	+61.5	−42.1	**+5.3**
	2080 (A2)	+10.9	+6.9	+15.6	+5.9	−42.2	**−21.6**	−17.2	+44.2	+44.7	+101.6	−71.5	**+18.2**
	2050 (B1)	+7.1	−4.1	+6.5	−3.3	−19.0	**−10.9**	−6.7	+14.5	+5.1	+68.0	−22.3	**+9.9**
	2080 (B1)	+8.0	−5.2	+10.3	−1.0	−22.7	**−12.0**	−9.4	+22.0	+17.6	+78.0	−38.6	**+12.2**

The column EI>10 (bold) indicates a composite analysis combining all area classified as suitable, favourable, and highly favourable for each species.

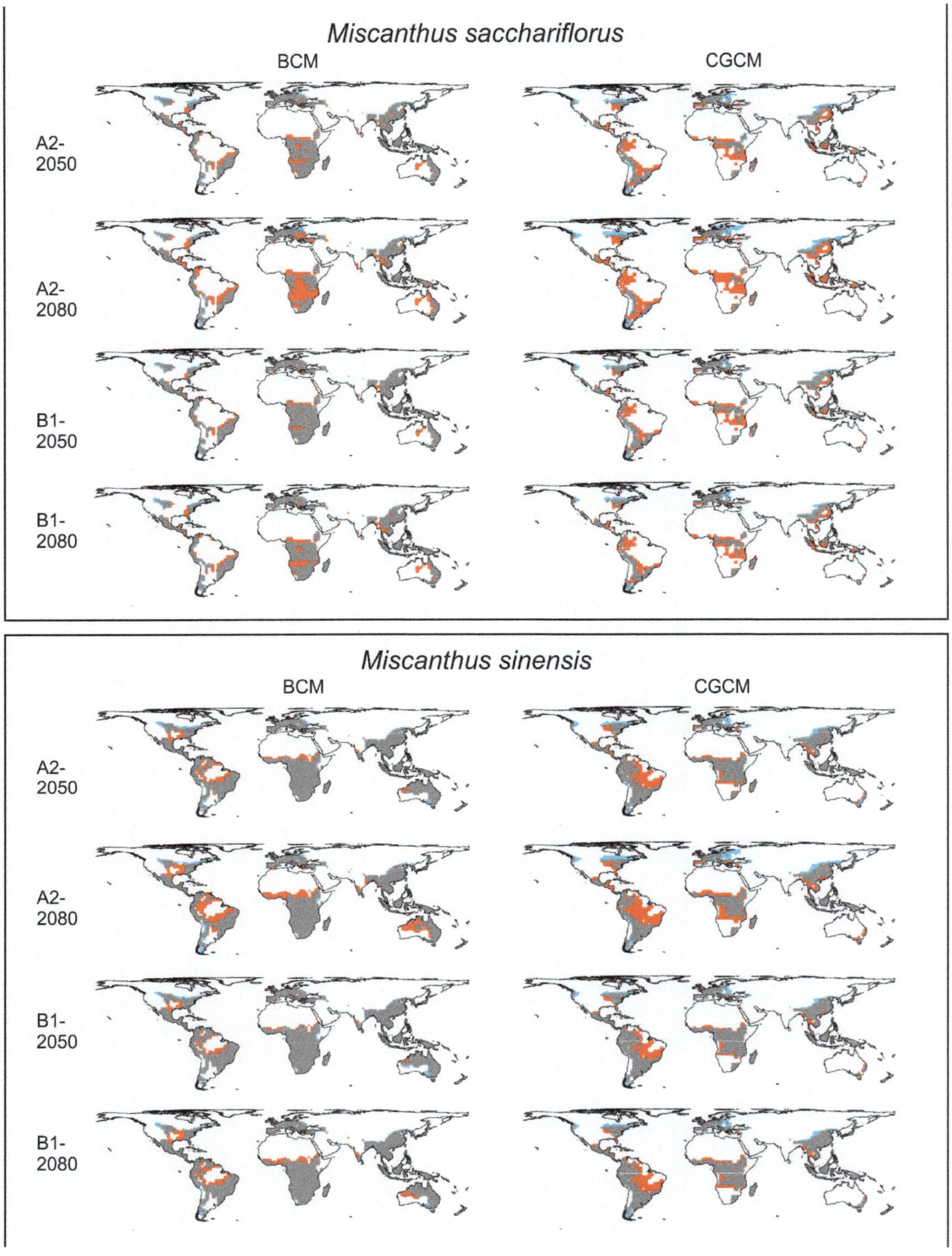

Figure 2. The potential future distribution of *Miscanthus sacchariflorus* and *M. sinensis* (for all area classified as suitable, favourable, and highly favourable, or EI >10) as modeled by CLIMEX under the BCM and CGCM models run using the A2 and B1 emissions scenarios. All potential ranges are shown relative to their baseline potential distributions. Areas shown in red indicate area where the range is projected to contract; grey indicates the species are potentially present under both baseline and future conditions; and blue indicates areas of range expansion.

In the non-native range, contraction is projected mainly in areas where the species have not yet been introduced such as South America, Africa, and parts of Australia for both species, as well as islands of Southeast Asia for *M. sacchariflorus*. Contraction of the potential range is also projected to occur in southern parts of the United States, but more so for *M. sinensis* than for *M. sacchariflorus*. The majority of range expansion is projected to occur in northern areas of North America, eastern Europe, and Scandinavia.

Table 4. Projected changes in the global climatically suitable area for *Miscanthus sacchariflorus* and *M. sinensis* under the BCM and CGCM models run using the A2 and B1 scenarios.

Model	Year, scenario	Expansion[a] (km×10⁶)	Contraction (km×10⁶)	No change (km×10⁶)	Overlap[b] (%)
		Miscanthus sacchariflorus			
BCM	2050 (A2)	1.54	5.74	34.81	86%
	2080 (A2)	2.28	13.99	26.56	65%
	2050 (B1)	1.10	3.88	36.67	90%
	2080 (B1)	1.35	7.16	33.38	82%
CGCM	2050 (A2)	3.08	11.94	16.51	58%
	2080 (A2)	5.08	18.20	10.25	36%
	2050 (B1)	2.33	9.56	18.89	66%
	2080 (B1)	2.73	11.41	17.04	60%
		Miscanthus sinensis			
BCM	2050 (A2)	2.08	5.85	48.85	89%
	2080 (A2)	2.78	11.79	42.91	78%
	2050 (B1)	1.95	3.85	50.84	93%
	2080 (B1)	1.60	6.72	47.98	88%
CGCM	2050 (A2)	3.68	9.36	33.10	78%
	2080 (A2)	6.01	15.17	27.29	64%
	2050 (B1)	2.84	7.47	35.00	82%
	2080 (B1)	3.51	8.60	33.86	80%

[a]Potential future suitable area is defined as EI>10.
[b]Overlap is calculated as the portion of the baseline range that overlaps with the projected range.

Sensitivity of Results to Choice of Climate Model and Scenario

By overlaying the results for the different models and scenarios, we can identify patterns of agreement and disagreement between the various projections (Fig. 3). This analysis helps to identify changes in the suitable range that are highly likely (e.g., robust to the selection of model or scenario), versus those that are more speculative (e.g., those that differ depending on the model or scenario chosen). This analysis also allows us to identify whether the projections differ because of differences between the climate models or between the scenarios. These results indicate that the choice of climate model accounts for more difference in the results than the choice of scenario and that the two *Miscanthus* species differ in their sensitivity to the selection of model and scenario (Table 5). For the model sensitivity analysis, agreement values are all either <1.0 (greater area of disagreement than agreement between models; five out of eight comparisons), or very slightly > 1.0 (maximum value 1.22). For the scenario sensitivity analysis, all agreement values are substantially >1.0. Different scenarios are more similar within models than are the same scenarios between models for these species.

Discussion

Potential Global Distribution

Niche and bioclimatic envelope models in general provide a coarse-scale indication of areas where the climate might be suitable for the species of interest to establish. The limitations of such models are well discussed elsewhere [34], [54], [55], [56]. Here, we note two associated factors that require consideration in interpreting our models: the species' ecology and their likely non-equilibrium distribution in the introduced ranges.

Little is known of the species' comparative ecology in the native and introduced ranges because most research to date has focused on determining optimal conditions for agricultural production (but see [57], [42], [50]). Parameterizing a model based solely on the native range distribution assumes similar biotic interactions in the introduced and native ranges. However, potential release from suppressive interactions such as competition, herbivory, parasitism, and disease (e.g., [58], [59], [60]), or a lack of mutualist organisms (e.g., [61]) could result in different realized distributions or niche shifts in the introduced compared to the native range (e.g., [62], [63], but see [64], [65]). We parameterized our model using both the native range distribution and some physiological data, which could improve model estimations compared to strictly correlative methods [66], [67]. Model validation using occurrences in the introduced range indicates that the models fit the current introduced distributions well. Determining whether an introduced species will establish beyond the modelled range requires further fine-scale assessment and/or field experiments [54], [68].

It is highly likely that *M. sinensis* and *M. sacchariflorus* are still spreading in the introduced ranges. Indeed, the species only became naturalized in North America in the mid-1900s [12], [11], and time since introduction is a well-known correlate of plant escape and abundance in the introduced range (e.g., [69], [70]). The bioclimatic models suggest that there are additional moderate to large amounts of climatically suitable area in Central America, South America, and Africa, as well as small parts of Australia and New Zealand, where these species have not yet established. The likelihood of spread beyond the modelled potential distributions is unknown, but given the increase in cultivation of these species as

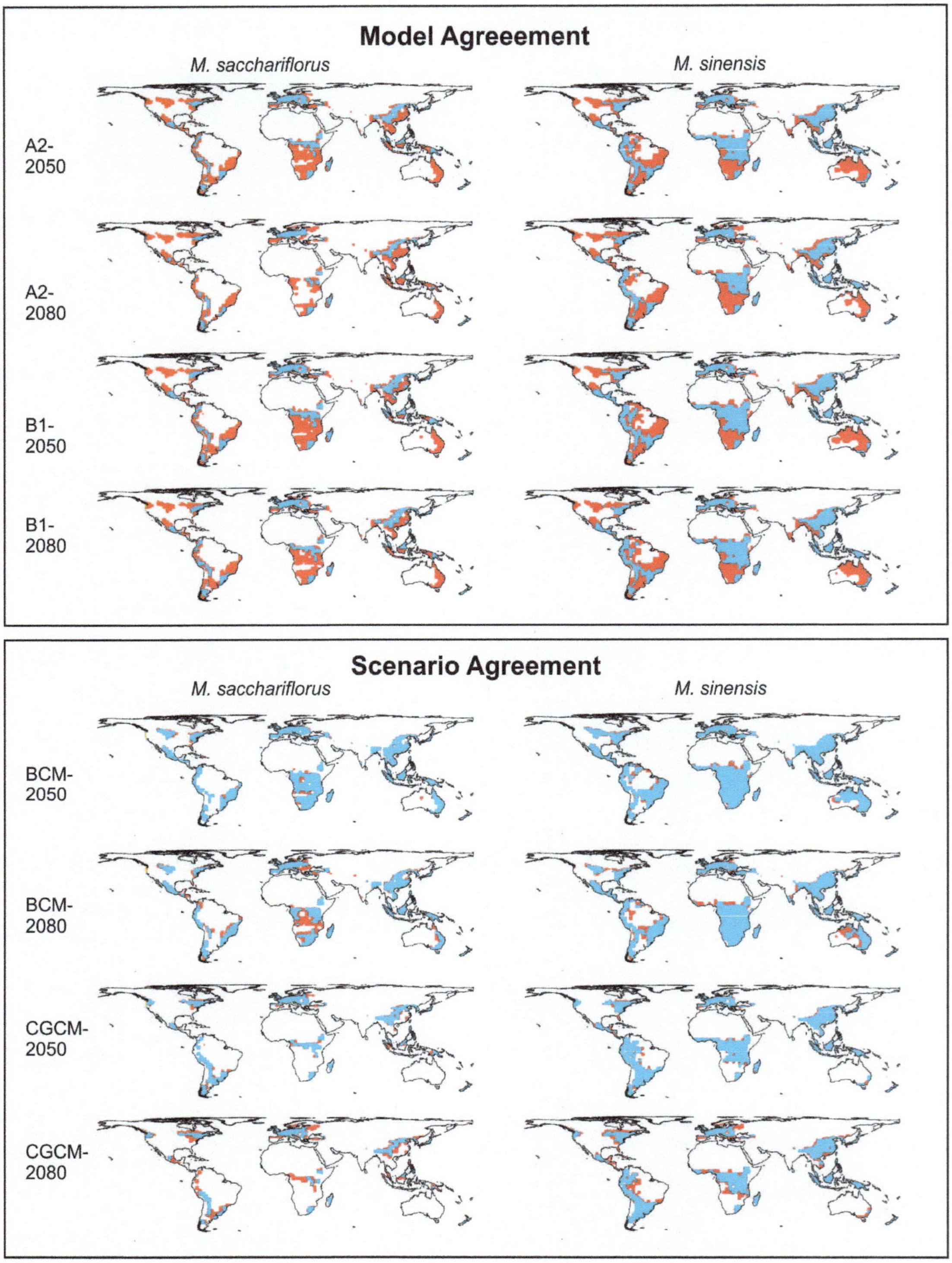

Figure 3. Areas of agreement and disagreement in the projected potential distribution of *Miscanthus sacchariflorus* and *M. sinensis* (all area classified as suitable, favourable, and highly favourable, or EI >10) as modeled by CLIMEX. The upper model shows areas of agreement between the BCM and CGCM models for a given scenario and year. The lower panel shows areas of agreement between the A2 and B1 scenarios for a given model and year. Areas shown in blue indicate suitable climate under both projections; areas shown in red indicate that only one of the two projections predicts suitable climate.

both horticultural and agricultural materials, which reduces dispersal limitations, and the potential for plant breeding programs to introduce new genetic material with a wider range of trait variation, proliferation within and beyond the introduced area should be monitored closely.

According to the native distributions and our models, *M. sinensis* has a wider range and greater climatically suitable area than *M. sacchariflorus* [37] (Fig. 1). Although these characteristics could make *M. sinensis* attractive for cultivation across a wide area, species that have wider native distributions and occur in more

Table 5. The index of agreement[a] between different models run under the same scenario and between different scenarios run within the same model, for projected distributions of *M. sacchariflorus* and *M. sinensis* in 2050 and 2080.

Model agreement:	***2050***	***2080***
M. sacchariflorus–A2	0.58	0.44
M. sacchariflorus–B1	0.64	0.63
M. sinensis–A2	1.17	0.96
M. sinensis–B1	1.21	1.22
Scenario agreement:		
M. sacchariflorus–BCM	13.68	3.23
M. sacchariflorus–CGCM	5.28	1.33
M. sinensis–BCM	16.35	6.25
M. sinensis–CGCM	11.74	3.25
Model average:	0.90	0.81
Scenario average:	11.76	3.51

[a]The index is calculated as the ratio of the area of overlap between the two projections to the area of non-overlap.

habitats and climate zones are likely to be more successful as invaders, regardless of other biological traits [71]. Thus, *M. sinensis* might also have greater potential to become a weedy or invasive plant than does *M. sacchariflorus*. Rather than developing one or two cultivars suitable for widespread production, a best management practice might be to develop regionally restricted cultivars to minimize widespread escape and invasion of novel species from the agriculture and horticulture trades.

Consequences of Parameter Sensitivity

The smaller native distribution and therefore narrower environmental tolerances of *M. sacchariflorus* likely contribute to its greater sensitivity to changes in model parameters than for *M. sinensis*. If the extent of the native distribution is correlated with model sensitivity across a range of species, this would have implications for modelling and interpreting models for both invasive species and rare species of conservation concern. For potential invaders, this might mean that some priority is given to species with wide distributions. For rare species, i.e., those with small native distributions that are not due to anthropogenically caused local extinctions, obtaining accurate model results could require greater accuracy in parameter estimation.

The two *Miscanthus* species models showed sensitivity to similar parameters, which might not be surprising, given that their distributions overlap in temperate areas. However, the main sets of parameters exhibiting sensitivity were those for which there are the least data. The most sensitive parameters were related to upper temperatures and heat tolerance, but most studies of temperature-related growth for these species have examined cold tolerance because of interest in their cultivation at northern latitudes (e.g., [39], [38]). Physiological heat thresholds remain to be explored for these species to improve the confidence of lower-latitude thresholds for growth in the northern hemisphere, where they have been introduced, as well as potential range contractions at lower latitudes under climate change.

Similarly, although weakly sensitive and thus potentially of lesser importance the upper temperature parameters, upper soil moistures and moisture tolerance have rarely been examined. Most studies of soil moisture effects for these species examine drought, rather than saturation (e.g., [45], but see [50]). The accuracy of these parameter estimates could be important in predicting potential invasion of these species into drainage ditches, riparian areas, and wetlands.

Most stress rate parameters were relatively insensitive to changes in value. CLIMEX determines stress as the annual exponential accumulation of weekly population reduction when a stress threshold is exceeded [34]. Stress accumulation rate is difficult to estimate empirically without extensive field or laboratory trials under various stress thresholds, and the magnitude of the accumulation rate could depend on the threshold value chosen; for many species, few data of this type likely exist [72]. However, the minimal sensitivity of the stress rate parameters implies that their accuracy is less influential than that of other, more easily estimated parameters.

Two previous tests of sensitivity using CLIMEX have some similarities. A study of the invasive tropical/subtropical shrub *Lantana camara* identified model sensitivity to limiting low and high temperatures and limiting low soil moisture [27]. A study of the invading pathogen *Phytophthora ramorum* identified model sensitivity to optimal high temperature and limiting and optimal low soil moisture [26]. However, neither study tested model sensitivity to the stress rate parameters. Nevertheless, both our and their models show high sensitivity to some of the limiting upper or lower temperature and moisture parameters. Sensitivity analyses should be performed for additional species to determine whether some parameters are consistently more sensitive than others. If sensitivity to specific parameters is consistent within biomes and species types (forb, shrub, etc.), researchers could focus their efforts on measuring those specific environmental tolerances to maximize model estimation accuracy.

Future Climate Projections

Although the climatically suitable area for the two *Miscanthus* species is projected to decrease globally with climate change, areas of North America, eastern Europe, and Scandinavia are projected to experience some future increase in suitable climate. This could be beneficial for cultivating these species as bioenergy crops in these regions if suitable habitat is available. However, it could also place these regions at greater risk of invasion through increases in the area of suitable climate outside of cultivation. These regions, in particular, are projected to be future hotspots of invasion for 99 of the worst invaders globally [73]. Areas of range contraction for the two *Miscanthus* species also coincide with areas where future invasion is projected to decrease [73].

Additionally, climate niche projections do not account for the potential that rapid evolution in introduced species and their recipient communities could allow species to become invasive beyond their current tolerances (e.g., [74], [75], [76], [77]). In these *Miscanthus* species, rapid evolution could be aided by the introduction of new horticultural genotypes from widely separated populations in the native range [78]. These species are obligate out-crossers [79], so isolated populations composed of a single clone do not produce seed. Introduction of different genotypes could increase the probability of sexual reproduction and long-distance spread via the wind-dispersed seed [80]. Similarly, if these plants are developed for biomass production, intensive plant breeding programs will aim to improve their performance under a variety of conditions, including resistance to pests and disease, drought and heat tolerance, cold tolerance, and possibly salinity tolerance [81]. These efforts will potentially expand the plants' realized distributions as well as their invasive potential.

Variation in the area of future projected climate was greater among climate models than among emissions scenarios. This has also been found previously, when quantified, for both plant (e.g., [82], [83], [84]) and insect (e.g., [31], [32]) species under a number of different model-scenario combinations. Coupled with the observation that some currently observed climate changes might be greater than those predicted by even the highest emissions scenarios [85], this result suggests that future bioclimatic envelope model projections should focus efforts on increasing the number of models compared using one high-emissions scenario to develop composite projections of future suitable climate areas.

Acknowledgments

We thank Emily Robinson-Berzitis and Filip Cybula for formulating the climate model data. We also thank two referees for helpful comments on an earlier version of the manuscript.

Author Contributions

Conceived and designed the experiments: HAH JAN. Performed the experiments: HAH SES. Analyzed the data: HAH SES ZG. Wrote the paper: HAH SES.

References

1. Mack R, Erneberg M (2002) The United States naturalized flora: largely the product of deliberate introductions. Ann Mo Bot Gard 89: 176–189.
2. Williamson M, Fitter A (1996) The varying success of invaders. Ecology 77: 1661–1666.
3. Pheloung PC, Williams PA, Halloy SR (1999) A weed risk assessment model for using as a biosecurity tool evaluating plant introductions. J Environ Manage 57: 239–251.
4. IPPC (2007) International Standards for Phytosanitary Measures: Framework for Pest Risk Analysis. ISPM No 2. Rome: Food and Agriculture Organization.
5. Koop A, Fowler L, Newton L, Caton B (2012) Development and validation of a weed screening tool for the United States. Biol Invasions 14: 273–294.
6. Baker RHA, Benninga J, Bremmer J, Brunel S, Dupin M, et al. (2012) A decision-support scheme for mapping endangered areas in pest risk analysis. Bulletin OEPP/EPPO Bulletin 42: 65–73.
7. Ewel JJ, O'Dowd DJ, Bergelson J, Daehler CC, D'Antonio CM, et al. (1999) Deliberate introductions of species: research needs - benefits can be reaped, but risks are high. Bioscience 49: 619–630.
8. Raghu S, Anderson RC, Daehler CC, Davis AS, Wiedenmann RN, et al. (2006) Adding biofuels to the invasive species fire? Science 313: 1742–1742.
9. Venette RC, Kriticos DJ, Magarey RD, Koch FH, Baker RHA, et al. (2010) Pest risk maps for invasive alien species: a roadmap for improvement. Bioscience 60: 349–362.
10. Atkinson CJ (2009) Establishing perennial grass energy crops in the UK: a review of current propagation options for *Miscanthus*. Biomass Bioenergy 33: 752–759.
11. Quinn LD, Allen DJ, Stewart JR (2010) Invasiveness potential of *Miscanthus sinensis:* implications for bioenergy production in the United States. Glob Change Biol Bioenergy 2: 310–320.
12. Pohl RW (1963) Phytogeographic notes on *Rottboellia, Paspalum,* and *Miscanthus* (Gramineae). Rhodora 65: 146–147.
13. Dore WG, McNeill J (1980) Grasses of Ontario. Agriculture Canada Monograph 26. Hull, Quebec, Canada: Canadian Government Publishing Centre.
14. Nishiwaki A, Mizuguti A, Kuwabara S, Toma Y, Ishigaki G, et al. (2011) Discovery of natural *Miscanthus* (Poaceae) triploid plants in sympatric populations of *Miscanthus sacchariflorus* and *Miscanthus sinensis* in southern Japan. Am J Bot 98: 154–159.
15. Dougherty RF, Quinn LD, Endres AB, Voigt TB, Barney JN (2014) Natural history survey of the ornamental grass *Miscanthus sinensis* in the introduced range. Invasive Plant Sci Manag 7: in press.
16. Clifton-Brown JC, Lewandowski I, Andersson B, Basch G, Christian DG, et al. (2001) Performance of 15 *Miscanthus* genotypes at five sites in Europe. Agron J 93: 1013–1019.
17. Jorgensen U, Mortensen J, Kjeldsen JB, Schwarz KU (2003) Establishment, development and yield quality of fifteen *Miscanthus* genotypes over three years in Denmark. Acta Agric Scand B Soil Plant Sci 53: 190–199.
18. Pyter R, Heaton E, Dohleman F, Voigt T, Long S (2009) Agronomic experiences with *Miscanthus x giganteus* in Illinois, USA. In Mielenz J, editor. Biofuels: methods and protocols. New York: Springer. 41–52.
19. Barney JN, DiTomaso JM (2008) Nonnative species and bioenergy: are we cultivating the next invader? Bioscience 58: 64–70.
20. Holdridge LR (1947) Determination of world plant formations from simple climatic data. Science 105: 367–368.
21. Woodward FI (1987) Climate and plant distribution. Cambridge: Cambridge University Press.
22. Gordon DR, Gantz CA (2008) Screening new plant introductions for potential invasiveness: a test of impacts for the United States. Conservation Letters 1: 227–235.
23. Scott JK, Panetta FD (1993) Predicting the Australian weed status of southern African plants. J Biogeog 20: 87–93.
24. Sutherst RW, Maywald GF (1985) A computerized system for matching climates in ecology. Agric Ecosyst Environ 13: 281–299.
25. Beerling DJ, Huntley B, Bailey JP (1995) Climate and the distribution of *Fallopia japonica:* use of an introduced species to test the predictive capacity of response surfaces. J Veg Sci 6: 269–282.
26. Venette RC, Cohen SD (2006) Potential climatic suitability for establishment of *Phytophthora ramorum* within the contiguous United States. For Ecol Manage 231: 18–26.
27. Taylor S, Kumar L (2012) Sensitivity analysis of CLIMEX parameters in modelling the potential distribution of *Lantana camara L.* Plos One 7(7): 1–16.
28. Sutherst RW, Baker RHA, Coakley SM, Harrington R, Kriticos DJ, et al. (2007) Pests under global change - meeting your future landlords? In: Canadell JG, Pataki DE, Pitelka LF, editors. Terrestrial ecosystems in a changing world. Berlin: Springer. 211–223.
29. Walther G, Roques A, Hulme PE, Sykes MT, Pysek P, et al. (2009) Alien species in a warmer world: risks and opportunities. Trends Ecol Evol 24: 686–693.
30. Dukes JS, Mooney HA (1999) Does global change increase the success of biological invaders? Trends Ecol Evol 14: 135–139.
31. Mika AM, Weiss RM, Olfert O, Hallett RH, Newman JA (2008) Will climate change be beneficial or detrimental to the invasive swede midge in North America? Contrasting predictions using climate projections from different general circulation models. Glob Change Biol 14: 1721–1733.
32. Mika AM, Newman JA (2010) Climate change scenarios and models yield conflicting predictions about the future risk of invasive species in North America. Agric For Entomol 12: 213–221.
33. Commonwealth Scientific and Industrial Research Organisation (2004) Dymex simulator application 2.0. South Yarra, Australia: Hearn Scientific Software.
34. Sutherst RW, Maywald GF, Kriticos DJ (2006) CLIMEX Version 3 User's Guide. Melbourne: CSIRO.
35. New M, Hulme M, Jones P (1999) Representing twentieth-century space-time climate variability. Part I: development of a 1961–90 mean monthly terrestrial climatology. J Clim 12: 829–856.
36. Sutherst RW (2003) Prediction of species geographical ranges. J Biogeog 30: 805–816.
37. Hodkinson TR, Chase MW, Lledo MD, Salamin N, Renvoize SA (2002) Phylogenetics of *Miscanthus, Saccharum* and related genera (*Saccharinae, Andropogoneae, Poaceae*) based on DNA sequences from ITS nuclear ribosomal DNA and plastic trnL intron and trnL-F intergenic spacers. J Plant Res 115: 381–392.
38. Yan J, Chen W, Lou F, Ma H, Meng A, et al. (2011) Variability and adaptability of *Miscanthus* species evaluated for energy crop domestication. Glob Change Biol Bioenergy 4: 49–60.

39. Farrell AD, Clifton-Brown JC, Lewandowski I, Jones MB (2006) Genotypic variation in cold tolerance influences the yield of *Miscanthus*. Ann Appl Biol 149: 337–345.
40. Naidu SL, Long SP (2004) Potential mechanisms of low-temperature tolerance of C4 photosynthesis in *Miscanthus* x *giganteus*: an in vivo analysis. Planta 220: 145–155.
41. Farage PK, Blowers D, Long SP, Baker NR (2006) Low growth temperatures modify the efficiency of light use by photosystem II for CO2 assimilation in leaves of two chilling-tolerant C4 species, *Cyperus longus* L. and *Miscanthus* x *giganteus*. Plant Cell Environ 29: 720–728.
42. Quinn LD, Stewart JR, Yamada T, Toma Y, Saito M, et al. (2012) Environmental tolerances of *Miscanthus sinensis* in invasive and native populations. Bioenerg Res 5(1): 139–148.
43. Naidu SL, Moose SP, Al-Shoaibi AK, Raines CA, Long SP (2003) Cold tolerance of C_4 photosynthesis in *Miscanthus* x *giganteus*: adaptation in amounts and sequence of C_4 photosynthetic enzymes. Plant Physiol 132: 1688–1697.
44. Clifton-Brown J, Lewandowski I (2000) Overwintering problems of newly established *Miscanthus* plantations can be overcome by identifying genotypes with improved rhizome cold tolerance. New Phytol 148: 287–294.
45. Clifton-Brown JC, Lewandowski I, Bangerth F, Jones MB (2002) Comparative responses to water stress in stay-green, rapid- and slow senescing genotypes of the biomass crop, *Miscanthus*. New Phytol 154: 335–345.
46. Daubenmire RF (1974) Plants and environment: a textbook of plant autecology. Sydney, Australia: Wiley.
47. Kriticos DJ, Sutherst RW, Brown JR, Adkins SW, Maywald GF (2003) Climate change and biotic invasions: a case history of a tropical woody vine. Biol Invasions 5: 145–165.
48. Hastings A, Clifton-Brown J, Wattenbach M, Mitchell CP, Smith P (2009) The development of MISCANFOR, a new *Miscanthus* crop growth model: towards more robust yield predictions under different climatic and soil conditions. Glob Change Biol Bioenergy 1: 154–170.
49. Yamasaki S (1990) Potential dynamics in overlapping zones of *Phragmites australis* and *Miscanthus sacchariflorus*. Aquat Bot 36: 367–377.
50. Li F, Qin X, Xie Y, Chen X, Hu J, et al. (2013) Physiological mechanisms for plant distribution pattern: responses to flooding and drought in three wetland plants from Dongting Lake, China. Limnology 14: 71–76.
51. Sutherst RW, Maywald GF (2005) A climate model of the red imported fire ant, *Solenopsis invicta* Buren (Hymenoptera: Formicidae): implications for invasion of new regions, particularly Oceania. Environ Entomol 34(2): 317–335.
52. Randall DA, Wood RA, Bony S, Colman R, Fichefet T, et al. (2007) Climate models and their evaluation. In Solomon S, Qin D, Manning M, Chen Z, Marquis M, Averyt KB, Tignor M, Miller HL, editors. Climate Change 2007: The Physical Science Basis. Contribution of Working Group I to the Fourth Assessment Report of the Intergovernmental Panel on Climate Change. Cambridge: Cambridge University Press. 589–662.
53. IPCC (2000) IPCC Special Report: emissions scenarios. Summary for policy makers. Geneva: Intergovernmental Panel on Climate Change.
54. Guisan A, Thuiller W (2005) Predicting species distribution: offering more than simple habitat models. Ecol Lett 8: 993–1009.
55. Heikkinen RK, Luoto M, Araujo MB, Virkkala R, Thuiller W, et al. (2006) Methods and uncertainties in bioclimatic envelope modelling under climate change. Prog Phys Geogr 30: 751–777.
56. Araújo MB, Peterson AT (2012) Uses and misuses of bioclimatic envelope modeling. Ecology 93: 1527–1539.
57. Horton JL, Fortner R, Goklany M (2010) Photosynthetic characteristics of the C4 invasive exotic grass *Miscanthus sinensis* Andersson growing along gradients of light intensity in the southeastern United States. Castanea 75: 52–66.
58. Keane R, Crawley M (2002) Exotic plant invasions and the enemy release hypothesis. Trends Ecol Evol 17: 164–170.
59. Ross JL, Ivanova ES, Severns PM, Wilson MJ (2010) The role of parasite release in invasion of the USA by European slugs. Biol Invasions 12: 603–610.
60. Schaffner U, Ridenour WM, Wolf VC, Bassett T, Mueller C, et al. (2011) Plant invasions, generalist herbivores, and novel defense weapons. Ecology 92: 829–835.
61. Parker MA (2001) Mutualism as a constraint on invasion success for legumes and rhizobia. Divers Distrib 7: 125–136.
62. Broennimann O, Treier UA, Mueller-Schaerer H, Thuiller W, Peterson AT, et al. (2007) Evidence of climatic niche shift during biological invasion. Ecol Lett 10: 701–709.
63. Gallagher RV, Beaumont LJ, Hughes L, Leishman MR (2010) Evidence for climatic niche and biome shifts between native and novel ranges in plant species introduced to Australia. J Ecol 98: 790–799.
64. Guo Q, Sax DF, Qian H, Early R (2012) Latitudinal shifts of introduced species: possible causes and implications. Biol Invasions 14: 547–556.
65. Petitpierre B, Kueffer C, Broennimann O, Randin C, Daehler C, et al. (2012) Climatic niche shifts are rare among terrestrial plant invaders. Science 335: 1344–1348.
66. Kearney M, Porter W (2009) Mechanistic niche modelling: combining physiological and spatial data to predict species' ranges. Ecol Lett 12: 334–350.
67. Kearney MR, Wintle BA, Porter WP (2010) Correlative and mechanistic models of species distribution provide congruent forecasts under climate change. Conserv Lett 3: 203–213.
68. Flory SL, Lorentz KA, Gordon DR, Sollenberger LE (2012) Experimental approaches for evaluating the invasion risk of biofuel crops. Environ Res Lett 7: 045904.
69. Pysek P, Sadlo J, Mandak B, Jarosik V (2003) Czech alien flora and the historical pattern of its formation: what came first to Central Europe? Oecologia 135: 122–130.
70. Huang QQ, Qian C, Wang Y, Jia X, Dai XF, et al. (2010) Determinants of the geographical extent of invasive plants in China: effects of biogeographical origin, life cycle and time since introduction. Biodivers Conserv 19: 1251–1259.
71. Pysek P, Jarosik V, Pergl J, Randall R, Chytry M, et al. (2009) The global invasion success of Central European plants is related to distribution characteristics in their native range and species traits. Divers Distrib 15: 891–903.
72. Sutherst RW, Maywald GF, Bourne AS (2007) Including species interactions in risk assessments for global change. Global Change Biol 13: 1843–1859.
73. Bellard C, Thuiller W, Leroy B, Genovesi P, Bakkenes M, et al. (2013) Will climate change promote future invasions? Global Change Biol 19: 3740–3748.
74. Broennimann O, Treier UA, Mueller-Schaerer H, Thuiller W, Peterson AT, et al. (2007) Evidence of climatic niche shift during biological invasion. Ecol Lett 10: 701–709.
75. Whitney KD, Gabler CA (2008) Rapid evolution in introduced species, 'invasive traits' and recipient communities: challenges for predicting invasive potential. Divers Distrib 14: 569–580.
76. Xu C, Julien MH, Fatemi M, Girod C, Van Klinken RD, et al. (2010) Phenotypic divergence during the invasion of *Phyla canescens* in Australia and France: evidence for selection-driven evolution. Ecol Lett 13: 32–44.
77. Sultan SE, Horgan-Kobelski T, Nichols LM, Riggs CE, Waples RK (2013) A resurrection study reveals rapid adaptive evolution within populations of an invasive plant. Evol Appl 6: 266–278.
78. Lavergne S, Molofsky J (2007) Increased genetic variation and evolutionary potential drive the success of an invasive grass. Proc Natl Acad Sci U S A 104: 3883–3888.
79. Heaton EA, Dohleman FG, Miguez AF, Juvik JA, Lozovaya V, et al. (2010) *Miscanthus*: a promising biomass crop. In: Kader JC, Delseny M, editors. Advances in botanical research. Volume 56. London: Academic Press. 76–138.
80. Quinn LD, Matlaga DP, Stewart JR, Davis AS (2011) Empirical evidence of long-distance dispersal in *Miscanthus sinensis* and *Miscanthus* x *giganteus*. Invasive Plant Sci Manag 4: 142–150.
81. Chapman SC, Chakraborty S, Dreccer MF, Howden SM (2012) Plant adaptation to climate change-opportunities and priorities in breeding. Crop Pasture Sci 63: 251–268.
82. Barney JN, DiTomaso JM (2010) Bioclimatic predictions of habitat suitability for the biofuel switchgrass in North America under current and future climate change scenarios. Biomass Bioenergy 34: 124–133.
83. Kriticos DJ, Watt MS, Potter KJB, Manning LK, Alexander NS, et al. (2011) Managing invasive weeds under climate change: considering the current and potential future distribution of *Buddleja davidii*. Weed Res 51: 85–96.
84. Bourdôt G, Lamoureaux S, Watt M, Manning L, Kriticos D (2012) The potential global distribution of the invasive weed *Nassella neesiana* under current and future climates. Biol Invasions 14: 1545–1556.
85. Rahmstorf S, Cazenave A, Church JA, Hansen JE, Keeling RF, et al. (2007) Recent climate observations compared to projections. Science 316: 709–709.

9

Gender Specific Reproductive Strategies of an Arctic Key Species (*Boreogadus saida*) and Implications of Climate Change

Jasmine Nahrgang[1,2]*, Øystein Varpe[2,3], Ekaterina Korshunova[1,3], Svetlana Murzina[4], Ingeborg G. Hallanger[1], Ireen Vieweg[1], Jørgen Berge[1,2]

1 Department of Arctic and Marine Biology, UiT The Arctic University of Norway, Tromsø, Norway, 2 University Centre in Svalbard, Longyearbyen, Norway, 3 Akvaplan-niva, Fram Centre, Tromsø, Norway, 4 Institute of Biology of the Karelian Research Centre, Russian Academy of Science, Petrozavodsk, Russia

Abstract

The Arctic climate is changing at an unprecedented rate. What consequences this may have on the Arctic marine ecosystem depends to a large degree on how its species will respond both directly to elevated temperatures and more indirectly through ecological interactions. But despite an alarming recent warming of the Arctic with accompanying sea ice loss, reports evaluating ecological impacts of climate change in the Arctic remain sparse. Here, based upon a large-scale field study, we present basic new knowledge regarding the life history traits for one of the most important species in the entire Arctic, the polar cod (*Boreogadus saida*). Furthermore, by comparing regions of contrasting climatic influence (domains), we present evidence as to how its growth and reproductive success is impaired in the warmer of the two domains. As the future Arctic is predicted to resemble today's Atlantic domains, we forecast changes in growth and life history characteristics of polar cod that will lead to alteration of its role as an Arctic keystone species. This will in turn affect community dynamics and energy transfer in the entire Arctic food chain.

Editor: Brian R. MacKenzie, Technical University of Denmark, Denmark

Funding: The work is partly financed by the Norwegian Research Council through the EWMA (nr 195160) and Polarisation (nr 214184/F20) projects, UiT - The Arctic University of Norway (Tromsø), Arctic Field Grant (RiS 5148, Svalbard Science Forum), Barents secretariat, Akvaplan-niva and the President of the Russian Federation (Grant NSh-1410.2014.4). Fieldwork and the time allocated for JB and ØV was supported by the NFR funded project CircA (nr 214271/F20). The funders had no role in study design, data collection and analysis, decision to publish, or preparation of the manuscript.

Competing Interests: The authors have declared that no competing interests exist.

* E-mail: jasmine.m.nahrgang@uit.no

Introduction

Climate variability and global warming have changed the Arctic, most notably seen in the abrupt decline in Arctic sea ice extent and thickness [1]. As the temperature is predicted to rise disproportionally more in the Arctic compared to global trends [2], these current and future perspectives are likely to pose a significant challenge for both the marine [3–5] as well as the terrestrial [6] Arctic ecosystems. Altered marine community structures [7–9] and a northward expansion of boreal species [10,11] are common expectations, to some extent also supported by observations [12]. When species expand their distribution ranges northwards, it may lead to either co-existence with, or replacement of their native Arctic counterparts [13]. Mechanisms involved in such range expansions are often directly associated with changes in water masses [14,15], but may also be more indirect through biological interactions [3,8]. Commonly predicted replacements include the high Arctic calanoid *Calanus glacialis* being replaced by its more boreal sister taxa, *Calanus finmarchicus* [16,17], as well as the replacement of the polar bear (*Ursus maritimus*) by killer whale (*Orcinus orca*) as a likely effect of increased temperatures and loss of sea ice in the Arctic [18].

In high Arctic marine ecosystems, the polar cod (*Boreogadus saida*) is regarded as a key link in the food web between lower and higher trophic levels [19,20]. Polar cod has been a focal point for studies of climate [11,21–23] and pollution impacts [24,25]. Yet, current knowledge regarding this important Arctic species is surprisingly scattered and inconclusive, preventing a holistic understanding of its life history and ecology. Understanding life history strategies, the schedules of growth, fecundity and mortality [26], is imperative for our ability to predict species' responses to environmental change. Importantly, there are tight couplings among altered energy acquisition, survival, and consequently responses in energy allocation [27]. We need to cover these dimensions to understand how polar cod will respond to environmental changes.

Through a large field study of wild polar cod populations from comparable shelf areas influenced by either Arctic or Atlantic water masses, hereafter called domains (Figure 1, Table S1), we established a natural large-scale climate experiment. We use this design to fill important gaps in knowledge regarding life history traits and reproductive strategies, and to test if they are stable across climatic domains and genders. Specifically, we ask if population structure, size-at-age, age-at-maturation, and reproductive strategy (including semelparity versus iteroparity) are similar across the two domains. Furthermore, we discuss potential direct and cascading ecological effects from a continued warming of the Arctic Ocean for arguably the most abundant and

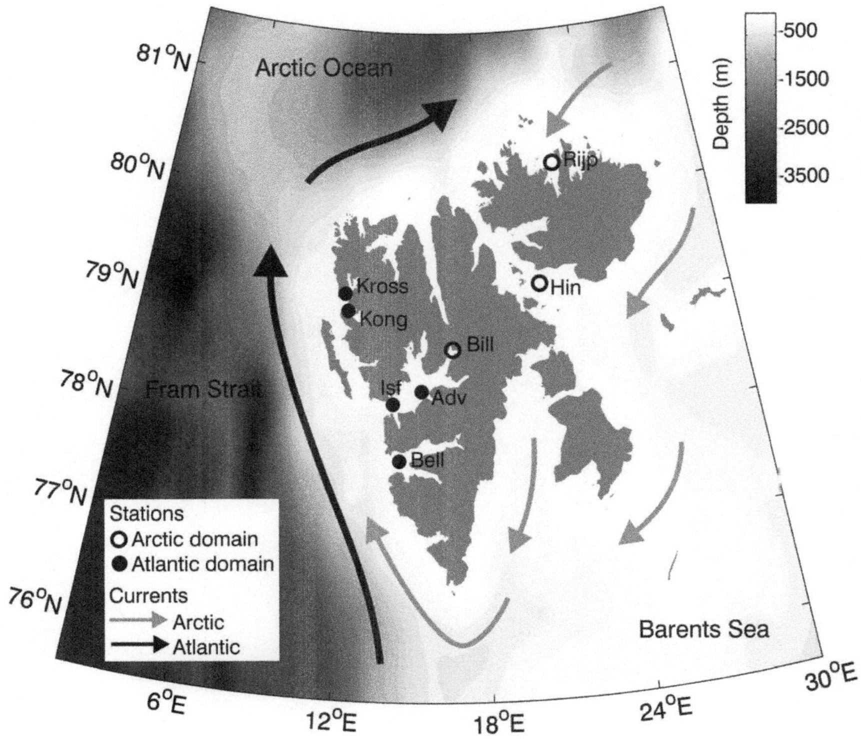

Figure 1. Sampling stations of Arctic and Atlantic domains. Map of the sampling stations with dominant current systems (Bell: Bellsund, Isf: Isfjorden, Adv: Adventfjorden, Bill: Billefjorden, Kong: Kongsfjorden, Kross: Krossfjorden, Hin: Hinlopen, Rijp: Rijpfjorden).

ecologically important fish species within the high Arctic food chain.

Material and Methods

Ethic statement

All work were performed according to and within the regulations enforced by the Norwegian Animal welfare authorities and no specific permissions were required. The R/V *Helmer Hanssen* is owned by the University of Tromsø and has all necessary autorization from the Norwegian Fisheries Directorate to use a bottom trawl to collect fish for scientific purposes. The lead author had all necessary training and certificates (FELASA C) to perform the work including sacrificing the sampled organisms. Furthermore, the organisms are neither protected nor endangered in the coastal waters of the Svalbard Archipelago. Upon trawling, polar cod were sacrificed by a sharp blow to the head and immediately dissected as specified below.

Study sites and polar cod sampling

Polar cod were collected using a Campelen bottom trawl from R/V *Helmer Hanssen* between 2010 and 2013 from 8 stations (Figure 1, see also Table S1 for a detailed overview). Trawling time was standardised to 15 min bottom time for all trawl hauls used in population estimates (Table 1), but reduced to 5 min in Adventfjorden. The opening of the trawl is 60 m across, the trawling speed was set to 2 kt. In order to make the catches comparable, trawling was carried out at a depth between 180 and 220 m bottom (with one exception in Bellsund, 130 m), depending on the stations and on comparable bottom types. For the samples taken in January 2012 (Rijpforden and Isfjorden) and in January and September 2013 (Rijpfjorden, Isfjorden, Krossfjorden and Kongsfjorden), 100% of the trawl catches (all species and specimens included) were considered. For all other stations (Table S1), only random sub-samples of polar cod were treated. Also, due to the failure to adequately collect pelagic early life stages with the bottom trawl [11], larvae and juvenile fish of total length <10 cm were removed from subsequent analyses.

Oceanographic data (Figure 2) were gathered from two moorings deployed in Kongsfjorden and Rijpfjorden, respectively (see [28] for details). The moorings were located in the immediate vicinity (1 km apart) to the trawling grounds in both fjords, and have been in operation since 2002 as part of a long term monitoring program.

The polar cod total length (TL, ±0.1 cm) and/or fork length (FL, ±0.1 cm), total wet weight (TW, ±0.1 g wet weight) and sex were recorded for all collected individuals. TW was recorded for a large fraction of the individuals (n = 1124, Figure S1). Somatic weight (0.1 g wwt), gonad and liver weight (±0.01 g wwt) were recorded when possible in order to calculate the gonado-somatic (GSI, n = 314, Table 2) and hepato-somatic index (HSI, n = 266, Table S2) respectively. Otoliths were collected for age determination from the dissected subsamples. In addition, otoliths were collected non-randomly to cover all size ranges (n = 296). Stomachs were dissected out from a subsample of specimens collected in Rijpfjorden and Kongsfjorden in September 2013 (Table S1) and the fresh material was examined under a dissecting microscope. Identification of prey items was made to the lowest possible taxon based on degree of digestion. Results are presented as frequency of occurrence (Table 3).

All presented data are based on total length measurements. When only fork length (FL) was available, TL was determined with help of a linear regression FL = 0.9641 TL+0.0113 (R^2 = 0.996), based on 1463 TL and FL records (see Figure S2). The GSI was calculated using the equation GSI = 100× gonad weight/somatic weight. Gonadal maturation was only considered from January

Table 1. Fish composition of benthic trawls taken in January 2012 and January and September 2013 in fjords from the Arctic and Atlantic domains.

	Arctic				Atlantic				
	Rijp	**Rijp**	**Rijp**	**Hin**	**Adv**	**Isf**	**Kross**	**Isf**	**Kong**
Taxon	Jan 2012	Jan 2013	Sept 2013	Sept 2013	Jan 2012	Jan 2013	Jan 2013	Sept 2013	Sept 2013
Boreogadus saida	***12.6 (1023)***	***14.4 (689)***	***22.3 (1325)***	***8.4 (310)***	***2.4 (306)***	0.6 (958)	1.1 (310)	23.4 (2358)	23.7 (2008)
Gadus morhua	5.5 (22)	0.4 (30)		0.07 (7)	***3.0 (300)***	***5.5 (89)***	***13.6 (163)***	***54 (22)***	***26.5 (7)***
Mallotus villosus	0.1 (6)	0.1 (45)	0.1 (5)	0.1 (7)	0.03 (3)	0.03 (7)	0.03 (2)		0.03 (3)
Leptoclinus maculatus	0.1 (6)	0.1 (28)	0.001 (1)	0.05 (10)	0.03 (3)	0.1 (26)		0.1 (19)	0.03 (5)
Sebastes sp		0.6 (138)	0.03 (4)	0.9 (16)	0.03 (6)	0.7 (26)	1.1 (51)	0.2 (1)	0.5 (7)
Hippoglossoides platessoides	1.3 (4)	0.3 (3)		0.8 (5)		3.8 (43)	0.1 (5)	0.8 (5)	2.4 (13)
Reinhardtius hippoglossoides	0.8 (1)			0.1 (5)	0.03 (9)	***5.3 (1)***	0.1 (1)	14 (24)	3.1 (7)
Liparis sp		0.3 (47)	0.4 (22)	0.02 (8)		0.04 (13)	0.1 (6)		0.02 (2)
Leptagonus decagonus		0.1 (2)		0.001 (1)		0.03 (4)	0.03 (6)		
Melanogrammus aeglefinus						0.1 (1)	1.6 (1)		0.01 (1)
Triglops spp		0.1 (23)		2.0 (2)					
Eumicrotremus spinosus		0.1 (3)		0.04 (4)					
Clupea harengus						0.5 (4)	0.01 (1)		
Lycodes spp						0.1 (4)		0.06 (1)	
Myxocephalus sp		0.1 (8)							

Total weight of each taxon (kg) and number of individuals in brackets. Dominant taxa according to biomass (>30% of total biomass, kg) in bold and italics. Trawl duration differed between stations, hence all numbers are standardized as catch per 15 min trawling time.

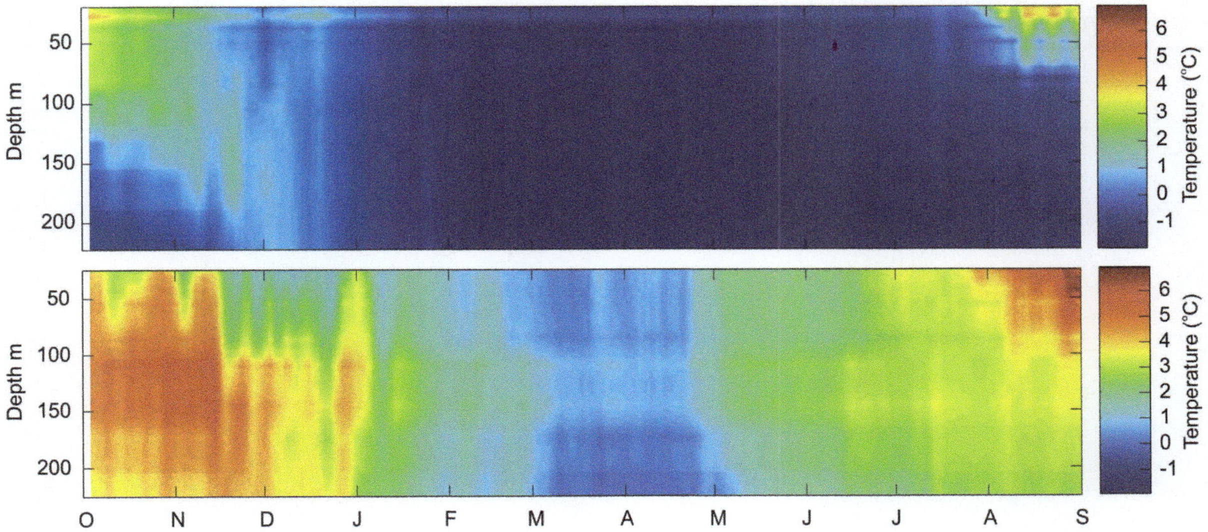

Figure 2. Seasonal temperature plots in an "Atlantic" and "Arctic" type fjord. Temperature plots from moored observatories between 4th of October 2012 and 1st of September 2013 in Rijpfjorden (upper) and Kongsfjorden (lower).

samples, as males are known to increase gonad size several months before females [46], making direct comparison of GSI between genders impossible except in the pre-spawning period in January. As a threshold level for maturity, mature polar cod were defined by GSI>10% (Table 2). For the sex-ratio distributions (Figure 3), the category "immature" were specimens with unidentifiable gonads.

Age estimation

Polar cod age (years) was determined based on otolith readings: for small transparent otoliths, white winter rings were counted in sub-surface light with a Leica M205 C stereo microscope and a Planapo 1.0× objective lens; for all larger otoliths cross sectioning and counting the rings under polarised light was necessary. For individuals from which otoliths were dissected, age was estimated using a length-age relationships (Figure 4). Length (and not weight) was considered the most reliable proxy for age, as total weight is highly depending on maturation and feeding status of each individual fish.

Table 2. Description of the maturing fraction of polar cod.

	Arctic			Atlantic		
TL (cm)	n	Female	Male	n	Female	Male
Percentage maturing						
]9–12]	26	14	83	38	41	65
]12–15]	106	81	95	49	58	73
]15–18]	53	97	100	2	100	100
]18–21]	23	100	100			
Gonado-somatic index of mature specimens						
]9–12]		17.7±7.9^{a}	22.1±2.8^{a}		14.7±1.8	21.4±5.5
]12–15]		20.4±3.9^{b}	26.2±7.2^{a,b}		17.0±5.3	20.9±4.5
]15–18]		20.6±4.6^{b}	28.2±4.6^{b,c}		23.0	25.0
]18–21]		25.5±5.1^{c}	34.4±6.8^{c}			
Estimation of amount eggs per female						
]9–12]		7738			5555	
]12–15]		11813			9220	
]15–18]		19335			24381	
]18–21]		36530				

Percentage maturing individuals (GSI>10%), their gonado-somatic index (mean ±SD) and the estimated number of eggs per female per size class (total length, cm) from the Arctic (n = 207) and Atlantic (n = 107) domains, combined for January 2011, 2012, and 2013. Different letters (a, b, c) represent significant differences (Welch-ANOVA and Dunnett's T3) in GSI between size classes for each sex and domain.

Table 3. Frequency of occurrence (%) of preys in polar cod stomachs.

	Arctic	Atlantic
n examined	298	157
n with content	287	132
Frequency of occurence of prey		
Calanoidae	5%	48%
Themisto libellula	100%	10%
Themisto abyssorum	0%	25%
Krill	1%	13%
Mysids	1%	13%
Fish	0.3%	5%
Polychaeta	0%	2%
Amphipoda	2%	5%
Cumacea	0%	14%
Pandalus borealis	1%	12%

Stomach contents of polar cod sampled in September 2013, from the Arctic (Rijpfjorden) and Atlantic (Kongsfjorden) domains. Only stomachs with content were included in the calculations.

Histological analysis of gonads and estimation of fecundity

Gonads of mature females from January 2011 (n = 5) and January 2012 (n = 4) were analysed histologically to determine oocyte stages, size (diameter and area), and density represented as the percentage of picture area covered by oocytes related to total picture area within the gonads (Table S3). Classical histological methods using hematoxylin and eosin stains were used. Histological sections were studied in the light microscope (50–1000×) Axioskop 40 (Carl Zeiss) with eye lens ×10 and objective lenses ×5, ×10, ×20, ×40, ×100. Sections on the slides were photographed (Pixera Pro 150ES) and analysed with the Videotest programme. No significant differences in oocyte size and developmental stage were found between domains, suggesting that polar cod from both domains were in the same stage of oocyte development. Oocytes in the vitellogenesis stage represented 98.8% of the weight percentage of all oocyte stages, hence fecundity calculations were based solely on this dominating stage (Table S3). Vitellogenic oocytes represented 49±19% (n = 24 slices) of the total volume of the gonads. Fecundity (amount of egg)

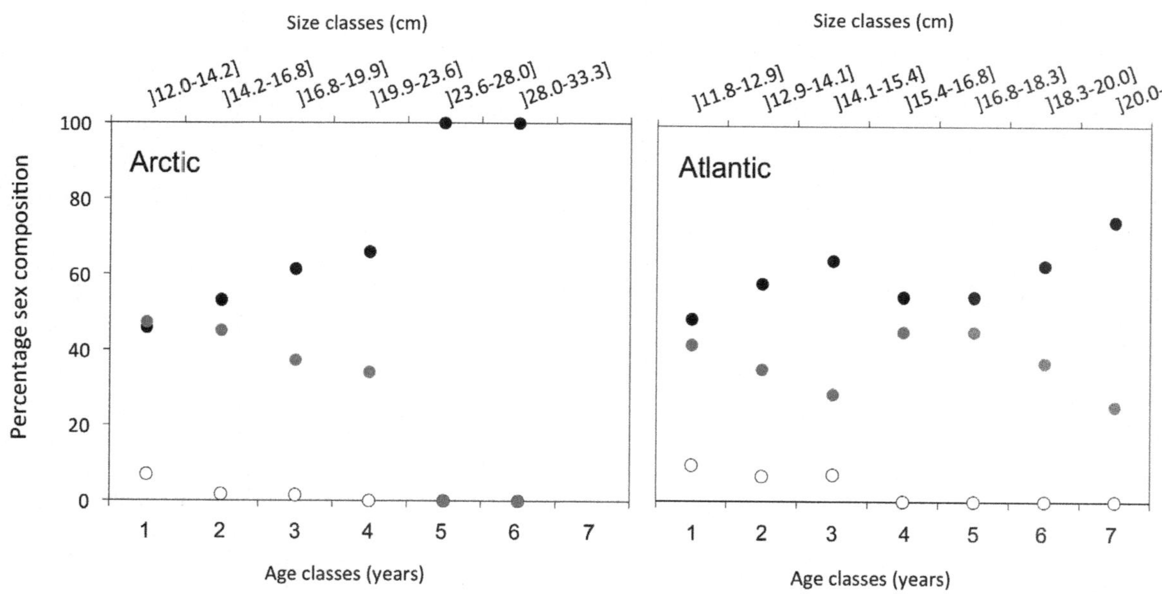

Figure 3. Sex ratio composition of polar cod in Arctic and Atlantic domains. Percentage female (black circles), male (grey circles) and immature (open circles) per age class of polar cod (lower x-axis) and size classes (cm, upper x-axis) from the Arctic (n = 1813) and Atlantic (n = 728) domains. Immature are individuals where gonads were not detectable. Age classes are derived from the linear regression presented in Figure 4, hence length ranges per age class differ for the two domains. Within the Arctic domain, a total of 74 females (max length 30.5 cm) and 30 males (max length 22.9 cm) were recorded larger than 20 cm. In the Atlantic domain only 53 females (max 26.5 cm) and 18 males (max length 25.0 cm) were recorded larger than 20 cm. For more details see Table S4.

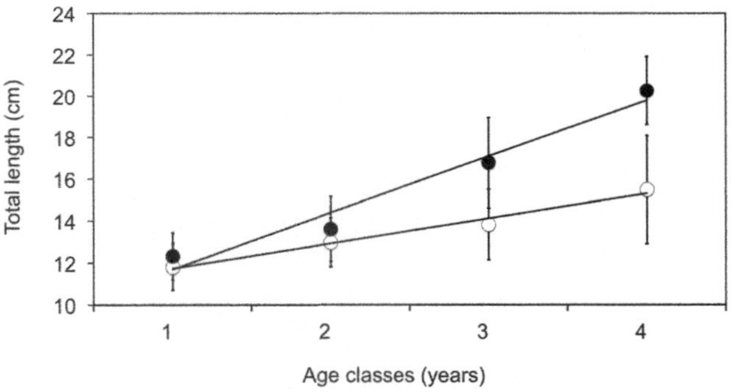

Figure 4. Relationship between total length (cm) and age (otolith based, years) of polar cod. Specimens from the Arctic (n = 97, closed) and Atlantic (n = 199, open) domains. Points indicate mean and wiskers are standard deviation, lines are predicted from a generalized linear model for the two domains where age was treated as a continuous variable. Arctic: TL = 2.710 Age+8.898 (R^2 = 0.96) and Atlantic: TL = 1.0825 Age+10.756 (R^2 = 0.98).

was then estimated based on the proportion of maturing females within size classes in the January 2011, 2012 and 2013 populations and their gonad weight (Table 2). Population fecundity was estimated based on the trawl catches (Figure S3) representing 5 examples of populations in the Arctic and Atlantic domains (see Table 1). The amount of maturing females per size class (N_{f_mat}) in these populations was calculated, using the proportion of females-at-size data (Figure 3) and the proportion of maturing females-at-size data (Table 2) for each domain. The total amount of eggs (Σ_{Neggs}) produced in the entire populations was then estimated based on the number of eggs (N_{eggs}) produced per maturing female-at-size. N_{eggs} assumes that 100% of the vitellogenic oocytes will mature into eggs. Vitellogenic oocyte weight (g) was estimated assuming neutral buoyancy:

$$N_{eggs} = \frac{\text{gonad weight (g)} - \text{at} - \text{size}\bullet 98.8\%\bullet 49\%}{\text{vitellogenic oocyte weight (g)}}$$

$$\text{Total population fecundity} \sum_{Neggs} = \sum_{l=]18-21\text{cm}]}^{l=]9-12\text{cm}]} N_{f\_mat}\bullet N_{eggs}$$

$$\text{Fecundity per female} - \text{at} - \text{size} = \frac{\text{Fecundity population} - \text{at} - \text{size}}{N_{f\_mat}}$$

Statistical analyses

A linear model (LM) was used to predict body length based on age, gender (male and female) and domain (factor with the two levels "Arctic" and "Atlantic", Table S4). Age was treated as a continuous variable. 281 individuals could enter this analyses, that is fish longer than 10 cm in length and younger than 5 years of age (Table S4). For the LM we started out with a full model, including all interactions, and performed model simplification by stepwise removal of non-significant terms as suggested by Crawley [29]. Abundance and HSI were not included as potential explanatory variables due to lack of sufficient data. The LM was performed in R [29].

For calculation of maturity data, we could only include the January samples (see above), reducing sample size to 170 individuals, 66 in the Arctic and 104 in the Atlantic. Instead of including maturity in the linear model we tested for differences in length between mature and immature fish within domain and age-group (T-tests without corrections for multiple comparisons, Figure 5). For mature fish we pooled males and females as gender did not contribute significantly in the larger linear model above.

Potential statistical differences in GSI between gender, size classes or domains (e.g. Table 2) were analysed using IBM SPSS statistics version 19.0. Requirements for normality and homogeneity of variances were not met, therefore the robust tests of equality of means (Welch-ANOVA) was used, followed by a multiple comparison test (Dunnett T3 test) to distinguish specific differences among groups. For all analyses the significance level was set to $p \leq 0.05$.

Results and Discussion

The total dataset consists of 9165 specimens of wild caught polar cod, all of which are included in analyses on population structure (Table 1). In all other analyses, only a subset of the entire data contained the variables of interest, hence sample sizes varies and are smaller than for the population structure analysis (see Table S1). It is important to point out that we have taken the opportunity to gather all existing data from polar cod made available from scientific cruises around Svalbard during the last 4 years to provide a more holistic and general prospective on life history characteristics of polar cod. Due to this, the sampling design was often heterogenous and not appropriate to explain the mechanisms behind the observed trends. We do, however, highlight potential hypotheses for the observed patterns.

Climatic domains and polar cod population structure

The two climatic domains of the present study were defined based on oceanographical characteristics as well as the diversity of potential prey and predators. The high Arctic archipelago of Svalbard is characterized by water masses of both Atlantic and Arctic origin (Figure 2) [30]. Rijpfjorden as well as Billefjorden [31] and the southern part of Hinlopen are predominantly influenced by Arctic water masses due to their location (Figure 2), and are therefore referred to as "Arctic-type" fjords [30]. In contrast, the fjords on the west coast of Svalbard (Bellsund, Isfjorden, Adventfjorden, Kongsfjorden and Krossfjorden) are largely dominated by warm water masses without formation of seasonal sea ice (referred to as 'Atlantic type' fjord [28,32], Figure 2). It is important to note that in our approach, we do not consider if temperature is directly influencing the populations of polar cod (see also below). Rather, temperature is used as an indicator of differences in both the physical and biological habitat of the two domains.

The Atlantic domains are characterized by higher presence of potential predators on polar cod (Table 1), first of all the Atlantic cod (*Gadus morhua*) [33,34]. Similarly, the composition of polar cod prey in the present study (Table 3) is markedly different between Atlantic and Arctic domains. Polar cod from the Arctic domain fed almost exclusively on the large Arctic amphipod *Themisto libellula*, while in the Atlantic domain, the diet was more diverse and included both mysids, cumaceans, polychaetes and fish (Table 3). The frequency of occurrence of *Themisto* spp. in the Atlantic domain was only 35%, of which more than two thirds were the Atlantic *T. abyssorum*. Similar observations have been made by previous authors [11,35], but never before interpreted at a regional (domain) scale.

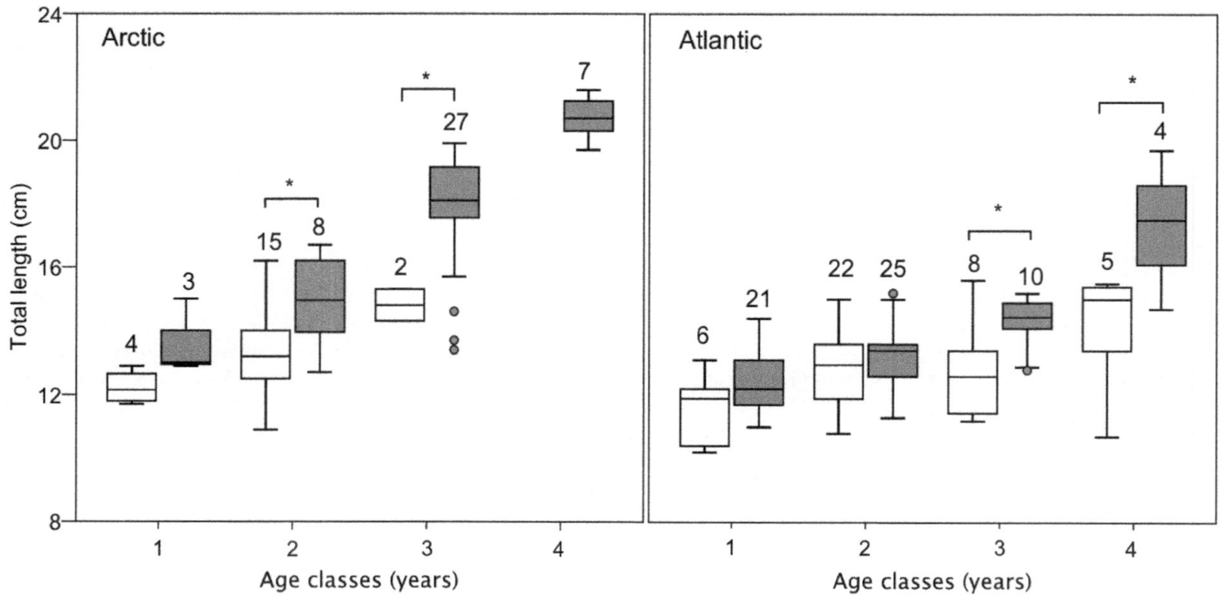

Figure 5. Total length (cm) of polar cod in relation to otolith based age (years) and maturity status. Mature (grey boxes) and immature (white boxes) individuals sampled in January 2011, 2012 and 2013 within the Arctic (n = 66) and Atlantic (n = 104) domains. For both domains, both genders were merged due to otherwise small sample size and because gender was not found to explain variability in length in the linear model operating on the larger dataset. Plots represent the median (line), 25%–75% percentiles (box), non-outlier range (wisker) and outliers (circle). Numbers above each box are n. Asterisks show significant (t-test without corrections for multiple comparisons, $p<0.05$) differences between mature and immature fish within domain and age-group.

Polar cod from Arctic domains generally showed a broader size range and a higher abundance of large specimens (Table 1, Figure S3). In accordance with previous anecdotal observations from the western coast of Spitsbergen (Atlantic domains) [36] and Cheshskaya Bay and Novaya Zemlya (Arctic domain) [37], the polar cod length-at-age was higher in the Arctic compared to the Atlantic domain (Figure 4). In addition, previous studies from the American and Canadian Arctic [38,39] have also reported differences in the size of polar cod between inshore and offshore areas. The differences in size were attributed to habitat differences, with inshore habitat providing favorable growing conditions (warmer coastal waters) as well as better protection against predation. Our analyses of body length variability suggested an increasing effect of domain with age and a starting point where young fish (age 1) did not differ between domains (Figure 4, Table S4). Furthermore, length-at-age did not depend on gender (Table S4). The difference in length-at-age between domains could not be explained by a shifted trade-off between growth in length versus body weight, as the relationship between total length and total weight was similar for both domains (Figure S1). The mechanisms resulting in a different length-at-age of polar cod between domains (Figure 4) are likely complex. A lower length-at-age, and earlier age-at-maturation (see discussion below) of polar cod from the Atlantic domain, can have several explanations including direct temperature effects on growth, higher predation risk, higher inter-specific species competition or density dependent effects of polar cod itself (such as intra-specific competition) [11,19,27,40]. Of these four potential explanations, we regard the latter two to be of least importance. A previous study has suggested that the potential interspecific competition for food is limited between gadoid species of equal size [11]. Also, based on our density estimates we find it unlikely that density dependent effects are at play, even at the highest abundances reported herein for Isfjorden and Kongsfjorden (September 2013, Table 1) where mean abundance did not exceed 40 individuals km^{-2}. Future studies should however apply acoustical methods for more precise density estimates. Finally, one of the most striking differences between the two domains, in terms of species composition is the general low numbers of large piscivorous fish predators in the Arctic domain ([11,41] and Table 1). Furthermore, the diet of polar cod in the Atlantic domain may represent an energetic drop in foraging efficiency of adult fishes that may explain the smaller length-at-age found in polar cod from the Atlantic domain from age 2 (Figure 4). Early life stages of polar cod (up to 1), however, preferentially forage on *Calanus* spp [20]. Such a relationship between growth and availability of large Arctic zooplankton (mainly *T. libellula*) have previously been demonstrated for capelin in the Barents Sea [42]. Finally, although there are marked differences in temperature between the two domains (Figure 2), we have not analysed how much temperature may account for the observed patterns between domains. Temperature, however, plays a critical role, both directly on the physiology of organisms [40], but also in shaping the ecosystem structures such as predator/prey composition. In the present study, it is not possible to extract adequate temperature data in order to discriminate between direct effects of temperature on polar cod physiology and effects of temperature on ecosystem structure. Our spatial resolution only allows for two point measurements, not a continuous variable to which changes in polar cod populations could be related. In addition, with sampling being performed in different seasons, temperature snapshots at each sampling point (temporal and spatial) can not be used without making unsupported assumptions.

Gender-specific reproductive strategy in the Arctic domain

In the Arctic domain, females have a longer life expectancy compared to males (Figure 3, Table S5), potentially reflecting a gender-specific trade-off between growth and reproduction.

Indeed, the sex ratio distribution against the age/size classes was relatively even in the lower age classes (age 1 to 3), while the relative proportion of females increased with age, reaching 100% at age class 5. Such a skewed sex ratio has been anecdotally reported for polar cod from other true Arctic regions [38] and references therein, [43–45], but without being discussed nor analysed. This repeatedly observed sex-ratio pattern in Arctic domains therefore appears to be a pan-Arctic phenomenon. Life history theory predicts trade-offs between growth and reproduction, especially for a species with a positive relationship between size and fecundity [26]. Body size may also limit the amount of resources that can be stored and thus the number of eggs produced [46], adding to the potential benefits of large size for reproduction. Accordingly, we found both an increasing percentage of maturing individuals per size class as well as an increasing gonadosomatic index (GSI) with size (Table 2). Furthermore, mature individuals of the Arctic domain were larger than immature individuals across all age classes (Figure 5). Assuming that female fecundity in polar cod is limited by body size [44], this factor in addition to the longer life expectancy of females (Figure 3) may indicate that there is a selective pressure favouring individuals to delay reproduction and invest more energy in somatic growth at young age. Males, on the other hand, for which body size does not restrict fecundity as much as in females [47], seemed to invest more energy (e.g. higher GSI, Table 2; earlier maturation in the season [44]) in reproduction at a young age. Indeed, a higher proportion of sexually mature males was observed in the smaller size classes compared to females (Table 2). Although eggs are more energy rich than sperm, a higher energy investment of males compared to females may translate as the sum of the costs to produce and maintain larger sized gonads for a longer period of time, including e.g. reduced motility and higher predation risks. Reproduction for males, therefore, may come with increased mortality costs (e.g. [47]) which may explain the absence of older males (Figure 3).

Polar cod have previously been considered a semelparous species, but it has been demonstrated that they are capable of reproducing in two consecutive years while in captivity [20,44]. Our study supports a more iteroparous life history due to the high percentage of mature individuals within each size class (Table 2). We hypothesise that males reproduce earlier in life to reduce risks of predation, but at the expense of a higher energy investment at a small size leading to higher post-spawning mortality. This trade-off further suggests a male reproductive strategy closer to semelparity than the iteroparous female's strategy. Iteroparity versus semelparity are outcomes of trade-offs between growth, fecundity and survival [26]. Semelparity is for instance suggested in cases where adult mortality is expected to be high [47,48]. Moreover, the degree of semelparity within a species may differ between males and females as suggested for capelin (*Mallotus villosus*) in the Barents Sea [47,49].

Altered reproductive strategy in the Atlantic domain

The gender-specific patterns were less pronounced within the Atlantic domain, although the sex ratio was still skewed towards females in all age/size classes (Figure 3, Table S4). Importantly, the difference between the two domains regarding sex ratios of older fish, is most likely an effect of Atlantic fishes not growing as large as within the Arctic domain. In the Atlantic domain however, a larger proportion of the females matured within the smallest size class compared to the Arctic domain (Table 2), and the energy investment in terms of GSI was reduced in both sexes compared to the Arctic domain (Table 2). Together with the altered sex-ratio pattern, these differences suggest a different reproductive strategy compared to the Arctic domain. Polar cod from the Atlantic domain appear to allocate more energy to reproduction at a young age, at the cost of somatic growth, possibly as an adaptation to higher predation risk (Table 1). Changes in reproductive strategies across latitudes and environments have been reported in other fish species. The most common example is that of the riverine American shad (*Alosa sapidissma*) which shows increased iteroparity with increasing latitudes and decreasing environmental predictability [50]. Similarly, capelin also expresses different reproductive strategies in different spawning habitats; iteroparity in beach spawning versus semelparity in open sea spawning capelin [49]. These differences are also paralleled by other life history traits of the two capelin populations; body size, distance of spawning migration and age at first maturity and are believed to result from differences in environmental characteristics of the spawning habitats (e.g. physical forcing and predation pressure [49]).

There was a trend in all age classes of mature individuals being longer than immature, although with a stronger pattern within the Arctic domain (Figure 5). Also, immature specimens showed a consistent but non-significant trend of a larger hepatosomatic index (HSI) compared to mature specimens (Table S2). These data suggest that mature specimens used their hepatic energy storage for gonadal growth [51]. We thus hypothesize that individuals with a reduced fitness, characterized by a reduced length-at-age and lower energy reserves (reduced HSI), would skip reproduction in a given year. This is supported by a recent study [52] on the Northeast Arctic cod (*Gadus morhua*) where skipped spawning is common for individuals that fail to build up sufficient energy reserves for spawning.

Our fecundity estimates (Table 4) are within the ranges of previously published estimates [38,39,42], and suggest that polar cod from the Atlantic domain are at a disadvantage compared to the Arctic domain. At the individual level (Table 2) and within each size class, the estimated fecundity varied only by 22–28% between domains. However, assuming iteroparity for females, the expected lifetime fecundity of each female (sum of all eggs produced) was twice as high in the Arctic compared to the Atlantic domain (75000 and 39000 eggs per female, respectively). Even if assuming semelparity (average number of eggs produced across all size classes) for females, the larger size-at-age in the Arctic leads to a higher lifetime fecundity for Arctic than for Atlantic females (19000 and 13000 eggs per female, respectively). This suggests lower fitness for Atlantic domain- compared to Arctic domain polar cod. The factors that varied between domains included probability of maturing, size- and GSI-at-age. Hence, the earlier age-at-maturation of Atlantic domain polar cod resulted in similar fecundity between domains in the lower size classes, but did not seem to fully compensate their reduced size and GSI compared to the Arctic domain. The total estimated egg production in the Arctic domain (Rijpfjorden) catches were 5.3 and 3.3 million eggs in 2012 and 2013 respectively compared to between 0.1 and 0.6 million eggs for the Atlantic domain catches (Table 4). Seen from a population perspective, there is thus an order of magnitude higher number of eggs produced in the Arctic versus the Atlantic domain. The factors controlling this pattern include both a larger size-at-age and a higher number of mature females in the Arctic domain. However, another important aspect to consider is the potential advantage of an increased growth season for eggs and larvae in the Atlantic domains [21,23].

Consequences of climate change for polar cod and the ecosystem

We document that polar cod exhibit different reproductive and growth patterns between the Arctic and Atlantic domains.

Table 4. Population fecundity estimates.

	Arctic		Atlantic		
	Rijp		Adv	Isf	Kong
TL (cm)	2012	2013	2012	2013	2013
Estimated amount of maturing females					
]9–12]	21.1	1.4	4.2	12.6	2.5
]12–15]	177.2	119.2	17.4	33.2	5.5
]15–18]	17.1	154.3	2.6	8.6	3.2
]18–21]	20.3	34.0			
Estimated total amount of eggs (10^3)					
]9–12]	163	10	23	70	14
]12–15]	2092	1407	160	305	50
]15–18]	331	2984	62	210	77
]18–21]	740	1243.			
Total eggs in population (10^6)					
]9–21]	3.3	5.6	0.3	0.6	0.1

Estimation of the total amount of maturing females and eggs produced in the catch of five trawl hauls (January 2013, 2013) in the Arctic domain (Rijpfjorden) and Atlantic domain (Adventfjorden, Isfjorden and Kongsfjorden). The numbers were estimated based on data from Figure 3, Table 2 and the total polar cod population of each trawl haul (Fig S3).

Although mechanistically not fully understood, our study suggests that a warmer climate will lead to changes in phenotypic traits, including earlier maturation, smaller size, increased investment in reproduction at early age, and in sum a reduced fecundity. These differences may be due to both phenotypic plasticity as well as differential selection acting on isolated populations. It is, however, at present not possible to fully discriminate between the two, and both explanations may to some degree be valid. However, previous studies have not been able to reveal a genetic isolation of populations between the Pechora Sea, Svalbard and Greenland [53]. As such, phenotypic plasticity may be the most important process. Importantly, however, regardless of the mechanisms at play, the observed differences in both population structure, gender balance and age-at-maturation are likely to have important ecological cascading effects, and are as such important to consider.

Representing a wide range of coastal habitats (open and silled fjords, straits and shelf regions on the western and northern side of Svalbard), and in line with previous studies from other parts of the Arctic (e.g. [38] and references therein, [45]), we argue that our observed patterns are relevant at a pan-Arctic scale. We therefore, hypothesize that a differential diet combined with increased predation on polar cod in the Atlantic domain may have caused the observed phenological differences. Although polar cod has been suggested as a central spawner, performing spawning migrations to key areas south of Svalbard and southeast of Novaya Zemlya [54], observations of both mature individuals (with GSI>20%) and young-of-the-year polar cod widely distributed in all fjords of Svalbard rather suggest that polar cod populations spawn locally [11]. Hence, polar cod is more likely to be affected by changes in the local environment than a species conducting large-scale spawning migrations. This hypothesis is further supported by Bouchard and Fortier [23], showing that water masses with differential temperatures and salinities may contribute to important phenological changes in the hatching season of polar cod larvae. We therefore hypothesise that climate change at the pan-Arctic scale may lead to similar patterns in polar cod populations, as those currently seen locally within Atlantic domains around Svalbard. These shifts seem to be a response to environmental forcing (e.g. predation pressure, prey availability and temperature) through changed trade-offs between somatic growth and reproduction among all size and age classes. Should polar cod be affected by climate change in this way, cascading effects to top predators are likely. Although mammalian predators may shift their preferences towards larger prey when available, Atlantic cod and other large temperate fish species are known to migrate seasonally, and are less available for non-migratory top predators such as the harbour seal [33,55,56] during winter and spring. During this time, such predators will be forced to shift diet back to polar cod. It is therefore likely that smaller sized polar cod will have significant impact on food web interactions and transfer of energy from the base of the food chain and up to top predators. Finally, a population fecundity reduced by one order of magnitude is likely to have significant implications for the global polar cod stock and further alter ecosystem structure and functioning.

In conclusion, we present unique evidence of a gender and domain specific reproduction strategy in polar cod, and how the predicted warming of the Arctic might alter these in the future. We argue that the differential predation pressure and prey availability, as well as a temperature difference between Arctic and Atlantic domains may play a substantial role in the observed reproductive patterns. Ultimately, we suggest that the findings from this study may provide a glimpse of what a future warmer Arctic may lead to. In a contiguous Arctic Ocean and shelf seas, a dramatic reduction in polar cod fecundity and size-at-maturation will have ecological cascading effects in the food web and play a significant role for future ecosystem services in Arctic areas affected by global warming.

Supporting Information

Figure S1 Relationship between total length (cm) and total weight (g) of polar cod. Cubic regression line from the Arctic ($n = 547$, $R^2 = 0.97$) and Atlantic ($n = 577$, $R^2 = 0.97$) domains.

Figure S2 Relationship between total length (cm) and fork length (cm) of polar cod (n = 1463). Data includes specimens from November 2010 (Isf), September 2011 (Hin), January 2011 (Adv, Isf, Bell), April 2012 (Adv) and September 2012 (Bill, Hin, Kong, Rijp).

Figure S3 Example of polar cod population size structure in January 2012 and January and September 2013. Arctic domain Rijpfjorden (black continuous line) and Hinlopen (dot-dashed, Sept 2013) and in the Atlantic domain Adventfjorden (dotted line, in Jan 2012), Isfjorden (dotted line in Sept 2012 and 2013) and Kongsfjorden (dashed line). Numbers in brackets indicate total amount of polar cod in each trawl haul.

Table S1 Overview of the sampling stations and number of specimens considered for each analyses. Fish <10 cm in length were excluded, except for the population structure data (Table 1). Number of specimens between analyses can differ due to lack of adequate information for some specimens. For instance, January 2012 Rijpfjorden, otoliths were available for 73 specimens (Fig 3), but GSI was only present for 66 specimens (Table 2 and Fig 5).

Table S2 Hepatosomatic index (HSI%, mean ±SD) of polar cod. Immature (I) and mature (M) polar cod were collected in January 2011, 2012 and 2013, from the Arctic (162) and Atlantic (104) domains. Numbers in bold and italics are significant differences (T-test) in means between immature and mature polar cod for a domain and size class (cm).

Table S3 Histological analysis of gonads of mature polar cod from January 2011 (Isfjorden) and 2012 (Rijpfjorden). The Table shows mean occurrence of oocyte stages (%) based on oocyte counts (n = 254 oocyte counts), mean diameter (mm) and weight (ug) of each oocyte stage: oogonia (Oo), previtellogenesis (PVit), cortical alveoli (CA), lipid inclusions formation (LIF), vitellogenesis (Vit), maturation (Mat), post-ovulatory follicles (POF), atretic oocytes (AO). There were no statistically significant differences in egg size (t-test, p<0.05) or maturation stages between domains and years.

Table S4 Summary output from a linear model for polar cod total length, explained by age, climatic domain as well as the interaction between age and domain. This model explained 65% of the variability in polar cod body length. Gender and all other interaction terms were omitted as explanatory variables after model simplification based on stepwise deletion of non-significant terms. P-values were smaller than 0.005 for all variables in the table. n = 281. Parameter estimate for intercept and slope in the table is for the Arctic domain. Intercept and slope for the Atlantic is obtained by adding the estimates for domain and the interaction term.

Table S5 Percentage female polar cod per age class (total n per age class) from the Arctic domain stations (upper panel) and Atlantic domain stations (lower panel). Age classes are derived from the linear regression presented in Fig 3. Fish <10 cm in length were excluded. Bill: Billefjorden, Hin: Hinlopen, Rijp: Rijpfjorden Isf: Isfjorden, Adv: Adventfjorden, Kong: Kongsfjorden, Kross: Krossfjorden, Bell: Bellsund.

Acknowledgments

We thank the University Centre In Svalbard and UiT The Arctic University of Norway for providing ship time. We thank Colin Griffiths and Malin Daase for providing the temperature plots and Paul Wassmann, Stig Falk-Petersen, Geir Johnsen and Finlo Cottier for reviewing the manuscript.

Author Contributions

Conceived and designed the experiments: JN JB. Performed the experiments: JN EK IGH IV JB. Analyzed the data: JN ØV EK SM JB. Contributed reagents/materials/analysis tools: JN ØV EK SM JB. Wrote the paper: JN ØV EK SM IGH IV JB.

References

1. Wang M, Overland JE (2009) A sea ice free summer Arctic within 30 years? Geophys Res Lett 36: L07502.
2. Anisimov OA, Vaughan DG, Callaghan TV, Furgal C, Marchant H, et al. (2007) Polar regions (Arctic and Antarctic). Climate Change 2007: Impacts, Adaptation and Vulnerability. Parry ML, Canziani OF, Palutikof JP, van der Linden PJ, Hanson CE, editors. Contribution of Working Group II to the Fourth Assessment Report of the Intergovernmental Panel on Climate Change, Cambridge University Press, Cambridge. pp. 653–685.
3. Grebmeier JM, Overland JE, Moore SE, Farley EV, Carmack EC, et al. (2006) A Major Ecosystem Shift in the Northern Bering Sea. Science 311: 1461–1464.
4. Bluhm BA, Gebruk AV, Gradinger R, Hopcroft RR, Huettmann F, et al. (2011) Arctic marine biodiversity: an update of species richness and examples of biodiversity change. Oceanography 24: 232.
5. Kovacs KM, Aguilar A, Aurioles D, Burkanov V, Campagna C, et al. (2012) Global threats to pinnipeds. Mar Mammal Sci 28: 414–436.
6. Hansen BB, Grotan V, Aanes R, Saether BE, Stien A, et al. (2013) Climate Events Synchronize the Dynamics of a Resident Vertebrate Community in the High Arctic. Science 339: 313–315.
7. Doney SC, Ruckelshaus M, Emmett Duffy J, Barry JP, Chan F, et al. (2012) Climate Change Impacts on Marine Ecosystems. Annu Rev Marine Sci 4: 11–37.
8. Kortsch S, Primicerio R, Beuchel F, Renaud PE, Rodrigues J, et al. (2012) Climate-driven regime shifts in Arctic marine benthos. Proc Natl Acad Sci USA 109: 14052–14057.
9. Alvain S, Le Quéré C, Bopp L, Racault M-F, Beaugrand G, et al. (2013) Rapid climatic driven shifts of diatoms at high latitudes. Remote Sens Environ 132: 195–201.
10. Perry AL, Low PJ, Ellis JR, Reynolds JD (2005) Climate Change and Distribution Shifts in Marine Fishes. Science 308: 1912–1915.
11. Renaud PE, Berge J, Varpe Ø, Lønne OJ, Nahrgang J, et al. (2012) Is the poleward expansion by Atlantic cod and haddock threatening native polar cod, *Boreogadus saida*? Polar Biol 35: 401–412.
12. Wassmann P, Duarte CM, Agustí S, Sejr MK (2011) Footprints of climate change in the Arctic marine ecosystem. Glob Change Biol 17: 1235–1249.
13. Huntley B, Collingham YC, Willis SG, Green RE (2008) Potential Impacts of Climatic Change on European Breeding Birds. PLoS ONE 3: e1439.
14. Beaugrand G, Reid PC, Ibanez FI, Lindley JA, Edwards M (2002) Reorganization of North Atlantic Marine Copepod Biodiversity and Climate. Science 296: 1692–1694.
15. Dalpadado P, Ingvaldsen RB, Stige LC, Bogstad B, Knutsen T, et al. (2012) Climate effects on Barents Sea ecosystem dynamics. ICES J Mar Sci 69: 1303–1316.
16. Slagstad D, Ellingsen IH, Wassmann P (2011) Evaluating primary and secondary production in an Arctic Ocean void of summer sea ice: An experimental simulation approach. Progr Oceanogr 90: 117–131.
17. Kjellerup S, Dünweber M, Swalethorp R, Nielsen TG, Møller EF, et al. (2012) Effects of a future warmer ocean on the coexisting copepods *Calanus finmarchicus* and *C glacialis* in Disko Bay, western Greenland. Mar Ecol Prog Ser 447: 87–108.
18. Ferguson SH, Higdon JW, Westdal KH (2012) Prey items and predation behavior of killer whales (*Orcinus orca*) in Nunavut, Canada based on Inuit hunter interviews. Aquat Biosyst 8: 3.
19. Christiansen JS, Hop H, Nilssen EM, Joensen J (2012) Trophic ecology of sympatric Arctic gadoids, *Arctogadus glacialis* (Peters, 1872) and *Boreogadus saida* (Lepechin, 1774), in NE Greenland. Polar Biol 35: 1247–1257.

20. Hop H, Gjøsæter H (2013) Polar cod (*Boreogadus saida*) and capelin (*Mallotus villosus*) as key species in marine food webs of the Arctic and the Barents Sea. Mar Biol Res 9: 878–894.
21. Fortier L, Ponton D, Gilbert M (1995) The match/mismatch hypothesis. Mar Ecol Prog Ser 120: 11–27.
22. Benoit D, Simard Y, Fortier L (2008) Hydroacoustic detection of large winter aggregations of Arctic cod (*Boreogadus saida*) at depth in ice-covered Franklin Bay (Beaufort Sea). J Geophys Res 113.
23. Bouchard C, Fortier L (2011) Circum-arctic comparison of the hatching season of polar cod *Boreogadus saida*: A test of the freshwater winter refuge hypothesis. Progr Oceanogr 90: 105–116.
24. Jonsson H, Sundt RC, Aas E, Sanni S (2010) The Arctic is no longer put on ice: evaluation of Polar cod (*Boreogadus saida*) as a monitoring species of oil pollution in cold waters. Mar Pollut Bullet 60: 390–395.
25. Nahrgang J, Camus L, Broms F, Christiansen JS, Hop H (2010) Seasonal baseline levels of physiological and biochemical parameters in polar cod (*Boreogadus saida*): Implications for environmental monitoring. Mar Pollut Bullet 60: 1336–1345.
26. Stearns SC (1992) The evolution of life histories. Oxford University Press, Oxford, 247 Pp.
27. Enberg K, Jørgensen C, Dunlop ES, Varpe Ø, Boukal DS, et al. (2011) Fishing-induced evolution of growth: concepts, mechanisms and the empirical evidence. Mar Ecol 33: 1–25.
28. Wallace MI, Cottier FR, Berge J, Tarling GA, Griffits C, et al. (2010) Comparison of zooplankton vertical migration in an ice-free and a seasonally ice-covered Arctic fjord: An insight into the influence of sea ice cover on zooplankton behavior. Limnol Oceanogr 55: 831–845.
29. Crawley MJ (2007). The R book. Wiley, West Sussex. R-Core-Team (2013) R: a language and environment for statistical computing, 877 pp.
30. Cottier F, Tverberg V, Inall M, Svendsen H, Nilsen F, et al. (2005) Water mass modification in an Arctic fjord through cross-shelf exchange: The seasonal hydrography of Kongsfjorden, Svalbard. J Geophys Res 110c12005.
31. Arnkværn G, Daase M, Eiane K (2005). Dynamics of coexisting *Calanus finmarchicus*, *Calanus glacialis* and *Calanus hyperboreus* populations in a high-Arctic fjord. Polar Biol 28: 528–538.
32. Nilsen F, Cottier F, Skogseth R, Mattsson S (2008) Fjord–shelf exchanges controlled by ice and brine production: The interannual variation of Atlantic Water in Isfjorden, Svalbard. Cont Shelf Res 28: 1838–1853.
33. Olsen E, Aanes S, Mehl S, Holst JC, Aglen A, et al. (2010) Cod, haddock, saithe, herring, and capelin in the Barents Sea and adjacent waters: a review of the biological value of the area. ICES J Mar Sci 67: 87–101.
34. Dolgov AV, Johannessen E, Høines Å (2011) Main species and ecological importance. In: Jakobsen TJ, Ozhigin VK, editors. The Barents Sea Ecosystem, Resources, Management: Tapir Academic Press, Trondheim. pp. 193–200.
35. Lønne OJ, Gulliksen B (1989) Size, age and diet of Polar cod. Polar Biol 9:187–191.
36. Falk-Petersen IB, Frivoll V, Gulliksen B, Haug T (1986) Occurrence and size/age relations of polar cod, *Boreogadus saida* (Lepechin), in Spitsbergen coastal waters. Sarsia 71: 235–245.
37. Andriashev AP (1954) Fishes of the northern seas of the USSR (English edn, 1964). Jerusalem: Israel Programme Scientific Translations.
38. Craig PC, Griffiths WB, Haldorson L, McElderry H (1982) Ecological studies of Arctic cod (*Boreogadus saida*) in Beaufort Sea coastal waters, Alaska. Can J Fish Aquat Sci 39: 395–406.
39. Bradstreet MSW (1982) Occurrence, habitat use, and behavior of seabirds, marine mammals, and Arctic cod at the Pond Inlet ice edge. Arctic 35: 28–40.
40. Pörtner HO, Knust R (2007) Climate change affects marine fishes through the oxygen limitation of thermal tolerance. Science 315: 95–97.
41. Wienerroither R, Johannesen E, Dolgov A, Byrkjedal I, Bjelland O, et al. (2011) Atlas of the Barents Sea fishes. IMR/PINRO joint report series 1-2011: 274 pp.
42. Gjøsæter H, Ajiad AM (1994) Growth of polar cod, *Boreogadus saida* (Lepechin), in the Barents Sea. ICES J Mar Sci 51: 115–120.
43. Gillispie JG, Smith RL, Barbour E, Barber WE (1997) Distribution, abundance, and growth of Arctic cod in the northeastern Chukchi Sea. Am Fish Soc Symp 19: 81–89.
44. Hop H, Trudeau VL, Graham M (1995) Spawning energetics of Arctic cod (*Boreogadus saida*) in relation to seasonal development of the ovary and plasma sex steroid levels. Can J Fish Aquat Sci 52: 541–550.
45. Lear WH (1979) Distribution, size and sexual maturity of Arctic cod (*Boreogadus saida*) in the northwest Atlantic during 1959–1978. CAFSAC Res Doc 79: 17–40.
46. Reznick DN, Braun B (1987) Fat cycling in the mosquitofish (*Gambusia affinis*): fat storage as a reproductive adaptation. Oecologia 73: 401–413.
47. Huse G (1998) Sex-specific life history strategies in capelin (*Mallotus villosus*)? Can J Fish Aquat Sci 55: 631–638.
48. Young TP (1981) A general model of comparative fecundity for semelparous and iteroparous life histories. Am Nat: 27–36.
49. Christiansen JS, Præbel K, Siikavuopio SI, Carscadden JE (2008) Facultative semelparity in capelin *Mallotus villosus* (Osmeridae) - an experimental test of a life history phenomenon in a sub-arctic fish. J Exp Mar Biol Ecol 360: 47–55.
50. Leggett WC, Carscadden JE (1978) Latitudinal variation in reproductive characteristics of American shad (*Alosa sapidissima*): evidence for population specific life history strategies in fish. J Fish Res Board Can 35: 1469–1478.
51. Marshall CT, Yaragina NA, Lambert Y, Kjesbu OS (1999) Total lipid energy as a proxy for total egg production by fish stocks. Nature 402: 288–290.
52. Skjæraasen JE, Nash RD, Korsbrekke K, Fonn M, Nilsen T, et al. (2012) Frequent skipped spawning in the world's largest cod population. PNAS 109: 8995–8999.
53. Fevolden S-E, Martinez I, Christiansen JS (1999) RAPD and scnDNA analyses of polar cod, *Boreogadus saida* (Pisces, Galidae), in the north Atlantic. Sarsia 84: 99–103.
54. Sameoto DD (1984) Review of current information on Arctic cod (*Boreogadus saida* Lepechin) and bibliography. Bedford Institute of Oceanography, Dartmouth, 72 pp.
55. Gjertz I, Lydersen C, Wiig Ø (2001) Distribution and diving of harbour seals (*Phoca vitulina*) in Svalbard. Polar Biol 24: 209–214.
56. Bjørge A, Desportes G, Waring GT, Rosing-Asvid A (2010) Introduction: the harbour seal (*Phoca vitulina*) - a global prespective. NAMMCO Scientific Publications 8: 7–14.

10

Relative Roles of Weather Variables and Change in Human Population in Malaria: Comparison over Different States of India

Prashant Goswami[1]*, Upadhayula Suryanarayana Murty[2], Srinivasa Rao Mutheneni[2], Swathi Trithala Krishnan[1]

1 CSIR Fourth Paradigm Institute, Bengaluru, India, **2** CSIR Indian Institute of Chemical Technology, Hyderabad, India

Abstract

Background: Pro-active and effective control as well as quantitative assessment of impact of climate change on malaria requires identification of the major drivers of the epidemic. Malaria depends on vector abundance which, in turn, depends on a combination of weather variables. However, there remain several gaps in our understanding and assessment of malaria in a changing climate. Most of the studies have considered weekly or even monthly mean values of weather variables, while the malaria vector is sensitive to daily variations. Secondly, rarely all the relevant meteorological variables have been considered together. An important question is the relative roles of weather variables (vector abundance) and change in host (human) population, in the change in disease load.

Method: We consider the 28 states of India, characterized by diverse climatic zones and changing population as well as complex variability in malaria, as a natural test bed. An annual vector load for each of the 28 states is defined based on the number of vector genesis days computed using daily values of temperature, rainfall and humidity from NCEP daily Reanalysis; a prediction of potential malaria load is defined by taking into consideration changes in the human population and compared with the reported number of malaria cases.

Results: For most states, the number of malaria cases is very well correlated with the vector load calculated with the combined conditions of daily values of temperature, rainfall and humidity; no single weather variable has any significant association with the observed disease prevalence.

Conclusion: The association between vector-load and daily values of weather variables is robust and holds for different climatic regions (states of India). Thus use of all the three weather variables provides a reliable means of pro-active and efficient vector sanitation and control as well as assessment of impact of climate change on malaria.

Editor: Thomas A. Smith, Swiss Tropical & Public Health Institute, Switzerland

Funding: This work was supported by the project Integrated Analysis for Impact, Mitigation and Sustainability funded by CSIR, India. The funders had no role in study design, data collection and analysis, decision to publish, or preparation of the manuscript.

Competing Interests: The authors have declared that no competing interests exist.

* Email: goswami@cmmacs.ernet.in

Introduction

A topic of growing concern, and debate, is the role of climate change in variability and trends in malaria [1–5]. Several countries and regions have reported emerging or growing threat of malaria due to change in climate [6–10]. While deaths due to malaria depend on a number of factors [11–13] like vector control, access to adequate medical facilities and immunological history of the people, infections can be considered to depend primarily on abundance of mosquito vector and exposure of the host to bites and rate of transmission. However, for a population with given immunological history, transmission can be assumed to be constant, at least at short time scales (~ years). Thus the abundance of vectors and the degree of exposure can be said to determine the vulnerability and the severity of the epidemic. However, in regions of rapidly changing human populations, like in many states of India, the relative roles of abundance and change in exposure due to changing human population are not easy to determine. A quantitative delineation of the role of weather variables in malaria load can enable design of pro-active mitigation based on vector control; implementation of such pro-active vector control is becoming increasingly important in view of reported resistance of mosquitoes to insecticides like DDT [14]. At the same time, such quantitative relations between weather variables and malaria can enable reliable projection of vulnerability in a changed climate.

The importance of meteorological variables in genesis and survival of mosquito have been known for a long time, although the relative roles of these variables and various aspects are still being investigated. However, fairly sharp and well defined ranges of meteorological variables, especially temperature, are required for genesis and survival of malaria vector, although these thresholds also depend on the mosquito species [15–19]. For example, there are well known thresholds of minimum and maximum of temperature for vector genesis and survival. Each weather variable has a range in which it affects the mosquito genesis and survival; a temperature above 16°C is required for breeding and development of the aquatic stages of Anopheline mosquitoes in the Tropics. Similarly, experiments indicate a threshold of minimum temperature between 16°C and 19°C for the development of *Plasmodium falciparum* (PF) malaria parasite. Temperature affects the rate of reproduction of the pathogens, and thus plays important role in the survival and transmission the vector [20]. Temperature is also known to affect the gonotrophic cycle (physiological process consisting of digestion of blood-meal and development of ovaries), and longevity. However, other weather variables like rainfall and humidity also play important roles [20–21]. For example, amount, intensity as well as duration of rainfall all play important roles in the dynamics of vector population. Besides, other environmental variables like land use and land cover also affect vector dynamics [22–23]. Rainfall can significantly reduce the vector survival by flushing the larval mosquitoes out of their aquatic habitat [17]; it can be logically assumed that reduced larval survival will also lead to a smaller population of adult mosquitoes. Both dryness and rainfall can influence larval survival significantly, as shown by experiments with simulated dry conditions [16–17]. Similarly, transmission can be inhibited by development time of the pathogen larger than the life span of the insect [18]. Higher temperature, on the other hand, can lead to quicker maturity leading to higher number of off springs during the transmission period [19]. Thus vector abundance can be assumed to be primarily determined by the climate (average weather regime) and the weather (variability around the climate) of the region [15–22].

It is also known that the effect of environmental parameters on vector population can be highly location-specific. The phenology and population dynamics of mosquito can vary significantly with altitude; a study based on weekly rainfall intensity in the Lake Victoria basin showed significant differences in larval abundance with altitudes [24]. It was found, for example, that for northern Australia, determination of the periodicity and the amplitude of mosquito population (abundance peaks) required consideration of the frequency and the amplitude of the tides [25]; a flood plain or a forested region is likely to require very different considerations. Over India, several studies have shown relationship between weather variables and malaria [26–28]. India with its complex climatic heterogeneity, growing population, land-use and urbanization poses a unique challenge to identify role of climate change in malaria load [29–31]. In particular, a region can experience both decrease and increase in malaria prevalence as the meteorological variables move in or out of the genesis window. A robust and reliable assessment of vulnerability to malaria due to climate change require quantitative, validated models of malaria epidemiology that involve all the relevant weather variables; assessments based on the trend of a single meteorological variable do not have much relevance. While the role of meteorological variables in malaria has been well recognized, and several attempts have been already made to use weather-driven models for malaria [32–33], most methods have used only a single meteorological variable or time-averaged (weekly/monthly mean) values to assess impact of climate change on malaria. However, dynamics of mosquito population (genesis as well as mortality) depends on daily variability of the weather variables [19]; in general, this daily variability will be much higher than that in monthly-averaged variables. Thus, vector abundance, and hence impact of malaria, is likely to depend on the number of daily genesis windows.

There have been attempts to quantitatively describe the prevalence of malaria relating entomological parameters to malaria transmission that include many parameters like age-at-infection, human blood index, entomological inclusion rate, vectorial capacity etc. However, most of these attempts often do not involve all the weather variables explicitly, and may involve climate variables only indirectly [34] A critical gap is the relative roles of the three meteorological variables (temperature, rainfall, humidity) in relation to the number malaria cases. Although models of population dynamics of vector that consider effects of variables like temperature and rainfall have been considered [35–36], they often lack validation against observed epidemiological data. A model of malaria based on weather and exposure has been validated against the observed malarial cases over Arunachal Pradesh in north-east India [37]. However, there are several aspects that were neither covered nor inferable from our previous study [37]. An important issue is the robustness of the results in terms of applicability to diverse climatic regions. This issue is addressed in the present manuscript by considering climatically and endemically diverse regions in terms of the 28 states of India; these states represent climates from the tropical to the higher latitudes. Another important issue is the delineation of the role of weather variables from those of other socio-economic variables; this issue is addressed in the present work by considering the impact of human population. In particular, analysis with constant and varying populations are carried out to quantify the impact of change in the number of hosts (as against change in vector population due to weather variables) in the disease load. While impact of human population may be minimal in regions like Arunachal Pradesh, the change (increase) in exposure due to growth in human population can be appreciable in many areas of India. Similarly, quantification of the roles of the three weather variables is expected to fill a critical gap in our understanding of the dynamics of malaria.

Methods

Ethics Statement

We declare that the data on malaria cases in this study was collected and compiled by the co-authors from CSIR-IICT based on records at the Public Health centres.

The collected epidemiological data from primary health centers was analyzed anonymously and no particular patient by name was involved. The study received clearance from the institutional Ethical Committee at CSIR-Indian Institute of Chemical Technology.

Study Area

We consider the 28 states of India, characterized by wide ranges of temperature, humidity and rainfall (Table 1). To apply our analysis to the 28 states, we use the annual malarial case data from the state health directorate. The population data was adopted from www.faostat.fao.org.

Collection of Malaria Data

The data was collected from the Directorate of Health, State Govts of India, based on cases reported from Primary Health Centers (PHCs) from all the states containing data on epidemiological aspects of Malaria number of blood samples collected (NBSC)and the number of blood samples that tested positive (NBSP) for either *Plasmodium vivax* (NPV) or *Plasmodium falciparum* (NPF) infection, or mixed infection; thus blood sample positive (NBSP) is the sum total of NPV, NPF and mixed type thus provides the total cases of Malaria. The procedure for collecting

Table 1. Showing Abbreviations used for State name, annual (1961–2010) mean minimum and maximum values of temperature, humidity and rainfall averaged over each state.

S.No	Abbreviations	States	Temperature (°C)		Humidity (%)		Rainfall (mm)
			Min	Max	Min	Max	Max
1	AP	Andra Pradesh	23	38	33	90	13
2	Ar P	Aruachal pradesh	12	30	60	88	27
3	AM	Assam	14	32	47	99	20
4	BR	Bihar	15	34	32	99	16
5	DL	Delhi	4	36	48	91	19
6	GA	Goa	18	37	53	97	18
7	GJ	Gujarat	19	34	30	92	11
8	HY	Haryana	3	31	38	91	23
9	HP	Himachal pradesh	7	29	70	96	16
10	JK	Jammu & kashmir	2	28	53	94	16
11	JH	Jarkhand	11	31	44	93	16
12	KA	Karnataka	16	34	54	96	13
13	KL	Kerala	20	35	31	91	27
14	MP	Madhya Pradesh	19	32	49	90	20
15	MH	Maharashtra	20	36	59	97	16
16	MN	Manipur	14	30	43	96	19
17	MG	Meghalaya	16	32	47	92	18
18	MZ	Mizoram	14	30	52	98	11
19	NG	Nagaland	16	29	30	93	23
20	OR	Orissa	14	32	38	91	16
21	PB	Punjab	12	34	70	97	16
22	RJ	Rajasthan	20	39	53	94	6
23	SK	Sikkim	18	31	41	93	13
24	TN	Tamil nadu	21	37	55	93	21
25	TP	Tripura	14	34	33	93	16
26	UP	Uttar Pradesh	8	35	45	91	17
27	UK	Uttar Khand	9	33	57	95	17
28	WB	West Bengal	13	34	45	96	14

epidemiological data has been already described in an earlier work [37]; in particular, the data did not involve any personal data or identification of the individuals.

The Epidemiological data on malaria is available online in the National Vector Borne Disease Control Programme(NVBDCP), Ministry of Health & Family Welfare, Govt. of India, (URL: www.nvbdcp.gov.in) and also at (http://nvbdcp.gov.in/Doc/mal-situation-Oct13.pdf).

Thresholds for vector genesis

The threshold values of meteorological variables were adopted from published literature. Although there have been reports of association between malaria incidence and variables like rainfall with large time lags like 2–3 months such long-term associations are not consistent with the time scales of mosquito life cycles. In our scenario, mosquito genesis takes place whenever favorable meteorological conditions occur; while this time scale in practice could be in hours, we consider daily variables for our description. Each episode of mosquito genesis at daily scale leads to a vector load through the process of survival in terms of biting mosquitoes; this is turn leads to malaria load at daily scale. The annual load of malaria is a sum of these daily values.

We first examine the relation between the annual vector loads, calculated as vector abundance based on the number of genesis days, to determine the relative roles of the meteorological variables. Next we quantify the relative roles of the weather variables (without the effect of size of the human population) and the potential epidemiology (with effect of size human population) and compare against the observed epidemiology.

Meteorological Data

A critical requirement for our analysis is a long-period, homogeneous data on all the three meteorological variables (temperature, rainfall and humidity) at daily (or shorter) time scale. In the present case these daily values of near surface temperature and near surface humidity were obtained from global NCEP Reanalysis data in a 2.5°*2.5° grid and the daily rainfall on a global 1.8° *1.8° grid [38,39]. The daily data at state level was created through appropriate interpolation or averaging. The advantage of NCEP Reanalysis is that it provides a comprehensive and consistent long-period data for all the three variables on a grid. It has been also shown that the interpolated NCEP Reanalysis possesses good correspondence with high resolution data from other sources over the region [40], and provides an effective data set for studying role of weather variables in malaria. Information on NCEP data is available on http://www.ncep.noaa.gov/

Days of Vector Genesis

We define a (vector) genesis day as one on which all the three variables: temperature, rainfall and humidity, are within their respective ranges for genesis of mosquitoes. In particular, a day is counted as a genesis day if

$$18^\circ C \leq T \leq 32^\circ C$$

$$20\% \leq Q \leq 80\%$$

$$1.5mm \leq R \leq 20mm$$

where T, Q and R represent, respectively, daily average temperature, daily average humidity and daily rainfall, extracted from daily gridded data from NCEP Reanalysis.

As vector dynamics takes place on a daily (or even shorter) time scales, we have considered daily values of weather variables in our analysis; the annual values are thus sum over these daily values. It would be ideal, and interesting, to compare the daily values of observed and computed malaria cases. However, as the current data availability on the number of malaria cases is at annual scale, the validation is carried out at annual scale; besides, analysis of trends due to change in climate and population (hosts) is really meaningful only with annual values.

Infection depends both on the number of encounters between human and mosquito (exposure) leading to bites as well as transmission of the parasite. The exposure is a complex function of life style, migration and movement of the human population as well as vector abundance. Naturally, only a fraction of the total human population present is eventually bitten by mosquito; however, it is expected that a larger vector density provides higher exposure leading to more bites. Assuming that larger genesis days allow larger prevalence of vectors, we consider the number of genesis days in a year as representative of vector abundance or vector load. To delineate the role of vector genesis days (and hence climate change) and change of host population, we compute two quantities, as defined below:

The annual vector load (E_V) for each state for year n is calculated as the sum of the days, $N_{VG}(n)$, in the year that satisfy criteria for genesis of mosquitoes.

$$E_V(n,k) = a_v(k) \times N_{VG}(n)$$

Thus E_V represents the number of vector genesis days with a constant host (human) population for the state k. It is assumed that the actual vector load will be a result of modulation of the number of vector genesis days by other parameters like land use and land type not explicitly included here; this modulation for a state is represented through the coefficient $\alpha_v(k)$ characteristic of the state.

The annual potential malaria load is calculated by including the effect of human population as

$$E_p(k,n) = a_p(k) \times N_{VG}(k,n) \times \frac{N_H(k,n)}{N_H(k,2000)}$$

Here $N_H(k,n)$ is the human population for the state k in the year n. The coefficient $\alpha_v(k)$ and $\alpha_p(k)$ characterize our model in terms of modulation of vector load and malaria load through processes not explicitly included. For example, $\alpha_p(k)$ can contain information specific to the state like immunological history (transmission coefficient), and land use. Since only a fraction of the bites will result in infection these coefficients generally would have values $\ll 1$, depending on the number of mosquitoes that carry malaria parasites and encounter human host. However, to isolate the effects of vector genesis we use a single value (0.007) of $\alpha_v(k)$ as a scaling factor to focus on the role of vector load; thus effects of variation due to exposure and transmission from state to state are not considered here. Similarly, a single value of (0.002) is used for $\alpha_p(k)$; the difference in the values of $\alpha_v(k)$ and $\alpha_p(k)$ is to ensure that the actual values of E_v and E_p are of similar magnitudes. These values of $\alpha_v(k)$ and $\alpha_p(k)$ were obtained through an iterative process to arrive at minimum differences between recorded malaria cases ($E_O(k,n)$) and calculated values of malaria for one year and kept fixed for the other years [37]. In other words, the values of α_v and α_v represent optimum values to obtain best fit between the simulated and the observed values in a statistical sense.

Results

The 28 states of India (Table 1) show substantial variations in variables like annual maximum and minimum temperature. The number (in %) of days satisfying genesis criteria, in terms of all the three weather variables can be very different from those based on a single variable. Indeed, percentages of genesis days based on individual weather variables (T, Q or R) show significant differences not only in their values but also in their variability; the percentage of days based on all the three variables is significantly smaller than that for any single variable, as expected. The inter-annual variability in the % of genesis days with all the three meteorological variables is also very different from the corresponding variability with only one meteorological variable (Figure 1).

A comparison of vector load, with vector abundance derived from all the three weather variables, shows excellent agreement (Figure 2) for most of the states, both with and without modulation by growth in human population (E_P and E_V, respectively). As expected from the linear growth in human population in many states, the potential malaria load is always larger than the vector load, with difference between the two generally increasing with time (Figure 2) due to normalization with respect to the population of 2001–2010. However, while correlation between Eo and E_P is generally better than that between Eo and E_V (Table 2), the effect is not really appreciable; in particular, against 21 states with correlation coefficient equal to or more than 0.8 for predicted potential malaria load the corresponding number for vector load is 18 (Table 2).

To examine the relative roles of the three weather variables, vector load was first calculated with each of these three variables separately. With only temperature as the condition for mosquito genesis, the calculated number of malarial cases has very little correspondence with the number of reported cases (Figure 3);

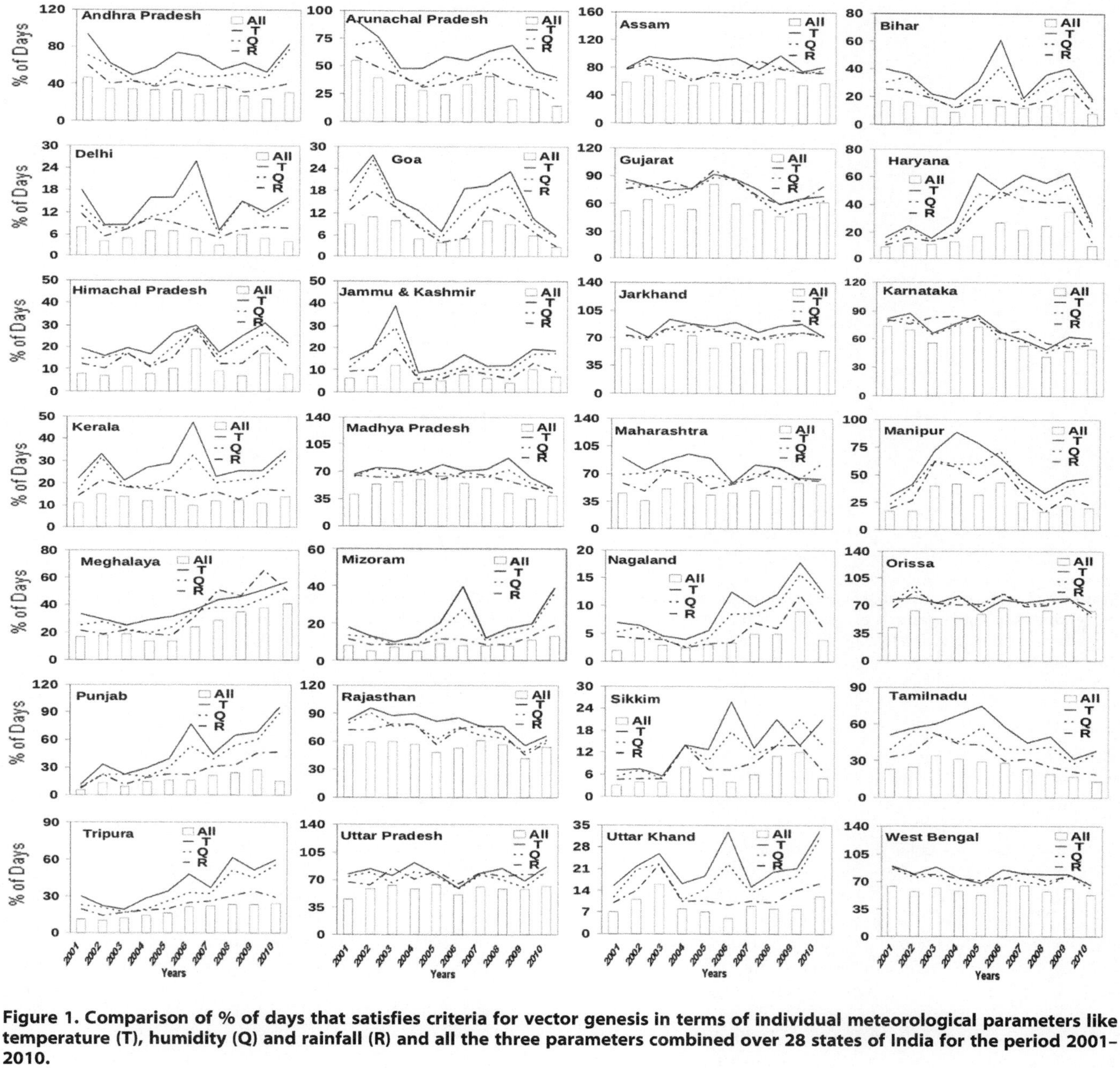

Figure 1. Comparison of % of days that satisfies criteria for vector genesis in terms of individual meteorological parameters like temperature (T), humidity (Q) and rainfall (R) and all the three parameters combined over 28 states of India for the period 2001–2010.

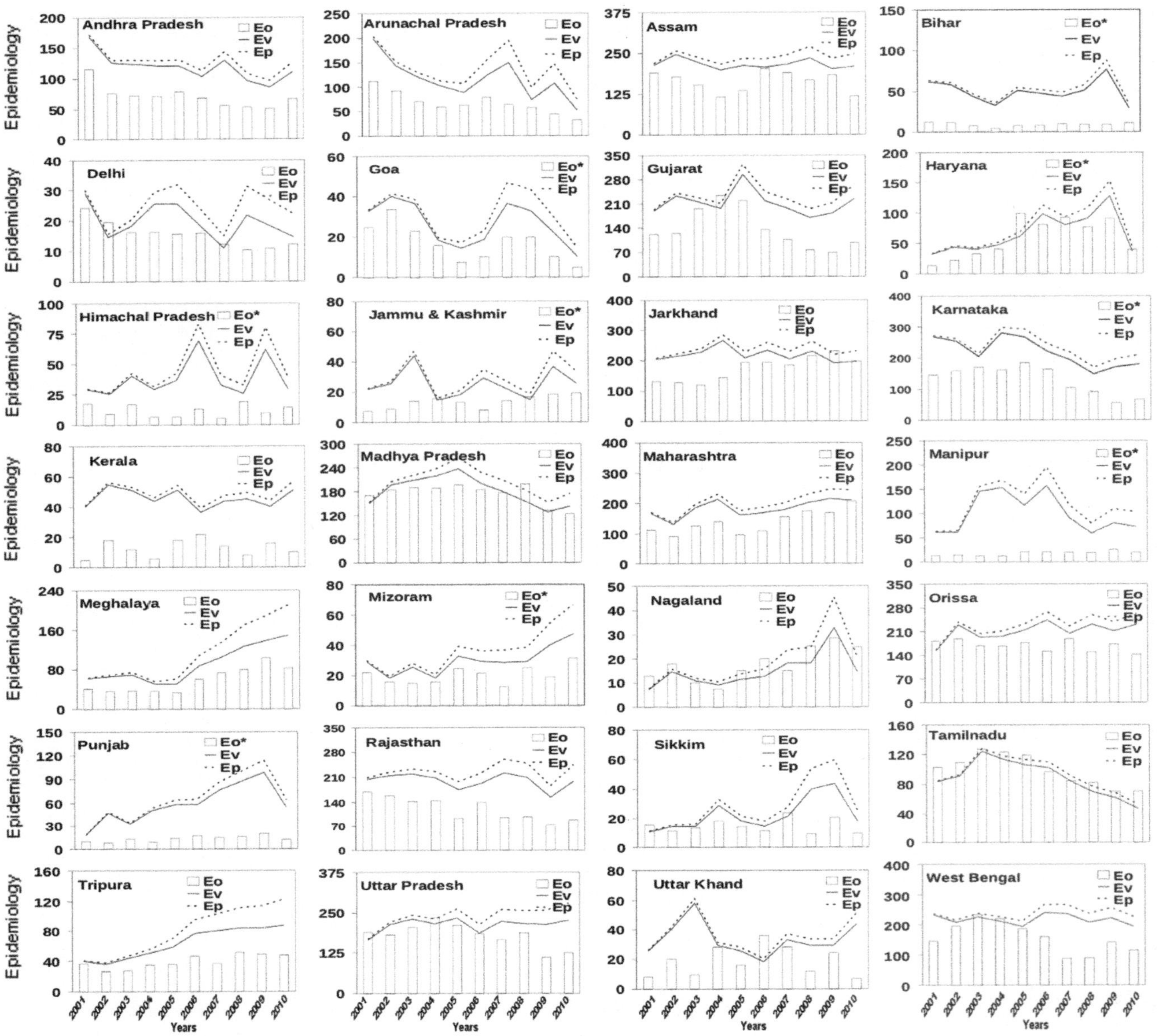

Figure 2. Comparison of observed annual epidemiology load (E_O) with epidemiology load based on vector load (E_V) calculated using days of vector genesis and constant human population with potential epidemiology load (E_P) (growth in human population included) over 28 states of India. The epidemiology is calculated as the number of blood samples that tested positive. The days of vector genesis here represent days in a year that fulfill combined meteorological conditions of temperature, humidity and rainfall for genesis of mosquitoes. The calculated epidemiology has been scaled by a factor (500 for which marked * and rest with 1000, as indicated) for easy comparison.

indeed, only five states are characterized by correlation coefficient between Eo and E_V significant at 99% confidence level for the degrees of freedom involved; the corresponding number for E_P is 8 (Table 2). Similar conclusions also hold for epidemiology load calculated with only rainfall as the condition for mosquito genesis (Figure 4), and only humidity as the condition for mosquito genesis (Figure 5); the numbers of states with correlation coefficient between Eo and either E_P or E_V significant at 99% confidence level are very few with either rainfall or humidity (Table 2).

It can be seen that the linear trends in the number of genesis days with individual weather variables can be quite different from those with all the three weather variables (Figure 6, top panel); in particular, the trends with the combined criteria can be either larger or smaller than those with a single weather variable. For easy assessment of significance, all trends have been expressed as percentage of respective standard deviation. More importantly, both appreciable positive and negative trends are present across the states (Figure 6, top panel). The linear trends calculated with either Eo and E_P, match well with those for observed number of malaria cases for most states (Figure 6, bottom panel). There is no significant or systematic improvement in the calculated trends against the observed trends, indicating the number of genesis days as the driving factor for the changes in malaria cases. There are only four states, Jharkhand (JH), Jammu and Kashmir (JK), Meghalaya (MG), and Maharashtra (MH), for which the trends are in opposite direction, but with marginal difference; out of these four states, the first three are mountainous. It is noteworthy; however, that inclusion of change in human population often reduces the trends in calculated malaria cases (Figure 6, bottom panel). While many states show decline in the number of genesis

Table 2. Correlation between observed (E_O) and estimated (E_P and E_V) epidemiology load with different combination of meteorological variables.

S.No	States	Correlation coefficient between epidemiology load and							
		All		T		Q		R	
		Ev	Ep	Ev	Ep	Ev	Ep	Ev	Ep
1	Andra Pradesh	**0.9**	**0.9**	**0.7**	**0.7**	**0.4**	**0.5**	**0.9**	**0.9**
2	Aruachal pradesh	**0.9**	**0.9**	**0.7**	**0.8**	**0.4**	**0.4**	**0.8**	**0.8**
3	Assam	0.3	**0.6**	0.3	0.2	0.2	0.1	**0.4**	**0.5**
4	Bihar	**0.7**	**0.8**	0.1	0.2	0.3	0.3	**0.4**	**0.4**
5	Delhi	**0.9**	**0.5**	0.2	0.3	0.3	0.3	**0.5**	**0.5**
6	Goa	**0.9**	**0.9**	**0.7**	**0.9**	**0.8**	**0.9**	**0.9**	1.0
7	Gujarat	0.3	**0.5**	**0.5**	**0.6**	**0.5**	**0.6**	**0.7**	**0.7**
8	Haryana	**0.9**	**0.9**	**0.9**	**0.9**	**0.9**	**0.9**	**0.9**	**0.9**
9	Himachal pradesh	**0.7**	**0.8**	0.1	0.1	0.1	0.1	0.1	0.1
10	Jammu & kashmir	**0.9**	**0.9**	0.3	0.2	**0.4**	**0.4**	**0.4**	**0.4**
11	Jharkhand	0.1	**0.6**	**0.5**	**0.4**	0.3	0.3	**0.4**	**0.4**
12	Karnataka	**0.9**	**0.9**	**0.6**	**0.7**	**0.7**	**0.7**	**0.9**	**0.9**
13	Kerala	**0.9**	**0.9**	**0.6**	**0.6**	**0.6**	**0.6**	0.2	0.2
14	Madhya Pradesh	**0.7**	**0.9**	**0.8**	**0.9**	**0.8**	**0.8**	**0.5**	**0.5**
15	Maharashtra	**0.9**	**0.9**	0.3	**0.4**	**0.6**	**0.6**	**0.9**	**0.8**
16	Manipur	**0.4**	**0.5**	0.2	0.2	0.3	0.2	0.2	0.3
17	Meghalaya	**0.9**	**0.8**	**0.9**	**0.9**	**0.9**	**0.9**	1.0	1.0
18	Mizoram	**0.9**	**0.9**	**0.7**	**0.7**	**0.8**	**0.8**	**0.8**	**0.8**
19	Nagaland	**0.8**	**0.7**	**0.9**	**0.9**	**0.9**	**0.9**	**0.8**	**0.7**
20	Orissa	**0.8**	**0.8**	0.2	0.2	**0.4**	**0.4**	0.2	0.2
21	Punjab	**0.9**	**0.9**	**0.6**	**0.6**	**0.6**	**0.6**	**0.6**	**0.6**
22	Rajasthan	**0.7**	**0.8**	**0.6**	**0.8**	**0.8**	**0.8**	**0.4**	**0.4**
23	Sikkim	**0.9**	**0.8**	0.0	0.0	0.0	0.0	**0.4**	**0.4**
24	Tamil nadu	**0.9**	**0.9**	0.5	**0.5**	**0.8**	**0.7**	**0.6**	**0.6**
25	Tripura	**0.9**	**0.9**	**0.9**	1.0	**0.9**	**0.9**	**0.9**	**0.9**
26	Uttar Pradesh	**0.6**	**0.7**	0.1	0.3	0.2	0.2	0.1	0.1
27	Uttar Khand	**0.9**	**0.9**	0.0	0.0	0.2	0.2	0.1	0.1
28	West Bengal	**0.7**	**0.9**	0.2	0.1	0.1	0.2	0.1	0.1
	No of states with correlation coefficient above 95% significance	25	28	15	16	18	18	20	19

Correlation coefficients of 0.8 are above 9.5% level of significance for the degrees of freedom The cases with correlation above 95% level of significance for the degrees of freedom involved are in bold.

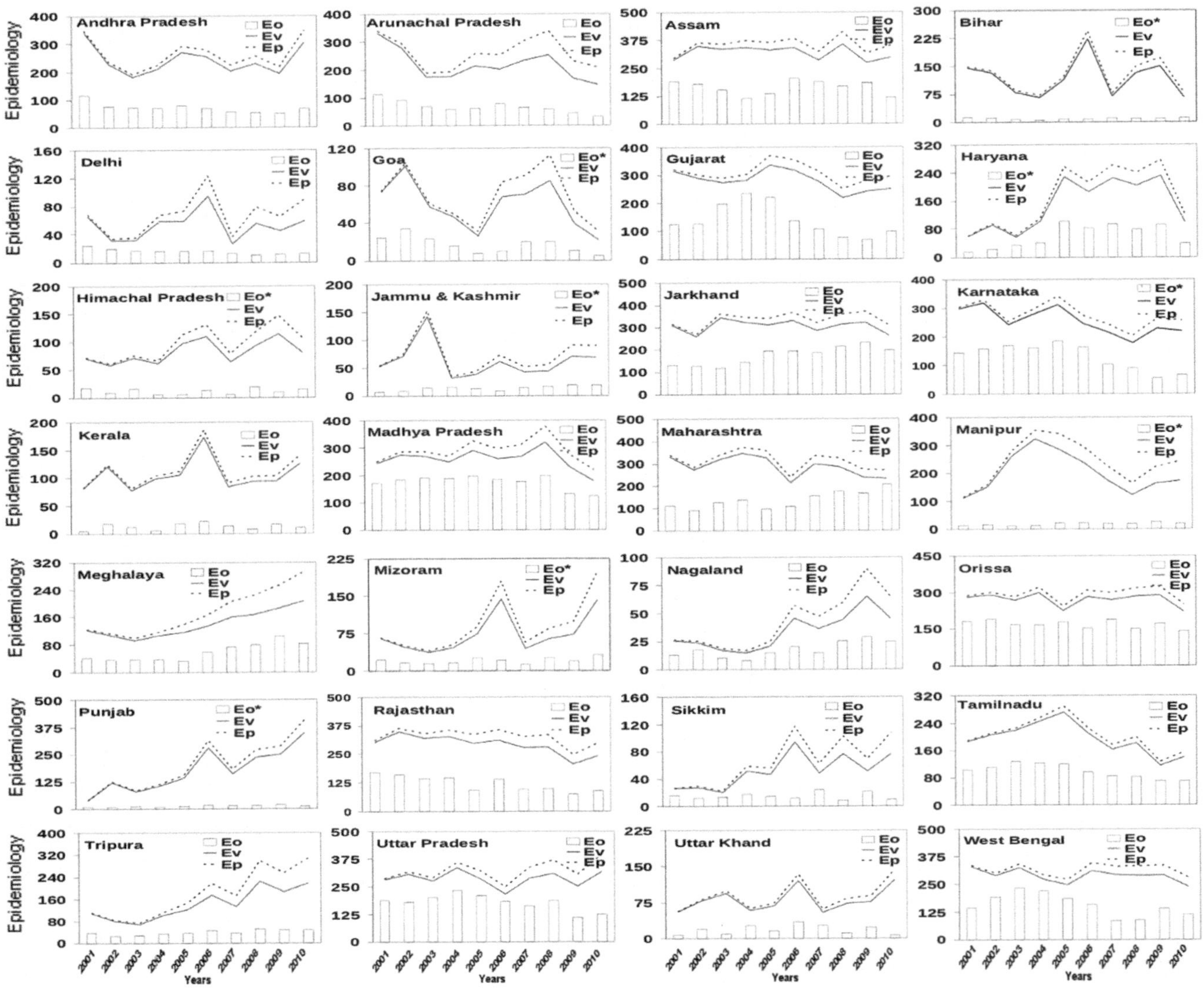

Figure 3. Comparison of observed annual epidemiology load (E_O) with epidemiology load based on vector load (E_V) calculated using days of vector genesis and constant human population with potential epidemiology load (E_P) (growth in human population included) over 28 states of India. The annual epidemiology is calculated as the number of blood samples that test positive. The days of vector genesis represent days in a year that fulfill only meteorological condition of temperature for genesis of mosquitoes. With only temperature as the condition for mosquito genesis, the calculated E_P has very little correspondence to the observed E_P (Figure 1); only a few (5–8) states show significant correlation (Table 2). The annual epidemiology has been scaled by a factor (500 for which marked * and rest with 1000, as indicated) for easy comparison.

days and corresponding decline in malaria load, a few states show significant positive trends, notably Maharashtra (MH) and Tripura (TP) (Figure 6, bottom panel). A few states, notably Himachal Pradesh (HP), Rajasthan (RJ) and Meghalaya (MG), show steep rise in both malaria prevalence and genesis days in recent years, indicating, once again, the driving role of the weather variables.

Discussions and Conclusions

The relative roles of weather variables in vector genesis, and increase in exposure due to growth in human population, are critical inputs for many issues like effectiveness of vector control and assessment of impact of climate change. Our results show that vector load defined in terms of genesis days calculated from the combined daily values of temperature, rainfall and humidity provides a close association with malaria; none of the weather variables alone provides significant skill.

The present results, along with our earlier results on Arunachal Pradesh [37], show that the association between weather variables and malaria is robust in terms of applicability to diverse climatic regions; the 28 states of India represent climates from the tropical to the higher latitudes and regions of varying endemicity. Another important insight is the delineation of the role of weather variables from the impact of human population in the disease occurrence; our results show the weather variables to be the primary drivers.

Death due to malaria depends on a host of processes like vector control, quality of health services, land management as well as public awareness. However, identification of factors responsible for the vector dynamics can reduce the uncertainties in designing control and mitigation strategies. Our identification of the strong association between vector loads computed based on combined

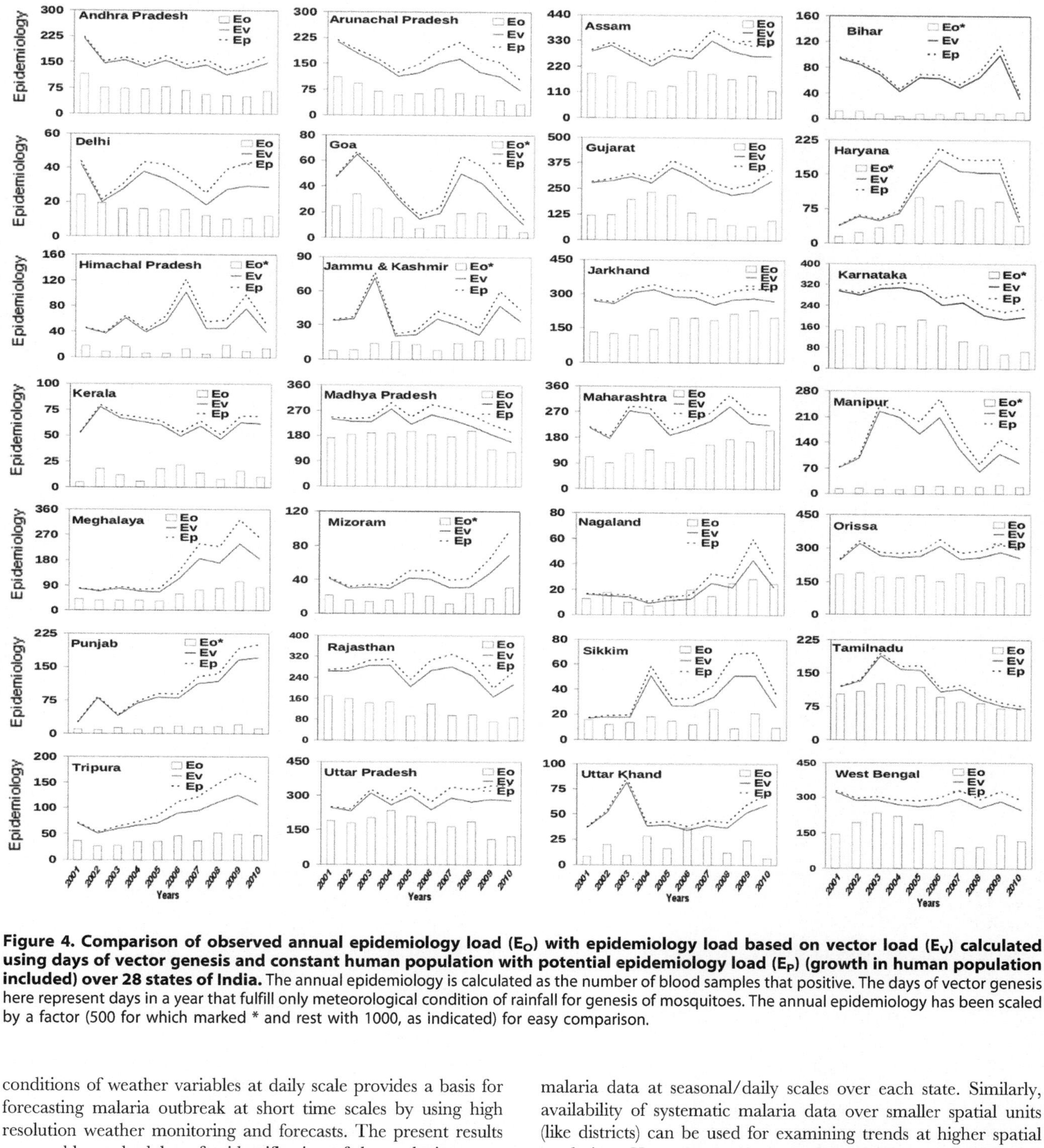

Figure 4. Comparison of observed annual epidemiology load (E_O) with epidemiology load based on vector load (E_V) calculated using days of vector genesis and constant human population with potential epidemiology load (E_P) (growth in human population included) over 28 states of India. The annual epidemiology is calculated as the number of blood samples that positive. The days of vector genesis here represent days in a year that fulfill only meteorological condition of rainfall for genesis of mosquitoes. The annual epidemiology has been scaled by a factor (500 for which marked * and rest with 1000, as indicated) for easy comparison.

conditions of weather variables at daily scale provides a basis for forecasting malaria outbreak at short time scales by using high resolution weather monitoring and forecasts. The present results can enable methodology for identification of the peaks in vector population which would precede the disease outbreak by a typical incubation time of the parasite in the human host [37]. It is worth emphasizing that the toll of the epidemic, in terms of deaths, will also depend critically on several other factors like access to health care, especially in remote areas [41]. To further quantify the association between vector load and disease prevalence it is necessary to calibrate the parameters for each region through extensive observational programs.

In the same way that we have considered trends in annual loads, it is possible to examine trends in seasonal or even daily vector loads. A primary requirement for such an analysis is reliable malaria data at seasonal/daily scales over each state. Similarly, availability of systematic malaria data over smaller spatial units (like districts) can be used for examining trends at higher spatial resolution. However, as the changes in climate as well as in population are slow processes, analysis at annual scale is logical.

The calculation of days of vector genesis in the present version is based on daily values of the meteorological variables. It is desirable, and possible, to include episodic (say 3-day window) values of the variables to include effects of the preceding meteorological conditions on vector genesis. This would ensure that flooding due to rains before genesis day can also affect the survival of the larvae. Similar arguments are also valid for extreme temperature and dryness. However, we have not included these effects here explicitly for simplicity; our results based on only the one-day criteria may be further improved due to multi-day

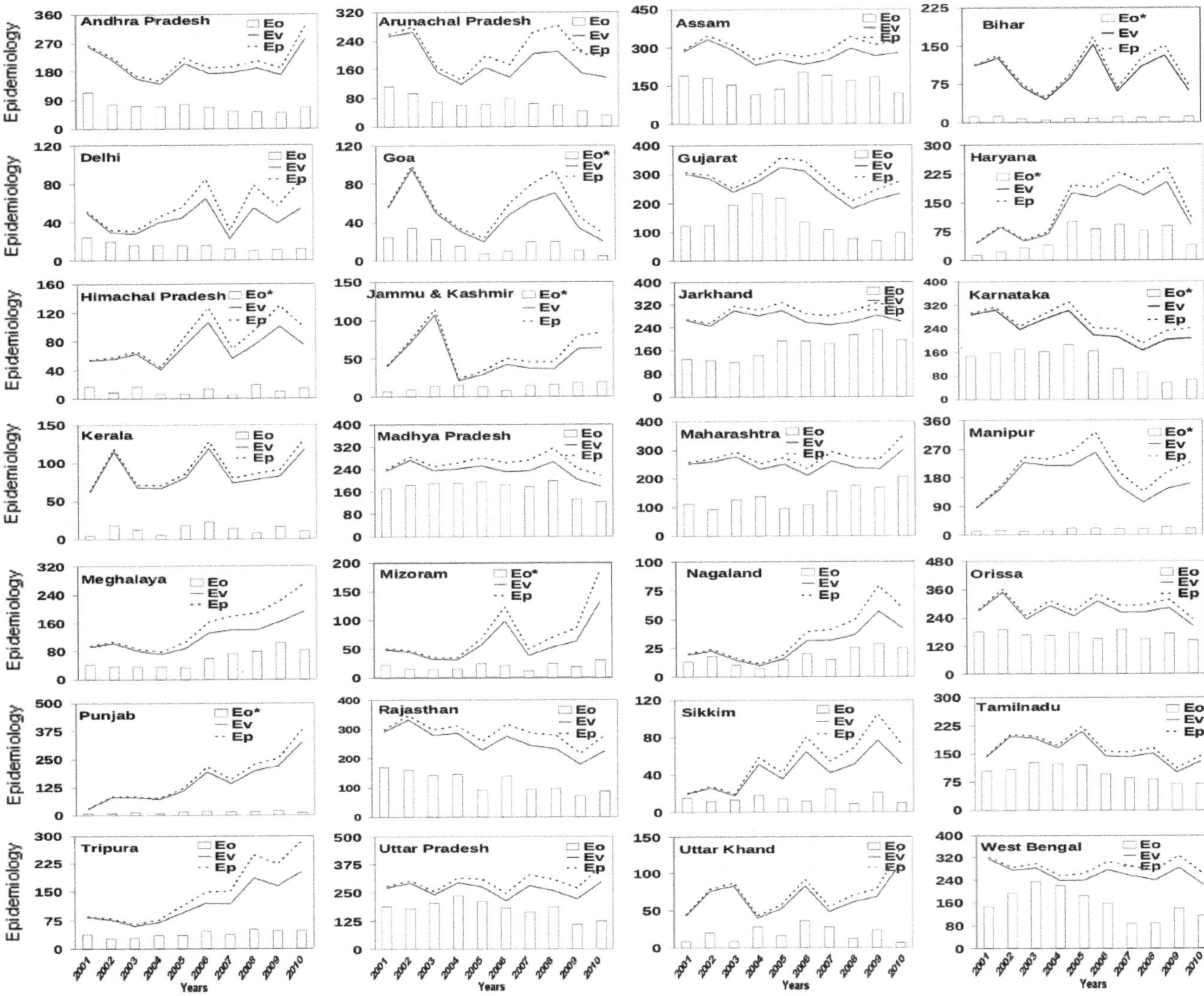

Figure 5. Comparison of observed annual epidemiology load (E_O) with epidemiology load based on vector load (E_V) calculated using days of vector genesis and constant human population with potential epidemiology load (E_P) (growth in human population included) over 28 states of India. The annual epidemiology is calculated as the number of blood samples that positive. The days of vector genesis here represent days in a year that fulfill only meteorological condition of humidity for genesis of mosquitoes. The annual epidemiology has been scaled by a factor (500 for which marked * and rest with 1000, as indicated) for easy comparison.

(episode) based criteria. Similarly, we have avoided detail effects of the meteorological variables on the life cycle of mosquito. For example, the effect of rainfall on mosquito dynamics is likely to be more at the larval stage. However, since reduction in the number of larvae due to (flushing by) rainfall is likely to reduce the number of adult mosquitoes, we have not separated the differential effect at different stages of the mosquito life cycle.

The results bring out clearly the minimum necessary analysis for investigating impact of climate change on malaria. At the same time, close relation between vector genesis days and the disease provides an excellent tool for investigating impacts of regional climate change on malaria, especially to investigate impact scenarios where the change, rather than the actual malaria load cases, is more important. In addition to genesis, there are other processes that may become important in situations of extreme climate change. For example, the life cycle of the malaria parasite inside the mosquito and the human host is important for transmission. Large changes in temperature can lead to reduction in the duration of gonotrophic cycle and in the extrinsic incubation period of malaria parasite and affect the rate of transmission [42]. Similarly, possible changes in the immunology profile of a people due to factors like migration may also become important in the long term. However, most of the processes in the post-larval stages become less important if vector sanitation is carried out at the genesis sites. The framework and the methodology are quite generic and can be applied in any geographical location with required calibration.

In practice, several species of mosquito are responsible for malaria in India [43–46]. The genesis and survival criteria may vary somewhat from species to species; similarly, the effects of local environmental conditions like run-off, land use and water stagnation may affect mosquito dynamics. However, as our results show, these are likely to be secondary effects.

The two parameters that define our model, α_V (k) and α_P (k), characterize various local factors (for a state) that modulate vector dynamics as well as malaria load. While we have considered a

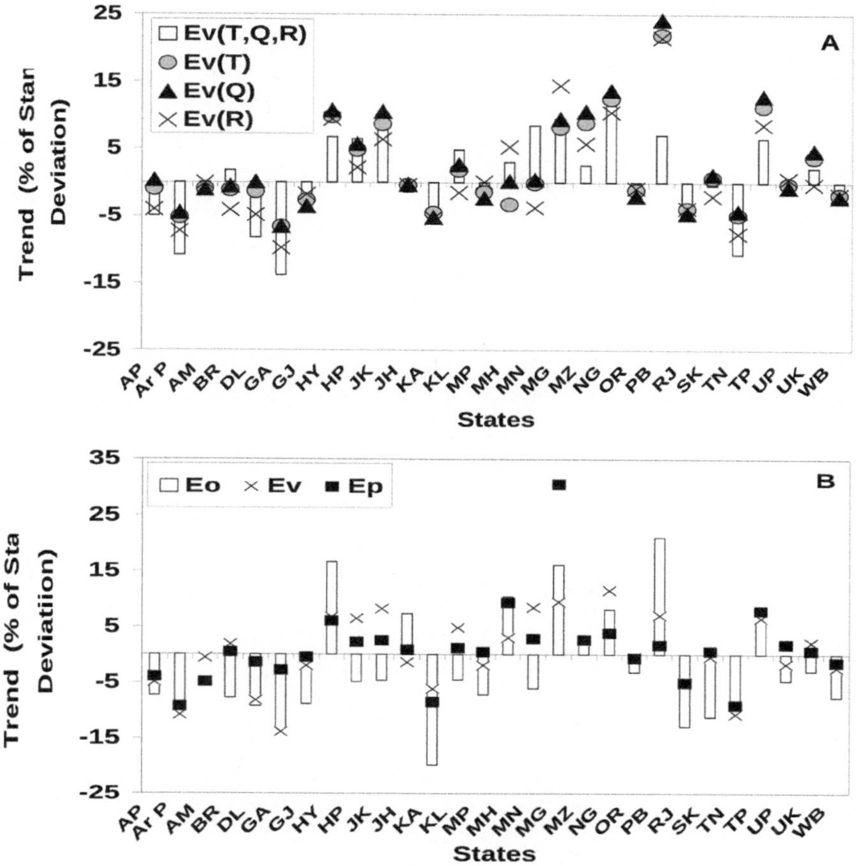

Figure 6. Linear trends in observed and calculated epidemiology (as % of respective standard deviation) (A) calculated based on genesis constraint of only temperature, only humidity, only rainfall and combined for the period of 1961–2010. (B) Observed epidemiology (E_O), vector load (E_V) and epidemiology potential (E_P) for the 28 states based on data for the period 2001–2010.

single value each for α_p and α_v, it is possible, and desirable, to consider values α_p and α_v that represent characteristics of a state. However, this may not change our results qualitatively, as it would only mean changing the values of α_p and α_v to calculate Ev and Ep for each state, while increasing the complexity of the model. For practical application, it would be necessary to determine the parameters like Ep for different states and different seasons. As discussed in our earlier work [37], this is possible by considering climatological conditions as well as factors like population and seasonablity of outdoor activity (exposure). The methodology for such an extension is, however, straight forward.

Author Contributions

Conceived and designed the experiments: PG USM SRM. Performed the experiments: PG USM SRM STK. Analyzed the data: PG STK. Wrote the paper: PG.

References

1. Epstein PR, Diaz HF, Elias S, Grabherr G, Graham NE, et al. (1998) Biological and physical signs of climate change: focus on mosquito-borne diseases. Bull Am Meteorol Soc 79: 3.
2. Lindsay SW, Martens WJ (1998) Malaria in the African highlands: past, present and future. Bull WHO 76: 33–45.
3. Hunter PR (2003) Climate change and waterborne and vector-borne disease. Journal of Applied Microbiology 94: 37–46.
4. Matsuoka Y, Kai K (1994) An estimation of climatic change effects on malaria. Journal of Global Environmental Engineering 1: 1–15.
5. Leaf A (1989) Potential health effects of global climate and environmental changes. New England Journal of Medicine 321: 1577–1583.
6. Patz JA, Epstein PR, Burke TA, Balbus JM (1996) Global climate change and emerging infectious diseases. JAMA 275: 217–223.
7. Githeko AK, Lindsay SW, Confalonieri UE, Patz JA (2000) Climate change and vector-borne diseases: a regional analysis. Bull WHO 78: 9.
8. Tanser FC, Brian S, David LS (2003) Potential effect of climate change on malaria transmission in Africa. The Lancet 362: 1792–1798.
9. Martens WJ, Niessen LW, Rotmans J, Jetten TH, McMichael AJ (1995) Potential impact of global climate change on malaria risk. Environment and Health Perspectives 103: 458–464.
10. Caminade C, Ndione JA, Kebe CMF, Jones AE, Danuor S, et al. (2010) Mapping rift valley fever and malaria risk over west africa using climatic indicators. Atmos Sci Let 12: 96–103.
11. Snow RW, Craig MH, Deichmannle U, Sueur DL (1999) A Preliminary Continental Risk Map for Malaria Mortality among African Children. Parasitol Today 15(3): 99–104.
12. Craig MH, Snow RW, Sueur DL (1999) Climate based distribution model of malaria transmission in sub-Saharan Africa. Parasitol Today 15: 105–111.
13. Macdonald G (1958) The epidemiology and control of malaria. Proc R Soc Med 51: 223–224.
14. Davidson G, Zahar AR (1973) The practical implications of resistance of malaria vectors to insecticides. Bull World Health Organ 49 (5): 475–483.
15. Paaijmans KP, Read AF, Thomas MB (2009) Understanding the link between malaria risk and climate. PNAS 106: 33.
16. Koenraadt CJ, Paaijmans KP, Githeko AK, Knols BG, Takken W (2003) Egg hatching, larval movement and larval survival of the malaria vector *Anopheles gambiae* in desiccating habitats. Malaria Journal 2: 20.
17. Paaijmans KP, Wandago MO, Githeko AK, Takken W (2007) Unexpected high losses of *Anopheles gambiae* larvae due to rainfall. PLoS ONE 2(11): e1146.

18. Leeson HS (1939) Longevity of *Anopheles maculipennis* race atroparvus, Van Theil, at controlled temperature and humidity after one blood meal. Bull Entom Res 30: 103–301.
19. Paaijmans KP, Blanford S, Bell AS, Blanford JL, Read AF, et al. (2010) Influence of climate on malaria transmission depends on daily temperature variation. PNAS 107(34): 15135–15139.
20. Gillett JD (1974) Direct and indirect influences of temperature on the transmission of parasites from insects to man. Symp British Soc Parasitol 12: 79.
21. Blanford JI, Blanford S, Crane RG, Mann ME, Paaijmans KP, et al. (2013) Implications of temperature variation for malaria parasite development across Africa. Scientific Reports 3: 1300.
22. Paaijmans KP, Thomas MB (2011) The influence of mosquito resting behavior and associated micro climate for malaria risk. Malaria Journal 10: 183.
23. Paaijmans KP, Imbahale SS, Thomas MB, Takken W (2010) Relevant microclimate for determining the development rate of malaria mosquitoes and possible implications of climate change. Malaria Journal 9: 196.
24. Imbahale SS, Paaijmans KP, Mukabana WR, Lammeren RV, Githeko AK, et al. (2011) A longitudinal study on *Anopheles* mosquito larval abundance in distinct geographical and environmental settings in western Kenya. Malaria Journal 10:81.
25. Yang GJ, Brook BW, Bradshaw CJA (2009) Predicting the Timing and Magnitude of Tropical Mosquito Population Peaks for Maximizing Control Efficiency. PLoS Negl Trop Dis 3(2): e385.
26. Gill CA (1920) The relationship between malaria and rainfall. Indian Journal of Medical Research 37: 618–632.
27. Gill CA (1921) The role of meteorology and malaria. Indian Journal of Medical Research 8: 633–693.
28. Dhiman RC, Sharmila P, Dash AP (2008) Climate change and malaria in India: Interplay between temperature and mosquitoes. Regional Health Forum 12(1): 27–31.
29. Mohapatra PK, Narain K, Prakash A, Bhattacharyya DR, Mahanta J (2001) Risk factors of malaria in the fringes of an evergreen monsoon forest of arunachal pradesh. Natl Med J India 14(3): 139–142.
30. Tiwari S, Ghosh SK, Ojha VP, Dash AP, Raghavendra K (2010) Reduced susceptibility to selected synthetic pyrethroids in urban malaria vector Anopheles stephensis: a case study in Mangalore city, south India. Malaria Journal 9: 179.
31. Kiszewski AE, Teklehaimanot A (2004) A review of the clinical and epidemiologic burdens of epidemic malaria. Am J Trop Med Hyg 71: 128–135.
32. Parham PE, Michael E (2010) Modeling the Effects of Weather and Climate Change on Malaria Transmission. Environ Health Perspect 118(5): 620–626.
33. Hoshen MB, Morse AP (2004) A weather-driven model of malaria transmission. Malaria Journal 3: 32.
34. Ermert V, Fink AH, Jones AE, Morse AP (2011) Development of a new version of the Liverpool Malaria Model. I. Refining the parameter settings and mathematical formulation of basic processes based on a literature review. Malaria Journal 10: 35.
35. Parham PE, Pople D, Celine CJ, Lindsay S, Hinsley W, et al. (2012) Modeling the role of environmental variables on the population dynamics of the malaria vector Anopheles gambiae sensu stricto. Malaria Journal 11: 271.
36. Loevinsohn ME (1994) Climatic warning and increased malaria incidence in Rwanda. The Lancet 19, 343(8899):714–718.
37. Goswami P, Suryanarayana UM, Srinivasa MR, Avinash K, Swathi TK (2012) Model of Malaria Epidemiology Involving Weather, Exposure and Transmission Applied to North East India. PLoS ONE 7(11): e49713.F.
38. Mo KC, Higgins RW (1996) Large-Scale Atmospheric Moisture Transport as Evaluated in the NCEP/NCAR and the NASA/DAO Reanalyses. J Climate 9: 1531–1545.
39. Kistler R, Kalnay E, Collins W, Saha S, White G, et al. (2001) The NCEP-NCAR 50-Year Reanalysis: Monthly Means CD-ROM and Documentation. Bull Amer Meteor Soc 82: 247–268.
40. Goswami P, Ramesh KV (2006) A comparison of interpolated NCEP (I-NCEP) rainfall with high resolution satellite observation. Geophys Res Lett 33(19): L19821.
41. Gaudart J, Toure O, Dessay N, Dicko AL, Ranque S, et al. (2009) Modelling malaria incidence with environmental dependency in a locality of Sudanese savannah area, Mali. Malaria Journal 8: 61.
42. Patz JA, Graczyk TK, Geller N, Vittor AY (2000) Effects of environmental change on emerging parasitic diseases. International Journal for Parasitology 30: 1395–1405.
43. Sharma SK, Tyagi PK, Padhan K, Adak T, Subbarao SK (2004) Malarial morbidity in tribal communities living in the forest and plain ecotypes of Orissa, India. Ann Trop Med Parasitol 98: 459–568.
44. Singh N, Nagpal AC, Saxena A, Singh MP (2004) Changing scenario of malaria in central India: The replacement of *P. vivax* by *P. falciparum* (1986–2000). Trop Med Int Hlth 9(3): 364–371.
45. Kumar A, Sharma VP, Sumodan PK, Thavaselvam D, Kamat RH (1994) Malaria control by utilizing *Bacillus sphaericus* (strain B-101, serotype H5a, H5b) against *Anopheles stephensi* in Panaji, Goa. J Am Mosq Cont Assoc 10(4): 534–539.
46. Vinayak S, Biswas S, Dev A, Kumar A, Ansari MA (2003) Prevalence of the K76T mutation in the Pfcrt gene of *Plasmodium falciparum* among chloroquine responders in India. Acta Tropica 87(2): 287–293.

Individualistic Population Responses of Five Frog Species in Two Changing Tropical Environments over Time

Mason J. Ryan[1,2]*, Michael M. Fuller[1], Norman J. Scott[1,2], Joseph A. Cook[1,2], Steven Poe[1,2], Beatriz Willink[3], Gerardo Chaves[3], Federico Bolaños[3]

1 Department of Biology, University of New Mexico, Albuquerque, New Mexico United States of America, **2** Museum of Southwestern Biology, University of New Mexico, Albuquerque, New Mexico United States of America, **3** Escuela de Biología, Universidad de Costa Rica, San Pedro, San José, Costa Rica

Abstract

Roughly 40% of amphibian species are in decline with habitat loss, disease, and climate change being the most cited threats. Heterogeneity of extrinsic (e.g. climate) and intrinsic (e.g. local adaptations) factors across a species' range should influence population response to climate change and other threats. Here we examine relative detectability changes for five direct-developing leaf litter frogs between 42-year sampling periods at one Lowland Tropical Forest site (51 m.a.s.l.) and one Premontane Wet Forest site (1100 m.a.s.l.) in southwest Costa Rica. We identify individualistic changes in relative detectability among populations between sampling periods at different elevations. Both common and rare species showed site-specific declines, and no species exhibited significant declines at both sites. Detection changes are correlated with changes in temperature, dry season rainfall, and leaf litter depth since1969. Our study species share Least Concern conservation status, life history traits, and close phylogenetic relationship, yet their populations changed individualistically both within and among species. These results counter current views of the uniformity or predictability of amphibian decline response and suggest additional complexity for conservation decisions.

Editor: Michael Sears, Clemson University, United States of America

Funding: A UNM Grove Scholarship, LAII PhD Fellowship, and IdeaWild grant to MJ Ryan and NSF grant DEB 0844624 to S. Poe funded this project. The funders had no role in study design, data collection and analysis, decision to publish, or preparation of the manuscript.

Competing Interests: The authors have declared that no competing interests exist.

* E-mail: mjryan42@gmail.com

Introduction

A primary focus of community ecology is to understand how species respond to environmental variation across space and time [1]. This focus has gained urgency as anthropogenic pressures alter species dynamics [2], pushing many species toward extinction [3]. Long-term studies of community ecology can link shifting population patterns to changes in climate and inform conservation efforts for imperiled species and communities [4,5].

Tropical amphibians are at the forefront of the current extinction crisis [4,6]. As a consequence of climate change, habitat loss, disease, and interactions among these factors, 40% of the 7,125 known amphibian species are at high risk of extinction in the near future [6,7]. Disease-induced population crashes caused by the fungal pathogen *Batrachochytrium dendrobatidis* (*Bd*) have been documented in many amphibian communities [8], but little is known about long-term population trends and non-disease threats in most species [4,7].

Identifying decadal-scale population trends for tropical amphibians has been difficult due to a dearth of historical baseline population data [9,10]. In one case where such data were available, an entire Neotropical leaf litter amphibian fauna was found to have declined over a 35-year period in the Atlantic lowlands of Costa Rica, with declines linked to climate change [11]. Although long-term site-specific studies provide valuable insights into local population responses to environmental change, they cannot assess interpopulation differences in susceptibility across species' ranges [12,13].

Measuring population changes in multiple environments is particularly important because climate change is not occurring uniformly across the landscape [14]. Instead, changes are localized due to variable conditions (e.g., topography, prevailing winds) resulting in a mosaic of novel climatic conditions at small spatial scales [14–16]. Species with broad ranges may include locally adapted populations that exhibit different tolerances to changing environmental conditions and disease [17–19]. Long-term assessments that incorporate population trends across more than one environment are needed to understand range-wide responses of species to change.

Characterizing interpopulation variation in the environmental sensitivity of amphibian species is critical for developing research and conservation priorities in a rapidly changing world [3,12]. Amphibians are especially sensitive to long-term global climate change because warming temperatures and altered hydrologic cycles are expected to increase thermal stress [20], affect disease susceptibility [21], desiccate breeding habitats [22], reduce availability of critical microhabitats [11,23], and alter foraging behavior and efficiency [24,25]. Yet, because climate change is occurring heterogeneously across the landscape [14], it is unclear if all populations of a species are at equal risk [13].

In the Neotropics, frogs of the clade Terrarana dominate leaf litter vertebrate community diversity and abundance [26–28]. Terrarana frogs reproduce by direct development, have no association with aquatic habitats, and depend on standing leaf litter for most aspects of their life, e.g., refugia, foraging, and egg laying sites [28,29]. Because of their strong leaf litter association and independence from aquatic habitats, many Terrarana frog species are expected to be directly or indirectly more vulnerable to climate change than disease [9,11,30]. There is a wide-range of *Bd*-susceptibility within Terrarana with riparian species being more susceptible to *Bd* compared to the terrestrial species studied here [30–32]. Following *Bd*-declines some strictly terrestrial Terrarana species increase in abundance and become dominant components of post-*Bd* assemblages [30]. By studying non-*Bd* vulnerable frog species it is possible to evaluate the effects of environmental change on the remaining species in a post-*Bd* world.

Here we expand the growing field of longitudinal population comparisons (e.g., [11,33,34]) by studying five wide–ranging Terrarana leaf litter frog species in two distinct tropical environments. The broad geographic distributions, syntopy, close phylogenetic relationship, and ecological similarities of these species make them ideal for comparative exploration of long–term population changes. We address relative detection changes within and among species, both within and between environments and over time. We used plot presence/absence data as an assay of relative detection probability from 1969 and 2009–2012 from one Lowland Tropical Forest (51 m.a.s.l.) and one Premontane Wet Forest (1100 m.a.s.l.) environment in southwestern Costa Rica.

Methods

Ethics Statement

The study was approved and conducted under animal care protocol 08UNM041 by the Institutional Animal Care and Use Committee at the University of New Mexico. Costa Rica Research Permits were granted through Javier Guevara at Ministerio de Ambiente y Energia (MINAE) to MJR. This study did not involve any endangered species.

Study Sites and Field Methods

Las Cruces Biological Station (LCBS) protects approximately 227 hectares of Premontane Wet Forest in the Coto Brus Valley (8.785778 N; 82.958889 W Decimal Degrees; 1100 m elevation) on the Pacific versant of the southern Talamanca Mountains, Puntarenas Province, Costa Rica. The 39–year mean annual rainfall is 3442 mm, with a distinct dry season from January–March, and a mean annual temperature of 20.7°C (Table S1; [35]). LCBS has been protected since 1962 and is surrounded by a matrix of smaller fragments and agricultural land [36].

Fundación Neotropica Station is located ~2 km southwest of Rincón de Osa (Rincón; 8.69602 N, – 83.50139 W, 51 m) on the Osa Peninsula in the southwest Pacific lowlands, Puntarenas Province, Costa Rica within the lowland Tropical Forest Zone. The region was forested until the late 1960s when a logging camp was established, and by the 1980s deforestation was nearly complete [37]. By 1996 the flatlands had been converted to pastureland, but the adjacent foothills and steep slopes remain largely forested, including approximately 300 hectares of primary and older secondary forest [37]. Because our surveys require relatively flat forest (see below), all studied plots were in the foothills of the forested mountain, approximately 200 m from cattle pastures. Plots were on the Fundación Neotropica Station or within 1 km of the station on adjacent private property. The 52–year mean annual rainfall for Rincón region is 4730 mm with a distinct dry season from January–March, and a mean annual temperature of 27.5°C (Table S1; [38]).

Participation of the original researcher, Norman J Scott, allowed us to replicate the field data collection and plot set-up techniques used for the 1969 baselines. Each plot was 25-foot square. The original plots were not resampled and each year new plots were sampled to minimize possible impacts from litter removal disturbances during sampling. Plots were placed haphazardly within the forest in flat areas away from trails, tree fall gaps and slopes; we could not completely randomize plot placement within the forest patches. We used Scott's [26] clearing techniques that require removal of all leaf litter to maximize frog observations per plot. After plots were sampled the leaf litter and debris were added back to the plots. During the dry season of March 1969, Scott [26] sampled 10 plots: five at LCBS and five at Rincón; from 2009–2012 we sampled 78 plots, 38 at LCBS and 40 at Rincón in March. We measured leaf litter depth using a ruler at the corners and center of each plot and averaged these measurements for an estimate of leaf litter depth/plot.

Statistical Methods for Relative Detection Probability

We measured changes in detection probability by scoring the proportion of plots occupied, e.g. presence/absence, for each species during a given sampling period (Summarized in Table S1). This is a statistically simplified approach to estimating detection probability, and we refer to this as relative detection probability. We used this approach because of limitations imposed by the original study design that precluded the use of robust algorithm based detection probability approaches such as Program PRESENCE [38]. Algorithm based methodologies have specific assumptions in model building that include primary and secondary sampling periods per field season [39]. Our study does not meet these assumptions because we lack a secondary sampling period. Instead, we attempted to exactly replicate the original study, which did not have a secondary sampling period [26].

Although we also collected abundance data, we analyzed relative detection probability rather than abundance because four of our five focal species were present in low numbers (see below) and such over-dispersion may cause problems for non-logistic approaches [40]. To assess changes in species presence/absence over time, we regarded 1969 as the initial sample period, and all samples recorded between 2009 and 2012 as secondary measurements of a repeated measures experiment. We compared the 1969 samples to the later period formed by pooling the 2009–2012 samples. We investigated temporal changes in presence/absence at each site separately and jointly (e.g., by including a time x site interaction term).

We converted the raw species counts recorded from each plot to presence/absence data, and analyzed the resulting occupancies by logistic regression. Statistical modeling was hindered by the condition of the data, which was characterized by low initial samples size and sparse occupancy (i.e., many empty samples). In addition, the repeated measures design, and probable spatial autocorrelation of samples collected from the same site [41], raised concerns that residual errors may be spatially and temporally correlated. The aforementioned data issues are problematic for standard logistic methods, which rely on maximum likelihood calculations to estimate model parameters. The Firth logistic method, which uses a penalized likelihood method, was developed to overcome computational challenges presented by small sample sizes, data sparsity, and non-independence [42]. Therefore, all of our logistic modeling was performed using the Firth method.

In evaluating the contribution of Site and Period to presence/absence, we constructed a separate model for each species. We considered an effect to be statistically significant if the probability of a non-zero coefficient (i.e., alpha) was less than 0.05. All statistical modeling was performed in R [43].

The year 2011 was a strong La Niña year, which resulted in significantly higher than normal rainfall in lower Central America [44] and at our study sites (unpublished data). The intent of our study was to focus on average trends and not exceptional events such as the strong La Niña. We included 2011 in preliminary analyses (which bolstered our current conclusions of population decline; see below), but because it differed from the other recent sampling years we excluded it from analyses presented here due to concerns of conflating long-term trends with changes due to anomalous climatic effects [44].

Climate Trends

We used meteorological data from the Loma Linda and LCBS meteorological stations to explore long-term climate patterns at LCBS. The Loma Linda (8.7385 N; – 82.922717; 1100 m) station is located 14 km south of LCBS and includes rainfall and temperature data from 1973–2007. The LCBS station has rainfall and temperature data from 2005–2012. Loma Linda and LCBS had comparable weather for the three years of data shared by these stations (2005–2007; Fig. S1 and S2). Combining data for these stations provided a continuous record from 1973–2012.

There are no complete and reliable meteorological records from any weather station in the vicinity of Rincón. To reconstruct the recent historical temperature and rainfall profile we relied on meteorological data from three regional lowland weather stations. Hacienda Barú National Wildlife Refuge (N 9.27152; – 83.88162 W; 24 m) is located 45 km north of Rincón and has documented monthly rainfall from 1981–2011 and temperature from 2001–2011. The Golfito weather station (8.39 N; – 83.11 W; 15 m) of the Instituto Meteorológico Nacional is located approximately 25 km south of Rincón and has kept monthly rainfall records from 1960–1983. We determined that Hacienda Barú and Golfito receive similar rainfall using the same methods as above for the four years of overlap, 1980–1983 (Fig. S3). The third station is in David, Panama (8.4 N, – 82.424167 W; 27 m) 117 km east of Rincón and the closest Pacific lowland station with long-term monthly temperature data from 1973–2000 [16]. We combined the weather data from these three stations to reconstruct general decadal climate change for the Pacific lowlands near Rincón [16].

We used linear regression to explore long-term changes in rainfall and temperature at each site. The binned wet and dry season rainfall and temperature data allowed us to test long-term, seasonal rainfall and temperature trends. We follow McDiarmid and Savage [38] in classifying months with less than 200 mm of precipitation as dry season (i.e., January–March).

Results

Changes in Relative Detection Probability

We were able to calculate relative detection probability changes for all five species from the plot presence/absence proportions for 1969 and 2009–2012 (excluding 2011) (Table 1). The general Firth logistic regression model results show that relative detection probability significantly changed for three species between sampling periods and one species showed a significant time Period X Site interaction (Table 2).

The site-specific Firth logistic regression results show individualistic changes in relative detection among species and sites. One species, *Craugastor stejnegerianus,* showed no change in relative detection at both sites over time, and the other species exhibited unique changes between sites over time (Table 3, Fig. 1). Three of five species showed a negative change in relative detection at the mid-elevation site, LCBS; two of these species, *C. rugosus* and *Diasporus vocator*, can be considered rare in our samples and were not detected in the later sampling period. At Rincón, two species showed a negative change, *C. crassidigitus* and *C. rugosus; C. rugosus* was rare in our samples and was not detected during the later sampling period (Fig. 1).

Climate Variables

We observed a significant increase in annual mean minimum temperature at Rincón of 0.059°C/year ($R^2 = 0.558$; $P = 0.0001$) and at LCBS of 0.064°C/year ($R^2 = 0.420$; $P = 0.0001$) since 1973 (Fig. 2). We found dry season precipitation changed divergently for each elevation. At Rincón, dry season precipitation significantly decreased by 8.47 mm/year on average since 1960 ($R^2 = 0.324$; DF = 55; $P = 0.0001$). There was no significant change in wet season ($R^2 = 0.044$; DF = 55; $P = 0.129$) or annual precipitation ($R^2 = 0.006$; DF = 55; $P = 0.554$). Conversely, at LCBS dry season precipitation significantly increased by 2.24 mm/year on average since 1973 ($R^2 = 0.148$; DF = 34; $P = 0.018$) with no significant change in wet season ($R^2 = 0.028$; DF = 34; $P = 0.339$) or annual precipitation ($R^2 = 0.042$; DF = 34; $P = 0.242$).

Litter Depth

At LCBS mean leaf litter depth decreased significantly from 7.02±2.11 cm (Std Dev) in 1969 to 4.88±1.69 cm in the 2000s (Kruskal-Wallis $\chi^2 = 4.425$; DF = 1; $P = 0.037$). At Rincón mean leaf litter depth exhibited no significant change between March 1969 and the 2000s (Kruskal-Wallis $\chi^2 = 1.553$; DF = 1; $P = 0.219$).

Discussion

Relative detection probability changes in the five frog species indicate two broad patterns in time and space that are associated with substantial changes in temperature, dry season precipitation and leaf litter depth. First, the site-specific results show relative detectability changes varied among species between sites, with neither site exhibiting uniform declines across all species between 42-year sampling periods (Fig. 1). Second, we observed substantial intraspecific variation in different environments between 42-year sampling periods. Our results suggest that these widely distributed leaf litter frogs show individualistic responses to environmental change, a pattern that fits the individualistic or Gleasonian ecological view [12,45].

The scope, design and results of our study differ from previous tropical amphibian decline studies in three main ways. First, previous amphibian population change studies documented community-wide declines within a single locality and environment (i.e. lowland rainforest or montane cloud forest) [11,46,47]. Our study measured simultaneous changes in the same five species across two distinct environments over time and found no uniform, community-wide decline. Our approach assessed inter- and intraspecific responses at broader geographic scales than previous work. Second, previous mid-elevation amphibian decline studies have focused on highly *Bd*-vulnerable riparian species that quickly declined (i.e., [30,47,48]) rather than terrestrial leaf litter species that are likely to be more vulnerable to climate change than *Bd* [9,11,30]. Our focus on non-*Bd*-vulnerable leaf litter frogs, 20-years after *Bd* arrived in the region (e.g. early 1990's) (32), allowed us to investigate responses associated with climate change rather than the confounding or direct effects of disease. Third, we rely on changes in relative detection, instead of the more commonly used count data [11,48], to measure if species have become more rare over time. This approach is more conservative than using count

Table 1. Plot presence/absence for 1969 and 2000s sampling periods.

Species	LCBS 1969	LCBS 2000s	Rincón 1969	Rincón 2000s
Craugastor crassidigitus	2/5 [20%]	9/28 [32%]	4/5 [80%]	1/30 [3%]
Craugastor rugosus	2/5 [20%]	0/28 [0%]	1/5 [20%]	0/30 [0%]
Craugastor stejnegerianus	5/5 [100%]	17/28 [60%]	5/5 [100%]	27/30 [90%]
Diasporus vocator	3/5 [60%]	0/28 [0%]	2/5 [40%]	2/30 [6%]
Pristimantis ridens	4/5 [80%]	4/28 [14%]	1/5 [20%]	5/30 [16%]

Proportions of plots occupied by each species at LCBS and Rincón between sampling periods. Values in brackets are proportion of plots occupied during that sampling period. These data were used to calculate relative detection probabilities used in the logistic regression analysis. See Table S1 for presence/absence for each individual year.

data to assess population change because it is less vulnerable to overestimating the magnitude of change for species with low sample sizes [49].

Amphibian populations are susceptible to stochastic variation and distinguishing natural amphibian population fluctuations from directional short- or long-term declines has been problematic [50,51]. Extreme population fluctuations tend to be driven by unpredictable changes to aquatic breeding habitats, droughts, or deluges, impacting aquatic breeding species more than terrestrial breeding species [51,52]. With no ties to running or standing water for reproduction, Terrarana frogs are not expected to exhibit short-term population fluctuations characteristic of aquatic species [11,52]. Because the sampling years in our analysis represent periods of relatively constant weather (Fig. 2) it is unlikely our patterns reflect short-term changes. During the 2011 sampling year that we removed from analysis, both sites received approximately 50% more rainfall than the long-term means. We excluded this year in order to maintain comparability with the early (1969) sampling period, which received an unexceptional amount of rain.

Site-specific Patterns

Because our study species are phylogenetically closely related and ecologically uniform, similar population responses to environmental change might be expected [30,53]. Yet we observed individualistic responses among species, with no general community wide trends at either site despite significant changes in temperature and dry season rainfall (Fig. 2). This result is consistent with patterns observed in North American small mammals and birds. Moritz et al [33] found elevational range shifts to be variable among small mammal congeners over a 100-year period [33]. Taper et al. [12] found differing patterns of decline among species of insectivorous songbirds. These examples of species-specific responses highlight the complexity of predicting individual species and community responses to climate change.

Why did we observe species-specific responses among closely related, ecologically similar frog species? We can only speculate on mechanisms. Body size is an important predictor of species extinction threat from disease or climate change with large species at greater risk than small species [32,54], but there is no relationship of body size and population trends among our species (Fig. 2, Table 1). We hypothesize that undetected micro–ecological differences (breeding phenology, diet, etc.) will emerge with future study of these species. For example, *Diasporus vocator* may have declined at LCBS because increased dry season rainfall saturated the soil leading to high rates of egg mortality [55]. Increased dry season rainfall was not evident at Rincón, and perhaps the eggs of *D. vocator* are especially sensitive to changes in rainfall patterns. This scenario is speculative, as we do not understand the ecology of *D. vocator* at this fine level. Autecological studies of our five frog species are needed to shed light on their varying responses.

Intraspecific Patterns

Observed differences in relative detection changes within species show that these leaf litter frogs respond to long-term environmental change individualistically. The majority of our species showed a decrease in detection at one site but not the other site between the 42-year sampling periods (Fig. 2). A similar result has been found in birds, where 77% of 47 species that occurred in more than one environment varied in degree of population change between environments [12].

These individualistic responses could be attributed to varying environmental stressors at each site differentially affecting traits that are constant across species (see previous section). Alternative-

Table 2. General logistic regression results.

Species	Time Period *P*-value	Coefficient ± STD Error	Site *P*-value	Coefficient ± STD Error	Time Period X Site *P*-value	Coefficient ± STD Error
Craugastor crassidigitus	0.082	−0.23±0.91	0.236	1.44±1.37	0.032*	−2.93±1.54
Craugastor rugosus	0.007**	−3.71±1.71	0.548	−0.76±1.37	0.770	0.69±2.47
Craugastor stejnegerianus	0.106	−1.98±1.66	1.00	0.00±2.29	0.468	1.64±2.39
Diasporus vocator	0.007**	−3.71±1.71	1.00	0.00±1.28	0.385	1.60±2.05
Pristimantis ridens	0.004**	−2.79±1.16	0.076	−2.20±1.46	0.099	2.36±1.62

Logistic regression results relative detection probability calculated from the plot occupancy.
See text for details on analyses. Significance levels: * = 0.05; ** = 0.01.

Table 3. Firth logistic regression results.

Species	Rincón *P*-value	Coefficient ± STD Error	LCBS *P*-value	Coefficient ± STD Error
Craugastor crassidigitus	0.001**	−3.16±1.18	0.802	−0.23±0.99
Craugastor rugosus	0.054	−3.01±1.78	0.007**	−3.71±1.71
Craugastor stejnegerianus	0.825	−0.34±1.72	0.106	−1.98±166
Diasporus vocator	0.052	−2.10±1.13	0.007**	−3.71±1.71
Pristimantis ridens	0.696	−0.44±1.14	0.004**	−2.79±1.16

Site-specific results of Firth logistic regression between sampling periods.
**denotes 0.01 significance level.

ly, local variation in decline susceptibility within species may produce our observed patterns. For example, *Craugastor crassidigitus* declined at Rincón but not LCBS. Temperature increased significantly at both sites, so perhaps the population of *C. crassidigitus* at Rincón is sensitive to warmer temperatures but the population of *C. crassidigitus* at LCBS is tolerant to increased temperatures. Such a scenario likely oversimplifies the complexities of ecological interactions that may be operating. Additional factors such as invertebrate predation [56] and physiological stresses associated with increased temperatures and altered rainfall patterns [20,57], or other dynamic interactions, may drive local changes. Regardless of the mechanisms, intraspecific variation in response to local environmental change is evident among both common and rare species.

One species, *Craugastor rugosus,* was rare at both sites in 1969 and not detected during the later sampling period (Table 1) and warrants special consideration. Because we did not detect *C. rugosus* during the later sampling period, we would infer that this species was locally extirpated. However, we conducted transect surveys to supplement the plot method and detected *C. rugosus*. Detection of this species using a secondary method contradicts the inference from our plot data. We suggest that when replicating historical population comparisons to assess declines, alternative survey methods should be employed to detect rare species. Such

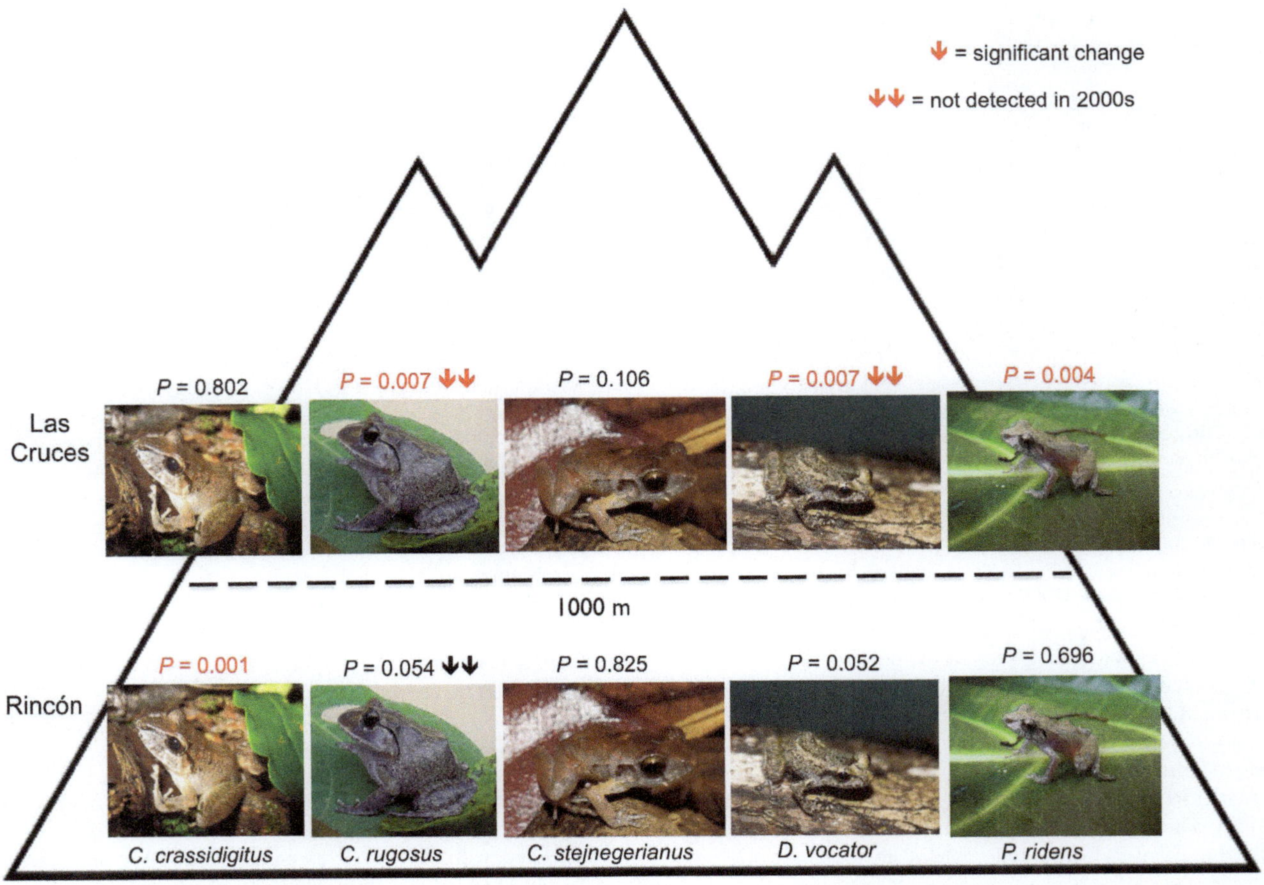

Figure 1. Changes in detection of the five species between the two sites. The site-specific Firth logistic regression *P*–values and direction of relative detection changes between sampling periods and elevation for each species. Upper row is LCBS and lower row is Rincón. See text for details of analysis and Table 1 for raw data.

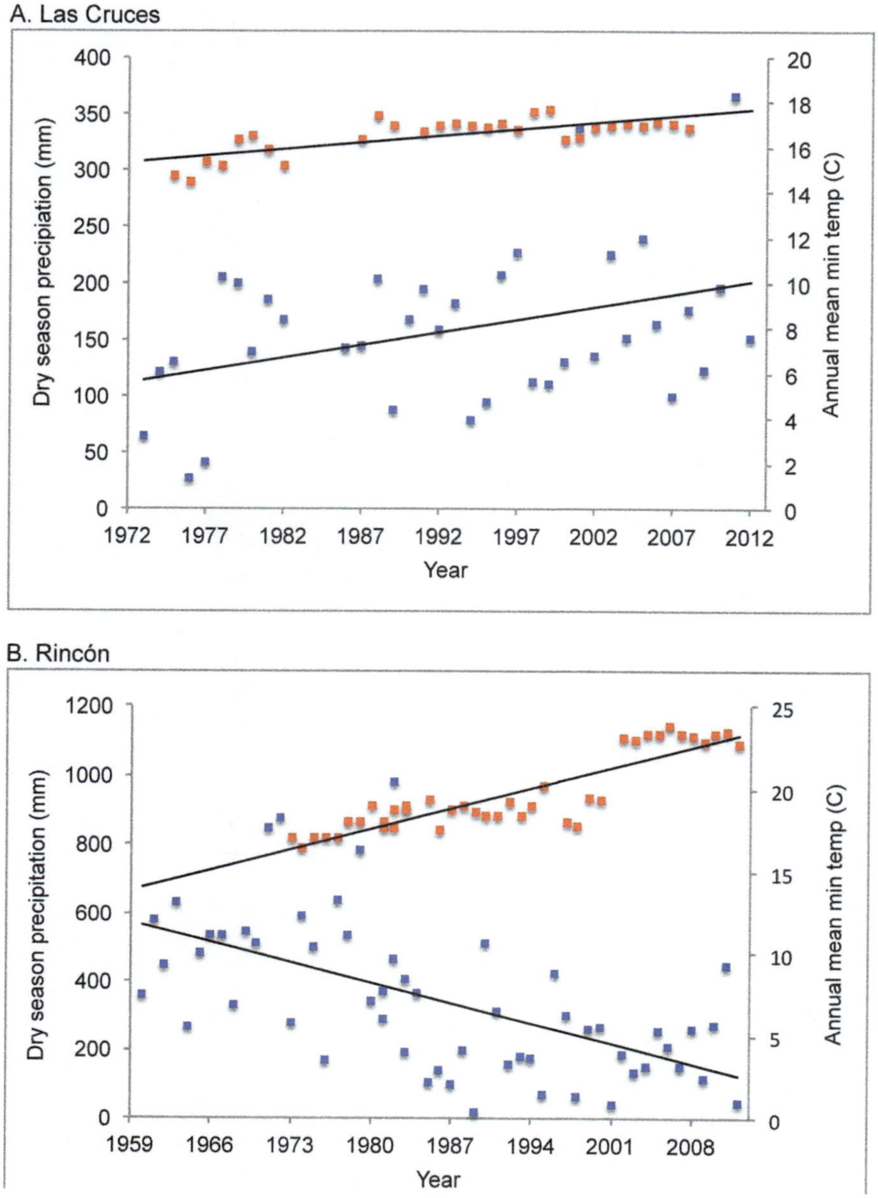

Figure 2. Climatic changes for LCBS and Rincón. Long-term trends in dry season precipitation and minimum annual temperature for A) LCBS and B) Rincón. Red squares represent temperature; blue squares represent dry season rainfall.

multifaceted approaches are especially needed when assessing population extirpations. Reliance on a single field survey method may overestimate a species' threatened status.

Finally, due to the ubiquity of *Bd* in Costa Rica and it's role in amphibian declines we cannot rule out the possibility that at least some of our observed changes are related to disease. This is more of a concern at LCBS than Rincón because no severe *Bd*-declines and die-offs have been reported at tropical lowland sites, despite *Bd* being detected at low elevations [58,59]. *Bd* arrived in the LCBS region in 1993, almost 20-years before our later sampling period [32]. There is no rigorous documentation of the *Bd* die-off at LCBS, but many Bd-susceptible species such as *Atelopus varius*, *Craugastor ranoides*, and others are now absent from the frog fauna [60]. It is not possible for us to determine whether *Bd* played a role in the decline of our study species at LCBS. However, we note that Picco and Collins [61] detected *Bd* at LCBS but did not detect *Bd* on any strictly terrestrial, direct-developing frog species there, including two of our study species. Furthermore, our study species and their relatives are known to increase in community dominance [30] and abundance [62] within four years of *Bd*-related faunal collapse. We suspect that *Bd* has had an impact on many amphibian species at LCBS. However there is no evidence that any of our study species has been affected.

Conclusions

We documented individualistic changes of frog populations between a 42–year period at two distinct sites. Observed changes are associated with increased temperatures, altered dry season rainfall, and changed leaf litter depth, all of which influence leaf litter amphibian populations [7,11].

Increased rarity in some populations and not others is both troubling and optimistic in terms of long-term persistence of these leaf litter frogs. Our results are optimistic because we did not observe local extirpations of rare or common species despite decades of environmental change and disease emergence. Instead, common species have remained relatively common and rare species have remained rare, albeit at much lower detectability than in the past. This result is not consistent with other Neotropical studies that found declines and extirpations in both common and rare species [11,46,48]. On the other hand, these results are troubling because we detected declines in species of Least Concern not previously reported to have declined [6]. All of our study species have been categorized as Least Concern by the International Union for Conservation of Nature [6]. Population declines in species of Least Concern may be subtler than those of endangered species [63]. The apparently slow population attrition we detected contrasts with the rapid population crashes characteristic of many endangered forms [48,64].

The above concerns clearly are pertinent to conservation decisions, but policy implications are not straightforward. Perhaps instead of focusing on a species as a whole, conservation actions should address local, geographically threatened or declining populations. The complicating factors we have identified, including decline of some populations of Least Concern species, interspecific variation among ostensibly ecologically uniform species, and differing intraspecific responses across space and time, should provide ample material for future discussions of conservation.

Acknowledgments

Thanks to the Organization of Tropical Studies Las Cruces Biological Station and Fundación Neotropica Research Station for allowing access to study sites. We thank Z. Zahawi from Las Cruces and L. Umaña from Fundación Neotropica for permission and logistical support. Permits for research in Costa Rica were granted from MINAE and J. Guevara. We thank Instituto Metereológico Nacional de Costa Rica. We thank the dozens of students in the UNM Tropical Biology Course for assistance in field data collection. This manuscript was greatly improved by suggestions by M. Workman, M. Balk, I. Latella, and three anonymous reviewers. Photograph of *C. stejnegerianus* was provided by K. Christenson.

Author Contributions

Conceived and designed the experiments: MJR NJS. Performed the experiments: MJR NJS JAC BW GC FB. Analyzed the data: MJR MMF. Wrote the paper: MJR MMF SP. Provided comments: NJS JAC BW GC FB.

References

1. MacArthur RH (1972) Geographical ecology, patterns in the distribution of species. Princeton: Princeton University Press. 269 p.
2. Parmesan C (2006) Ecological and evolutionary responses to recent climate change. Annu Rev Ecol Evol Syst 37: 637–669.
3. Pereira HM, Leadley PW, Proença V, Alkemade R, Scharlemann JPW, et al. (2010) Scenarios for global biodiversity in the 21st century. Science 330: 1496–1501.
4. Wake DB, Vredenburg VT (2008) Are we in the midst of the sixth mass extinction? A view from the world of amphibians. Proc Natl Acad Sci USA 105: 11466–11473.
5. Blaustein AR, Han BA, Relyea RA, Johnson PTJ, Buck JC, et al. (2011) The complexity of amphibian population declines: understanding the role of cofactors in driving amphibian losses. Ann NY Acad Sci 1223: 108–119.
6. Stuart SM, Chanson JS, Cox NA, Young BE, Rodrigues ASL, et al. (2004) Status and trends of amphibian declines and extinctions worldwide. Science. 306: 1783–1786.
7. Hof C, Araújo MB, Jetz W, Rahbek C (2011) Additive threats from pathogens, climate and land-use change for global amphibian diversity. Nature 480: 516–519.
8. Kilpatrick AM, Briggs CJ, Daszak P (2009) The ecology and impact of. chytridiomycosis: an emerging disease of amphibians. Trends Ecol Evol 25: 109–118.
9. Donnelly MA, Crump ML (1998) Potential effects of climate change on two Neotropical amphibian assemblages. Clim Change 39: 541–561.
10. Balmford A, Green RE, Jenkins M (2003) Measuring the changing state of nature. Trends Ecol Evol 18: 326–330.
11. Whitfield SM, Bell KE, Philippi T, Sasa M, Bolaños F, et al. (2007) Amphibian and reptile declines over 35 years at La Selva, Costa Rica. Proc Natl Acad Sci USA 104: 8352–8356.
12. Taper ML, Böhning-Gaese K, Brown JH (1995) Individualistic responses of bird species to environmental change. Oecologia 101: 478–486.
13. Hanski I, Mononen T (2011) Eco-evolutionary dynamics of dispersal in spatially heterogeneous environments. Ecol Lett 14: 1025–1034.
14. Loarie SR, Duffy PB, Hamilton H, Asner GP, Field CB, et al. (2009) The velocity of climate change. Nature 462: 1052–1055.
15. Hulme M, Viner D (1998) A climate change scenario for the tropics. Clim Change 39: 5–36.
16. Aguilar E, Peterson TC, Frutos R, Retana JA, Solera M, et al. (2005) Changes in precipitation and temperature extremes in Central America and northern South America, 1961–2003. J Geophys Res Biogeosci 110, D23107. doi:10.1029/2005JD006119.
17. Puschendorf R, Hoskin CJ, Cashins SD, McDonald K, Skerratt LF, et al. (2011) Environmental refuge from disease-driven amphibian extinction. Conserv Biol 25: 956–964.
18. Moritz C, Langham G, Kearney M, Krockenberger A, VanDerWal J, et al. (2012) Integrating phylogeography and physiology reveals divergence of thermal traits between central and peripheral lineages of tropical rainforest lizards. Philos Trans R Soc Lond B Biol Sci 367: 1680–1987.
19. Gordo I, Campos PRA (2013) Evolution of clonal populations approaching a fitness peak. Biol Lett doi:10.1098/rsbl.2012.0945.
20. Angilleta MJ (2010) Thermal adaptation. Oxford: Oxford University Press. 290 p.
21. Rohr JR, Raffel TR (2010) Linking global climate and temperature variability to widespread amphibian declines putatively caused by disease. Proc Natl Acad Sci USA 107: 8269–8274.
22. McMenamin SK, Hadly EA, Wright CK (2008) Climatic change and wetland desiccation cause amphibian decline in Yellowstone National Park. Proc Natl Acad Sci USA 105: 16988–16993.
23. Rouifed S, Handa TI, David JF, Hättenschwiler S (2010) The importance of biotic factors in predicting global change effects on decomposition of temperate forest leaf litter. Oecologia 163: 247–256.
24. Lensing JR, Wise DH (2006) Predicted climate change alters a trophic cascade that affects an ecosystem process. Proc Natl Acad Sci USA 103: 15502–15505.
25. Stuble KL, Pelini SL, Diamond SE, Fowler DA, Dunn RR, et al. (2013) Foraging by forest ants under experimental climatic warming: a test at two sites. Ecol Evol 3: 482–491.

26. Scott NJ (1976) The abundance and diversity of the herpetofauna of tropical forest litter. Biotropica 8: 41–58.
27. Toft CA (1981) Feeding ecology of Panamanian litter anurans: patterns in diet and foraging mode. J Herpetol 15: 139–144.
28. Stewart MM, Woolbright LL (1996) Amphibians. In: Reagan DP, Waide RB, editors. The food web of a tropical rain forest. Chicago: Chicago University Press. 273–320.
29. Savage JM (2002) The amphibians and reptiles of Costa Rica, a herpetofauna between two seas, between two continents. Chicago: University of Chicago Press. 934 p.
30. Smith KG, Lips KR, Chase JM (2009) Selecting for extinction: nonrandom disease-associated extinction homogenizes amphibian biotas. Ecol Lett 12: 1069–1078.
31. Ryan MJ, Lips KR, Eichholtz ME (2008) Decline and extirpation of an endangered Panamanian stream frog population (*Craugastor punctariolus*) due to an outbreak of chytridiomycosis. Biol Conserv 141: 1636–1647.
32. Lips KR, Reeve JD, Witters LR (2003) Ecological traits predicting amphibian population declines in Central America. Conserv Biol 17: 1078–1088.
33. Moritz C, Patton JL, Conroy CJ, Parra, White GC, et al. (2008) Impact of climate change on small-mammal communities in Yosemite National Park, USA. Science 322: 261–264.
34. Tingley MW, Monahan WB, Beissimger SR, Moritz C (2009) Birds track their Grinnellian niche through a century of climate change. Proc Natl Acad Sci USA 106: 19637–19643.
35. Dosch JJ, Peterson CJ, Haines BL (2007) Seed rain during initial colonization of replicate abandoned pastures in the premontane wet forest zone of southern Costa Rica. J Trop Ecol 23: 1–9.
36. Daily GC, Ceballos G, Pacheco J, Suzán G, Sánchez-Azofeifa A (2003) Countryside biogeography of Neotropical mammals: conservation opportunities in agricultural landscapes of Costa Rica. Conserv Biol 17: 1814–1826.
37. Joyce AT (2006) Land use change in Costa Rica: 1966–2006. Litografía e Imprente LIL S.A. 279 p.
38. McDiarmid RW, Savage JM (2005) The herpertofauna of the Rincón area, Península de Osa, Costa Rica, a Central American lowland evergreen forest site. In: Donnelly MA, Crother BI, Guyer C, Wake MH, White ME, editors. Ecology and evolution in the tropics, a herpetological perspective. Chicago: University of Chicago Press. 366–427.
39. MacKenzie DI, Nichols JD, Royle JD, Pollock KH, Bailey LA, et al. (2006) Occupancy estimation and modeling. Boston: Academic Press. 324 p.
40. Cameron AC, Trivedi PK (1998) Regression Analysis of Count Data. Cambridge: Cambridge University Press. 436 p.
41. Legendre P, Fortin M-J (1989) Spatial pattern and ecological analysis. Vegetatio 80: 107–138.
42. Heinze G, Schemper M (2002) A solution to the problem of separation in logistic regression. Stat Med 21: 2409–2419.
43. R Development Core Team (2008) R: A language and environment for statistical computing. R Foundation for Statistical Computing, Vienna, Austria ISBN 3-900051-07-0.
44. Boening C, Willis JK, Landerer FW, Nerem RS, Fasullo J (2012) The 2011 La Niña: so strong, the oceans fell. Geophys Res Lett 39: L19602. doi:10.1029/2012GL053055.
45. Mac Nally R, Nerenberg S, Thomson JR, Lada H, Clarke RH (2014) Do frogs bounce, and if so, by how much? Responses to the 'big wet' following the 'big dry' in south-eastern Australia. Global Ecol Biogeogr 23: 223–234.
46. Pounds JA, Fogden MPL, Campbell JH (1999) Biological response to climate change on a tropical mountain. Nature 398: 611–615.
47. Lips KR, Brem F, Brenes R, Reeve JD, Alford RA, et al. (2006) Emerging infectious disease and the loss of biodiversity in a Neotropical amphibian community. Proc Natl Acad Sci USA 103: 3165–3170.
48. Crawford AJ, Lips KR, Birmingham E (2010) Epidemic disease decimates amphibian abundance, species diversity, and evolutionary history in the highlands of central Panama. Proc Natl Acad Sci USA 107: 13777–13782.
49. Dorazio RM, Martin J, Edwards HH (2013) Estimating abundance while accounting for rarity correlated behavior, and other sources of variation in counts. Ecology 94: 1472–1479.
50. Pechmann JHK, Scott DE, Semlitsch RD, Caldwell JP, Vitt LJ, et al. (1991) Declining amphibian populations: the problem of separating human impacts from natural fluctuations. Science 253: 892–895.
51. Alford RA, Richards SJ (1999) Global amphibian declines: a problem in applied ecology. Annu Rev Ecol Evol Syst 30: 133–165.
52. Green DM (2003) The ecology of extinction: population fluctuation and decline in amphibians. Biol Conserv 111: 331–343.
53. Cooper N, Bielby J, Thomas GH, Purvis A (2007) Macroecology and extinction risk correlates of frogs. Global Ecol Biogeogr doi:10.1111/j.1466-8238.2007.00355.x.
54. Sodhi NS, Bickford D, Diesmos AC, Lee TM, Koh LP, et al. (2008) Measuring the Meltdown: Drivers of Global Amphibian Extinction and Decline. PLoS ONE 3(2): e1636. doi:10.1371/journal.pone.0001636.
55. Socci AM, Schlaepfer MA, Gavin TA (2005) The importance of soil moisture and leaf cover in a females lizard's (*Norops polylepis*) evaluation of potential oviposition sites. Herpetologica 61: 233–240.
56. Chalcraft DR, Andrews RM (1999) Predation on lizard eggs by ants: species interactions in a variable physical environment. Oecologia 119: 285–292.
57. McCain CM, Colwell RK (2011) Assessing the threat to montane biodiversity from discordant shifts in temperature and precipitation in a changing climate. Ecol Lett 14: 1236–1245.
58. Kilburn VL, Ibáñez R, Green DM (2011) Reptiles as potential vectors and hosts of the amphibian pathogen *Batrachochytrium dendrobatidis* in Panama. Dis Aquat Organ 97: 127–134.
59. Whitfield SM, Kerby J, Gentry LR, Donnelly MA (2012) Temporal variation in infection prevalence by the amphibian chytrid fungus in three species of frogs at La Selva, Costa Rica. Biotropica 44: 779–784.
60. Santos-Barrera G, Pacheco J, Mendoca-Quijano F, Bolaños F, Chaves G (2008) Diversity, natural history and conservation of amphibians and reptiles from the San Vito Region, southwestern Costa Rica. Rev Biol Trop 56: 755–778.
61. Picco AM, Collins JP (2006) Fungal and viral pathogen occurrence in Costa Rica amphibians. J Herpetol 41: 746–749.
62. Pounds JA, Fogden MPL, Savage JM, Gorman GC (1997) Tests of null models of amphibian declines on a tropical mountain. Conserv Biol 11: 1307–1322.
63. Adams MJ, Miller DAW, Muths E, Corn PS, Grant EHC, et al. (2013) Trends in amphibian occupancy in the United States. PLoS ONE 8: e64347. doi:10.1371/journal.pone.0064347.
64. Lips KR, Diffendorfer J, Mendelson JR III, Sears MW (2008) Riding the wave: Reconciling the roles of disease and climate change in amphibian declines. PLoS Biol 6(3): e72. doi:10.1371/journal.pbio.0060072.

Stochasticity and Determinism: How Density-Independent and Density-Dependent Processes Affect Population Variability

Jan Ohlberger*, Lauren A. Rogers, Nils Chr. Stenseth

Centre for Ecological and Evolutionary Synthesis (CEES), Department of Biosciences, University of Oslo, Oslo, Norway

Abstract

A persistent debate in population ecology concerns the relative importance of environmental stochasticity and density dependence in determining variability in adult year-class strength, which contributes to future reproduction as well as potential yield in exploited populations. Apart from the strength of the processes, the timing of density regulation may affect how stochastic variation, for instance through climate, translates into changes in adult abundance. In this study, we develop a life-cycle model for the population dynamics of a large marine fish population, Northeast Arctic cod, to disentangle the effects of density-independent and density-dependent processes on early life-stages, and to quantify the strength of compensatory density dependence in the population. The model incorporates information from scientific surveys and commercial harvest, and dynamically links multiple effects of intrinsic and extrinsic factors on all life-stages, from eggs to spawners. Using a state-space approach we account for observation error and stochasticity in the population dynamics. Our findings highlight the importance of density-dependent survival in juveniles, indicating that this period of the life cycle largely determines the compensatory capacity of the population. Density regulation at the juvenile life-stage dampens the impact of stochastic processes operating earlier in life such as environmental impacts on the production of eggs and climate-dependent survival of larvae. The timing of stochastic versus regulatory processes thus plays a crucial role in determining variability in adult abundance. Quantifying the contribution of environmental stochasticity and compensatory mechanisms in determining population abundance is essential for assessing population responses to climate change and exploitation by humans.

Editor: Christopher Joseph Salice, Texas Tech University, United States of America

Funding: JO was supported by VISTA (http://www.vista.no), a basic research program funded by Statoil and conducted in close collaboration with The Norwegian Academy of Science and Letters (http://english.dnva.no). LAR acknowledges support by the Norden Top-Level Research Initiative through the Nordic Centre for Research on Marine Ecosystems and Resources under Climate Change (NorMER, http//www.normer.uio.no). The funders had no role in study design, data collection and analysis, decision to publish, or preparation of the manuscript.

Competing Interests: The lead author received financial support through a basic research programme (VISTA), which is coordinated by the Norwegian Academy of Science and Letters, and funded by a commercial source (Statoil). This does not alter the authors' adherence to PLOS ONE policies on sharing data and materials.

* E-mail: jan.ohlberger@ibv.uio.no

Introduction

The relative importance of density-independent and density-dependent processes has long been a matter of debate in population ecology [1,2]. Both stochastic density-independent factors, for instance related to climate variation, and density-dependent processes (e.g. competition for food or habitat) influence the survival of individuals in a population and ultimately determine the variation in population abundance through time. The number of individuals at any life-history stage depends on the survival experienced by all life-stages preceding the given stage and the initial number of offspring, both of which may vary considerably from year to year. Processes occurring at early life stages are thus of great importance as they shape the dynamics of the reproductive life-stage. The question of how stochastic and density-dependent processes affect early life-stages of an organism is therefore of general interest to population ecologists, and has recently attracted increased attention due to our efforts to predict the biological consequences of climate change.

While there generally is consensus that density dependence is a feature of population dynamics for most species [3,4], it is crucial to understand how it operates and when during the life cycle it occurs. The relative timing of density dependence determines how stochastic variation (e.g. through climate) translates into changes in adult abundance. This is of particular interest in fisheries, because harvested fish stocks often show large fluctuations in recruitment [5–7]. While climate is expected to strongly affect survival of eggs and larvae in marine fish [8,9], density dependence often occurs later in life, particularly among juvenile fish [10,11]. Compensatory density dependence ('compensation'), which occurs when population growth is reduced at high abundance, tends to reduce population variability and partially offsets the loss of individuals that are harvested [3,11,12]. Compensation is implied in stock-recruitment relationships commonly used in models of fisheries population dynamics (e.g. Beverton-Holt). Density-dependent survival in juveniles modifies the impact of climate factors operating early in life, and leads to lower variability in year-class strength of adults compared to early life-stages [10]. Quantifying density dependence in different phases of the life cycle is thus important for conservation and fisheries management.

Evidence of density-dependent survival generally comes from time-series analysis [4], suggesting a causal relationship between population abundance and survival. Assessing consequences of this process for a population is however difficult without a model that incorporates all life-stages of an organism and thus covers the entire life cycle, including a relationship between spawners and recruitment. Life-cycle models dynamically link the multiple effects of intrinsic and/or extrinsic factors on the different life-stages from one time period to the next, thereby propagating information and uncertainty about influences on one life-stage to inform processes affecting other life-stages [13]. This is crucial for evaluating population-level effects of multiple intrinsic and extrinsic factors that operate at different times during a life cycle, such as density dependence, climate variation, and harvesting.

Here, we develop a life-cycle model for the population dynamics of a large marine fish, Atlantic cod (*Gadus morhua*). The Northeast Arctic (NEA) stock is currently the largest cod stock in the world and sustains a large and profitable fishery. Using a state-space approach we are able to incorporate catch statistics and scientific survey data on all cod life-stages from different time periods (covering >50 years), while accounting for a stochastic environment, inaccurate sampling, and non-independence of data. Our goal is to disentangle the effects of stochasticity and density regulation on the survival of pre-recruit life-stages, and to quantify the strength of density dependence and its consequences for population variability.

Methods

Northeast Arctic cod

Northeast Arctic cod are distributed along the coast of Norway and throughout the Barents Sea. Mature cod spawn from February to April along the Norwegian coast, particularly around the Lofoten islands. Spawned eggs drift northeastward along the coast in the Norwegian Coastal Current, and hatch into larvae less than a month after spawning and before reaching the feeding grounds in the Barents Sea. By late summer the 0-group fish (pelagic juveniles) are distributed over large areas of the Barents Sea, where they settle to the bottom in late autumn (demersal juveniles). The fish remain in the Barents Sea until they reach maturity and start their annual migrations to the spawning grounds, typically at 6-8 years old [14]. Fig. 1 shows the general life cycle and ecology of NEA cod and the observational data used in this study. Over the past 65 years, total stock biomass of NEA cod has fluctuated between 1 and 4 million tons, and annual fishery landings of 0.2–1.3 million tons have been reported [15].

Data sources

We used two types of observational data: indices of age-specific abundances from scientific surveys and age-specific fisheries landings. Fisheries landings data (1959-present) were obtained from international statistics based on NEA cod landings in the Norwegian and Barents Seas [15]. Indices of the abundance of cod eggs and larvae (1959–1990) were obtained from ichthyoplankton surveys in spring (April-May) and summer (June-July) in the Norwegian and Barents Seas by the Knipovich Polar Research Institute of Marine Fisheries and Oceanography (PINRO), Murmansk, Russia [16,17]. An abundance index for 0-group cod (1966-present) was obtained from autumn samplings (August-September) by the international 0-group survey in the Barents Sea [15,18] as described in [19]. Abundance indices of juvenile and adult cod of age 1–9 (1981-present) were obtained from Norwegian bottom trawl surveys in winter in the Barents Sea [15]. The above-mentioned surveys have good spatial coverage in those areas inhabited by the different cod life-stages. We used annual data (1959-present) on weight-at-age and age-specific probabilities of being mature [15] to calculate spawning stock biomass from the estimated abundances. Temperature records (1959-present) were obtained from the Kola meridian transect (33°30′E, 70°30′N to 72°30′N), a good indicator of the thermal conditions in the Barents Sea. Monthly values were calculated by averaging temporally along the transect and vertically from 0–200 m water depth (historical data from [20]; recent values provided by PINRO: www.pinro.ru).

Process model

The process model describes changes in age-specific abundance over time. It consists of a dynamic age-structured population model with lognormal process noise, which refers to stochasticity in the age-/stage-specific survival due to environmental effects. Spawning stock biomass (SSB) in a given year (y) is calculated from age-specific abundance ($N_{a,y}$), weight-at-age ($W_{a,y}$), and the probability of being mature ($P_{a,y}$):

$$SSB_y = \sum_{a=1}^{a=A} N_{a,y} P_{a,y} W_{a,y}$$

Here, a denotes age and A is maximum age. The number of eggs (N_e) is then calculated from spawning stock biomass and fecundity:

$$N_{e,y} = SSB_y \tau e^{\varepsilon sp}$$

Here, τ is mean fecundity (eggs kg^{-1}), and $\varepsilon_{sp} \sim N(0,\sigma_{sp})$ is a Gaussian error term with mean zero and variance σ_{sp} describing stochasticity in the spawning process, that is, process error associated with mean fecundity, fertilization of eggs, and egg survival prior to the egg survey (for estimates of annual egg production see [21]). The number of larvae (N_l) is calculated from egg abundance and mortality (M_e, day^{-1}) during the period between egg and larval surveys (t_e, days):

$$N_{l,y} = N_{e,y} e^{-t_e M_e}$$

Previous studies have shown that warmer temperature is associated with higher survival of early life-stages of cod, particularly larvae [22,23]. We tested for temperature dependence of egg and larval survival, and found that the effect on egg survival was highly correlated with and weaker than the effect on larval survival. We thus model temperature effects on larval but not egg survival. The number of 0-group fish (N_z) is calculated from larval abundance and mortality (M_l, day^{-1}) during the period between larval and 0-group survey (t_l, days), and a temperature effect ($\varphi_l T_l$) based on a proxy of the thermal conditions during the larval stage (June-August):

$$N_{z,y} = N_{l,y} e^{-t_l M_l + \varphi_l T_l}$$

It is worth noting that during the period between the larval and 0-group surveys most of the larvae undergo metamorphosis after

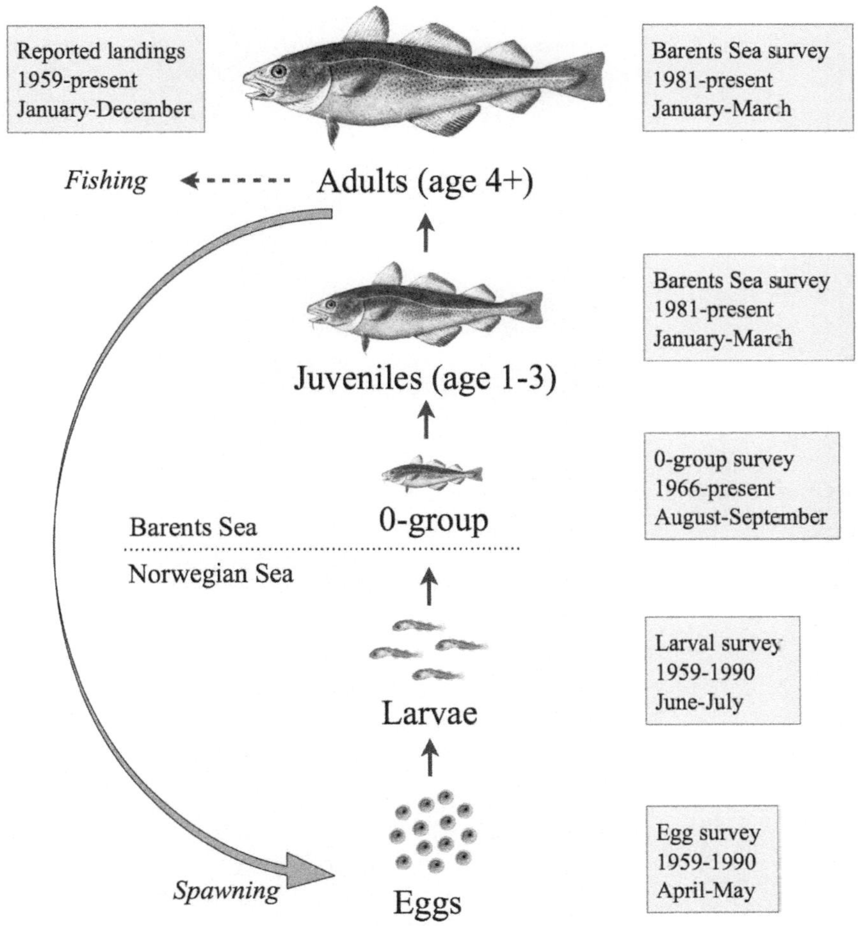

Figure 1. Schematic illustration of the NEA cod life cycle and available observational data.

which they are considered 'post-larvae' or 'early juveniles'. For simplicity, we refer to the mortality experienced during this period as the 'larval mortality'.

We considered density-dependent survival among age-classes 0–3. Intercohort predation (i.e. cannibalism) from older cod is known to decrease the survival of juveniles [15]. Because mortality due to cannibalism is expected to be particularly strong at the 0-group stage after the cod have entered the Barents Sea and settled to the bottom [24], we modeled cannibalism between the 0-group and cannibalistic cod of age 3–4. This choice was based on a correlation analysis between age-class abundances and 0-group process error using a model without cannibalism. Furthermore, intracohort density dependence was modeled for ages 1–3 based on an initial analysis of the observational data, which indicated density-dependent survival within these age-classes (Fig. 2). This analysis consisted of calculating the correlation between age-class abundance and a survival index, which was derived from the abundances of the same cohort the previous and subsequent years. This method ensures that the correlation is not biased due to correlated errors, as the estimation errors for the survival and abundance indices are independent [10]. Furthermore, based on a preliminary test of the form of density dependence, we used a Beverton-Holt relationship (see Text S1 in File S1), that is, we model compensation in survival, but no overcompensation (as implied by the Ricker model). The number of age 1 fish (N_1) is thus calculated from 0-group abundance the previous year, an intercohort interaction ($\gamma N_{3+4,y-1}$) describing the increase in mortality with abundance of older cannibalistic cod, and mortality (M_z, day^{-1}) during the period between 0-group and age 1 survey (t_z, days):

$$N_{1,y} = N_{z,y-1} e^{-t_z M_z} \frac{1}{1+\gamma N_{3+4,y-1}} e^{\varepsilon_{M_z}}, \text{ where } \frac{1}{1+\gamma N_{3+4,y-1}} e^{\varepsilon_{M_z}}$$

was restricted to be <1.

Here, the term $\varepsilon_{M_z} \sim N(0,\sigma_{M_z})$ accounts for variability in survival cf the 0-group. Accordingly, the number of age 2–4 fish is calculated from the abundance of the previous age-class in the previous year, juvenile mortality (M_j), and an intracohort interaction ($\beta_a N_{a-1,y-1}$) describing the increase in mortality with abundance:

$$N_{a,y} = N_{a-1,y-1} \frac{1}{1+\beta_a N_{a-1,y-1}} e^{-M_j}, \text{ for } 2 \leq a \leq 4$$

In our model, cod are recruited to the fishery at age 4, and fishing mortality on age 3 is assumed to be negligible [15]. The number of age 5–9 fish ($N_{a,y}$) is thus calculated from the abundance of the previous age-class in the previous year, natural mortality of adults (M_{4+}), and fishing mortality on that age-class the year before ($F_{a-1,y-1}$):

$$N_{a,y} = N_{a-1,y-1} e^{-(F_{a-1,y-1}+M_{4+})}, \text{ for } 5 \leq a \leq 9$$

This formulation assumes that natural and fishing mortality both occur throughout the year. Fishing mortality for a given age and year is modeled based on the age-selectivity of the fishery (f_a), and a year effect to account for temporal changes in fishing effort (f_y):

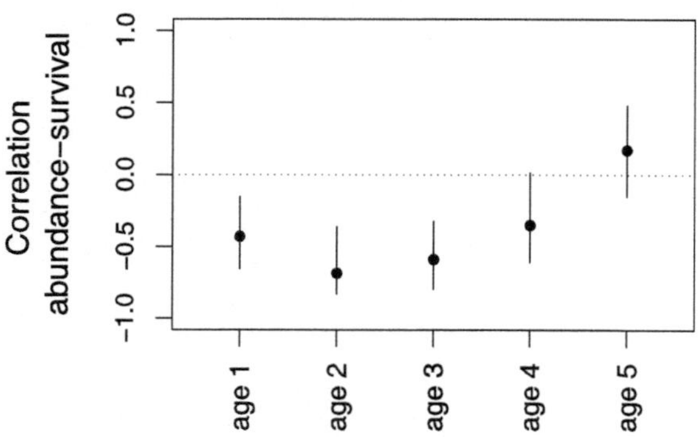

Figure 2. Correlations between log-abundances and survival indices for cod age-classes 1–5. The survival index was based on observational data and calculated as the difference between the log numbers of the subsequent and previous age-classes. Observation errors for the survival and abundance indices are thus independent and the correlation is not biased due to correlated errors. Circles and error bars indicate medians and 95% confidence intervals from bootstrapped samples, which were constructed by randomly sampling from the available abundance-survival pairs with replacement (repeated 10000 times).

$$F_{a,y}=e^{f_a+f_y}$$

Here, f_a represents the mean relative proportion by age of fish caught in the fishery, and f_y is a measure of fishing effort that is unknown and estimated in the model [25]. Separable year and age effects imply that total effort may have changed over time, but that no systematic changes in the age-selectivity of the fishery have occurred during the time period considered here. We allowed for effort to change gradually over time due to changes in fisher's behavior, investments or regulations [25]. It was modeled as a random walk process following a normal distribution according to $\varepsilon_{f_y} \sim N(0,\sigma_{f_y})$:

$$f_y=f_{y-1}+\varepsilon_{f_y}$$

For identifiability, effort in the first year was set to 0 [25,26]. Using an additional error term $W_{a,y} \sim N(0,\sigma_F)$ to account for stochastic variation in fishing mortality by age and year did not qualitatively change our results, but significantly increased computation time, and was thus not included in the model (see Text S1 in File S1). Finally, the catch for a given age and year ($C_{a,y}$) is related to age-specific abundances through:

$$C_{a,y}=\frac{F_{a,y}}{F_{a,y}+M_{4+}}N_{a,y}(1-e^{-(F_{a,y}+M_{4+})})$$

This equation (Baranov catch equation, see [27]) defines the true catch as the fraction of the total removal of individuals from the population due to fishing.

Observation model

The observation model links the true numbers to the observations. The true age-specific *abundances* and *catches* were treated as latent (unobserved) variables and were related to the survey *indices* and reported *landings* using lognormal errors. Errors in reported landings ($L_{a,y}$) were assumed to be log-normally distributed and independent between ages and years:

$$L_{a,y}=C_{a,y}e^{\varepsilon_{L_{a,y}}-0.5\sigma^2_{L_a}}$$

Here, $\varepsilon_{L_{a,y}}$ is a Gaussian error term with mean zero and variance σ_{L_a}, and the term $-0.5\sigma^2_{L_a}$ adjusts the mean of the catch error to one for any value of σ_{L_a} [27]. For the abundance indices, such a correction term was not explicitly included as it only affects the values of the age-specific catchability estimates [25]. Furthermore, the indices of age-specific abundance from the scientific surveys were assumed to be associated with observation error due to the inaccuracy of the scientific sampling. Abundance indices from the Barents Sea bottom trawl survey in winter (age-classes 1–9) were modeled according to:

$$I_{a,y}=q_aN_{a,y}e^{\varepsilon_{I_{a,y}}}\xi_y$$

Here, $\varepsilon_{I_{a,y}}$ is a Gaussian error term with mean zero and variance σ_{I_a}, and $\xi_y \sim N(0,\sigma_\xi)$ is a random year effect that accounts for correlated observation errors within years [28,29]. Hence, we estimate one variance parameter (σ_ξ) in addition to the nine age-specific observation error variances (σ_{I_a}). The q_a are proportionality constants reflecting the age-specific catchability of the survey. Because in 1993/1994 the sampling area of the Barents Sea bottom trawl survey and the mesh size in the codend were changed [30], survey data for ages 3+ were corrected according to [31]. Since no correction factors were available for ages 1 and 2, the catchability of these age-classes was modeled independently for the two periods before ($q_{a^{pre}}$) and after ($q_{a^{post}}$) 1993.

Because eggs, larvae, and the 0-group were sampled by different surveys, we used stage-specific error terms ($\varepsilon_{I_{s,y}}$) and proportionality constants (q_s) for each of these three life-stages:

$$I_{s,y}=q_sN_{s,y}e^{\varepsilon_{I_{s,y}}}$$

Thus, the constants q_e, q_l and q_z are the catchabilities, and $\varepsilon_{I_{e,y}}$, $\varepsilon_{I_{l,y}}$, and $\varepsilon_{I_{z,y}}$ are the Gaussian error terms with variances σ_{I_e}, σ_{I_l}, and σ_{I_z} for the egg, larval and 0-group indices, respectively.

Bayesian parameter estimation

We use a Bayesian framework for estimating the model parameters, which facilitates the incorporation of prior knowledge about parameter values. The priors are updated by the likelihood functions of the model to estimate the joint posterior distribution of all parameters given the observations. The use of this framework allows for the quantification of uncertainty in the parameter estimates and the model predictions. The Bayesian analysis was performed using JAGS [32], a program that uses Gibbs sampling [33], a variety of Markov Chain Monte Carlo (MCMC), to estimate the joint posterior distribution of the model parameters. We ran JAGS from within R (ver. 3.0.2, [34]) using the package R2jags [35]. MCMC runs consisted of three chains with a burn-in of 250 000 samples, and a posterior distribution based on 250 000 samples for each chain. We used a thinning of 1000 to reduce autocorrelation in the chains. This gave a total of 750 samples per parameter for the posterior distributions.

Parameter values and prior distributions

We fix the values for some of the parameters for which independent estimates were available, specifically mean fecundity and natural mortality of eggs and larvae. We used the following values: $\tau = 235$ eggs g^{-1} for mean fecundity [21], $M_e = 0.169$ day^{-1} for egg mortality, and $M_l = 0.058$ day^{-1} for larval mortality [36,37].

Mortality during the period from larval to 0-group survey was calculated assuming 60 days of larval stage (0.075 day^{-1}) and 60 days of early juvenile stage (0.04 day^{-1}), thus accounting for differences in mortality before and after metamorphosis [36,37]. We used informative priors for the density-independent natural mortality of the 0-group (M_z) and the demersal juveniles (M_j) ($year^{-1}$), which were set to be log-normally distributed with medians 0.425 [36] and 0.2 [38,39], and a variance of 0.5. Since the relative differences in catchability of the survey are not known, estimates of total mortality in these ages will be confounded with the catchability estimates [25]. We thus used informative priors to restrict the parameter estimates to a biologically reasonable range, and tested the sensitivity of our results to this assumption (see Supporting Information). Natural mortality of adults was set to $M_{4+} = 0.2$ $year^{-1}$ following [39]. This value was set in order to achieve model convergence, as is commonly done in non-Bayesian and VPA-type analyses [15]. Although a fixed natural mortality of adults poses a general limitation, our approach was developed to investigate mechanisms operating at the early life-stages before the fish are recruited to the fishery (see [25] on estimating temporal trends in natural mortality). Vague priors were used for the age-specific ($\log(q_a)$) and the 0-group ($\log(q_z)$) log-catchabilities [25], which were set to be normally distributed with variance 1000 and means 0. The log-catchabilities of eggs ($\log(q_e)$) and larvae ($\log(q_l)$) were given normal priors with variance 1000 and means -25. The age-specific fishing selectivities (f_a) were given normal priors with variance 1000 and means -1.6, and restricted to negative values. These mean values of the vague priors reflect informed yet arbitrary choices, and the posterior distributions do not depend upon them. Uniform priors [0, upper limit] were used for the density dependence parameters (β_a), and the observation and process errors (σ_L, σ_ξ, σ_{I_a}, σ_{I_z}, σ_{I_l}, σ_{I_e}, σ_{sp}, σ_{M_z}, σ_{f_y}). The upper limits of uniform priors were selected such that they were not approached during the sampling process.

Model diagnostics

We assessed model convergence using common diagnostics [40–42], which confirmed convergence of our model with the number of samples used for the posterior distributions. In addition to these formal diagnostics we used informal criteria to ensure model convergence: chains were initialized with different starting values to address problems related to meta-stability in parameter space, trace and density plots were used to detect improper mixing and asymptotic behavior of chains (Fig. S1 in File S1). Furthermore, autocorrelation in the chains was consistently below 0.2 for all parameters (Fig. S2 in File S1), and the posterior samples generally showed low cross-correlations, with the notable exception of the age-effects in fishing mortality (Fig. S3 in File S1).

Results

The variation associated with the spawning process (σ_{sp}) and 0-group survival (σ_{M_z}) exemplifies the strong impact of environmental stochasticity on the cod population dynamics (Fig. S4 in File S1). A process error standard deviation of 0.7 roughly corresponds to a 2-fold variation in egg abundance. Furthermore, higher environmental temperatures were associated with higher larval survival (but note that the 95% credible interval includes the zero value). Based on the median value of the posterior distribution, mean survival over the entire period from larval to 0-group survey varied between 0.08% and 0.12% depending on temperature during larval development. While these numbers seem small in absolute terms, they imply a 1.5-fold increase in the number of surviving larvae between the coldest and warmest years. We did not include a temperature effect on egg survival, because this parameter was estimated to be close to zero and was highly correlated with the effect on larval survival.

The density-independent natural mortality rates of the 0-group fish ($M_z = 0.35$ $year^{-1}$), and older juveniles ($M_j = 0.15$ $year^{-1}$), had median estimates that were similar but not identical to the median values of the prior distributions. The effect of cannibalistic older cod (ages 3–4) on the survival of the 0-group was clearly estimated (median [95% credible interval]: $\gamma = 1.82^{-9}$ [7.27e^{-10}–3.28e^{-9}]), and intracohort density-dependent survival was detected in juveniles of age 1 to 3 (medians [95% credible intervals]: $\beta_1 = 9.48^{-11}$ [5.37e^{-12}–2.69e^{-10}], $\beta_2 = 1.69^{-10}$ [8.12e^{-12}–4.26e^{-10}], $\beta_3 = 2.27^{-10}$ [1.56e^{-11}–4.75e^{-10}]). The posterior distributions show that an intracohort density term was well estimated for ages 1, 2, and 3 (Fig. S5 in File S1). The resulting overall mean survival rates of ages 1–3 were 0.77, 0.75, and 0.76, respectively, while the mean 0-group survival was 0.41. Using the estimates of the density-dependent and density-independent survival rates, we can predict the relationships between subsequent juvenile age-classes and compare them to the observed data (Fig. 3). This validation shows that the functional form of density dependence (Beverton-Holt) predicts the relationship between age-class abundances well, and that the model estimates accurately reflect the uncertainty in these relationships.

The strength of density dependence in juvenile survival is evidenced by the estimated time series of overall survival from age 1 to age 4, which depends on the combined intracohort density-dependent mortality rates in addition to the density-independent mortality (Fig. 4A). The number of recruits (age 4) per 0-group cod decreases with increasing abundance (Fig. 4B). To further quantify the degree of density dependence among juveniles and its dampening effect on the variability in year-class strength, we calculated the ratio between the largest and smallest year-classes before and after intracohort density dependence at age 1 and age 4. The difference between the largest and smallest year-class is more than 43-fold at age 1 and about 11-fold at age 4. Overall, the inter-annual variance in the estimated time series increases from about 0.51 in SSB to 0.95 at age 1 and thereafter decreases again to 0.40 at age 4 (Fig. 4C).

The age-specific fishing mortality shows a smooth hump-shaped relationship, with the lowest average mortality at age 4 and the highest average mortality at age 8–9 (Fig. S6 in File S1). The year-effect in fishing mortality, which represents gradual changes in effort over time, shows a temporal trend that reflects the historic changes in fishing pressure (Fig. S7 in File S1). Fishing mortality slightly increased until the late 1980s, then dropped significantly within a few years, returned to previous levels in the late 1990s, and has been declining continuously ever since. The resulting temporal dynamics in spawning biomass as predicted by the model (Fig. 5A) are in close correspondence with estimates from a virtual population analysis [15]. This is not surprising given that the same data and natural mortality of adults (age 4+) were used in our model. Abundance trends for the different life-stages and age-classes as predicted by the model are further in general agreement with the temporal dynamics seen in the observational data (Fig. S8 in File S1). The age-specific observation errors show a U-shaped relationship, with the lowest observation error on 6-year-old cod

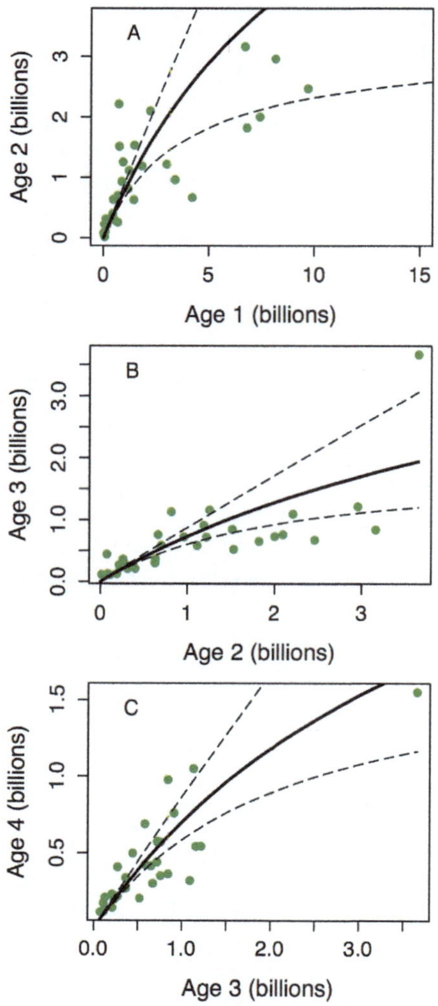

Figure 3. Relationships between cod age-class abundances. Observations (circles) were corrected for estimated survey catchability. Model predictions are shown as medians (solid lines) and 95% credible intervals (dashed lines), which were obtained by sampling parameter values independently for each MCMC draw to account for pairwise correlations between parameters.

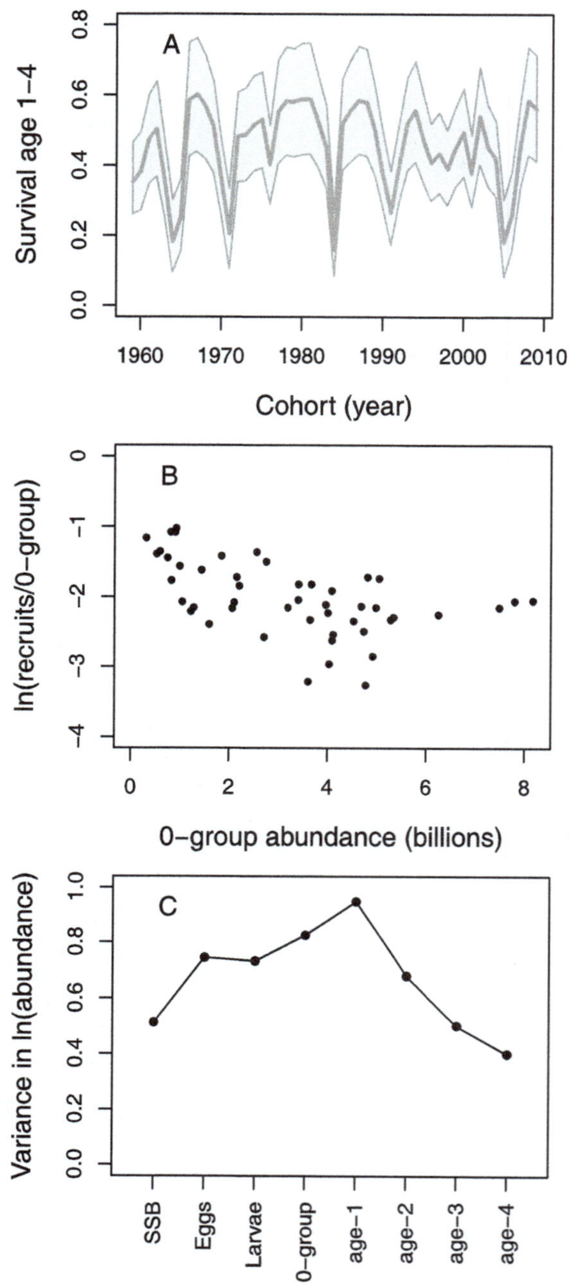

Figure 4. Juvenile survival rate and variance in life-stage or age-class specific abundance. Shown are (A) survival rate from age 1 to 4, (B) relationship between the number of recruits per 0-group and 0-group abundance, and (C) interannual variance of the log-abundance time-series. Lines in (A) represents median values (thick) and 95% credible intervals (thin), obtained by calculating survival from age 1 to 4 independently for each MCMC draw in order to account for parameter correlations. In (C) variance is shown for log-transformed estimates of spawning stock biomass (SSB) and abundances of the early life-stages and age-classes 1–4.

(Fig. S9 in File S1) and the highest observation error on the 0-group. The catchability estimates for the survey suggest that surveyability of age-classes 1–2 increased after 1993, and that surveyability continuously declines from age 3 to 9 (Fig. S10 in File S1). All parameter estimates are presented in Table S1 in File S1.

Discussion

We studied the impacts of density-independent and density-dependent factors on early life-stage survival and the consequences of these processes for the variability in adult year-class strength. We did this by developing a dynamic population model for NEA cod based on catch statistics and scientific survey data for all life-stages, from eggs to spawners, while accounting for the inaccuracy of the observations and stochasticity in the population dynamics. The parameter estimates highlight the importance of density dependence for population regulation. Intracohort density dependence in survival occurs among juveniles of age 1–3, indicating that this period of the life cycle determines the compensatory capacity of the stock. This compensation dampens the variation in early life-stage abundances (egg, larvae, and 0-group) and thus reduces the impact of environmental stochasticity on the recruitment to the fishery.

Environmental stochasticity has a strong impact on the production of eggs and the survival of larvae and 0-group cod. First, egg production as well as 0-group survival are highly stochastic processes as evidenced by the large variances of the

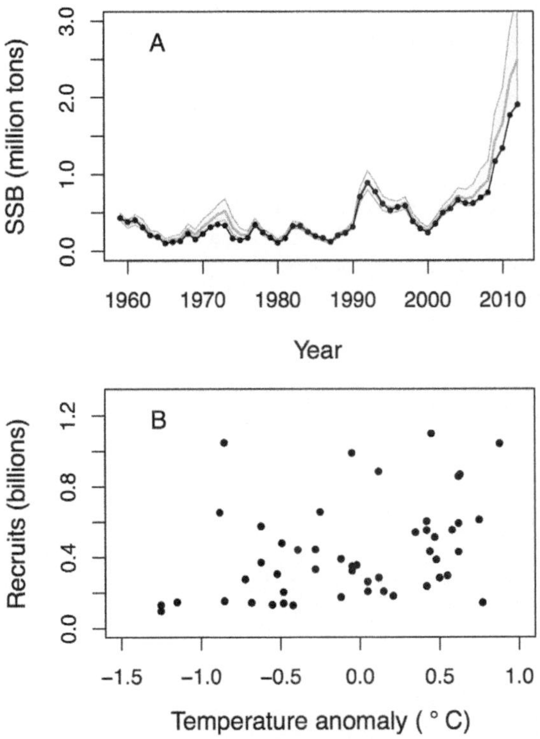

Figure 5. Temporal trend in spawning stock biomass and temperature-recruit relationship. (A) Predicted spawning stock biomass (grey lines, median values with 95% credible intervals) in comparison with the VPA assessment (black line and circles, ICES 2013). (B) Relationship between predicted number of recruits (age 4) and temperature anomaly during the larval stage.

process errors. Second, temperature has a positive effect on the survival of larval cod, which is in line with other studies suggesting that growth and survival of cod early life-stages depend on temperature [22,23]. Temperature-dependent survival may be caused by direct and/or indirect effects. Higher temperatures may either directly increase survival, for instance through faster development, which reduces the duration and thus total mortality of a life-stage, or indirectly increase survival through an increase in food supply. Interestingly, low temperatures consistently result in low recruit abundance, whereas high temperatures may result in either small, medium or large year-classes (Fig. 5), suggesting that warm temperature is a necessary but not a sufficient condition for good recruitment [43].

Density dependence in the juvenile life-stage was found to be compensatory, that is, increased densities reduce the survival of the juvenile age-classes. Although density-dependent processes can be compensatory or depensatory, previous studies have suggested compensatory dynamics in NEA cod [10,36], whereas depensation does not seem to occur in this stock within the range of historically observed abundances [44]. Recently, it was shown that the variance in cohort abundance of NEA cod increases up to age 1 and subsequently decreases until age 3, indicating a dampening effect of density dependence on cohort fluctuations [37]. Another study found evidence for density dependence at ages 1–3 by correlating estimates of age-specific process errors (using a similar approach as adopted here) to the corresponding abundances [38].

Both competition and cannibalism from older cod have previously been suggested as a cause of density-dependent survival of juvenile cod [24,45–48]. In this study, cannibalism from older cod was found to affect the survival of 0-group cod, while intracohort density dependence was found to be a significantly better predictor of survival among age 1–3 juveniles than cannibalism (intra- and intercohort terms for a given age-class could not be estimated at the same time). Cannibalism is expected to affect juvenile survival after the cod have entered the Barents Sea and settled to the seafloor. In fact, the most abundant cod length group in the stomachs of larger cannibalistic cod is about 10–14 cm [24], which roughly corresponds to the 0-group stage. During the period from the 0-group survey in August-September to the bottom-trawl survey in January-March the cod grow from ~6–9 cm [22] to ~12–15 cm [49]. Competition is also expected to become significant when the cod settle to the seabed and assume a demersal existence. However, it has also been reported that older juveniles (ages 2–3) are more concentrated in the southeastern Barents Sea, thereby indicating higher densities [14]. In line with these observations, our results suggest that density dependence occurs in ages 1 to 3. Separating this effect by age group is difficult due to pairwise correlations among parameter estimates. However, as illustrated by Fig. 4, overall juvenile survival is clearly reduced at high cohort densities. On the other hand, we were not able to estimate an intracohort density dependence term for the 0-group, possibly due to the high observation error of these data [17,19], which is reflected in our estimated variance of the 0-group observation error (Table S1 in File S1). In summary, both cannibalism and competition affect the survival of juveniles. Importantly, the intracohort density dependence among juveniles of age 1–3 considerably dampens the variability in cohort abundance.

The relatively strong compensation within the juvenile life-stage dampens the large variation in abundance of the early life-stages caused by environmental stochasticity. The condition of the spawning stock, here included through the weight-at-age and probability of being mature, as well as climatic and oceanographic conditions during the spawning period presumably have a strong effect on egg production. In addition, temperature is associated with changes in survival of cod larvae. Due to the density regulation in the demersal juvenile stage (ages 1–3), the variation in the abundance of cod entering the fishery is considerably lower than for the preceding age-classes, and similar to levels of recruitment variation that are commonly observed for marine fishes [7]. Higher abundances at the larval and post-larval stage (0-group) imply lower juvenile survival due to density-dependent survival, which operates after the stochastic impacts on the spawning process and early life-stages. The negative slope of the relationship between survival rate from 0-group to recruitment at age 4 and abundance reflects the degree of compensation in this population (Fig. 4B). Consequently, one implication of density-dependent survival is that it reduces the impact of climate variability on the year-class strength of recruits. The fishery is thus less affected by climate than would otherwise be expected in the absence of density regulation in the juvenile life-stage. It is further known that recruitment is positively linked to spawner abundance in NEA cod [50,51], and that the incidence of poor recruitment increases at spawner biomasses below about half a million tons [52]. In this study, we assumed a linear relationship between spawner biomass and egg production with a lognormal error term. With its current total stock biomass of ~3.5 million tons, the NEA cod continues to sustain an enormous fishery with recent annual catches around 1 million tons.

In conclusion, the timing of stochastic versus regulatory events plays a crucial role in determining adult abundance in marine fish populations. Using a life-cycle model for NEA cod that dynamically links the population life-stages, we have demonstrated that environmental stochasticity, including climatic variation, affects the survival of early life-stages of cod, while density-dependent survival occurs at the juvenile stage (age 1–3). Compensatory density dependence at the juvenile stage reduces the impact of climatic and other stochastic processes operating earlier in life, as well as changes in spawning stock biomass due to harvesting, thus acting to stabilize temporal variation in the abundance of older life-stages. This type of density regulation in demographic rates is the basis of sustainable harvest as it causes population growth to increase at low abundance, and is therefore fundamental to population persistence.

Supporting Information

File S1 Contains the files: Text S1: Model tests and extensions. Table S1: Posterior medians and 95% credible intervals for all model parameters. Figure S1: Posterior distributions of all model parameters. Figure S2: Temporal autocorrelation of all model parameters. Figure S3: Cross-correlations of all model parameters. Figure S4: Estimated process errors. Shown are the process errors for spawning (ε_{sp}) and early juvenile mortality (ε_{M_z}). **Figure S5: Prior and posterior distributions of mortality rates.** Prior (grey) and posterior (black) distributions are shown for the density-independent mortality of the 0-group (M_z) and ages 1–3 (M_j) and posterior distributions (uniform priors) are shown for the density-dependent mortality rates of ages 1–3 (β_1, β_2, β_3). **Figure S6: Parameter estimates for the age-specific fishing mortality. Figure S7: Estimated year-effect of the fishing mortality. Figure S8: Abundance trends for eggs, larvae, 0-group cod, and age-classes 1–9.** Shown are the observations (black), corrected for age-specific catchability, and the model predictions (grey) with 95% credible intervals. Note the log-scale and the different periods for the abundance time-series corresponding to the actual observations (eggs/larvae: 1959–1990; 0-group: 1966–2010; ages 1–9: 1981–2010). **Figure S9: Estimates of age-specific observation errors of the Barents Sea survey. Figure S10: Estimates of age-specific catchabilities of the Barents Sea survey.** For age-classes 1 and 2 surveyability was independently estimated for the period before 1993 (open circles).

Acknowledgments

We thank Anne M. Eikeset, Dag Ø. Hjermann, Øystein Langangen, Geir Ottersen, Rémi Patin, and Leif Chr. Stige for inspiring discussions. Bjarte Bogstad provided scientific survey data on age-specific abundances of cod. B. Bogstad, Ø. Langangen, and L. C. Stige provided insightful comments on previous versions of this manuscript. We further thank staff and scientists at the Knipovich Polar Research Institute of Marine Fisheries and Oceanography (PINRO) for collecting data on cod eggs and larvae as well as the temperature records of the Kola section.

Author Contributions

Analyzed the data: JO LAR. Wrote the paper: JO LAR NCS. Conceived the study: JO.

References

1. Andrewartha HG, Birch LC (1954) The distribution and abundance of animals. Chicago: University of Chicago Press.
2. Hanski I (1990) Density dependence, regulation and variability in animal populations. Phil Trans R Soc B 330: 141–150.
3. Hixon MA, Pacala SW, Sandin SA (2002) Population regulation: Historical context and contemporary challenges of open vs. closed systems. Ecology 83: 1490–1508.
4. Brook B, Bradshaw C (2006) Strength of evidence for density dependence in abundance time series of 1198 species. Ecology 87: 1445–1451.
5. Hjort J (1914) Fluctuations in the great fisheries of Northern Europe viewed in the light of biological research. Conseil Permanent International Pour L'Exploration De La Mer 20: 1–237.
6. Shepherd JG, Cushing DH (1990) Regulation in fish populations: myth or mirage? Phil Trans R Soc B 330: 151–164.
7. Myers RA (2001) Stock and recruitment: generalizations about maximum reproductive rate, density dependence, and variability using meta-analytic approaches. ICES J Mar Sci 58: 937–951.
8. Houde ED (2008) Emerging from Hjort's Shadow. J Northw Atl Fish Sci 41: 53–70.
9. Rijnsdorp AD, Peck MA, Engelhard GH, Moellmann C, Pinnegar JK (2009) Resolving the effect of climate change on fish populations. ICES J Mar Sci 66: 1570–1583.
10. Myers RA, Cadigan NG (1993) Density-dependent juvenile mortality in marine demersal fish. Can J Fish Aquat Sci 50: 1576–1590.
11. Rose KA, Cowan JH Jr, Winemiller KO, Myers RA, Hilborn R (2001) Compensatory density dependence in fish populations: importance, controversy, understanding and prognosis. Fish Fish 2: 293–327.
12. Walters CJ, Martell SJD (2004) Fisheries Ecology and Management. Princeton, New Jersey, USA: Princeton University Press.
13. Maunder MN, Deriso RB (2011) A state–space multistage life cycle model to evaluate population impacts in the presence of density dependence: illustrated with application to delta smelt (*Hyposmesus transpacificus*). Can J Fish Aquat Sci 68: 1285–1306.
14. Olsen E, Aanes S, Mehl S, Holst JC, Aglen A, et al. (2010) Cod, haddock, saithe, herring, and capelin in the Barents Sea and adjacent waters: a review of the biological value of the area. ICES J Mar Sci 67: 87–101.
15. ICES (2013) Report of the Arctic Fisheries Working Group (AFWG).
16. Mukhina NV (1992) Results of the ichthyoplankton investigations in the Norwegian and Barents Seas in 1959–1990. In: Belikov SV, editor. Ekologicheskie problemy Barentseva moria. Murmansk, Russia (In Russian): PINRO.
17. Mukhina NV, Marshall CT, Yaragina NA (2003) Tracking the signal in year-class strength of Northeast Arctic cod through multiple survey estimates of egg, larval and juvenile abundance. J Sea Res 50: 57–75.
18. ICES (2007) Report of the Arctic Fisheries Working Group (AFWG).
19. Stige LC, Hunsicker ME, Bailey KM, Yaragina NA, Hunt GL J Jr (2013) Predicting fish recruitment from juvenile abundance and environmental indices. Mar Ecol Progr Ser 480: 245–261.
20. Tereshchenko VV (1996) Seasonal and year-to-year variations of temperature and salinity along the Kola meridian transect. Reykjavik, Iceland: ICES CM 1996/C:11.
21. Marshall CT, Needle CL, Thorsen A, Kjesbu OS, Yaragina NA (2006) Systematic bias in estimates of reproductive potential of an Atlantic cod (*Gadus morhua*) stock: implications for stock-recruit theory and management. Can J Fish Aquat Sci 63: 980–994.
22. Ottersen G, Loeng H (2000) Covariability in early growth and year-class strength of Barents Sea cod, haddock, and herring: the environmental link. ICES J Mar Sci 57: 339–348.
23. Stige LC, Ottersen G, Dalpadado P, Chan K-S, Hjermann DØ, et al. (2010) Direct and indirect climate forcing in a multi-species marine system. Proc Roy Soc B 277: 3411–3420.
24. Yaragina NA, Bogstad B, Kovalev YA (2009) Variability in cannibalism in Northeast Arctic cod (*Gadus morhua*) during the period 1947–2006. Marine Biology Research 5: 75–85.
25. Aanes S, Engen S, Saether BE, Aanes R (2007) Estimation of the parameters of fish stock dynamics from catch-at-age data and indices of abundance: can natural and fishing mortality be separated? Can J Fish Aquat Sci 64: 1130–1142.
26. Gudmundsson G (1994) Time series analysis of catch-at-age observations. J R Stat Ser C, Appl Stat 43: 117–126.
27. Quinn TJ, Deriso RB (1999) Quantitative Fish Dynamics. New York: Oxford University Press.
28. Myers RA, Cadigan NG (1995) The statistical analysis of catch-at-age data with correlated errors. Can J Fish Aquat Sci 52: 1265–1273.
29. Millar R, Meyer R (2000) Bayesian state-space modeling of age-structured data: fitting a model is just the beginning. Can J Fish Aquat Sci 57: 43–50.
30. Jakobsen T, Korsbrekke K, Mehl S, Nakken O (1997) Norwegian combined acoustic and bottom trawl surveys for demersal fish in the Barents Sea during winter.ICES CM 1997/Y:17.
31. Korsbrekke K, Mehl S, Nakken O, Pennington M (2001) A survey-based assessment of the Northeast Arctic cod stock. ICES J Mar Sci 58: 763–769.
32. Plummer M (2003) JAGS: A program for analysis of Bayesian graphical models using Gibbs sampling. Proceedings of the 3rd International Workshop on Distributed Statistical Computing: 1–10.

33. Gelman A, Carlin JB, Stern HS, Rubin DB (2003) Bayesian data analysis. 2nd ed. Boca Raton, FL: Chapman and Hall/CRC.
34. R Development Core Team (2013) R: A language and environment for statistical computing. Vienna, Austria: R Foundation for Statistical Computing.
35. Su Y-S, Yajima M (2013) Package "R2jags".
36. Sundby S, Bjørke H, Soldal AV, Olsen S (1989) Mortality rates during the early life stages and year-class strength of northeast Arctic cod (*Gadus morhua* L.). Rapports et Proces-verbaux des Réunions Conseil International pour l'Éxploration de la Mer 191: 351–358.
37. Langangen Ø, Stige LC, Yaragina NA, Vikebø F, Bogstad B, et al. (2013) Egg mortality of northeast Arctic cod (*Gadus morhua*) and haddock (*Melanogrammus aeglefinus*). ICES J Mar Sci: AdvanceAccess.
38. Eikeset AM (2010) The ecological and evolutionary effects of harvesting Northeast Arctic cod - insights from economics and implications for management. University of Oslo.
39. Brinch CN, Eikeset AM, Stenseth NC (2011) Maximum likelihood estimation in nonlinear structured fisheries models using survey and catch-at-age data. Can J Fish Aquat Sci 68: 1717–1731.
40. Gelman A, Rubin DB (1992) Inference from iterative simulation using multiple sequence. Statist Sci 7: 457–511.
41. Raftery AE, Lewis S (1992) How many iterations in the Gibbs sampler. In: Bernardo JM, Berger JO, Dawid AP, Smith AFM, editors. Bayesian Statistics. Oxford, UK: Oxford University Press, Vol. 4. pp. 763–773.
42. Geweke J (1992) Evaluating the accuracy of sampling-based approaches to the calculation of posterior moments. In: Bernardo JM, Berger JO, Dawid AP, Smith AFM, editors. Bayesian Statistics. Oxford, UK: Oxford University Press, Vol. 4. pp. 169–193.
43. Ellertsen B, Fossum P, Solemdal P, Sundby S (1989) Relation between temperature and survival of eggs and first-feeding larvae of northeast Arctic cod (*Gadus morhua* L.). Rapports et Proces-verbaux des Réunions Conseil International pour l'Éxploration de la Mer 191: 209–219.
44. Keith DM, Hutchings JA (2012) Population dynamics of marine fishes at low abundance. Can J Fish Aquat Sci 69: 1150–1163.
45. Bogstad B, Lilly GR, Mehl S, Palsson OK, Stefansson G (1994) Cannibalism and year-class strength in Atlantic cod (*Gadus morhua* L.) in Arcto-boreal ecosystems (Barents Sea, Iceland, and eastern Newfoundland). ICES Marine Science Symposia 198: 576–599.
46. Hjermann DØ, Stenseth NC, Ottersen G (2004) The population dynamics of Northeast Arctic cod (*Gadus morhua*) through two decades: an analysis based on survey data. Can J Fish Aquat Sci 61: 1747–1755.
47. Ciannelli L, Dingsør GE, Bogstad B, Ottersen G, Chan K-S, et al. (2007) Spatial anatomy of species survival: effects of predation and climate-driven environmental variability. Ecology 88: 635–646.
48. Dingsør GE, Ciannelli L, Chan K-S, Ottersen G, Stenseth NC (2007) Density dependence and density independence during the early life stages of four marine fish stocks. Ecology 88: 625–634.
49. Michalsen K, Ottersen G, Nakken O (1998) Growth of North-east Arctic cod (*Gadus morhua* L.) in relation to ambient temperature. ICES J Mar Sci 55: 863–877.
50. Ottersen G, Sundby S (1995) Effects of temperature, wind and spawning stock biomass on recruitment of Arcto-Norwegian cod. Fish Oceanogr 4: 278–292.
51. Myers RA, Barrowman NJ (1996) Is fish recruitment related to spawner abundance? Fishery Bulletin 94: 707–724.
52. Hylen A (2002) Fluctuations in abundance of northeast arctic cod during the 20th century. ICES Marine Science Symposia 215: 543–550.

Co-Authorship and Bibliographic Coupling Network Effects on Citations

Claudio Biscaro[1,2,3]*, **Carlo Giupponi**[1]

1 Department of Economics, Ca' Foscari University of Venice, Venice, Italy, **2** Department of Management, Ca' Foscari University of Venice, Venice, Italy, **3** Institut für Organization und Globale Managementstudien, Johannes Kepler Universität, Linz, Austria

Abstract

This paper analyzes the effects of the co-authorship and bibliographic coupling networks on the citations received by scientific articles. It expands prior research that limited its focus on the position of co-authors and incorporates the effects of the use of knowledge sources within articles: references. By creating a network on the basis of shared references, we propose a way to understand whether an article bridges among extant strands of literature and infer the size of its research community and its embeddedness. Thus, we map onto the article – our unit of analysis – the metrics of authors' position in the co-authorship network and of the use of knowledge on which the scientific article is grounded. Specifically, we adopt centrality measures – *degree, betweenneess*, and *closeness centrality* – in the co-authorship network and *degree, betweenness centrality* and *clustering coefficient* in the bibliographic coupling and show their influence on the citations received in first two years after the year of publication. Findings show that authors' *degree* positively impacts citations. Also *closeness centrality* has a positive effect manifested only when the giant component is relevant. Author's *betweenness centrality* has instead a negative effect that persists until the giant component - largest component of the network in which all nodes can be linked by a path - is relevant. Moreover, articles that draw on fragmented strands of literature tend to be cited more, whereas the size of the scientific research community and the embeddedness of the article in a cohesive cluster of literature have no effect.

Editor: Christos A. Ouzounis, The Centre for Research and Technology, Hellas, Greece

Funding: The authors gratefully acknowledge the financial support of the KULTURisk Project (FP7-ENV.2010.1.3.2-1-265280). The funders had no role in study design, data collection and analysis, decision to publish, or preparation of the manuscript.

Competing Interests: The authors have declared that no competing interests exist.

* E-mail: claudio.biscaro@jku.at

Introduction

Generating scientific knowledge is as a social activity in which scientists find problems to tackle, become aware of connections between elements, elaborate on existing ideas to produce new or refined answers. It is a recipe with both social and knowledge ingredients. However, studies on the impact of scientific knowledge have mostly stressed either the social or the knowledge part of the story, neglecting their interaction.

Studies on co-authorship networks have gathered the attention of many scholars, not just because of their descriptive and synthetic power to describe the evolution of research communities, but also because social networks play a significant role on the generation of knowledge [1–3]. Social interactions still play a major role even in an era in which knowledge is accessible on-line [4,5], in contrast of what common sense would suggest. Co-authorship networks (hereafter CA) are a type of social network based on co-author relationships that are built over time by scientists.

The increasing number of researches that link CA and scientific impact – citations – reminds us that collaboration and its structure profoundly affect the quality of work. In this respect, scholarly attention increased towards the properties of CA and their effects on citations [2,6,7]. Newman explored the advantage of the first publications in a field [8], Mazloumian and colleagues documented the bandwagon effect of accolades, world-class recognitions, and landmark papers on prior works of the rewarded scientists [9]. With respect to the network position of co-authors, centrality measures significantly correlate with the article citation count [7]. In particular, degree centrality – that counts the number of personal contacts of an author – and betweenness centrality [2] – that measures how many shortest paths connecting any two authors in the dataset run through any single node.

Co-authorship relations represent the social side of the generative activity of scientists. For individuals, whose main capital and product is knowledge, social interactions are a crucial to improve the quality of their work. They are so relevant that opportunities of unintentional interaction are favored (or even forced) in some contexts oriented to knowledge generation, such as high-tech firms [10,11]. The reason is that the quality of ideas and knowledge work is given not just by the number of people who create it, but mainly by the knowledge diversity to which they are exposed. Establishing social connections with diverse groups enables the exposure to multiple and different intellectual domains, methods, perspectives and techniques [12,13] and the inclusion of "whole domains of elements … into the combinative hopper" (Simonton, 1995: 473). Recent works showed that also in different fields of science the attitude towards interdisciplinary is rewarded with more citations [2,7,15].

We need a different lens to overcome the paradox in which studies on knowledge generation are mainly carried out by studying the social component, neglecting the most fundamental one: knowledge itself. After all, even the social part of the theory assumes that it is by combining distant or multiple domains of knowledge that new and impactful ideas are generated. Yet, knowledge remains just an outcome. We think that the bibliographic coupling network is the appropriate lens to look at how knowledge is combined into a scientific work. The bibliographic coupling network (hereafter BC) is constructed on the shared references among publications [16], thus it provides deeper insights on the scientific activity, as it reveals information on how authors use and construct links among the existing literature. It can be used to analyze the articles' position within the literature, to infer the size of its research community and it may help answer conjectures that have not yet been addressed.

Indeed, the idea of analyzing the knowledge used by authors is not revolutionary, and it could be traced back at least to the intuition of John of Salisbury, then quoted by Newton who wrote that he was able to see further as he stood on giants' shoulders. This inspiring portrait of a scientist who builds on other scientists' work reveals that scientific work is in an ongoing evolution and there exist a real (and a spiritual) connection among scientists. Along the thread of scientific evolution, Polanyi [17] noticed that scientists adjust their efforts on the basis of "the hitherto achieved results of the others" (p. 2) thus creating a continuum of different works that results in a continuous progress that links different scientific domains and whereby the key knowledge sources change over time [18–22].

For this reason, we want to incorporate the analysis of the use of knowledge sources – the references – into the more traditional variables that look at the social interaction in order to study the effects of both social and knowledge networks on the scientific impact of knowledge production. Thus, we integrate metrics of BC and CA to see their effect on articles' citations. Thereby, as we follow this intent, we align with the recent work of Uddin, Hossain and Rasmussen [2], as our unit of analysis is the publication whereby we map metrics from the two network.

To better capture network effects, we isolate papers of a specific literature of vulnerability in climatic change and exclude those outside that literature, whose citations could have a different distribution [23,24]. We do that by performing an unsupervised textual analysis and categorization of all papers into topics [25–27]: this step also allows us to identify elements of originality and innovation in order to control for advantages coming from the introduction of new research topics or new combinations of topics. Then, we construct the CA and BC network [28], to answer the following research questions: (1) How does the co-authorship network structure influence the scientific impact of an article in terms of citations? (2) How do the article's knowledge sources influence its scientific impact?

In this work we use the terms paper, article, publication and work interchangeably. Sometimes the term node or vertex will be referred to the author/co-author or to the paper/article according to the network under scrutiny is the CA or the BC, and, similarly, the term tie or edge will be referred to the co-authorship relation or to the fact that two articles share at least one reference.

The rest of our article proceeds with the description of the methods with which we constructed the dataset and addressed our research questions, than we describe the measures we used and the theoretical reasons underpinning their adoption. Next, we will discuss our methods and research setting. After that we will present the results of our analysis and we will conclude deriving general theoretical contributions.

Methods

To test our arguments, we need a set of articles that originated within a coherent body of literature. Thus, after downloading the dataset, our first intent is consolidate the set by screening out those publications which entered our search due to the fact that the multiple keyword combinations adopted could in part be used also in contexts far from our interests, like the medical one. We carry out a machine learning classification task [25,27,29] on the textual information (more information both on method and on the process is contained in File S1), and isolate only the publications coherent with the identified literature. Only then, we generate two types of networks based on scholarly collaborations, and on articles' references.

The bibliographic coupling and co-authorship networks

The structure of a scientific literature and that one of collaboration between authors can be analyzed through networks that are based on the mathematical mapping of relations (edges) between dyads of elements (vertices or nodes).

From a set of scientific articles, it is possible to establish relations among different attributes, e.g., references, co-authors, keywords. In this research, we focus on two types of networks: the co-authorship network (CA), and the bibliographic coupling network (BC). In the BC, the articles are the vertices and an edge is established when they have at least one shared reference. Analogously in the co-authorship (CA) network, vertices are the authors and the edges are established between vertices who co-authored an article. We treat the BC as unweighted: the weight of the edge is not affected by the number of shared references.

In Figure 1, we represent how to construct the two networks starting from a set of three articles i, j, and k. With respect to the CA, the activity of co-authoring both in i and j, enables author α to span across two sets of co-authors β, ω and τ (through article i) and γ (through article j).

Generally the choice on networks of citations privileges the choice of visualizing the links among citing and cited articles. This type of network, called co-citation, provides a picture that changes in time and in which there are as many nodes as the number of citing and cited articles. To have a more compact and time-independent network, we choose to establish relations among those articles in the dataset that share references, as shown in Figure 1, where two same references are contained in articles i,j, while j and k share one. Thus, the BC contains only three nodes and two edges. Node j is linked with and i, and with k on the basis of the shared references, while there is no edge between i and k.

The choice of BC has also another key advantage with respect to the co-citation network. Other than compactness of the network (3343 nodes versus over 150,000), the relations that the node (article) establishes with the rest of the network (with the extant literature) is grounded in what the authors decided to include in their reference list. Thereby it is established by means of authors' choices. One of the disadvantages of the BC is that the number of relations depends on the size of the reference list, giving a positive bias towards those articles that have a larger reference list. However, we will control for it as described in the § Measures.

Data Source

As the scientific impact is dependent on the stream of literature, we build the dataset through a sequence of steps aimed at gathering a homogenous population of articles in terms of thematic structure, information on references and co-authors. The first step consists in the identification of literature streams and it is done through an accurate selection of keywords reported in

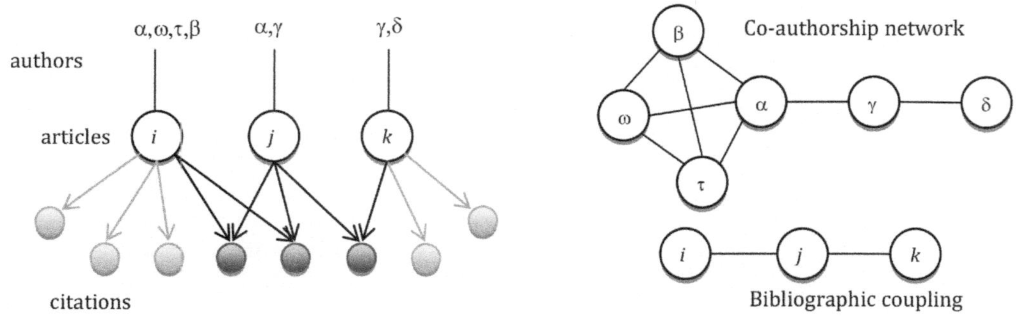

Figure 1. Example of bibliographic coupling network and co-authorship network stemming from a set of three articles.

Table 1. The choice of keywords has dealt with the concept of *vulnerability* that is a crucial notion for several research communities: particularly for climate change adaptation and disaster risk reduction. A positive note on the literature we chose is that the meaning of vulnerability varies in the diverse research streams, which developed in isolation. Only recently, under the push of international bodies (such as IPCC and UNISDR), these fragmented scientific streams are unifying their glossary. Thereby, this is a rich context whereby to test our approach. The focus of scientists spans over a multitude of aspects that vary from strategies to adapt and anticipate possible consequences of a natural extreme event to strategies to better cope with their effects in a multitude of different social or ecological systems, affected by diverse types of natural events (floods, draughts, storms, heat waves, and extreme wind, among the others), and mediated by various morphological, geological, and social conditions.

The search, performed on April 30 2013, in the ISI Web of Knowledge returned 5585 papers published between 1985 and 2013.

We refine the dataset in three steps. We select articles that have at least two entire years of forward citations (thus we exclude articles published after 2010), belong to the same scientific 'macro-discipline' (the entire process of data analysis is summarized in Figure 2), and have references.

First, we discard the articles published after the 2010, because they do not have two complete years of citation history that is the measure of scientific impact chosen. We choose a 2 year time window as all articles belong to the same disciplinary area and we know that they should have analogous citation patterns.

Second, using the Stanford Topic Modeling suite, we classify articles based on the latent structure of topics extracted from the textual information available from the search: abstracts and titles (we do not consider keywords, as not all journals require them). The algorithm needs to arbitrarily set the number of topics. It is set to 75, a number that takes into account the large variability of the themes discussed in the field: types of environments, receptors, and units of analysis. Topics are coded by the second author of this paper, who is expert in the field (with more than three publications), and labeled on the basis of their 20 most likely words (see the SI for details). Such coding is then validated by two other experts, who also identify nine non-relevant topics. This leads to the exclusion of 270 articles whose content – topic proportion >0.5 – falls in one of these nine topics (see the SI for details).

Third, we exclude articles for which it is not possible to compute the BC network for having no references.

This process brings the dataset to a population of 3343 articles published between 1989 and 2010. We then create a series of BC and CA networks on temporal slices that keep fixed the first observation in the dataset and move forward by one year (e.g., 1989–2002, 1989–2003, 1989–2004).The temporal slices enable us to observe the position of the authors in the CA and the article in the BC at the time of the article publication. For each network slice, we extract our independent variables by means of the library 'igraph'[30] running in the statistical software R. Last, we performed the statistical analysis by means of negative binomial regressions. In Figure 3, we provide a visual conceptualization of our research questions.

Table 1. Keywords.

Vulnerability & risk assessment
Vulnerability & risk management
Vulnerability & adaptive management
Vulnerability & water resource management
Vulnerability & climate
Vulnerability & climate change
Vulnerability & climate change adaptation
Vulnerability & disaster risk
Vulnerability & disaster risk reduction

Measures

Scientific impact. The dependent variable is the articles' scientific impact measured by the number of citations received (hereafter *citation count*) within a two-year fixed time window starting after the year of publication. We decide to exclude the citations received during the year of publication, as that would create a bias in favor of the articles published early in the year. A fixed time window to measure citations has been chosen for several reasons: it is shown to be a good measure to determine the scientific impact within a discipline [23]; it is not biased in favor old articles that can be cited for a longer period as it does not fully reflect the ranking of the most cited overall (0.68 spearman correlation coefficient with the cumulative citations received in April 2013); and with respect to the yearly citation average, efficiency [31], it has the advantage to be limited in time, therefore it focus only on the immediate impact of the social activity of co-authors and positioning of the article, while being not affected by phenomena such as drifts in the literature that could impact citations but are also slower to occur. However, the two-year time window is correlated with efficiency (0.84 with $p<0.001$).

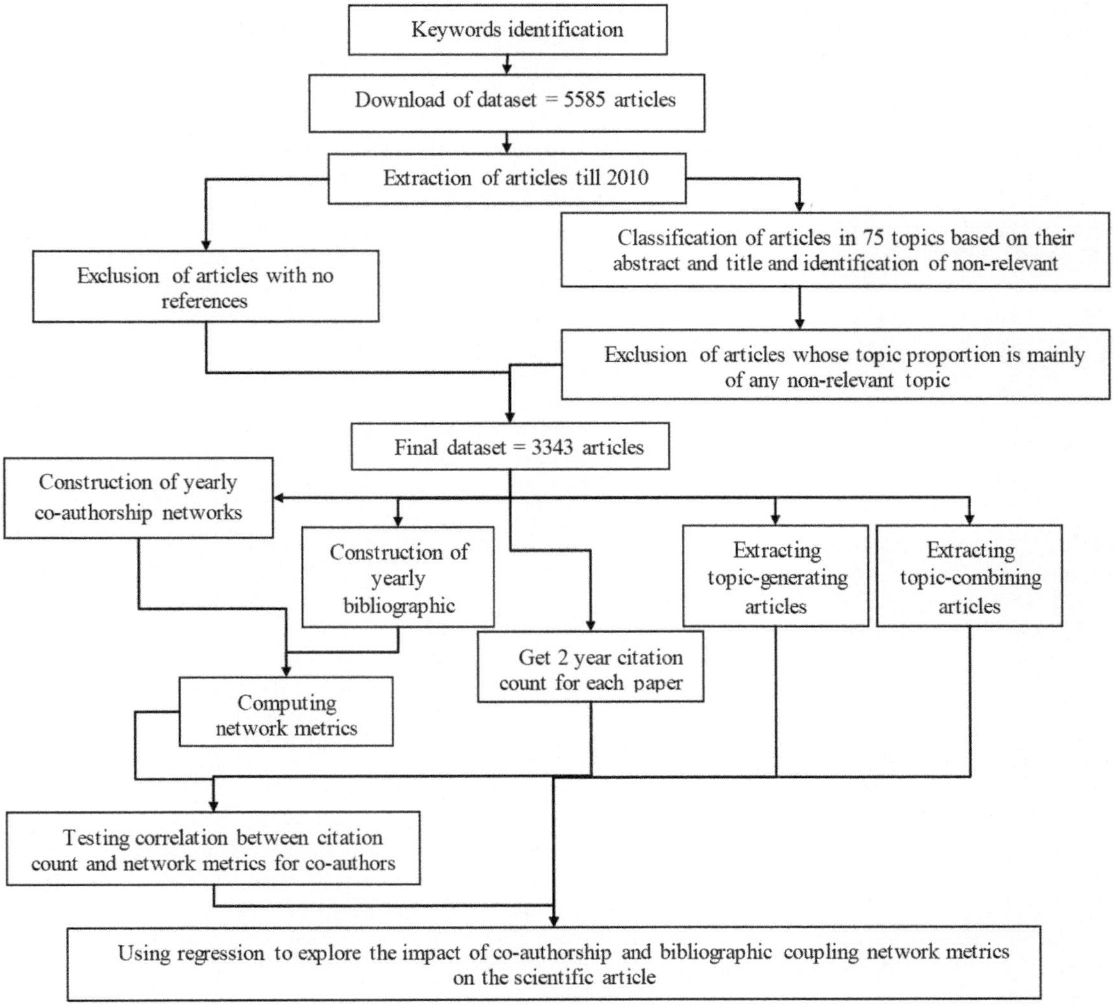

Figure 2. Mapping the process of data analysis.

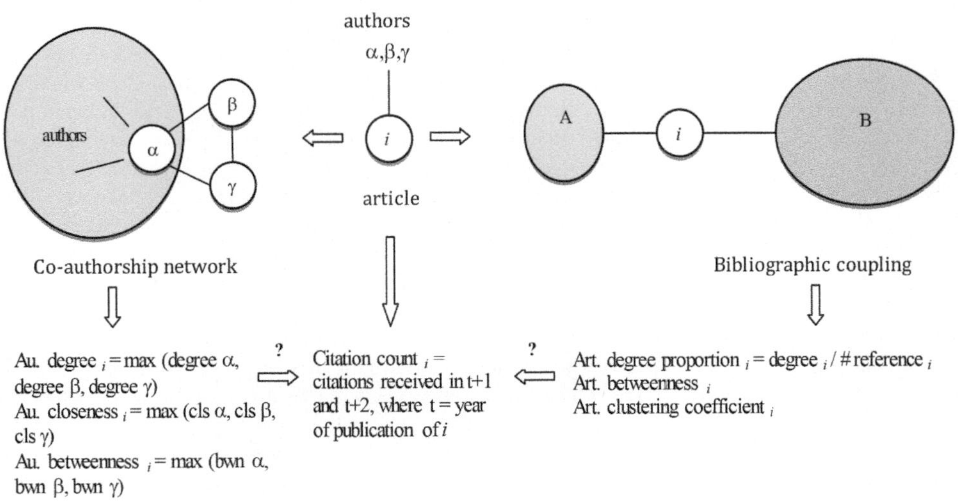

Figure 3. Conceptualization of the research questions. Starting from the article *i* written by α,β and γ and published in year t, two networks are derived. On the left side, the co-authorship network shows that α is already part of a group of other authors and had joint works with two of them (represented by the two edges inside the author set). β and γ are connected to the set of authors through α by means of their new joint-work *i*. We then extract three measures, degree, closeness and betweenness for each co-author of *i*, and map only the maximum value into *i*. On the right side, the article *i* have common references with articles of the sets A and B. We measure the degree, betweenness centralities and clustering coefficient of the article *i*. We construct the dependent variable cumulating the citations received by article *i* in the two years following the year of its publication. Last, we verify the effects the co-authorship and bibliographic coupling networks measures on the citation count of article *i*.

Degree centrality is the count of the first neighbors of a node. In the CA, it equates with the number of co-authors with whom any author has collaborated at the end of year t (hereafter *author degree*). We normalize the count in the BC network, as we use the *degree proportion* that is the degree centrality of article i divided by the number articles cited by i. Thereby keeping constant the degree of article i, its *degree proportion* will decrease with the size of its reference list.

Closeness centrality considers how close the node is to any other node in the network. Therefore a high closeness score demonstrates a short distance between the node – on which it is computed – and any other node. For a single node i, it is the inverse of the mean distance of the geodesic (shortest) path g to any other.

$$\text{Closenesscentrality}_i = \frac{n}{\sum_j g_{ij}}$$

Betweenness centrality of a node i is computed summing the number n^i_{st} of geodesic paths between any two nodes s and t that pass through i over the total number g_{st} of geodesic path between the two nodes. This is normalized by dividing by the number of ties between any other two nodes.

$$\text{Betweennesscentrality}_i = \frac{\frac{\sum_{st} n^i_{st}}{g_{st}}}{(n-1)(n-2)}$$

An author i with a high betweenness centrality score acts as the shortest path between many other actors, thus potentially benefiting from brokering advantages, especially when the other actors are disconnected if node i is removed.

Clustering coefficient of a node i is the number of f neighbors of i that are connected between each other divided by the number of pairs of neighbors of i. The measure, computed only for the BC networks, captures the embeddedness of an article in the existing literature. High values of clustering coefficient show that the articles with which i shares references, also share references among them.

Our unit of analysis is the article, therefore we retrieve the only the maximum value for each measure of the CA networks as they represent the value of the most influential co-author who transfers the highest value of authority to the paper [9].

All metrics are computed on yearly slices to capture the values at the year of publication.

Control variables

Topic generation. An article generates a topic when the proportion of text attributed to a topic is for the first time larger than 25%. When in the same year and in the same topic, the threshold is surpassed by multiple articles, all of them are classified as topic-generating. For example, Strzepek et al. [32] and Jose et al. [33] brought in the same year the concept of vulnerability to the context of *water resources and river basins* and we classify both as topic generating. Their proportion of words related to the topic of *water resources and river basins* are beyond the threshold and there is no prior article that accomplishes that. A similar operationalization has been performed by Kaplan and Vakili [34] who considered a less conservative threshold of 20%. Topic generation is a dummy variable that assumes value 1 when an article generates a new topic and 0 otherwise. With this variable, we control for those papers that introduce a certain topic in the field that could be the most important and be rewarded by higher citations [35], almost regardless of their quality [8]. By opening a new research stream, they benefit from a citation advantage over the followers as they lead future research that will necessarily cite them.

New topic combination. Similarly to topic generation, articles introduce a new topic combination when they establish a combination of topics that was not present in the dataset until the year of publication. We adopt the same rule used to attribute topic generation: a topic proportion beyond the threshold of 25% determines the presence of that topic in the article and the rule of the multiple attribution of a topic combination applies similarly to *topic generation*: all those articles that introduce the same topic-combination in the same year are identified as introducing a new topic combination. Topic combination controls for the effect of those articles which generate a combinatorial type of innovation [36] at the level of topic.

For robustness, results do not significantly change when checked with thresholds of .20 and .30.

Size of the citing literature. Citations depend on the size of the universe of article from which citations are drawn; therefore there is a need to control for the size of this expanding universe. As a proxy of the expanding universe of articles, we take the number of articles in the dataset two years after the year of publication of each article. Although we recognize that this universe must not be exact universe of citing articles, it provides the sense of growing attention towards the topic of vulnerability and, secondly, it is a monotonically growing body of scientific literature, such as the entire universe of scientific literature. For robustness, we also tried two different versions of the metric: (1) we inflated the measure with a relatively large fixed number representing the articles outside the dataset (10,000) that could be interested in referring to articles within; (2) we increased that fixed number (10,000) by a 4.1% each year as it is a plausible rate of growth of scientific articles in the period between 1990–2007 [37]. Results proved to be robust.

Number of authors. The number of authors is positively correlated with the *citation count*, see Table 2, and there may be several reasons: a paper with multiple co-authors is more likely to be more complex in terms of knowledge sources, as it required the work of multiple actors; the quality of the content can also be enhanced by the *labor limae* that can be performed by multiple hands; furthermore, the dissemination of the ideas included in the article can spread in the co-authorship network starting from multiple starting nodes. Thus, we sift out the effect by including the number of authors in the subsequent statistical analysis.

Experience in the field. Experience is associated with a higher level of specialization, knowledge of the relevant problems in the field and with a deeper ability in publishing and diffusing ideas [38]. Thereby expert authors have both cognitive and reputation advantages. They are better known within the field, have consolidated relationships in the research community, and know how to make networking better than their less-expert peers. This should translate into higher quality production and more effective dissemination. We operationalize the experience in the field for any article i by measuring the number of articles in the dataset published by each co-author prior to the publication of i. Among these figures, we chose the maximum, as we think that it is the most expert author who has the highest influence on the paper's impact.

Citing review bias. In several scientific disciplines there is a high concentration of reviews among the most cited articles [1]. Notwithstanding the difficulty to discriminate between review and non-review papers based solely on the number of references,

Table 2. Spearman correlation coefficients with significance level.

	Citations	2	3	4	5	6	7	8	9
(1) Citations									
(2) Author degree	0.34†								
(3) Author betweenness	0.31†	0.81†							
(4) Author closeness	−0.11†	−0.09†	−0.03						
(5) Article degree proportion	0.22†	0.19†	0.28†	0.11†					
(6) Article betweenness	0.30†	0.25†	0.29†	−0.11†	0.70†				
(7) Clustering coefficient	0.15†	0.18†	0.23†	−0.30†	0.56†	0.39†			
(8) Number of authors	0.22†	0.70†	0.38†	−0.07†	−0.02	0.08†	0.03		
(9) Size of Literature	0.18†	0.29†	0.28†	−0.93†	−0.02	0.17†	0.34†	0.15†	
(10) Author's experience	0.28†	0.77†	0.88†	−0.07†	0.32†	0.32†	0.27†	0.25†	0.28†

Levels of significance: † $p < .001$. All measures are computed at the year of publication of each paper.

because to introduce a new conceptual framework to analyze new data, often authors draw on multiple contributions in the extant literature, yet, to control for the citing bias towards articles with a large number of references, we clustered articles in two groups (review, non-review) based on the number of article cited in their reference list. The two groups have profoundly different means ($\mu_{review} \approx 37$, ($\mu_{non\text{-}review} \approx 106$), thus we created a dummy variable with value 1 if the article is a review, 0 otherwise.

Bandwagon effect. Citations are boosted by the peer recognition of the author. As already noticed, there is an effect of world-class recognitions and landmark papers on the citations of both prior and subsequent works of the authors [9]. With no availability of data on scientific prizes and accolades for the authors, we decided to control for the effect of peer-recognition by identifying the 309 authors of the articles that were most cited at the end of 2012 (top 1%, 33 articles), and put a dummy variable on the 199 articles written by them in the period starting two years before their landmark paper, with the exclusion of their landmark one. The two year time-window, in which we count citations, makes articles written before that period unaffected by posterior success, as the citing authors should not be probably aware of subsequent success (except for circulating working papers that we cannot control).

To understand the impact of the variables on the *citations count*, we perform negative binomial regressions – a regression model specific for count data in which the dependent variable has overdispersion – with nested sets of regressors. Results are presented in the next section.

Results

In this section, we first provide an overview of the data, and then we answer the research questions and show how the *citation count* is affected by the structure and positions of co-authors within the co-authorship network, and by the position of the article in the literature.

Summary statistics

After a handful of publication in the early 1990s, the literature on vulnerability increases steadily over the years, as displayed in Figure 4 (left). From 2004, the number of papers increases in a steep-log phase. Analogous is the trend of co-authors displayed in Figure 4 (right). In dark grey the cumulative number of authors in the dataset, while in light grey the number of new authors entering in the dataset at any year. Most articles are written by 3 authors ($\mu = 3.404$, $\sigma = 3.021$, min $= 1$, max $= 57$), they have on average 48.65 references ($\sigma = 34.537$, min $= 1$, max $= 398$), and receive on average 4.95 citations in the first two years after the publication ($\sigma = 9.890$, min $= 0$, max $= 273$), however the distribution of citations is skewed to the right.

For the non-normal distribution of most variables in the study, we use Spearman correlation to compute the correlation coefficients between the *citation count* and the independent variables, as shown in Table 2. Correlations show that citation count is positively associated with most independent variables with the expection of *author closeness*. The negative correlation between *Author closeness* is particularly interesting and needs further analysis, because it means that the impact of the paper seems to be negatively related to the proximity of its authors to all other authors in the network. As expected, instead, *author degree* and *author betweenness* co-vary with the citations received by the articles, thereby connecting distant authors and having a large number of co-authors co-occur with higher citations, and they are also

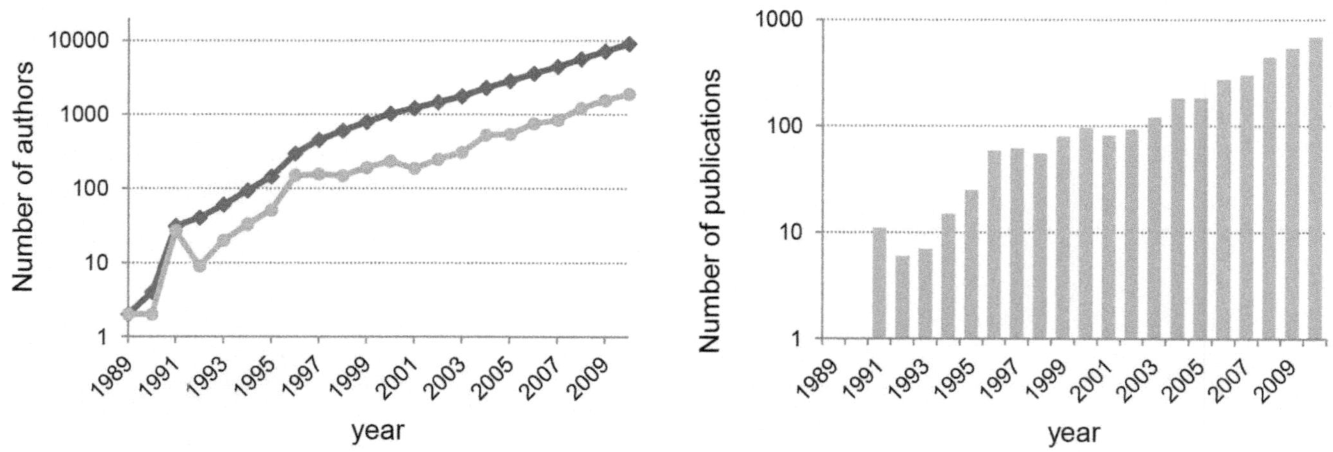

Figure 4. The expansion of the Vulnerability literature: number of authors (left) and publications (right) per year. On the left part of the figure, the dark line represents the cumulative number of authors who at least have one publication in the dataset, while the light line represents the number of new authors.

positively correlated (0.81 with $p<0.001$), as it occurs in most networks [39].

We also find that *article degree proportion*, *article betweenness* and *article clustering coefficient* are positively correlated with the citations received (.22 and .30 respectively with $p<.001$) and between each other (.70 with $p<.001$). The *article degree proportion*'s value show that there is a link between size of the research community to which the article belongs and its *citation count*, and *article betweenness* correlation show that bridging fragmented strands of literature usually signals an increase in the *citation count*. The positive value between *article clustering coefficient* and *citation count* shows that being embedded in a literature co-varies with citations. Also control variables such as the *number of authors*, the *size of literature* and the *author's experience* are positively and significantly correlated with the citation count.

In the next paragraph we present the results of the regressions computed on three models. In model 1, we replicate a part of a recent study that analyzed the structural effect of the co-authorship network on the *citation count* [2]. In model 2, we add the BC measures, whereas model 3 is generated on a smaller dataset that considers articles published between 2008 and 2010, a period in which the giant component – the largest component of the network in which all nodes can be linked by a path – in co-authorship network connects a relevant portion of the nodes. To control for outliers, we exclude the first three observation for their abnormally large score of *Author closeness* that is due to the number of nodes in the CA (the first three observations take values of 1 and .33, whereas the fourth largest observation of .038 – as shown in figures S1 and S2 in File S1).

All regressions are performed with the negative binomial regression model that is appropriate for count-data models and has no specific assumption on the dependent variable, unlike the Poisson and the zero-inflated Poisson. Poisson assumes that the mean and the variance of the dependent variable should be equal, while in our dataset they differ significantly ($\mu=4.952$, $\sigma^2=97.8116$). Nonetheless the large number of articles in the dataset whose *citation count* takes value 0, we still prefer the negative binomial model to a zero-inflated Poisson, because the latter model assumes that many of the observations that take value zero are drawn from a different distribution in which articles will never be cited. In our case, there is no theoretical reason to assume that non cited articles come from a different distribution. Results are displayed in Table 3.

The impact of co-authors' network position on the article's citation

Author degree has a positive highly significant but very small coefficient (between 0.02 and 0.01) throughout the three models. Such a positive association on citations remains when we add also the network metrics involving the shared references (in model 2), and is robust also in the smaller dataset comprising the articles published between the 2008 and 2010. This result shows that articles written by authors who have established more co-authorship relations (with other authors in the dataset) tend to be cited more. It must be noted that this cannot be a strategy pursued by authors, as the increase in citations generated by each relation is extremely small. Surprisingly, and unlike other studies [2,7], *author betweenness* is negatively and significantly associated with the citations received by the paper as long as the giant component connects a relevant part of the authors in the dataset (model 1 and 2 versus model 3). The coefficient is large, because the normalized betweenness scores take extremely small values (min 0, max 0.025, with a right skewed distribution in which 75% of the values are below $6 \cdot 10^{-7}$). The intuition is that bridging between groups of co-authors is negatively associated to *citation count*. It is not clear the reason underpinning such a result: it could be due to the experience of the bridging author and to the mathematical construction of the measure of betweenness. Regarding the experience of the bridging author, research in cognitive science says that a long and vast experience is necessary to successfully put together different pieces of knowledge [40,41], thereby we could think there may be some interaction between betweenness and authors' experience. Moreover the mathematical representation of the metric of betweenness overemphasizes the size of the population groups that are linked, while neglecting the redundancy of edges, thereby not capturing the number of different knowledge bases. For example, assuming that different groups have different expertise, betweenness does not distinguish if an author creates a bridge between two numerous groups (two knowledge domains), or alternatively more groups (more knowledge domains) but less numerous. Furthermore betweenness is equally sensitive to nodes directly and indirectly connected to the author, thereby it is also dependent to the size of the component.

Table 3. Results of regression analysis.

	model 1	model 2	model 3
	Estimate	**Estimate**	**Estimate**
Intercept	0.73†	0.64†	0.45
	0.06	*0.06*	*0.31*
Author degree	0.02†	0.02†	0.01†
	0.00	*0.00*	*0.00*
Author betweenness	−52.88†	−51.98†	−21.60
	14.61	*14.59*	*14.86*
Author closenesss	−9.03	−4.63	3161.00†
	10.74	*10.93*	*891.30*
Article degree proportion		−0.93	0.36
		0.81	*0.97*
Article betweenness		0.82**	0.17
		0.25	*0.41*
Article clustering coeff.		0.30*	0.24
		0.12	*0.16*
New topic generation	−0.29.	−0.26	
	0.16	*0.16*	
New topic combination	0.08	0.09	0.00
	0.07	*0.07*	*0.10*
Number of authors	0.08†	0.08†	0.07†
	0.01	*0.01*	*0.01*
Size of Literature	3.91E-05**	3.530E-05*	-3.29E-06
	1.36E-05	*1.409E-05*	*4.67E-05*
Review	0.72†	0.69†	0.64†
	0.05	*0.06*	*0.07*
Author's experience	0.03*	0.02.	0.00
	0.01	*0.01*	*0.01*
Bandwagon effect	0.08	0.05	0.02
	0.10	*0.10*	*0.13*
Observations	3340	3340	1684
Log-likelihood	−8505.26	−8498.22	−4509.65
McFadden R2	0.04	0.04	0.04

Signif. Codes: '†' p<0.001; '**' p<0.01; '*' p<0.05; '.' p<0.1. Standard errors in italics. McFadden r squared is 1 - (loglikelihood (model)/loglikelihood (intercept))

Thereby one author's betweenness score is high if she sits in a large network, despite that only a very limited number of them are directly in contact. For this reasons, we believe that *author betweenness* becomes not significant when we reduce the observation to those in presence of a relevant giant component, see Table 4. Although the score is high, it does not reflect the personal benefit that a person can acquire from sitting in a position more 'in between' within the network. Results then suggests us not to be conclusive in our conjectures and instead suggest carrying out further analysis to understand whether this or other measures should be used to capture the idea of creating bridges across groups.

Results of model 1 and 2 suggest that there is no association between *author closeness*, i.e., the proximity of an author to all others in the network, and the *citation count*. However, *author closeness* is sensitive to the structure of the CA. Initially the lack of collaboration between groups does not generate a unified network structure and authors cannot use their network of relations to usefully disseminate their work. Closeness plays a positive role in the presence of a tangible giant component (see Table 4). Model 3 shows the results of the regression on the subset of 1683 articles published between 2008 and 2010 where *author closeness*'s coefficient becomes positive and significant (3161 with p-value = 0.0004), proving that closeness is positively associated with *citation count*, and this may be due to the fact that direct forms of social interaction – co-authorships – facilitate the diffusion of ideas.

In summary, data support the idea that authors benefit from their accumulated co-authorship relations, and that being embedded (better when in the core rather than in the periphery) in a large network boosts the citation count of the paper. Betweenness centrality instead appears to be a problematic measure to analyze the combination of knowledge occurring in the single paper.

Table 4. Number of authors in the Giant Component (GC) with respect to the total authors in the network.

	1998	2000	2002	2004	2006	2008	2010
Authors in the field	594	1003	1429	2222	3461	5398	8590
Authors in the GC	79	94	98	231	723	1565	3589
Percentage in the GC	13.30%	9.37%	6.86%	10.40%	20.89%	28.99%	41.78%

The impact of the bibliographic coupling network

Model 2 shows that *article betweenness* is positively associated to the *citation count* (0.82, p-value <0.01) and highlights the scientific value of those works that find connections among others that are already present in the literature. It draws the attention to the fact that looking for relationships among theoretical arguments and finding connections with remotely connected or yet disconnected knowledge domains and theories is rewarded in terms of citations, even in the short term of two years. When we restrict the analysis to the articles written from 2008 to 2010, and the giant component comprises over 90% of nodes, it seems less obvious the advantage to find new ties with the existing literature, moreover there are fewer groups of nodes to incorporate in the main component. Thus the effect of *article betweenness* on citations becomes not significant when the giant component in the BC absorbs most nodes, as shown both in the regression of model 3 and in Figure 5.

Similarly, *article clustering coefficient* shows that being embedded in a literature benefit the *citation count* of the article (0.30 with p-value <0.05), but the effect disappears when computed for the articles written after 2008. One possible reason is that an increasing homogeneity of articles' references could be forced by exogenous pressures thus reducing the positive effect of the embeddedness. This hypothesis could be reinforced by the call for integration of the concept of vulnerability among various streams of literatures made by international bodies of research such as UNISDR and IPCC.

Article degree proportion, a proxy of the size of the research community, does not significantly impact the *citation count*. Thus we can conclude that, at least in this dataset, sharing references with a large number of other articles in the extant literature is not a practice that increases the number of citations received, suggesting that recognizing arguments already well spread in the literature by acknowledging very popular prior research, as well as the size of the research community, have no apparent relation with the scientific impact. Therefore, these practices that are typical of positioning a paper within a literature sort no significant effect at the citation side.

Curiously, among the control variables, *topic generation* is negatively associated to the number of citations, although in model 2 the p-value is slightly larger than 10%, while there is no effect for the *new topic combination*. The result of a new topic is not due to a different distribution of citations over the years (see SI), but it may be due to the fact that when vulnerability was brought into different topics, the formalization of the construct was still fuzzy, thus articles did not benefit from this inclusion. Only after the year 2002, there have been the most prominent theoretical advancements that established the concept and provided useful frameworks[42,43]. However, topic generating articles are concentrated before the year 2000.

The *number of authors*, the *type of article* (review or non-review) and the *size of literature* are all positively and significantly associated with citations, and especially review articles have a large positive effect on citations. When we control for the position of the article in the literature, the *author's experience* and the *size of literature* lose significance, showing that, at least in this setting, the number of articles published in the field is not a good predictor for the success of the next article, nor is the size of the citing community. In our setting, there is no *bandwagon effect* or, if there is, it is not extended to all co-authors of the most cited papers. This property is more likely to reside in a handful of them (perhaps the first and last one) who can on the one hand produce outstanding articles, and on the other benefit from peer-recognition.

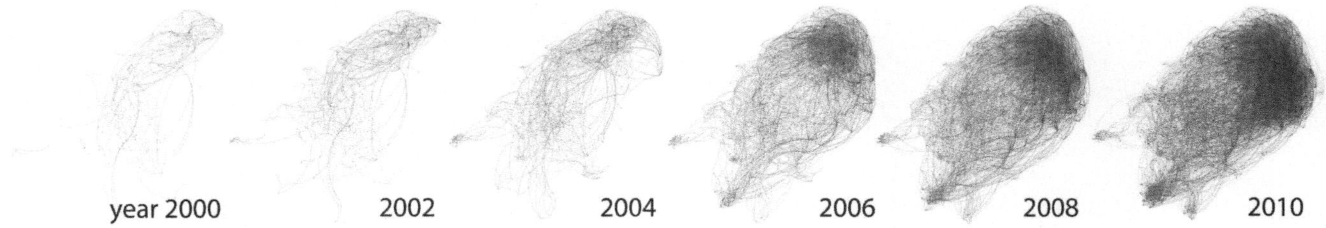

year	2000	2001	2002	2003	2004	2005	2006	2007	2008	2009	2010
number of nodes	420	501	594	715	898	1081	1356	1659	2104	2649	3343
size of the GC	310	377	466	576	739	915	1183	1468	1911	2452	3144
% of nodes in the GC	73.8%	75.2%	78.4%	80.6%	82.3%	84.5%	87.2%	88.5%	90.9%	92.6%	94.1%
avr. path distance	3.48	3.39	3.27	3.24	3.27	3.15	3.07	3.07	2.89	2.82	2.70
avr. clustering coeff.	0.44	0.44	0.48	0.46	0.43	0.42	0.45	0.44	0.43	0.42	0.42

Figure 5. The evolution of the Vulnerability literature with the Bibliographic Coupling network. Nodes are represented by articles in the dataset and the edges link articles which share one or more references. In this network, we show only articles with at least 10 citations and that are connected to the giant component (1164 nodes in 2010 that are 72,48% of amount of articles with more than 10 citations).

Discussion and Conclusion

In this work, we analyzed the effects on citations of the social and knowledge networks on which scientific articles are grounded. This work extends the knowledge accumulated on the effects of the co-authorship network on the scientific impact [2,3,7,15], and tries to ground it better in the theory of knowledge combination upon which it is based [12,14,41,44], by bringing in a second type of network that analyzes the knowledge sources. In particular, we felt compelled to answer two research questions: (1) how does the co-authorship network structure influence the scientific impact of an article in terms of citations? And (2) how do the article's knowledge sources influence its scientific impact? To answer these questions, we retrieve the articles published in the scientific literature on vulnerability of social and natural environment due to climate change and natural hazards. Then we construct the dynamics of the co-authorship and bibliographic coupling networks, based on co-authorship relations among authors and shared references among articles respectively.

With no surprise, we find that the structure of scientific collaborations matters. The cumulative number of co-authors has a positive – yet slight – impact on the citations of the article, while a larger and positive effect is given by the proximity of authors to all others in the field. However, the effect due to the proximity manifests itself only when it is possible to trace a collaboration link among a relevant share of authors in the field. This is consistent with the idea that knowledge diffusion is aided by personal relationships which could be costlessly and effortlessly tapped. Surprisingly, and in contrast to other studies [2,7], we found that bridging among groups of authors is penalized in terms of citations. Such result is counterintuitive, because the intuition would suggest that authors who create a bridge among different groups, also bridge among their knowledge bases, thus their ideas could benefit from the eventual distance among knowledge domains. We believe that the intuition, and the theory, still hold, but we propose two reasons that could explain such a result. A first reason that could moderate the negative impact of bridging could be due to experience of the author who establishes ties with other groups, as studies in cognitive science claim [14,40,41]. The second reason is regards the measure of betweenness centrality, adopted in this and similar studies: betweenness centrality on the one hand overemphasizes the effect of the indirect connections and thus depends on the size of the component and on the other neglects the redundancy of edges among nodes. Therefore it does not reflect the number of diverse knowledge domains a single author can benefit from. Other measures could be explored in future research and a promising perspective could be given by an adaptation of the clustering coefficient computed over sets of nodes (the papers' co-authors), instead on individual ones. In other words, when two authors decide to work together, their respective networks get together, but they are still separated by the edge that links the two coauthors. Therefore, knowledge diversity must be computed counting the common ties between every possible dyad among all prior co-authors of the authors.

With regard to knowledge sources, we found that articles receive more citations when the authors are aware of what happens in different strands of literature, demonstrated by the references included in their article, and are able to make a synthesis between the preexisting and disconnected ideas. Thereby we claim that articles that find ways to tie together fragmented pieces of literature obtain more citations. Also we found that articles receive more citations when they are positioned in a literature that builds on a common base of articles.

Neither a specific benefit, nor a disadvantage comes from the size of the research community. We identified the size of the research community by looking at the practice adopted by authors who signal their belonging to a community by citing what 'similar' papers cite, thus boosting the degree centrality in the bibliographic coupling network.

Besides the use of the bibliographic coupling network, the adoption of topic modeling is a second innovative methodological component of this work. We used this tool to sort a large set of literature and control for innovative papers in terms of their content. We want to stress the utility of the tool in describing the findings – some unexpected – related to the topic variables. We saw that topic generating articles are concentrated in the early years (1989–2002) when the literature is still fragmented: the co-authorship network and bibliographic coupling network are made of many components. These are signs that the body of literature on vulnerability does neither cohesively grow upon seminal contributions nor as a unified body of literature whereby authors speak to each other. Instead, articles belong to different and separate domains (probably pertinent to the topics as pre-existing and different strands of literature) that introduce the concept of vulnerability in different years. We may conjecture that they do not benefit from importing the concept of vulnerability in their literature, perhaps because the concept is still fuzzy, differently defined in different domains, and seminal contributions had yet to come. As regard topic combination, the vast nature of the dataset allowed the identification of relatively distant topics such as drought, river basins and arctic, whose combination in the same article may signal a generic approach and wide scope rather than a provision of specific and innovative contributions. We believe that peer recognition comes from deep analyses and novel results: features that still elude our algorithms.

Indeed, we recognize that this work comes with limitations, which drive also ideas for future research. As already mentioned, the negative impact of the bridging author is yet unclear and future research should be carried out to discover why there seems to be no advantage to broker. We propose that future research may adopt team-adjusted measures of network constraints [45], or clustering coefficient computed on the redundancies of the ties of sets of nodes.

We are aware that a second limitation is due to the large number of journals and edited books (1079 sources) that impeded us to control for possible journal effects on citations. We think that having a control for journals would absorb some of the variability in the data. However, such a limitation does not affect the validity of results that focused on the effect of co-authorship and bibliographic coupling network structures on the early citation received.

Another limitation comes with the simplification. We attributed the value given by the co-author with highest centrality scores to each article, and this prevented us from analyzing the information given by the heterogeneity of the set of co-authors. However, in this work, we think we have accomplished a first step to reconcile the often concealed knowledge aspect of generation of scientific knowledge with the more studied social one.

Supporting Information

File S1 Contains the files: Figure S1 – Scatterplot of the Bibliographic Coupling network data. Figure S2 – Scatterplot of the Co-authorship network data.

Acknowledgments

This work benefitted from the comments by Massimo Warglien on previous drafts of the paper and those of the participants at a seminar at the Laboratory of Experimental Economics at Ca' Foscari University of Venice, Venice, Italy.

Author Contributions

Conceived and designed the experiments: CB. Performed the experiments: CB. Analyzed the data: CB CG. Wrote the paper: CB. Designed the software used in the analysis: CB.

References

1. Ioannidis JP (2006) Concentration of the most-cited papers in the scientific literature: analysis of journal ecosystems. Plos One 1: e5.
2. Uddin S, Hossain L, Rasmussen K (2013) Network Effects on Scientific Collaborations. Plos One 8.
3. Wallace ML, Larivière V, Gingras Y (2012) A small world of citations? The influence of collaboration networks on citation practices. Plos One 7: e33339.
4. Moretti E (2012) The new geography of jobs: Houghton Mifflin Harcourt.
5. Bettencourt LM, Lobo J, Helbing D, Kühnert C, West GB (2007) Growth, innovation, scaling, and the pace of life in cities. Proceedings of the national academy of sciences 104: 7301–7306.
6. Heinze T, Bauer G (2007) Characterizing creative scientists in nano-S&T: Productivity, multidisciplinarity, and network brokerage in a longitudinal perspective. Scientometrics 70: 811–830.
7. Yan EJ, Ding Y (2009) Applying Centrality Measures to Impact Analysis: A Coauthorship Network Analysis. Journal of the American Society for Information Science and Technology 60: 2107–2118.
8. Newman MEJ (2009) The first-mover advantage in scientific publication. Epl 86.
9. Mazloumian A, Eom YH, Helbing D, Lozano S, Fortunato S (2011) How Citation Boosts Promote Scientific Paradigm Shifts and Nobel Prizes. Plos One 6.
10. Catmull E (2008) How Pixar fosters collective creativity: Harvard Business School Publishing.
11. Dekas KH, Bauer TN, Welle B, Kurkoski J, Sullivan S (2013) Organizational Citizenship Behaviour, version 2.0: A Review and Qualitative Investigation of OCBs for Knowledge Workers at Google and Beyond. Academy of Management Perspectives 27: 219–237.
12. Hargadon A, Sutton RI (1997) Technology brokering and innovation in a product development firm. Administrative Science Quarterly: 716–749.
13. Perry-Smith JE (2006) Social yet creative: The role of social relationships in facilitating individual creativity. Academy of Management Journal 49: 85–101.
14. Simonton DK (1995) Foresight in insight? A Darwinian answer. In: Sternberg RJ, Davidson JE, editors. The nature of insight. Cambridge, MA: MIT Press. pp. 465–494.
15. Yin L-c, Kretschmer H, Hanneman RA, Liu Z-y (2006) Connection and stratification in research collaboration: An analysis of the COLLNET network. Information Processing & Management 42: 1599–1613.
16. Kessler MM (1963) Bibliographic coupling between scientific papers. American documentation 14: 10–25.
17. Polanyi M (2000) The republic of science: its political and economic theory. Minerva 38: 1–21.
18. Sun XL, Kaur J, Milojevic S, Flammini A, Menczer F (2013) Social Dynamics of Science. Scientific Reports 3.
19. Newman MEJ (2001) The structure of scientific collaboration networks. Proceedings of the National Academy of Sciences of the United States of America 98: 404–409.
20. Chen CM (2004) Searching for intellectual turning points: Progressive knowledge domain visualization. Proceedings of the National Academy of Sciences of the United States of America 101: 5303–5310.
21. Janssen MA, Schoon ML, Ke WM, Borner K (2006) Scholarly networks on resilience, vulnerability and adaptation within the human dimensions of global environmental change. Global Environmental Change-Human and Policy Dimensions 16: 240–252.
22. Janssen MA (2007) An update on the scholarly networks on resilience, vulnerability, and adaptation within the human dimensions of global environmental change. Ecology and Society 12.
23. Dorta-Gonzalez P, Dorta-Gonzalez MI (2013) Impact maturity times and citation time windows: The 2-year maximum journal impact factor. Journal of Informetrics 7: 593–602.
24. Redner S (1998) How popular is your paper? An empirical study of the citation distribution. European Physical Journal B 4: 131–134.
25. Blei DM (2012) Probabilistic Topic Models. Communications of the Acm 55: 77–84.
26. Blei DM, Lafferty JD (2007) A correlated topic model of science. Annals of Applied Statistics 1: 17–35.
27. Blei DM, Ng AY, Jordan MI (2003) Latent Dirichlet allocation. Journal of Machine Learning Research 3: 993–1022.
28. Newman M (2010) Networks: An Introduction: OUP Oxford.
29. Chang J, Blei DM (2010) Hierarchical relational models for document networks. The Annals of Applied Statistics 4: 124–150.
30. Csardi G, Nepusz T (2006) The igraph software package for complex network research. InterJournal, Complex Systems 1695.
31. Uddin S, Hossain L, Abbasi A, Rasmussen K (2012) Trend and efficiency analysis of co-authorship network. Scientometrics 90: 687–699.
32. Strzepek KM, Yates D, El Quosy D (1996) Vulnerability assessment of water resources in Egypt to climatic change in the Nile Basin. Climate Research 6: 89–95.
33. Jose A, Sosa L, Cruz N (1996) Vulnerability assessment of Angat Water Reservoir to climate change. Water, Air, and Soil Pollution 92: 191–201.
34. Kaplan S, Vakili K (2012) Identifying Breakthroughs: Cognitive vs. Economic.
35. Redner S (2005) Citation Statistics from 110 Years of Physical Review. Physics Today 58: 49–54.
36. Frigotto ML, Riccaboni M (2011) A few special cases: scientific creativity and network dynamics in the field of rare diseases. Scientometrics 89: 397–420.
37. Larsen PO, von Ins M (2010) The rate of growth in scientific publication and the decline in coverage provided by Science Citation Index. Scientometrics 84: 575–603.
38. Maske KL, Durden GC, Gaynor PE (2003) Determinants of scholarly productivity among male and female economists. Economic Inquiry 41: 555–564.
39. Goh KI, Oh E, Kahng B, Kim D (2003) Betweenness centrality correlation in social networks. Physical Review E 67: 017101.
40. Simon HA, Chase WG (1973) Skill in chess: Experiments with chess-playing tasks and computer simulation of skilled performance throw light on some human perceptual and memory processes. American scientist 61: 394–403.
41. Simonton DK (1999) Creativity as blind variation and selective retention: Is the creative process Darwinian? Psychological Inquiry 10: 309–328.

42. Turner BL, Kasperson RE, Matson PA, McCarthy JJ, Corell RW, et al. (2003) A framework for vulnerability analysis in sustainability science. Proceedings of the national academy of sciences 100: 8074–8079.
43. Adger WN (2006) Vulnerability. Global environmental change 16: 268–281.
44. Schilling MA (2005) A" small-world" network model of cognitive insight. Creativity Research Journal 17: 131–154.
45. Burt RS (2004) Structural holes and good ideas. American Journal of Sociology 110: 349–399.

Phylogeny Predicts Future Habitat Shifts Due to Climate Change

Matjaž Kuntner[1,2,3]*, Magdalena Năpăruş[4,5], Daiqin Li[2,6], Jonathan A. Coddington[3]

1 Institute of Biology, Scientific Research Centre, Slovenian Academy of Sciences and Arts, Ljubljana, Slovenia, 2 Centre for Behavioural Ecology and Evolution, College of Life Sciences, Hubei University, Wuhan, Hubei, China, 3 National Museum of Natural History, Smithsonian Institution, Washington, D. C., United States of America, 4 Centre of Landscape–Territory–Information Systems - CeLTIS, University of Bucharest, Bucharest, Romania, 5 Tular Cave Laboratory, Kranj, Slovenia, 6 Department of Biological Sciences, National University of Singapore, Singapore, Singapore

Abstract

Background: Taxa may respond differently to climatic changes, depending on phylogenetic or ecological effects, but studies that discern among these alternatives are scarce. Here, we use two species pairs from globally distributed spider clades, each pair representing two lifestyles (generalist, specialist) to test the relative importance of phylogeny versus ecology in predicted responses to climate change.

Methodology: We used a recent phylogenetic hypothesis for nephilid spiders to select four species from two genera (*Nephilingis* and *Nephilengys*) that match the above criteria, are fully allopatric but combined occupy all subtropical-tropical regions. Based on their records, we modeled each species niche spaces and predicted their ecological shifts 20, 40, 60, and 80 years into the future using customized GIS tools and projected climatic changes.

Conclusions: Phylogeny better predicts the species current ecological preferences than do lifestyles. By 2080 all species face dramatic reductions in suitable habitat (54.8–77.1%) and adapt by moving towards higher altitudes and latitudes, although at different tempos. Phylogeny and life style explain simulated habitat shifts in altitude, but phylogeny is the sole best predictor of latitudinal shifts. Models incorporating phylogenetic relatedness are an important additional tool to predict accurately biotic responses to global change.

Editor: Mónica Medina, Pennsylvania State University, United States of America

Funding: This research was supported in part by a Raffles Museum for Biodiversity Research (RMBR) Short-term Fellowship and the grants P10236, BI-US/09-12-016 and MU-PROM/12-001 from the Slovenian Research Agency to M.K. and by the NSFC grant (31272324) and Singapore Ministry of Education (MOE) AcRF grant (R-154-000-476-112) to D.L. The funders had no role in study design, data collection and analysis, decision to publish, or preparation of the manuscript.

Competing Interests: The authors have declared that no competing interests exist.

* E-mail: kuntner@gmail.com

Introduction

Biotic change due to global warming is increasing dramatically. Although climatic changes are increasingly well understood and future predictions are improving, it is much more difficult to predict biotic responses to climate change [1,2]. Models predict global climates to become warmer in the following decades (Fig. S1), and precipitation patterns will change accordingly, but how organisms will respond to such changes is less clear. Naturally, different life histories and even clades of organisms may respond in various ways [3]. For example, they may respond with a shift in time, e.g. changing their phenology [2,4], or a shift in space, such as moving to different habitats, latitudes, or altitudes [2,5,6,7], insofar as habitats are available [8]. Crucial species interactions may be broken by both types of shifts [9]. Tropical species, in particular, may be severely affected [4,10]. Actual adaptive evolutionary change within a few decades may be possible for some organisms with extraordinary genotypic and phenotypic plasticity [11,12] but is unlikely for most as it would require over 10,000 times faster rates of adaptive change than those estimated [13]. Finally, the species unable to cope with anthropogenic global changes, and those whose dispersal ability is low, are likely to become extinct either globally or locally [1,3,14].

Communities, ecosystems and biomes are also likely to be forced to shift in space [2], become increasingly destabilized due to biotic loss [15], or be subject to thermophilization [16] and invasion [17]. In addition, food web and trophic interactions may be altered [18]. While biomes, notably tropical and subtropical forests, may expand in range due to increased rainfall and atmospheric CO_2 levels [19], such trends are expected to be countered by human habitat degradation, probably resulting in net habitat loss [20]. In some cases, temperate assemblages may increase in species richness or exhibit a higher species turnover [21]. However, at the global level with all factors combined, most biomes and communities are predicted to deteriorate, and thus may simply shrink or disappear. Their space then may be taken over by exotic and invasive species and communities, whose success can sometimes be directly attributed to global change [17].

Terrestrial organisms may be most prone to global climate changes (but see freshwater and marine reviews [22,23]), yet

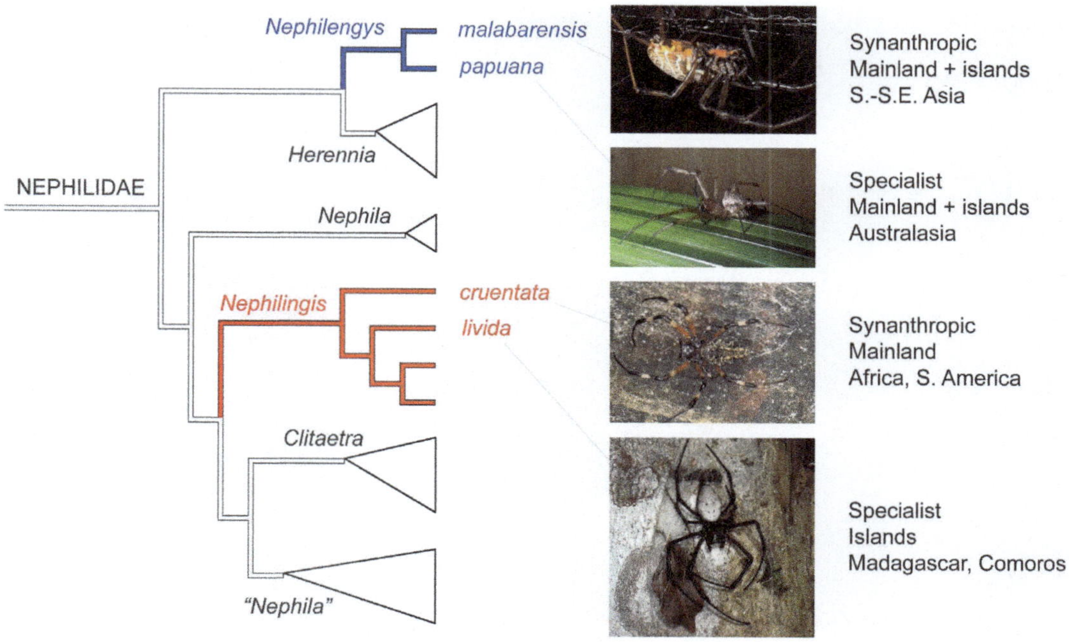

Figure 1. Two nephilid species pairs, their phylogeny, and basic ecology. One species from each clade is synanthropic and the other one a habitat specialist. This sample tests the relative importance of phylogeny versus life history on species responses to climatic changes. The phylogenetic hypothesis builds on a nephilid species level study that used 4kB of nucleotide data in addition to morphology, and an array of analytical approaches and sensitivity analyses [28].

surprisingly few studies have modeled the future of closely related terrestrial lineages—clades—in order to discern among phylogenetic and ecological effects while studies modelling single species continue to abound [24,25]. As an example of clade predictions, Diamond et al. [26] examined models for thermal tolerance in ants based on current and future climates and found that tropical ants had lower tolerances to warming. Tropical ant faunas, thus, are more susceptible to global warming, and that is precisely where the most diversity lies. However, is this necessarily true for all tropical ants? One would expect that phylogeny may affect biotic responses to climate change at least as much as ecology or natural history [27]. Fine grained studies discerning among these factors are currently lacking [3].

To test the relative importance of phylogeny versus ecology in predicted biotic responses to climate change, we selected two clades within which species have occupied two distinct ecological niches. The spider genera *Nephilengys and Nephilingis* [28] are large terrestrial invertebrate predators. The Asian *Nephilengys malabarensis* is synanthropic and hence relatively generalized, and the Australasian *Ny. papuana* is a forest species, thus relatively specialized [29]. Likewise the African-South American *Nephilingis cruentata* is synanthropic, whereas the Madagascan-Comoroan *Ni. livida* is a forest species (Fig. 1) [29,30,31]. The four species are fully allopatric, but combined they occupy all subtropical-tropical regions [29]. Ecologically the two specialists, *Ni. livida* and *Ny. papuana* should be more prone to habitat loss than the generalists *Ny. malabarensis* and *Ni. cruentata*. This prediction derives from the assertions that species with greater ecological generalization should be more likely to shift to higher elevation and latitude [3] and that niche breath positively correlates with species geographic ranges [32]. Alternatively, phylogenetic relatedness may better explain response trends. We first used predicted temperature and precipitation changes (Fig. S1) and a GIS simulation tool to model habitat degradation, gain, and loss in the four species (Fig. 2, Fig. S2) at 20 year intervals from 2000–2080. We then investigated the extent to which phylogeny versus ecology explained habitat changes. A priori predictions are difficult: if commonly derived species undergo life history adaptation to reduce competition, then ecology may be a better predictor. If not, phylogeny may be used to predict habitat shifts due to climate change.

Materials and Methods

Modeling current habitat suitability

We used all available locality data for the four species [29]: *Ni. cruentata* (specimen records N = 436), *Ni. livida* (N = 138), *Ny. malabarensis* (N = 138) and *Ny. papuana* (N = 39). Since our previous study phylogenetic evidence has emerged placing *Ni. cruentata* and *Ni. livida* in the genus *Nephilingis*, which is phylogenetically distant from *Nephilengys* containing *N. malabarensis* and *Ny. papuana* [28]. This is currently the best available phylogenetic hypothesis of nephilid species relationships, having utilized 4 kB of nucleotide data in addition to morphology, and an array of analytical approaches and sensitivity analyses [28]. That the previously congeneric species in fact belong to distinct clades, which nevertheless contain ecologically dissimilar species (one synanthropic and one habitat specialist within each genus), adds power to our testing of phylogenetic versus life history effects on habitat suitability. Our previous model selected best fit pairs of ecological parameters for each species current habitat suitability assessment based on the outcomes of backward linear regression: annual mean temperature and elevation was used for the *Ni. cruentata* model, global land cover and elevation for the *Ni. livida* model, and global land cover and annual mean precipitation for both *Ny. malabarensis* and *Ny. papuana* models [29]. Here, we used the same pairs of parameters for all four species models, which are indicative of global climate change: annual mean temperature (TMA) and annual mean precipitation (PMA). Based on these two parameters,

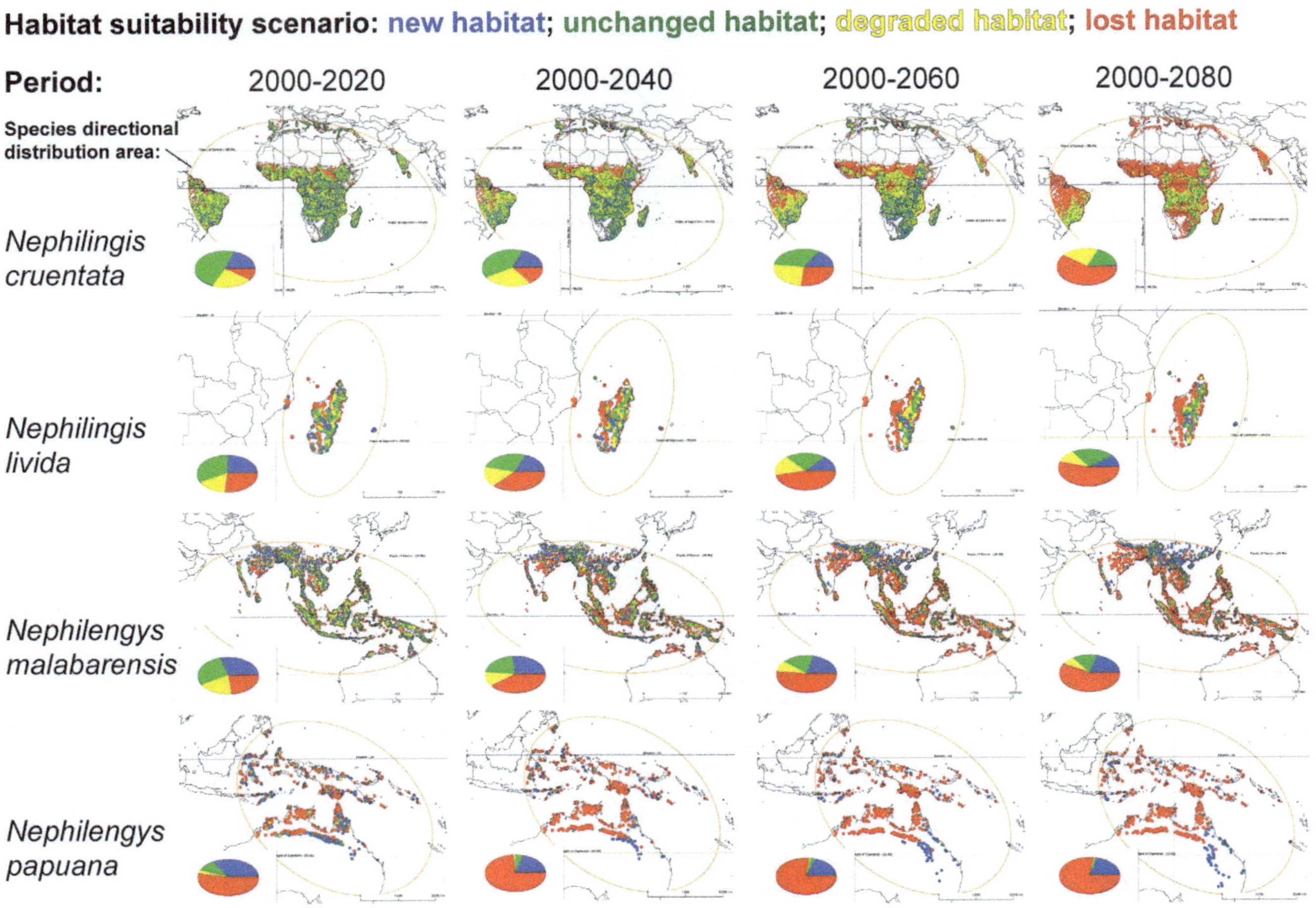

Figure 2. Predicted habitat suitability changes of two *Nephilingis* and two *Nephilengys* species over increasing time ranges. The predictions are based on the IPCC scenario A1B for temperature and precipitation (see Figs S1–S2) and are modeled for the time ranges 2000 vs. 2020, 2040, 2060, and 2080. Future habitat is classified as new, unchanged, degraded or lost.

we modeled the starting point for the year 2000, taken as each species' current habitat suitability.

Modeling future habitat suitability

We downloaded the raster data for TMA and PMA for the years 2020 to 2080 from the Downscaled Global Circulation Models data portal (http://www.ccafs-climate.org), and the current distribution (1950–2000) of climates from the WorldClim Database [33]. The raster layers are based on the IPCC Special Report on Emissions Scenarios (SRES) [34], where several scenario families of predicted climate change were proposed. The scenario A1 builds on the assumption of maximum energy requirements, and the sub-choices are based on emissions expected from the use of different fuel sources: fossil intensive (A1F1), technologically developed non-fossil sources (A1T) and a balance across sources (A1B). Among the worst case scenario (A1) we chose the balanced use of fuel sources (A1B) as the basis for our models. The remaining three scenarios build on minimum energy requirements and emissions (B1), on high energy requirements (A1F1) and on lower energy requirements; however, as the scenarios A2 and B2 only predict a limited number of years (2020, 2050 and 2080), and the A1 scenario predicts each decade, our choice of scenario (A1B) seemed justified. For global modeling use, IPCC recommends the use of A1 and B1 [35], among which we chose to model a more pessimistic version due to the availability of predicted climate change (for B1 no associated datasets exist). The A1B scenario uses 24 Global Circulation Models [36]. All rasters have a spatial resolution of 2.5 arc-minutes, using a WGS84 datum.

The A1 group of scenarios predicts a global average surface warming until 2100 from 1.4 to 6.4°C [35]. Fig. S1 shows projected global changes in TMA and PMA for the period covered by our models (current = 2000; projected 2020, 2040, 2060, 2080). Based on these, we modeled habitat suitability of each of the four species following our GIS modeling methodology [29]. These maps define each species' directional distribution as its potential target area and thus potential dispersal range [29], and within it, we used the frequency distribution values for TMA and PMA for current specimen records [29]. We here explore the changes for each species in three habitat categories: high, moderate and low habitat suitability (Fig. S2). Several data sources exist that model global changes according to the A1B scenario. We choose the data from the Canadian Centre for Climate Modelling and Analysis - CGCM3.1 (T47), 2005 [37]. This data model version is the basis for the suite of model simulations in the IPCC Fourth Assessment Report [35]. T47 version of CGCM3.1 data model has a surface grid whose spatial resolution is roughly 3.75 degrees latitude/longitude and 31 levels vertically, and an ocean grid resolution of roughly 1.85 degrees, with 29 levels vertically [38,39].

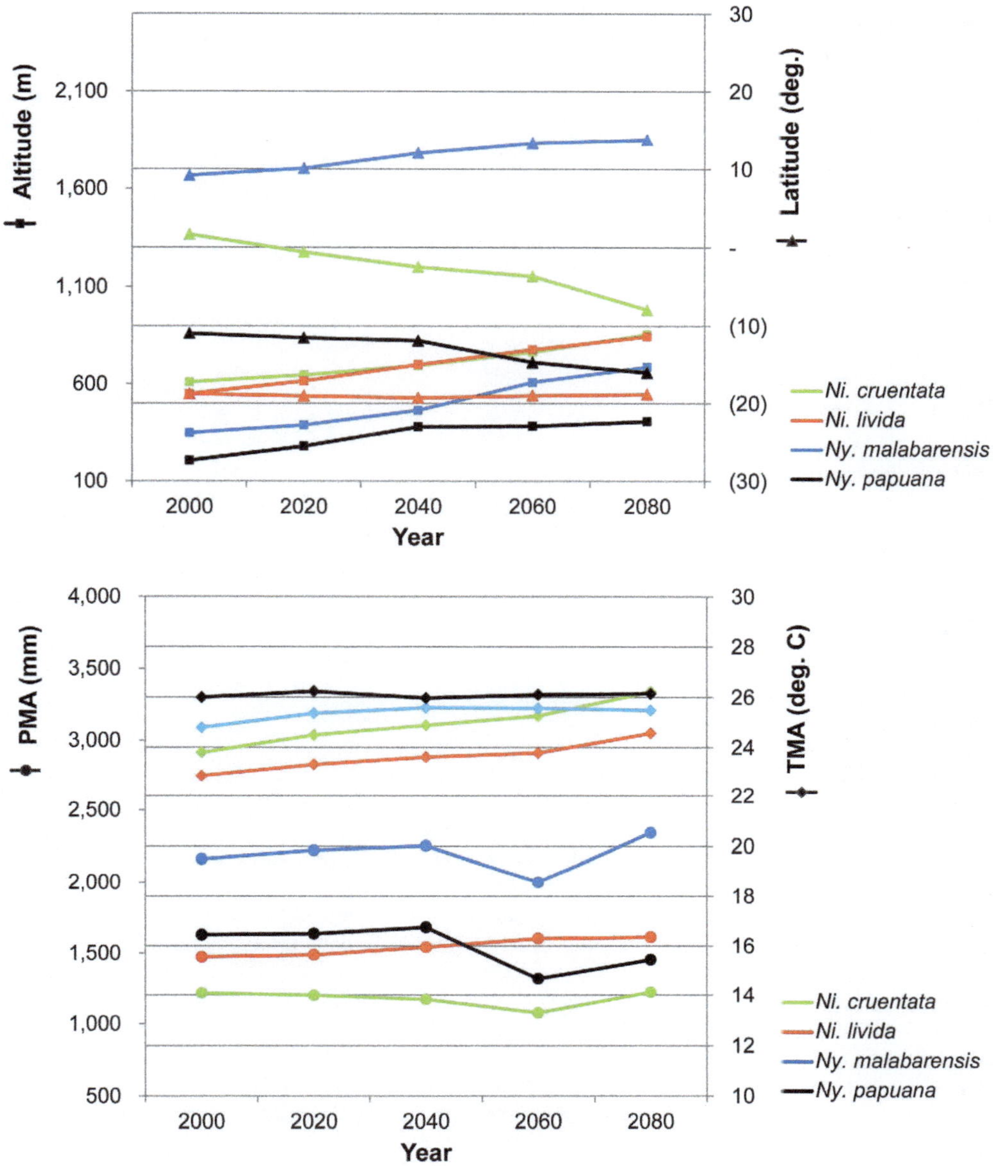

Figure 3. Changes for all species in altitude and latitude (above), and temperature and precipitation (below). Species maintain their characteristic climatic conditions by shifting towards higher altitudes and latitudes. Geographical and ecological averages for the modeled suitable habitats combined all categories (low, medium and high suitability). NI = *Nephilingis*, NG = *Nephilengys*, PMA = mean annual precipitation, TMA = mean annual temperature.

Assessing species habitat loss

To assess the degrees of projected habitat losses or gains due to climatic changes, we compared the differences in predicted habitat suitability categories for the following time periods: 2000–2020, 2000–2040, 2000–2060, and 2000–2080. We customized our previous GIS model [29] with additional tools in ArcGIS [40]: i) First, we reclassified each year's habitat suitability raster data where no habitat suitability was shown from 'NoData' to '99' using the tool 'Reclassify'; those values that showed a habitat suitability value were left as they were (high = 3; moderate = 2; low = 1); ii) We then compared using the 'Combine' tool the habitat category changes between the year 2000 and other time periods and reclassified the resulting combinations into the following nominal values: 1 = new habitat (indicating projected habitat suitability gain); 2 = unchanged habitat (indicating no change in habitat suitability); 3 = degraded habitat (indicating a loss in habitat suitability value, but not below 1) and 4 = lost habitat (indicating a complete loss of projected habitat suitability, i.e. resulting in a change to 99).

Statistical analyses

Since the TMA and PMA data were not normally distributed, we employed standard non-parametric statistics to test the effects of phylogenetic relatedness (*Nephilingis* vs. *Nephilengys*) and life history (synanthropic vs. habitat specialist) on current species ecological preferences. To determine whether phylogenetic or ecological factors are good predictors of the simulated habitat shifts, we first devised a randomized dataset of 10,000 simulated georeferenced records per period per species (totaling 200,000 records; Table S1), calculated latitudinal and altitudinal shifts and coded them as categorical data (i.e., shift or not) based on the shifted values. We performed generalized linear model (GLM)

Table 1. Predicted percentages of future habitat change per species relative to the year 2000.

Species	Habitat categories	2020	2040	2060	2080
Nephilingis cruentata	new	19.9%	19.2%	16.1%	1.2%
	unchanged	48.5%	40.0%	31.7%	17.8%
	degraded	20.9%	25.1%	25.6%	20.4%
	lost	10.7%	15.7%	26.6%	60.6%
Nephilingis livida	new	24.5%	19.1%	14.2%	10.6%
	unchanged	33.9%	26.6%	21.6%	23.3%
	degraded	15.2%	17.8%	17.7%	11.3%
	lost	26.4%	36.5%	46.5%	54.8%
Nephilengys malabarensis	new	29.3%	26.8%	21.5%	21.0%
	unchanged	29.4%	21.6%	16.8%	13.2%
	degraded	17.8%	12.5%	9.7%	7.2%
	lost	23.5%	39.1%	52.0%	58.6%
Nephilengys papuana	new	33.3%	20.8%	20.2%	20.9%
	unchanged	10.5%	4.2%	2.5%	1.5%
	degraded	3.2%	1.0%	0.7%	0.5%
	lost	53.0%	74.0%	76.6%	77.1%

using ordinal logistical model twice, once for latitudinal shifts and once for altitudinal shifts, with phylogeny, life style, period (2020, 2040, 2060 and 2080) and habitat suitability (high = 3; moderate = 2; low = 1) as predictors. Both period and habitat suitability showed the same effects on shifts in the initial GLM tests, thus being excluded from the final models, which tested the effects of only two factors (phylogeny and life style) and the 2-way interaction on the categorical latitude and altitude shifts. The model was fitted using multinominal distribution with cumulative logit link error. Akaike information criterion (AIC: the smallest is the best) was used to select the best model. All statistical analyses were done using IBM SPSS Statistics 21 (IBM Corporation, USA). Reported p-values are two-tailed tests, with $\alpha = 0.05$.

Results

The global habitat suitability models for 2000, 2020, 2040, 2060 and 2080 predict overall shrinking or disappearance of suitable habitat for all four species (Fig. S2), although minor amounts of newly appropriate habitat emerge over time. Categorizing habitat change as unchanged, degraded, lost, or new, the major trend across all time ranges is dramatic increase in either lost or degraded habitat for all species (Fig. 2; Table 1). Losses and gains can take different tempos. *Ni. cruentata* shows a dramatic loss by 2080, but in *Ny. papuana* the most dramatic loss occurs by 2040. Others are intermediate. In no case does new habitat tend to increase (Table 1). By 2080, habitat loss ranges from 54.8% (*Ni. livida*) to 77.1% (*Ny. papuana*). Even the widespread, generalist species *Ni. cruentata* and *Ny. malabarensis* lose 60.6% and 58.6% of their habitat. According to the model, all four species roughly maintain their known temperature and precipitation preferences (Fig. 3). However in order to do so, all species shift towards higher altitudes and latitudes (except *Ni. livida*, whose island habitats are obviously fixed) (Fig. 3).

Phylogeny significantly explains variation in current temperature preferences (Mann-Whitney $U = 24{,}965$, $df = 1$, $p < 0.0001$). Both phylogeny and lifestyle significantly explain current precipitation preferences, but phylogeny much more strongly than

Table 2. Results from a generalized linear model (GLM) testing the effects of two explorative factors (phylogeny and lifestyle) on how spiders respond to climate conditions (based on the IPCC scenario A1B for temperature and precipitation) by shifting latitude or altitude.

Shift	Exploratory factors	Wald χ^2	*df*	*p*
Latitude	Phylogeny (genus-pair)	310.711	1	<0.0001
	Life style	3.65	1	0.056
	Phylogeny × Life style	13.730	1	<0.0001
Altitude	Phylogeny (genus-pair)	30.825	1	<0.0001
	Life style	17.350	1	<0.0001
	Phylogeny × Life style	0.082	1	0.775

The model that included two factors and 2-way interaction was fitted using multinominal distribution with cumulative logit link error.
GLM test for latitude shift: Goodness of fit: AIC = 116.261; Omnibus test: $\chi^2 = 328.2$, $df = 3$, $p < 0.0001$; GLM test for altitude shift: Goodness of fit: AIC = 107.247; Omnibus test: $\chi^2 = 48.242$, $df = 3$, $p < 0.0001$.

lifestyle (Mann-Whitney U = 25,803, *df* = 1, $p<0.0001$ and U = 13,503, *df* = 1, $p<0.016$).

GLM analyses on simulated data show that phylogeny predicts latitudinal shifts better than life style does (Goodness of fit: AIC = 116.261; Omnibus test: $\chi^2 = 328.2$, *df* = 3, $p<0.0001$; Table 2): *Nephilengys* is more likely to shift latitude than *Nephilingis*. Both phylogeny and life style are good predictors of the simulated altitudinal shifts (Goodness of fit: AIC = 107.247; Omnibus test: $\chi^2 = 48.242$, *df* = 3, $p<0.0001$; Table 2): altitudinal shifts are more likely in *Nephilingis* than *Nephilengys* and in synanthropic species than in habitat specialists.

Discussion

Models that forecast biodiversity patterns have been grouped into four categories [41]: 1) those considering individual species, 2) those grouping species by niche, 3) those based on general circulation or coupled ocean-atmosphere-biosphere theories, and 4) those based on species-area curves [41]. We argue that models using phylogenetic relatedness should also be considered because our results have shown that phylogeny can strongly predict climatic and habitat preference.

Studies considering individual species fail to account for species interactions and phylogenetic factors and are therefore of limited general use. In one case, the already limited range of the golden striped salamander decreased even more [42]; in another an invasive species, the Australian redback spider, increased its already substantial distribution [43]. Outside of a phylogenetic context, these patterns make no general predictions for evolutionarily or ecologically closely related organisms.

More taxonomically inclusive studies often predict changes in large clades of organisms, notably amphibians [44,45] or ants [26], but again, fine-grained clade-based studies are too preliminary. For example, the ranges of European amphibians and reptiles are generally predicted to shrink by 2050 [44], but this study was based on two opposite, and equally unlikely assumptions of unlimited versus no dispersal. A study of plant, bird and mammal assemblages in Europe only found a weak relationship between phylogeny and climate change vulnerability [46]. However, Willis et al. [47] investigated the changes in phenology and abundance of a temperate flora over 150 years and found that different clades are quite differently affected by climate change.

Niche modelling is limited as it fails to account for differences in species natural histories [3]. Only recently have studies taken into account species trait variability in assessing predicted responses to climate change. Angert et al. [3] examined to what extent species' traits are predictive of expanding their ranges. Predicting ecologically general species with greater dispersal abilities to more easily shift to higher latitudes and altitudes in response to climatic changes, they did find the expected relationships in passeriform birds and odonates, but their models yielded low explanatory power.

Our study extended this logic on a phylogenetically fine-grained taxonomic sample, by testing the relative importance of phylogeny versus species traits in predicted biotic responses to climate change in two species pairs from globally distributed spider clades, each pair representing two lifestyles. Despite the prediction that the unrelated specialists (forest-dwelling species) would cope with projected global changes worse than the unrelated generalists (synanthropic species), our results project habitat shrinkage patterns that may better be predicted by phylogeny than life style.

Our modeling approach assumes genotypic and phenotypic species homogeneity, perhaps an unwarranted assumption considering wide geographical ranges these species occupy. In reality, even widespread generalists show considerable genetic and phenotypic adaptations, spiders being no exception [48]. The predicted habitat shifts therefore may not be equally likely throughout the entire range of these taxa, and it may be possible that levels of adaptive change could take place at short temporal scales such as these presented in our study. The modes of habitat loss and extinction may depend on many additional factors, such as geography or basic natural history that are too complex to model accurately [3]. Phylogeny—which does broadly predict many life history traits—can simplify the prediction of biotic responses to future climate change. For accurate modelling, of course, many other factors are important as well, such as species interactions [49,50], persistence abilities [51], population dynamics and changes in genetic diversity [52]. However, more work assessing phylogeny versus ecology and life history as explanations for responses to climate change is needed.

Supporting Information

Figure S1 Predicted global changes in temperature and precipitation based on the IPCC scenario A1B. These predicted changes were used as bases for modeling species distribution 2000–2020, 2000–2040, 2000–2060 and 2000–2080.

Figure S2 Models predicting future habitat suitability for two *Nephilingis* and two *Nephilengys* species. The models for the time periods 2000, 2020, 2040, 2060 and 2080 are based on the IPCC scenario A1B for temperature and precipitation changes (see Fig. S1).

Acknowledgments

We thank Mitch Aide, Ingi Agnarsson and Gregor Aljančič for discussions and three anonymous referees for their constructive comments.

Author Contributions

Conceived and designed the experiments: MK MN. Performed the experiments: MK MN. Analyzed the data: MK MN DL JAC. Contributed reagents/materials/analysis tools: MK MN DL JAC. Wrote the paper: MK MN DL JAC.

References

1. Sandel B, Arge L, Dalsgaard B, Davies RG, Gaston KJ, et al. (2011) The Influence of Late Quaternary Climate-Change Velocity on Species Endemism. Science 334: 660–664.
2. Bellard C, Bertelsmeier C, Leadley P, Thuiller W, Courchamp F (2012) Impacts of climate change on the future of biodiversity. Ecology Letters 15: 365–377.
3. Angert AL, Crozier LG, Rissler LJ, Gilman SE, Tewksbury JJ, et al. (2011) Do species' traits predict recent shifts at expanding range edges? Ecology Letters 14: 677–689.
4. Pau S, Wolkovich EM, Cook BI, Davies TJ, Kraft NJB, et al. (2011) Predicting phenology by integrating ecology, evolution and climate science. Global Change Biology 17: 3633–3643.
5. Franco AMA, Hill JK, Kitschke C, Collingham YC, Roy DB, et al. (2006) Impacts of climate warming and habitat loss on extinctions at species' low-latitude range boundaries. Global Change Biology 12: 1545–1553.
6. Saikkonen K, Taulavuori K, Hyvonen T, Gundel PE, Hamilton CE, et al. (2012) Climate change-driven species' range shifts filtered by photoperiodism. Nature Climate Change 2: 239–242.

7. Beale CM, Baker NE, Brewer MJ, Lennon JJ (2013) Protected area networks and savannah bird biodiversity in the face of climate change and land degradation. Ecology Letters 16: 1061–1068.
8. Lawler JJ, Ruesch AS, Olden JD, McRae BH (2013) Projected climate-driven faunal movement routes. Ecology Letters 16: 1014–1022.
9. Hegland SJ, Nielsen A, Lazaro A, Bjerknes AL, Totland O (2009) How does climate warming affect plant-pollinator interactions? Ecology Letters 12: 184–195.
10. Colwell RK, Brehm G, Cardelus CL, Gilman AC, Longino JT (2008) Global warming, elevational range shifts, and lowland biotic attrition in the wet tropics. Science 322: 258–261.
11. Erwin DH (2009) Climate as a Driver of Evolutionary Change. Current Biology 19: R575–R583.
12. Munday PL, Warner RR, Monro K, Pandolfi JM, Marshall DJ (2013) Predicting evolutionary responses to climate change in the sea. Ecology Letters 16: 1488–1500.
13. Quintero I, Wiens JJ (2013) Rates of projected climate change dramatically exceed past rates of climatic niche evolution among vertebrate species. Ecology Letters 16: 1095–1103.
14. Jump AS, Penuelas J (2005) Running to stand still: adaptation and the response of plants to rapid climate change. Ecology Letters 8: 1010–1020.
15. Post E (2013) Erosion of community diversity and stability by herbivore removal under warming. Proceedings of the Royal Society B-Biological Sciences 280.
16. Gottfried M, Pauli H, Futschik A, Akhalkatsi M, Barancok P, et al. (2012) Continent-wide response of mountain vegetation to climate change. Nature Climate Change 2: 111–115.
17. Sandel B, Dangremond EM (2012) Climate change and the invasion of California by grasses. Global Change Biology 18: 277–289.
18. Barton BT, Schmitz OJ (2009) Experimental warming transforms multiple predator effects in a grassland food web. Ecology Letters 12: 1317–1325.
19. Banfai DS, Bowman DM (2006) Forty years of lowland monsoon rainforest expansion in Kakadu National Park, Northern Australia. Biological Conservation 131: 553–565.
20. Heubes J, Kuhn I, Konig K, Wittig R, Zizka G, et al. (2011) Modelling biome shifts and tree cover change for 2050 in West Africa. Journal of Biogeography 38: 2248–2258.
21. Hillebrand H, Soininen J, Snoeijs P (2010) Warming leads to higher species turnover in a coastal ecosystem. Global Change Biology 16: 1181–1193.
22. Heino J, Virkkala R, Toivonen H (2009) Climate change and freshwater biodiversity: detected patterns, future trends and adaptations in northern regions. Biological Reviews 84: 39–54.
23. Brierley AS, Kingsford MJ (2009) Impacts of Climate Change on Marine Organisms and Ecosystems. Current Biology 19: R602–R614.
24. Razgour O, Juste J, Ibanez C, Kiefer A, Rebelo H, et al. (2013) The shaping of genetic variation in edge-of-range populations under past and future climate change. Ecology Letters 16: 1258–1266.
25. Bennie J, Hodgson JA, Lawson CR, Holloway CTR, Roy DB, et al. (2013) Range expansion through fragmented landscapes under a variable climate. Ecology Letters 16: 921–929.
26. Diamond SE, Sorger DM, Hulcr J, Pelini SL, Del Toro I, et al. (2012) Who likes it hot? A global analysis of the climatic, ecological, and evolutionary determinants of warming tolerance in ants. Global Change Biology 18: 448–456.
27. Kearney MR (2013) Activity restriction and the mechanistic basis for extinctions under climate warming. Ecology Letters 16: 1470–1479.
28. Kuntner M, Arnedo MA, Trontelj P, Lokovšek T, Agnarsson I (2013) A molecular phylogeny of nephilid spiders: Evolutionary history of a model lineage. Molecular Phylogenetics and Evolution 69: 961–979.
29. Năpăruş M, Kuntner M (2012) A GIS model predicting potential distributions of a lineage: A test case on hermit spiders (Nephilidae: Nephilengys). PLoS ONE 7: e30047.
30. Kuntner M, Agnarsson I (2011) Biogeography and diversification of hermit spiders on Indian Ocean islands (Nephilidae: Nephilengys). Molecular Phylogenetics and Evolution 59: 477–488.
31. Kuntner M (2007) A monograph of Nephilengys, the pantropical 'hermit spiders' (Araneae, Nephilidae, Nephilinae). Systematic Entomology 32: 95–135.
32. Slatyer RA, Hirst M, Sexton JP (2013) Niche breadth predicts geographical range size: a general ecological pattern. Ecology Letters 16: 1104–1114.
33. Hijmans RJ, Cameron, S E., Parra, J L., Jones, P G. and Jarvis A. (2005) Very high resolution interpolated climate surfaces for global land areas. International Journal of Climatology 25: 1965–1978.
34. IPCC (2000) Special Report Emissions Scenarios - Summary for Policymakers.
35. IPCC (2007) A Report of Working Group I of the Intergovernmental Panel on Climate Change - Summary for Policymakers.
36. Ramirez-Villegas J, Jarvis A (2010) Downscaling Global Circulation Model outputs: the Delta method decision and policy analysis working paper no. 1 Cali, Colombia International Center for Tropical Agriculture, CIAT. 1–18 p.
37. Scinocca JF, McFarlane NA, Lazare M, Li J, Plummer D (2008) Technical Note: The CCCma third generation AGCM and its extension into the middle atmosphere. Atmospheric Chemistry and Physics 8: 7055–7074.
38. Randall DA, Wood RA, Bony S, Colman R, Fichefet T, et al. (2007) Cilmate Models and Their Evaluation. In:Climate Change 2007: The Physical Science Basis. Contribution of Working Group I to the Fourth Assessment Report of the Intergovernmental Panel on Climate Change [Solomon, S., D. . Qin, M. Manning, Z. Chen, M. Marquis, K.B. Averyt., M. Tignor and H.L. . Miller (eds.)]. Cambridge: Cambridge University Press.
39. Flato GM, Boer DY (2001) Warming Asymmetry in Climate Change Simulations. Geophys Res Lett 28: 195–198.
40. ESRI (2010) ArcGIS 9.3.1 Environmental Systems Research Institute, Inc., USA.
41. Botkin DB, Saxe H, Araujo MB, Betts R, Bradshaw RHW, et al. (2007) Forecasting the effects of global warming on biodiversity. Bioscience 57: 227–236.
42. Teixeira J, Arntzen JW (2002) Potential impact of climate warming on the distribution of the Golden-striped salamander, Chioglossa lusitanica, on the Iberian Peninsula. Biodiversity and Conservation 11: 2167–2176.
43. Vink CJ, Derraik JGB, Phillips CB, Sirvid PJ (2011) The invasive Australian redback spider, Latrodectus hasseltii Thorell 1870 (Araneae: Theridiidae): current and potential distributions, and likely impacts. Biological Invasions 13: 1003–1019.
44. Araujo MB, Thuiller W, Pearson RG (2006) Climate warming and the decline of amphibians and reptiles in Europe. Journal of Biogeography 33: 1712–1728.
45. Hof C, Araujo MB, Jetz W, Rahbek C (2011) Additive threats from pathogens, climate and land-use change for global amphibian diversity. Nature 480: 516–U137.
46. Thuiller W, Lavergne S, Roquet C, Boulangeat I, Lafourcade B, et al. (2011) Consequences of climate change on the tree of life in Europe. Nature 470: 531–534.
47. Willis CG, Ruhfel B, Primack RB, Miller-Rushing AJ, Davis CC (2008) Phylogenetic patterns of species loss in Thoreau's woods are driven by climate change. Proceedings of the National Academy of Sciences of the United States of America 105: 17029–17033.
48. Kuntner M, Agnarsson I (2011) Phylogeography of a successful aerial disperser: the golden orb spider Nephila on Indian Ocean islands. Bmc Evolutionary Biology 11.
49. Pigot AL, Tobias JA (2013) Species interactions constrain geographic range expansion over evolutionary time. Ecology Letters 16: 330–338.
50. Araujo MB, Luoto M (2007) The importance of biotic interactions for modelling species distributions under climate change. Global Ecology and Biogeography 16: 743–753.
51. Hof C, Levinsky I, Araujo MB, Rahbek C (2011) Rethinking species' ability to cope with rapid climate change. Global Change Biology 17: 2987–2990.
52. Habel JC, Rodder D, Schmitt T, Neve G (2011) Global warming will affect the genetic diversity and uniqueness of Lycaena helle populations. Global Change Biology 17: 194–205.

15

Evidence of Climate-Induced Range Contractions in Bull Trout *Salvelinus confluentus* in a Rocky Mountain Watershed, U.S.A.

Lisa A. Eby[1], Olga Helmy[1], Lisa M. Holsinger[2], Michael K. Young[3]*

1 University of Montana, Wildlife Biology Program and Department of Ecosystem and Conservation Sciences, College of Forestry and Conservation, Missoula, Montana, United States of America, **2** U.S. Department of Agriculture Forest Service, Rocky Mountain Research Station, Fire Sciences Laboratory, Missoula, Montana, United States of America, **3** U.S. Department of Agriculture Forest Service, Rocky Mountain Research Station, Forestry Sciences Laboratory, Missoula, Montana, United States of America

Abstract

Many freshwater fish species are considered vulnerable to stream temperature warming associated with climate change because they are ectothermic, yet there are surprisingly few studies documenting changes in distributions. Streams and rivers in the U.S. Rocky Mountains have been warming for several decades. At the same time these systems have been experiencing an increase in the severity and frequency of wildfires, which often results in habitat changes including increased water temperatures. We resampled 74 sites across a Rocky Mountain watershed 17 to 20 years after initial samples to determine whether there were trends in bull trout occurrence associated with temperature, wildfire, or other habitat variables. We found that site abandonment probabilities (0.36) were significantly higher than colonization probabilities (0.13), which indicated a reduction in the number of occupied sites. Site abandonment probabilities were greater at low elevations with warm temperatures. Other covariates, such as the presence of wildfire, nonnative brook trout, proximity to areas with many adults, and various stream habitat descriptors, were not associated with changes in probability of occupancy. Higher abandonment probabilities at low elevation for bull trout provide initial evidence validating the predictions made by bioclimatic models that bull trout populations will retreat to higher, cooler thermal refuges as water temperatures increase. The geographic breadth of these declines across the region is unknown but the approach of revisiting historical sites using an occupancy framework provides a useful template for additional assessments.

Editor: Marcus Sheaves, James Cook University, Australia

Funding: This project was primarily funded by the Joint Fire Sciences Project (JFSP-09-1002-9). MT NSF EPSCoR (EPS-0701906) and the Longwell Foundation also provided fellowship support to O.H. The funders had no role in study design, data collection and analysis, decision to publish, or preparation of the manuscript.

Competing Interests: The authors have declared that no competing interests exist.

* E-mail: mkyoung@fs.fed.us

Introduction

Freshwater ecosystems host a disproportionately large amount of the Earth's biodiversity, including many fish species of economic and cultural value [1], yet account for an outsize share of globally imperiled species [2], [3]. Aquatic organisms in freshwater ecosystems are expected to be particularly sensitive to climate shifts because most are ectothermic and have a relatively narrow thermal range for growth and survival [4], [5]. Bioclimatic models accounting for climate change predict an array of phenological changes and range shifts in freshwater aquatic species [6], [7]. Alteration in the timing of life history events has been relatively widely observed [8], [9]. In contrast, confirmation of predictions that stenothermic cold-water fishes should be undergoing distributional shifts to cooler, high-elevation refuges has been elusive [10], particularly in North America [11].

The northern Rocky Mountains, U.S.A. is undergoing climate-mediated shifts e.g., reduced annual snowpack, earlier annual peak snowmelt, and winter precipitation switching from snow to rain, that are contributing to changes in hydrologic and thermal regimes [12], [13], [14]. Summer water temperatures have increased up to 0.3°C/decade [15] and summer base flows are declining [16]. The most stenothermic coldwater fish in the northern Rocky Mountains is the bull trout (*Salvelinus confluentus*) [17], which is listed as threatened under the U.S. Endangered Species Act. Juvenile bull trout rear in cold stream reaches across the upper elevations of river networks, with the upstream extent of individual populations limited by channel size and gradient [18]. Regional temperature increases associated with climate change have led to dire predictions about the persistence of this species in the U.S. [17], [19], but there is little empirical evidence of climate-related shifts. An additional complexity is attributing changes in occupancy directly to climate change [20]. Wildfire is a frequent natural disturbance that can lead to the decades-long elevation of summer stream temperatures because of the loss of shade from riparian vegetation [21], [22]. A recent increase in fire severity and size in the western U.S. has been linked to climate change [23]. Similarly, climate projections favor headwater invasions by less thermally restricted nonnative species such as brook trout (*S. fontinalis*) [24] that can reduce bull trout occupancy [25].

In this study, we repeated a late 20th-century inventory of bull trout occupancy within a river network that encompasses a broad temperature and elevation gradient. Our objective was to compare site-scale abandonment and colonization probabilities to determine whether they differed and if they were associated with landscape features such as wildfire occurrence and severity, habitat attributes including gradient, width, large wood, temperature, and elevation, and biotic variables such as proximity to strongholds of migratory adults and brook trout presence. If the range of bull trout contracted in response to climate change, we expected site-level abandonment probabilities to be greatest at the warmest sites and to exceed those for colonization.

Materials and Methods

Study area

The East Fork Bitterroot River basin is a 1,055-km^2 watershed in west-central Montana, U.S.A. (Figure 1). The basin is mainly a forested landscape with lower elevations dominated by ponderosa pine (*Pinus ponderosa*) and Douglas-fir (*Pseudotsuga menziesii*) and higher elevations by lodgepole pine (*Pinus contorta* var. *latifolia*), subalpine fir (*Abies lasiocarpa*), and Engelmann spruce (*Picea engelmannii*). The watershed is a temperate, snowmelt-dominated system with a range of elevations from 1,220 to 2,887 m. In 2000, wildfires burned 52.0% (29.2% at moderate to high severity) of the basin and 3.8% (2.5% at moderate to high severity) in 2007. Maximum summer stream temperatures in reaches where moderate- to high-severity fires burned in riparian stands remain elevated 1.4 to 2.2°C above those from reaches adjacent to unburned stands [22]. Over a comparable interval (1994–2007) maximum summer stream temperatures at some unburned sites also increased 1.9–2.6°C [22], which is higher than the July/August 0.24°C/decade increase described for the Greater Yellowstone area [26] and 0.22°C/decade increase across the U.S. Northwest [15]. Average daily maximum summer water temperature have been increasing in recent years in the main-stem East Fork Bitterroot River, as have summer air temperatures at the weather station nearest our study area (Sula, MT; Figure 1).

The East Fork Bitterroot River is a core conservation area for bull trout [27]. This watershed consists primarily of public land administered by the U.S. Forest Service (USFS) and has no known barriers to fish movement within our study area [27]. Bull trout in this basin exhibit partial migration, with resident and migratory individuals in most spawning tributaries. Resident individuals spend their entire lives within their natal stream or tributary, moving only short distances (e.g., <2 km), whereas migratory individuals spawn in headwater tributaries but migrate to the river to forage [28], [29]. Other native fish in the basin include westslope cutthroat trout *Oncorhynchus clarkii lewisi*, slimy sculpin (*Cottus cognatus*), longnose dace (*Rhinichthys cataractae*), mountain whitefish (*Prosopium williamsoni*), and longnose suckers (*Catostomus catostomus*). Non-native brown trout (*Salmo trutta*), brook trout, and rainbow trout (*O. mykiss*) are present throughout the main stem and in several tributaries in the basin.

A number of factors influence bull trout habitat occupancy in its U.S. range. As noted earlier, site occupancy of juvenile or resident bull trout is strongly correlated with maximum temperatures [21], [30]. However large individuals with migratory life histories are not restricted to cold thermal environments and can move through much warmer waters (e.g., 21°C 7-day average daily maximum temperature; [31]) before reaching spawning areas. Bull trout are also associated with relatively large patches of connected, complex habitat [17], [32]. In the Bitterroot River basin, the probability of bull trout presence in stream reaches was positively correlated with large wood, stream width, and relative abundance of main-stem bull trout at a tributary mouth, and negatively correlated with stream gradient and the presence of brook trout [33]. Bull trout in this basin tend to occupy streams to their headwaters (until stream width < 2 m; [33]), thus increases in occupancy are only likely at downstream locations or in previously unoccupied streams.

Data collection

First- through 4th-order streams were sampled between 1992 and 1995 to determine bull trout occupancy patterns in the Bitterroot River basin [33]. In this sampling, three 500-m study sites were equally spaced over the estimated length of suitable habitat in each tributary. Between 2009 and 2011, we revisited 74 sites on streams sampled previously [33] within the connected portions of the East Fork Bitterroot River basin to examine whether bull trout occupancy had changed. We relocated the sites and replicated the sampling methods of the earlier study. Fish were collected with a single-pass survey using a backpack electrofishing unit during the summer low-flow period. Care was taken to electroshock slowly and inclusively through all areas of cover. Our resampling was confined to 1st- through 3rd-order streams that were small enough to effectively sample with these techniques. As in the previous study, we divided each site into five sequential 100-m sections for sampling. All fish were identified to species, counted, and measured (total length). As in the earlier study, we avoided basing bull trout occupancy on the ephemeral presence of a large, migratory adult; presence in a section was defined by the capture of ≥ 2 bull trout, at least one of which was less than 250 mm (and thus likely to be a juvenile or small resident adult of local origin) [21]. All sampling was performed in accordance with guidelines specified under scientific collection permits issued to Lisa Eby by Montana Fish, Wildlife and Parks and under the protocols approved in the University of Montana Institutional Animal Care and Use Committee AUP 031-09.

We used field measures and GIS spatial data layers to assess covariates potentially related to bull trout occupancy. We counted large wood in the first 100-m section of every 500-m site. We estimated bankfull width (m) for the section by measuring it at three representative locations. Elevation (m) was noted in the field from the GPS unit (Garmin 60CSx) and validated from the 30-m cell size National Elevation Dataset [34]. Gradient at each site was derived from this dataset using TauDEM software [35]. We used estimates of bull trout abundance in the main-stem East Fork Bitterroot River (http://fwp.mt.gov/fishing/mFish/newSearch.html; accessed 8/2/13) to assess the proximity of sites to main-stem locations where adults were commonly captured. We categorized sites into three groups: (0) bull trout common in the main-stem East Fork Bitterroot River at the tributary mouth; (1) bull trout common in the main-stem within 2.5 km of the tributary mouth; and (2) bull trout common in the main-stem > 2.5 km from the tributary mouth. Because we did not have a single year with temperature data at every site and annual variation in temperature is large, we used a locally calibrated stream temperature model that allowed standardized representation of relative temperatures among sites (Text S1).

We obtained fire severity GIS layers from the Bitterroot National Forest (Hamilton, MT). Burn severity is used to describe the amount of fire-related change including overstory vegetation mortality, soil heating, and fuel consumption [36], [37]. Our sites within moderate- and high-severity riparian burns had the majority of the riparian area (and watershed) burned, thus fire severity and proportion of site burned were positively related. We grouped burn severity into two categories and at each site we indicated whether the riparian area experienced no-to-low severity burns or medium-to-high severity burns.

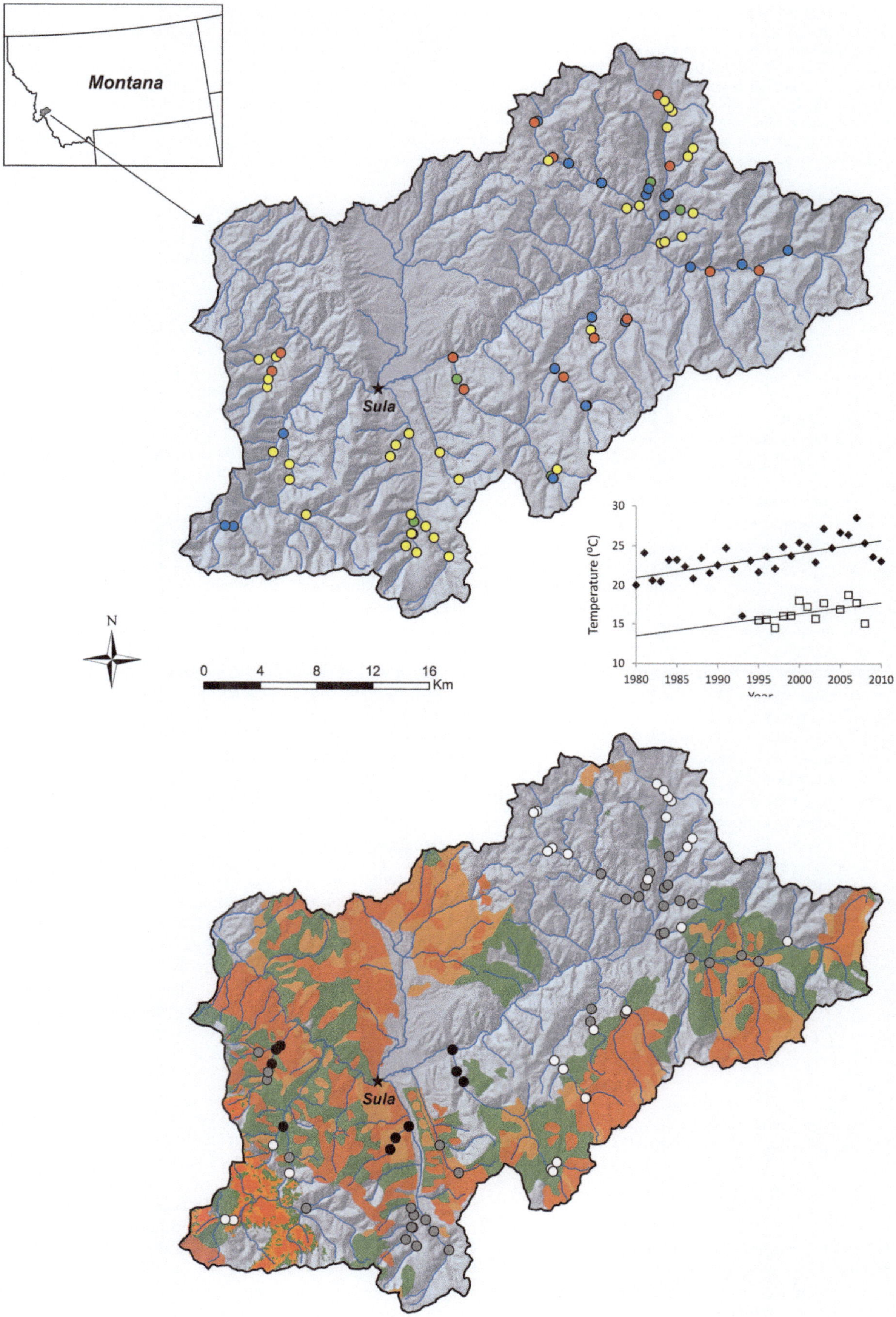
Montana
Sula
N
0 4 8 12 16
Km
Temperature (°C)
30
25
20
15
10
1980 1985 1990 1995 2000 2005 2010
Sula

Figure 1. Study area. Sampling locations (500-m sites that were initially visited in 1992–1995 and resampled in 2009–2011) in the East Fork Bitterroot River watershed. Top panel: patterns in occupancy (yellow, not occupied in either period; blue, occupied in both periods; red, occupied in first but not second period; green, occupied in second but not first period). These reflect observed patterns not corrected for probability of detection. Water and air temperature patterns within the East Fork Bitterroot River basin are inset. Black diamond symbols are average daily summertime (July and August air temperature recorded over the study period at the closest weather station at Sula, MT ($y = 0.1567x - 289.27$, $R^2 = 0.34$, $p = 0.0006$). Hollow squares are average daily maximum water temperatures over the summer season (July 15 to September 30) from the East Fork Bitterroot River main stem 28.6 km upstream of the confluence with the West Fork Bitterroot River ($y = 0.1441x - 271.96$, $r^2 = 0.22$, $p = 0.10$). Bottom panel: sampling locations shaded to indicate estimates of abandonment probability (white: 0.21–0.32, grey: 0.32–0.47, and black: 0.47–0.62). Burn severity for fires in the watershed is indicated by low severity in green, moderate severity in orange, and high severity in red. Grey is outside of the fire perimeters.

Data analyses

We used program PRESENCE 4.1 [38] to estimate detection probability, occupancy, and abandonment and colonization probabilities. We constructed a survey history based on the five, 100-m sections for each site across the 1992–1995 and 2009–2011 surveys to estimate these parameters. These separate survey intervals were regarded as seasons. Estimates for probability of detection were modeled as a function of standardized values for season, large wood, fire, width, and gradient. Because large wood and the occurrence of medium- to high-severity fire at the site change over time, we treated these as survey-specific covariates. Given that both survey events had imperfect detectability and different crews (but the same field protocol), we compared detection probability between the earlier and more recent surveys. In addition, we explored all possible combinations of covariates in competing models to examine which covariates best described the probability of detection across sites (based on maximum likelihood estimators and Akaike Information Criterion (AIC) [39].

We used a multi-season model to estimate site abandonment and colonization probabilities of bull trout. The model uses initial occupancy estimates for the first sampling period and derives estimates for the abandonment and colonization probabilities that determine whether a species occurs at a site during the second sampling period [40]. To examine changes in occupancy in our data set, we first determined whether estimates of bull trout abandonment or colonization probabilities were significantly different from zero and from one another. Any changes were regarded as apparent because occupancy was only estimated twice, not at repeated intervals. Finally, we tested whether large wood, gradient, width, brook trout presence, proximity to where bull trout were common in the main-stem, occurrence of medium- or high-severity fire, or relative temperature co-varied with these probabilities. In addition, we examined how elevation, an occasional surrogate for water temperature [21], degree of connectivity, channel gradient, and human land use, individually co-varied with colonization and abandonment probabilities. Models were fit using maximum likelihood estimators and ranked based on AIC scores. We considered all models within 2 AIC units of the top model, but disregarded uninformative parameters i.e., covariates for which approximate 85% confidence intervals overlapped zero [41]. We also performed stepwise variable removal, based on the minimum absolute value of β/SE, stopping when variable exclusion led to a decrease in AIC score for the model [41].

Results

Presence of bull trout in previously occupied sites declined (from 33 sites in 24 streams to 20 sites in 22 streams), and absence from previously unoccupied sites decreased (from 41 sites in 26 streams to 36 sites in 26 streams). Brook trout occupancy declined between the earlier (12 sites in 6 streams) and later (7 sites in 5 streams) surveys. In addition, thirty-one of the 74 sites had adjacent riparian burns during the period between surveys, 12 of which were from medium- to high-severity fires.

Significant correlations among covariates used to explain probability of detection were weak or absent (Table S1), so all were considered in models of detection probability. There were three top models (within 2 AIC units of each other; Table 1). Our naïve (no covariates) probability of detection estimate was 0.54 (SE, 0.03). Naïve detection probabilities did not differ between seasons (season 1: 0.54, SE 0.04; season 2: 0.54, SE 0.06) and season did not increase the model AIC score. Therefore, the top model for estimating probability of detection included only large wood and width as covariates. Based on this model, site-specific probability of detection varied from 0.09 to 0.99 (Table S2).

The overall (no site or survey covariates) abandonment probability (0.36, SE 0.07) was almost 3-fold greater than that for colonization (0.13, SE 0.07). Most covariates in models for estimating colonization or abandonment probabilities were uninformative because of small effect sizes or large standard errors. The three top models with informative covariates for abandonment probabilities included either no covariate or the single covariates of elevation or temperature (Table 2). We model-averaged the top three models with informative parameters. Estimated abandonment probabilities increased approximately three-fold from cooler to warmer sites (Figure 2) and high- to low-elevation sites (Figure 3). Elevation and temperature were negatively correlated (−0.59). No informative covariates were retained in the top model for estimating colonization.

Discussion

By revisiting historically sampled sites within a river network that encompasses a broad temperature and elevation gradient, we demonstrated that site abandonment probabilities of bull trout were highest at warmer, low-elevation sites over the last two decades. This coincided with increases in summer stream temperatures in the East Fork Bitterroot River basin. Neither colonization nor abandonment probabilities were related to variables reflecting habitat, biotic interactions, or recent disturbance, and probabilities of abandonment were three-fold greater than those of colonization. Collectively, these results represent the first empirical evidence supporting predicted declines in the distribution of bull trout as a consequence of climate change [17], [26], [42]. We acknowledge that the observed relation between climate-related warming and reduction in bull trout occupancy is correlative and that other covariates we did not consider may have influenced this outcome, but the effects of temperature on bull trout distributions are consistent across its historical range [17]. A complete mechanistic understanding of the effects of warming temperatures on bull trout has not been realized, but the restricted scope for growth of juvenile life history stages at warm temperatures is clear [4], [43]. Given that much of the historical range of bull trout is undergoing relatively rapid warming as a consequence of climate change [14], [26], testing the generality of

Table 1. Models within two Akaike Information Criterion (AIC) units of the top model for estimating probability of detection of bull trout with probability of occupancy, colonization, and abandonment held constant.

Model[a,b]	AIC	DeltaAIC	AIC Weight	Model likelihood	Parameters	-2Loglikelihood
$p(W_{4.22}, LW_{4.47})^c$	479.22	0.00	0.5467	1.0000	6	467.22
$p(S_{0.55}\ W_{4.22}, LW_{4.47})$	480.88	1.66	0.2384	0.436	7	468.04
$p(W_{4.22}, LW_{4.47}, G_{0.35})$	481.09	1.87	0.2146	0.393	7	467.09

[a]Variables subscripted with β/SE absolute values; variables with values < 1.4 are regarded as uninformative (Arnold 2010).
[b]Abbreviations: W, width; LW, large wood; S, season; G, gradient.
[c]Only this model lacks uninformative variables.

our observations will be straightforward if comparable long-term data sets become available.

Only those covariates most closely related to the decreased thermal suitability of bull trout habitat—water temperature and elevation—appeared in the top models for estimating abandonment probabilities. This may seem surprising, given that occupancy models for bull trout have included an array of habitat and biotic variables [18], [33]. We did not, however, model where bull trout are currently found, but examined what influenced recent changes in that distribution. Consequently, it might be expected that elevation (represented in a linear function across a broad range) would not explain the distribution of bull trout in the Bitterroot River basin [33], but was a top predictor of locations abandoned by bull trout. We attribute most of the explanatory power of elevation in our model to its relation to water temperature, but acknowledge that it can, in part, represent effects of other variables, such as the presence of nonnative species, the effects of fire, or proximity to population strongholds. In this study, however, none of these were informative contributors to models of changes in bull trout occupancy. At the sites we examined, brook trout occupancy was low and appeared to decline during the study. Elsewhere, replacement of bull trout by brook trout appears to be associated with particular valley morphologies [18], [25], that may not be prevalent in the study area. We regard it as unlikely that warming temperatures also reduced brook trout occupancy [24] because this species prefers warmer temperatures than does bull trout [4]. More plausible is that brook trout were declining in response to fire effects [44] but this species was too poorly represented in the data to evaluate this trend.

Stand-replacing fires tend to cause warmer stream temperatures [22], [45], thus it might be expected that reductions in bull trout occupancy would be associated with fire directly, or indirectly via fire's relation to elevation. Although low-elevation sites adjacent to burned areas were some of our warmest sites, there was no significant correlation ($r = 0.45$) between relative temperature and fire because wildfires burned across the entire watershed. Moreover, changes in water temperature may be ameliorated by other fire-related changes in habitat, such as increased autochthonous productivity, macroinvertebrate community shifts, or channel alteration [46], and previous observations of bull trout in the study area did not reveal population declines following fire [44]. Nevertheless, because stand-replacing fire in riparian zones leads to chronic increases in summer stream temperatures it has the potential to contribute to local, site-specific changes in occupancy by bull trout. In addition, anticipated increases in fire extent or frequency attributable to climate change [23], [47] may lead to more profound shifts, or outright extirpations, of populations across the landscape, where populations are isolated or landscapes are prone to large, post-fire debris flows [21].

Connectivity has long been thought to influence the persistence of salmonid populations because these fishes can be highly mobile and frequently form metapopulations [48]. In our analyses, rank distance to the main-stem river sections where bull trout are

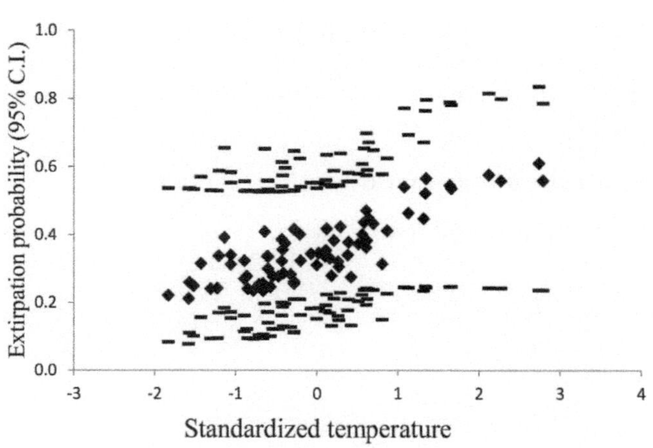

Figure 2. Effect of temperature on abandonment probabilities. Model-averaged abandonment probabilities (filled diamonds) from the top three informative models (Table 2) with their upper and lower 95% confidence intervals (dashes) versus standardized relative temperature across sites in the East Fork Bitterroot River basin.

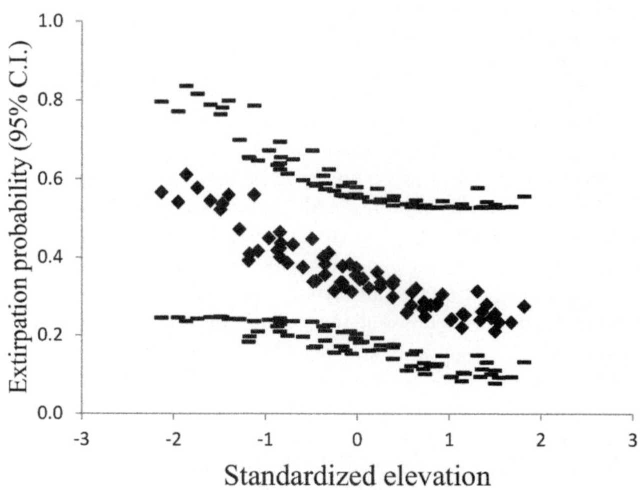

Figure 3. Effect of elevation on abandonment probabilities. Model-averaged abandonment probabilities (filled diamonds) from the top three informative models (Table 2) with their upper and lower 95% confidence intervals (dashes) versus standardized elevation.

Table 2. Models within two AIC units of the top model for using changes in occupancy (ψ) to estimate colonization (γ) and abandonment (ε) probabilities of bull trout.

Model[a,b,c]	AIC	DeltaAIC	AIC weight	Model likelihood	Parameters	-2Loglikelihood
$\psi, \gamma, \varepsilon(E_{1.59})^d$	478.38	0	0.10	1	7	464.38
$\psi, \gamma, \varepsilon(B_{0.5})$	478.62	0.24	0.088	0.88	7	464.62
$\psi, \gamma, \varepsilon(T_{1.43})^d$	478.94	0.56	0.075	0.76	7	464.94
$\psi, \gamma, \varepsilon^d$	479.22	0.84	0.065	0.66	6	467.22
$\psi, \gamma, \varepsilon(B_{0.47}, D_{0.73})$	480.11	1.73	0.042	0.42	8	464.11
$\psi, \gamma, \varepsilon(B_{0.46}, T_{0.72})$	480.12	1.74	0.042	0.42	8	464.12
$\psi, \gamma, \varepsilon(B_{0.66}, F_{0.73})$	480.12	1.74	0.042	0.42	8	464.12
$\psi, \gamma, \varepsilon(D_{1.04})$	480.17	1.79	0.040	0.41	7	466.17
$\psi, \gamma, \varepsilon(B_{0.24}, G_{0.62})$	480.24	1.86	0.039	0.40	8	464.24
$\psi, \gamma, \varepsilon(T_{1.43}, LW_{0.68})$	480.45	2.0	0.035	0.36	8	464.45

[a]All models contained the probability of detection function p(W, LW).
[b]Variables subscripted with β/SE absolute values.
[c]Abbreviations: E, elevation; T, temperature; LW, large wood; W, stream width; G, gradient; B, brook trout presence at site; D, distance to where bull trout are common in the main-stem East Fork Bitterroot River; F, occurrence of moderate- to high-severity fire.
[d]Models without uninformative variables. Model-averaged parameter estimates for the untransformed coefficients: E, −0.73 (0.46 SE); T, 0.63 (0.44 SE).

common did not account for the probability of abandonment or colonization at tributary sites. It may have been that the distances involved (a few tens of kilometers at most) do not represent meaningful levels of isolation for bull trout, migratory forms of which often traverse much longer distances [31], [49]. Genetic evidence indicates that despite declines in the abundance of migratory bull trout in the East Fork Bitterroot River basin over the last few decades, dispersal by bull trout among tributaries and between tributaries and the main-stem East Fork Bitterroot River remains common [50]. If climate change or anthropogenic habitat alteration increases the energetic and demographic costs of migration beyond some threshold, the influence of connectivity may become more evident [51].

An ongoing paradox is that demographic shifts among freshwater species have been difficult to detect despite that these taxa may be among the most sensitive to climate change [11]. Although a few studies have suggested that declines in freshwater fish abundance could be related to climate change [52], [53], [54], beyond the present study only one other [55] has reported changes in their distributions. This could be partly attributable to a paucity of adequately georeferenced historical data sets, particularly those that permit detectability estimation. Nevertheless, the enormity of current and historical fish monitoring efforts by state, tribal, and federal agencies throughout North America suggests that many such data sets exist. For areas lacking historical temperature data, recent advances in modeling dendritic ecological networks [56] can facilitate accurate hindcasting and prediction of stream temperatures for basins well represented by recent temperature records [21].

Other obstacles to detecting the effects of climate change on cold-water fishes reflect their habitat and biology. Although rising water temperature appears to be a consistent trend in many portions of the historical range of bull trout and other western North American salmonids [15], [26], uncertainty about the response of particular watersheds [57] or certain species [55] remains high. Fish abundance is exceptionally temporally variable and sometimes requires decades of sampling before statistically significant trends emerge [58], [59]. In addition, such trends may be superimposed on long-term variation in abundance dictated by climate cycles such as the Pacific Decadal Oscillation [60]. Nevertheless, correlations between bull trout abundance and broad-scale climate cycles (or this species' abundance and that of other salmonids) are weak [61], [62], [63]. Analyses that rely on occupancy, rather than abundance, may be less vulnerable to temporal fluctuations.

In summary, we found that in this core, connected conservation area for bull trout, patterns in site abandonment were consistent with the predicted effects of stream temperature warming. Extending the current time series of observation of bull trout in the East Fork Bitterroot River basin, as well as similar studies in other basins, is essential to evaluating the generality of this trend. Monitoring designs that focus on stenothermic species and on locations most likely to undergo rapid change offer the greatest power for detecting responses of aquatic species to climate change [11]. For bull trout, these include low-elevation reaches that are warming rapidly and high-elevation reaches undergoing flow reductions from declining snowpacks [64]. In the absence of such targeted designs, however, revisiting historically sampled sites across a range of elevations within a stream network to examine changes in occupancy constitutes a practical alternative [10], [65], [66].

Supporting Information

Table S1 Correlation coefficients. Pearson correlation coefficients for standardized variables in analyses including elevation (E), large wood (LW), bank-full width (W), gradient at site (G), relative temperature (T), the presence of medium to high severity burns at the site (F), the presence of brook trout (B), and the distance from the tributary confluence to where bull trout are common in the main-stem (D). An asterisk indicates a significant correlation ($P \leq 0.05$).

Table S2 Detection probabilities. Detection probabilities associated with each site and each survey considering large wood during each survey and stream width at each site. Probability of detection without covariates was 0.54 (SE, 0.03).

Acknowledgments

We thank T. Smith, W. Glenn, L. Hall, J. Leppi, and T. Moua for help with fieldwork and T. McMahon and C. Sestrich for sharing their previous dataset. D. Isaak, S. Wenger, R. Al-Chokhachy, and A. Ruesch provided helpful comments on drafts of this manuscript.

Author Contributions

Conceived and designed the experiments: LAE OH LMH MKY. Performed the experiments: OH. Analyzed the data: LAE OH LMH. Wrote the paper: LAE MKY. Manuscript comments: OH LMH.

References

1. Dudgeon D, Arthington AH, Gessner MO, Kawabata ZI, Knowler DJ, et al. (2006) Freshwater biodiversity: importance, threats, status and conservation challenges. Biol Rev Camb Philos Soc 81: 163–182.
2. Williams JE, Williams RN, Thurow RF, Elwell L, Philipp DP, et al. (2011) Native fish conservation areas: a vision for large-scale conservation of native fish communities. Fisheries 36: 267–277.
3. Burkhead NM (2012) Extinction rates in North American freshwater fishes, 1900–2010. BioScience 62: 798–808.
4. McMahon T, Zale A, Barrows F, Selong J, Danehy R (2007) Temperature and competition between bull trout and brook trout: a test of the elevation refuge hypothesis. Trans Am Fish Soc 136: 1313–1326.
5. Elliott JM, Elliott JA (2010) Temperature requirements of Atlantic salmon *Salmo salar*, brown trout *Salmo trutta* and Arctic charr *Salvelinus alpinus*: predicting the effects of climate change. J Fish Biol 77: 1793–1817.
6. Ficke AD, Myrick CA, Hansen LJ (2007) Potential impacts of global climate change on freshwater fisheries. Rev Fish Biol Fish 17: 581–613.
7. Beer WN, Anderson JJ (2013) Sensitivity of salmonid freshwater life history in western US streams to future climate conditions. Glob Chang Biol 8: 2547–2556.
8. Crozier LG, Scheuerell MD, Zabel RW (2011) Using time series analysis to characterize evolutionary and plastic responses to environmental change: a case study of a shift toward earlier migration date in sockeye salmon. Am Nat 178: 755–773.
9. Warren DR, Robinson JM, Josephson DC, Sheldon DR, Kraft CE (2012) Elevated summer temperatures delay spawning and reduce redd construction for resident brook trout (*Salvelinus fontinalis*). Glob Chang Biol 18: 1804–1811.
10. Comte L, Buisson L, Daufresne M, Grenouillet G (2012) Climate-induced changes in the distribution of freshwater fish: observed and predicted trends. Freshw Biol 58: 625–639.
11. Isaak DJ, Rieman BE (2013) Stream isotherm shifts from climate change and implications for distributions of ectothermic organisms. Glob Chang Biol 19: 742–751.
12. Mote PW, Hamlet AF, Clark MP, Lettenmaier DP (2005) Declining mountain snowpack in western North America. Bull Am Meteorol Soc 86: 39–49.
13. Knowles N, Dettinger MD, Cayan DR (2006) Trends in snowfall versus rainfall in the western United States. J Clim 19: 4545–4559.
14. Pederson GT, Graumlich LJ, Fagre DB, Kipfer T, Muhlfeld CC (2010) A century of climate and ecosystem change in western Montana: what do temperature trends portend? Clim Change 98: 133–154.
15. Isaak DJ, Wollrab S, Horan D, Chandler G (2012b) Climate change effects on stream and river temperatures across the northwest US from 1980–2009 and implications for salmonid fishes. Clim Change 113: 499–524.
16. Leppi JC, DeLuca TH, Harrar SW, Running SW (2012) Impacts of climate change on August stream discharge in the Central-Rocky Mountains. Clim Change 112: 997–1014.
17. Rieman BE, Isaak D, Adams S, Horan D, Nagel D, et al. (2007) Anticipated climate warming effects on bull trout habitats and populations across the interior Columbia River basin. Trans Am Fish Soc 136: 1552–1565.
18. Wenger SJ, Isaak DJ, Dunham JB, Fausch KD, Luce CH, et al. (2011a) Role of climate and invasive species in structuring trout distributions in the interior Columbia River Basin, USA. Can J Fish Aquat Sci 68: 988–1008.
19. Wenger SJ, Som NA, Dauwalter DC, Isaak DJ, Neville HM, et al. (2013) Probabilistic accounting of uncertainty in forecasts of species distributions under climate change. Glob Chang Biol 10.1111/gcb.12294.
20. Parmesan C, Burrows MT, Duarte CM, Poloczanska ES, Richardson AJ, et al. (2013) Beyond climate change attribution in conservation and ecological research. Ecol Lett 16: 58–71.
21. Isaak DJ, Luce CH, Rieman BE, Nagel DE, Peterson EE, et al. (2010) Effects of climate change and wildfire on stream temperatures and salmonid thermal habitat in a mountain river network. Ecol Appl 20: 1350–1371.
22. Mahlum SK, Eby LA, Young MK, Clancy CG, Jakober M (2011) Effects of wildfire on stream temperatures in the Bitterroot River Basin, Montana. Int J Wildl Fire 20: 240–247.
23. Westerling AL, Hidalgo HG, Cayan DR, Swetnam TW (2006) Warming and earlier spring increase western U.S. forest wildfire activity. Science 313: 940–943.
24. Wenger SJ, Isaak DJ, Luce CH, Neville HM, Fausch KD, et al. (2011b) Flow regime, temperature, and biotic interactions drive differential declines of trout species under climate change. Proc Natl Acad Sci U S A 108: 14175–14180.
25. Rieman BE, Peterson JT, Myers DL (2006) Have brook trout (*Salvelinus fontinalis*) displaced bull trout (*Salvelinus confluentus*) along longitudinal gradients in central Idaho streams? Can J Fish Aquat Sci 63: 63–78.
26. Isaak DJ, Muhlfeld CC, Todd AS, Al-Chokhachy R, Roberts J, et al. (2012a) The past as prelude to the future for understanding 21st-century climate effects on Rocky Mountain trout. Fisheries 37: 542–556.
27. Montana Bull Trout Restoration Team (2000) Restoration plan for bull trout in the Clark Fork River basin and Kootenai River basin Montana. Montana Department of Fish, Wildlife and Parks, Helena. 114 pp.
28. Nelson ML, McMahon TE, Thurow RF (2002) Decline of the migratory form in bull charr, *Salvelinus confluentus*, and implications for conservation. Env Biol Fish 64: 321–332.
29. Al-Chokhachy R, Budy P (2008) Demographic characteristics, population structure, and vital rates of a fluvial population of bull trout in Oregon. Trans Am Fish Soc 137: 1709–1722.
30. Dunham J, Rieman B, Chandler G (2003) Influences of temperature and environmental variables on the distribution of bull trout within streams at the southern margin of its range. N Am J Fish Manage 23: 894–904.
31. Howell PJ, Dunham JB, Sankovich PM (2010) Relationships between water temperatures and upstream migration, cold water refuge use, and spawning of adult bull trout from the Lostine River, Oregon, USA. Ecol Freshw Fish 19: 96–106.
32. Dunham JB, Rieman BE (1999) Metapopulation structure of bull trout: Influences of physical, biotic, and geometrical landscape characteristics. Ecol Appl 9: 642–655.
33. Rich CF, McMahon TE, Rieman BE, Thompson WL (2003) Local-habitat, watershed, and biotic features associated with bull trout occurrence in Montana streams. Trans Am Fish Soc 132: 1053–1064.
34. U.S. Geological Survey (2006) National elevation dataset (http://ned.usgs.gov/).
35. Tarboton DG (2008) Terrain analysis using digital elevation models (TauDEM). (http://hydrology.neng.usu.edu/taudem/).
36. Keane RE, Agee JK, Fule P, Keeley JE, Key C, et al. (2008) Ecological effects of large fires on US landscapes: benefit or catastrophe? Int J Wildl Fire 17: 696–712.
37. Keeley JE (2009) Fire intensity, fire severity, and burn severity: a brief review and suggested usage. Int J Wildl Fire 18: 116–126.
38. Hines JE (2006) PRESENCE2 (v4.1). Software to estimate patch occupancy and related parameters. USGS-PWRC. http://www.mbr-pwrc.usgs.gov/software/presence.html.
39. Burnham KP, Anderson DA (2002) Model selection and multimodel inference: a practical information-theoretic approach. Springer-Verlag, New York.
40. Donovan TM, Hines J (2007) Exercises in occupancy modeling and estimation. <http://www.uvm.edu/envnr/vtcfwru/spreadsheets/occupancy.htm>
41. Arnold T (2010) Uninformative parameters and model selection using Akaike's Information Criterion. J Wildl Manage 74: 1175–1178.
42. Ruesch AS, Torgersen CE, Lawler JJ, Olden JD, Peterson EE, et al. (2012) Projected climate-induced habitat loss for salmonids in the John Day River network, Oregon, U.S.A. Con Biol 26: 873–882.
43. Selong JH, McMahon TE, Zale AV, Barrows FT (2001) Effect of temperature on growth and survival of bull trout, with application of an improved method for determining thermal tolerance in fishes. Trans Am Fish Soc 130: 1026–1037.
44. Sestrich CM, McMahon TE, Young MK (2011) Influence of fire on native and nonnative salmonid populations and habitat is a western Montana basin. Trans Am Fish Soc 140: 136–146.
45. Dunham JB, Rosenberger AE, Luce CH, Rieman BE (2007) Influence of wildfire and channel reorganization on spatial and temporal variation in stream temperature and the distribution of fish and amphibians. Ecosystems 10: 335–346.
46. Gresswell RE (1999) Fire and aquatic ecosystems in forested biomes in North America. Trans Am Fish Soc 128: 193–221.
47. McKenzie D, Gedalof Z, Peterson DL, Mote P (2004) Climatic change, wildfire, and conservation. Con Biol 18: 890–902.
48. Rieman BE, Dunham JB (2000) Metapopulations and salmonids: a synthesis of life history patterns and empirical observations. Ecol Freshw Fish 9: 51–64.
49. Bjornn TC, Mallet J (1964) Movements of planted and wild trout in an Idaho river system. Trans Am Fish Soc 93: 70–76.
50. Nyce LG, Eby LA, Clancy CG, Painter S, Leary RF (2013) Genetic population structure of bull trout in the East Fork Bitterroot River drainage, Montana. N Am J Fish Manage 33: 432–445.

51. Starcevich SJ, Howell PJ, Jacobs SE, Sankovich PM (2012) Seasonal movement and distribution of fluvial adult bull trout in selected watersheds in the mid-Columbia River and Snake River basins. PLoS ONE 7: e37257.
52. Hari RE, Livingstone DM, Siber R, Burkhart-Holm P, Guettinger H (2006) Consequences of climatic change for water temperature and brown trout populations in alpine rivers and streams. Glob Chang Biol 12: 10–26.
53. Winfield IJ, Hateley J, Fletcher JM, James JB, Bean CW, et al. (2010) Population trends of Arctic charr (*Salvelinus alpinus*) in the UK: assessing the evidence for a widespread decline in response to climate change. Hydrobiologia 650: 55–65.
54. Almodóvar A, Nicola GG, Ayllón D, Elvira B (2011) Global warming threatens the persistence of Mediterranean brown trout. Glob Chang Biol 18: 1549–1560.
55. Comte L, Grenouillet G (2013) Do stream fish track climate change? Assessing distribution shifts in recent decades. Ecography 39: 1–11.
56. Peterson EE, Ver Hoef JM, Isaak DJ, Falke JA, Fortin MJ, et al. (2013) Modelling dendritic ecological networks in space: an integrated network perspective. Ecol Lett 16: 707–719.
57. Arismendi I, Johnson SL, Dunham JB, Haggerty R (2013) Descriptors of natural thermal regimes in streams and their responsiveness to change in the Pacific Northwest of North America. Freshw Biol 58: 880–894.
58. Maxwell D, Jennings S (2005) Power of monitoring programmes to detect decline and recovery of rare and vulnerable fish. J Appl Ecol 42: 25–37,
59. Al-Chokhachy R, Budy P, Conner M (2009) Detecting declines in the abundance of a bull trout (*Salvelinus confluentus*) population: understanding the accuracy, precision, and costs of our efforts. Can J Fish Aquat Sci 66: 649–658.
60. Mantua NJ, Hare SR, Zhang Y, Wallace JM, Francis RC (1997) A Pacific interdecadal climate oscillation with impacts on salmon production. Bull Am Meteorol Soc 78: 1069–1079.
61. High B, Meyer KA, Schill DJ, Mamer ERJ (2008) Distribution, abundance, and population trends of bull trout in Idaho. N Am J Fish Manage 28: 1687–1701.
62. Copeland T, Meyer KA (2011) Interspecies synchrony in salmonid densities associated with large-scale bioclimatic conditions in central Idaho. Trans Am Fish Soc 140: 928–942.
63. Meyer KA, Garton EO, Schill DJ (2014) Bull trout trends in abundance and probabilities of persistence in Idaho. N Amer J Fish Manage 34: 202–214.
64. Luce CH, Abatzoglou JT, Holden ZA (2013) The missing mountain water: slower westerlies decrease orographic enhancement in the Pacific Northwest USA. Science. DOI:10.1126/science.1242335
65. Altwegg R, Wheeler M, Erni B (2008) Climate and the range dynamics of species with imperfect detection. Biol Lett 4: 581–584.
66. Tingley MW, Beissinger SR (2009) Detecting range shifts from historical species occurrences: new perspectives on old data. Trends Ecol Evol 24: 625–633.

Plastic Responses to Elevated Temperature in Low and High Elevation Populations of Three Grassland Species

Esther R. Frei*¤, Jaboury Ghazoul, Andrea R. Pluess

Department of Environmental Systems Science, Ecosystem Management, ETH Zurich, Zurich, Switzerland

Abstract

Local persistence of plant species in the face of climate change is largely mediated by genetic adaptation and phenotypic plasticity. In species with a wide altitudinal range, population responses to global warming are likely to differ at contrasting elevations. In controlled climate chambers, we investigated the responses of low and high elevation populations (1200 and 1800 m a.s.l.) of three nutrient-poor grassland species, *Trifolium montanum*, *Ranunculus bulbosus,* and *Briza media*, to ambient and elevated temperature. We measured growth-related, reproductive and phenological traits, evaluated differences in trait plasticity and examined whether trait values or plasticities were positively related to approximate fitness and thus under selection. Elevated temperature induced plastic responses in several growth-related traits of all three species. Although flowering phenology was advanced in *T. montanum* and *R. bulbosus*, number of flowers and reproductive allocation were not increased under elevated temperature. Plasticity differed between low and high elevation populations only in leaf traits of *T. montanum* and *B. media*. Some growth-related and phenological traits were under selection. Moreover, plasticities were not correlated with approximate fitness indicating selectively neutral plastic responses to elevated temperature. The observed plasticity in growth-related and phenological traits, albeit variable among species, suggests that plasticity is an important mechanism in mediating plant responses to elevated temperature. However, the capacity of species to respond to climate change through phenotypic plasticity is limited suggesting that the species additionally need evolutionary adaptation to adjust to climate change. The observed selection on several growth-related and phenological traits indicates that the study species have the potential for future evolution in the context of a warming climate.

Editor: Ben Bond-Lamberty, DOE Pacific Northwest National Laboratory, United States of America

Funding: This study was funded by the Swiss National Science Foundation No. 3100A0-116277 (www.snf.ch) to J.G. and A.R.P., the Competence Center Environment and Sustainability (CCES) BioChange (http://www.cces.ethz.ch/projects/clench/BioChange) to J.G., and the Basler Stiftung für biologische Forschung to E.R.F. The funders had no role in study design, data collection and analysis, decision to publish, or preparation of the manuscript.

Competing Interests: The authors have declared that no competing interests exist.

* E-mail: esther.frei@slf.ch

¤ Current address: WSL Institute for Snow and Avalanche Research SLF, Davos, Switzerland

Introduction

Global annual mean temperature has risen by approximately 0.85°C over the past century and climate scenarios project a further warming of 2 to 4°C for the 21st century [1]. This rapid temperature increase and the concomitant alterations of other environmental factors, such as higher levels of atmospheric CO_2 and changing precipitation regimes, will drastically alter living conditions for plant and animal species across the globe and affect individuals, populations and communities [2–5]. Temperature is one of the main factors determining plant physiology and performance, and as a consequence, strongly influences the geographic distribution of plants [6,7]. Therefore, climate warming has caused latitudinal and altitudinal range shifts of many plant species worldwide and further shifts are expected [2,3]. Along elevation gradients, a considerable increase in number of plant species on alpine summits indicates an upward shift of species' ranges [8–12]. However, the migration potential might be inferior to the rapid rate of current climate change, which is expected to render many species unable to track the climate they are currently adapted to [13,14]. Continuing habitat loss and fragmentation further exacerbate migration by impeding gene flow [15,16] and threaten the persistence of many species [17,18].

Alternatively to shifts in abundance and distribution, plants may persist in a changing climate through evolutionary adaptation and phenotypic plasticity [19–22]. Several studies provide evidence of climate driven population differentiation [23–26]. Thus, temperature and climate warming might be strong selective agents leading to the adaptive evolution of key plant traits [27,28]. In contrast to evolutionary adaptation, which requires several generations, current phenotypic plasticity in plants, the ability of a genotype to change its phenotype in response to different environments [29], allows short-term responses to rapid warming [21,30,31]. Among the best documented plastic responses to climate change are shifts in phenology observed in many species across the globe [2,3,32]. Although these meta-analyses are mainly based on genetically unstructured samples, which could not disentangle genetic and plastic changes, studies in phenological garden networks on structured samples show similar results [33,34]. Besides plastic shifts in phenology, several studies have documented morphological and physiological trait plasticity to temperature changes, reviewed in [35], and specifically plastic responses to elevated temperature, such as decreased seed dormancy [36,37] or reduced seed longevity [38].

Adaptive plasticity is the ability of an individual genotype to express phenotypes, which enhance fitness in response to variation in selection [21,31,39]. This is likely facilitating species persistence under climate change and assisting rapid evolutionary adaptation [40–44]. While numerous studies reported adaptive plasticity to different environmental conditions, such as shade [45], drought [46,47] and flooding [48], empirical evidence for adaptive plastic responses to climate change is still scarce, but see [49,50]. The capacity for plastic responses is supposed to vary due to different physiological thresholds, such as temperature and precipitation tolerance. Consequently, trait plasticity generally differs among species, and might even vary among populations from contrasting habitats across a species' range [50,51]. Higher trait plasticity would be an advantage for plants experiencing greater spatial and temporal habitat heterogeneity because it allows them to maximize their fitness under different environmental conditions [39,52,53].

Mountain ecosystems are characterised by steep climate gradients and by increased environmental variation over short spatial and temporal scales at high as compared to nearby low elevation sites [54]. Because of the greater habitat heterogeneity we hypothesize that high elevation plants exhibit greater trait plasticity than their low elevation congeners. Varying climate conditions along altitudinal gradients might also result in different selection pressures that shape intraspecific trait adaptation to elevation. Little is known about how the adaptive potential of plants and their ability to respond to climate change through phenotypic plasticity shape phenotypic variation along altitudinal gradients. Transplant experiments in the Pyrenees and the Australian Alps revealed genetic differentiation but also trait plasticity [55,56]. In the Swiss Alps Frei et al. [57] described plasticity in reproductive phenology as an important plant response to transplantation to warmer low elevation sites. However, since these experiments were performed along natural elevation gradients, they were not able to separate the effects of temperature and warming from the influence of concomitant climate factors.

In this study we grew plants in controlled climate chambers and investigated responses of low (1200 m a.s.l.) and high (1800 m a.s.l.) elevation populations of three montane grassland species, *Trifolium montanum* L., *Ranunculus bulbosus* L. and *Briza media* L. to experimental warming. Plants were grown under elevated temperature, corresponding to predicted future conditions under climate change, and in control treatments corresponding to current ambient conditions at the plant origin whereas all other environmental parameters were kept identical. We measured growth-related, reproductive and phenological traits of the experimental plants and assessed differences in trait plasticity of low and high elevation populations. Using phenotypic selection analyses [58], we examined whether trait values and plasticities are under selection. Specifically, we addressed the following questions: (i) How are plant traits and fitness affected by elevated temperature? (ii) Does the degree of trait plasticity vary between low and high elevation populations? (iii) Are traits under selection? Does the strength or direction of selection differ between ambient and elevated temperature conditions? (iv) Does variation in temperature impose selection on trait plasticity? Studying current levels of genetic differentiation and variation in trait plasticity in response to elevated temperature will lead to a better understanding of future plant responses to predicted climate change.

Materials and Methods

Study Species

The common perennial species *Trifolium montanum* L., *Ranunculus bulbosus* L. and *Briza media* L. were selected for this study based on their co-occurrence in semi-dry nutrient-poor calcareous grasslands and their similar altitudinal range from 400 to 2000 m a.s.l [59]. The three species differ in phenology, pollination syndromes and reproduction: *Trifolium montanum* is an insect-pollinated late-flowering herb, *R. bulbosus* an insect-pollinated early-flowering herb and *B. media* is a wind-pollinated grass.

The leguminous *Trifolium montanum*, which forms a symbiosis with soil rhizobia, produces 1–5 flowering shoots from June to July. A shoot bears 1–6 inflorescences with c. 150 white flowers each, which are pollinated by bumblebees and bees. Seeds are small (c. 0.7 mg), but dispersal distance is very limited as they are primarily gravity-dispersed [60].

Ranunculus bulbosus produces a corm that serves as nutrient storage and as summer perennating organ [61]. In late winter, *R. bulbosus* begins to mobilize corm reserves to add new leaves to the overwintering rosette and starts to form a new corm. From late April to July, the plant produces one or more flower stalks of c. 30 cm height bearing 8–15 yellow flowers [62]. The seeds lack dispersal aids and are therefore expected to be mainly gravity dispersed. After a period of aestivation, a new rosette is formed in early autumn [63,64].

Briza media reproduces clonally and sexually with each ramet potentially forming one flower stalk [65]. The flowers emerge from June to July, have large feathery stigmas, and mature into indehiscent fruits that disperse next to the mother plant mainly by gravity [65]. Long-distance seed dispersal by grazing animals has also been observed [66].

Plant Material

In summer 2008, seeds were collected in seven to thirteen population pairs per species across the Swiss Alps (Table S1). For seed sampling, permits were obtained from the managers of the respective sites. Each population pair consisted of one population in the centre of the species range at c. 1200 m a.s.l. (1095–1275 m a.s.l.) and another close to the upper range limit at c. 1800 m a.s.l. (1720–1860 m a.s.l.), hereafter referred to as low and high elevation populations respectively. The vertical distance of 600 m between these elevations approximates a difference in annual mean temperature of 4°C [54] corresponding to the expected temperature rise of a typical climate change scenario for the year 2100 [1]. The horizontal distance between the two populations of a pair was 1–18 kilometres. Seeds were sampled from at least ten maternal plants in each population, air-dried and stored in paper bags at 4°C.

In spring 2009, seeds were germinated in a greenhouse. After four to six weeks, seedlings were planted into individual pots (800 cm^3), filled with a 3:2 mixture of nutrient-poor commercial soil and sand. Pot positions were randomized weekly. From October 2009 to April 2010, the plants were overwintered outdoors at Davos (1500 m a.s.l.) to support vernalisation processes.

Climate Chamber Experiment

In May 2010, a controlled climate chamber experiment was established to test the responses of plants to current ambient temperature, corresponding to the temperature at their altitude of origin, and to elevated temperature, corresponding to 4°C higher temperature as compared to current temperature at their altitude of origin. The lowest temperature treatment (hereafter 'T_{min}') was ambient temperature for plants of high elevation origin (1800 m a.s.l.). Elevated temperature for high elevation plants was equivalent to ambient temperature for plants of low elevation origin (1200 m a.s.l.), and was designated as 'T'. Elevated temperature for low elevation plants was denoted as 'T_{plus}' and corresponded to ambient temperature at 600 m a.s.l. Each temperature treatment was replicated in two climate chambers. To separate temperature from other climate effects, only temperature was varied in this experiment while humidity and illumination treatments were kept the same in all chambers. Temperature, humidity and light regime followed both a daily as well as a seasonal cycle. To simulate the daily temperature cycle with lower temperature, higher humidity and dark conditions at night and opposed conditions during the day, seven step cycles were programmed according to climate station measurements [67]. For example in July, temperatures, relative humidity and light were varied diurnally between 5°C/80%/0 klx as the night's minimum and 18°C/48%/21 klx as the daily maximum (Table 1). To account for seasonal changes of the climate regime, program cycles were adapted monthly.

The plants of each species and population were assigned equally to ambient and elevated temperature treatments and to both climate chambers per treatment. They were rotated weekly within each chamber and monthly between the two chambers of the same treatment. The experiment included 174 plants from ten population pairs of *T. montanum*, 88 plants from seven population pairs of *R. bulbosus* and 312 plants from 13 population pairs of *B. media*.

Assessment of Plant Traits

In May 2010 and at the end of the growing season (October 2010), number of leaves (respectively ramets for *B. media*) and length of the longest leaf of each plant were measured. To obtain a measure for growth rate, the increment of number of leaves (respectively ramets for *B. media*) was calculated. In September 2010, one mature leaf per individual plant was sampled to assess specific leaf area (SLA). Immediately after sampling, the leaf blades were scanned (HP AllInOne colour Scanner, Hewlett-Packard GmbH, Dübendorf, Schweiz) and their areas were determined with LAMINA version 1.0.2 [68]. The scanned leaf blades were oven-dried at 60°C for 48 h and weighed. To determine SLA, we divided the leaf area by the dry weight of the leaf blade [69].

From May to July 2010, 75% of *T. montanum* and 84% of *R. bulbosus* individuals flowered. Due to the low flowering rate in *B. media* (8%), reproductive and phenological traits were only assessed for the two herbs. The phenological development was monitored twice a week to record the Julian Day (JD) of appearance of the first flower bud and the first open flower of each individual. In August, flowers of *R. bulbosus* and inflorescences of *T. montanum* were counted and the reproductive biomass (i.e. flower stalks and flowers) was harvested.

In October 2010, all plants were harvested and oven-dried. Above-ground (i.e. reproductive organs and leaves) and below-ground biomass of *T. montanum* and *R. bulbosus* were measured separately. Subsequently, biomass partitioning (i.e. above-ground biomass as a proportion of total biomass) and reproductive allocation (i.e. the proportion of reproductive biomass on total above-ground biomass) were calculated. In *B. media*, only above-ground biomass was assessed because we were not able to remove all soil sticking to the fine root system without considerable loss of root biomass.

Statistical Analyses

We first tested for effects of elevated temperature and genetic differentiation on phenotypic trait variation. Then, we conducted two sets of selection gradient analyses where we tested for selection on plant traits as well as for selection on trait plasticities in response to elevated temperature. All statistical analyses were performed with the statistical software R [70]. The residuals were checked for deviations from the model assumptions and the data was transformed if necessary [71]. To account for multiple tests we applied a Bonferroni correction within species whenever applicable.

We performed nested mixed-model analyses of covariance (ANCOVA) for each species separately to analyse the effects of elevated temperature and altitude of origin. The model consisted of the two fixed factors 'altitude of origin' and 'temperature treatment', the random factor 'population' and the two interaction terms 'temperature treatment × altitude of origin' and 'temperature treatment × population'. A significant 'altitude of origin' effect (G_{Alt}) indicates genetic differentiation between low and high elevation populations. A significant 'temperature treatment' effect (E_{Temp}) indicates trait variation due to different environmental conditions i.e. phenotypic plasticity in a measured trait. A significant 'temperature treatment × altitude of origin' interaction ($G_{Alt} \times E_{Temp}$) indicates differences in plasticity between the altitudes of origin. The 'altitude of origin' was tested on the population level and the 'temperature treatment' on the interac-

Table 1. Daily cyles of the climate chamber program for the three temperature treatments T_{min}, T and T_{plus} reflecting July outdoor temperatures at 1800 m, 1200 m and 600 m a.s.l.

Interval	Running time (h)	Temperature (°C)			Relative humidity (%)	Illumination (klx)
		T_{min}	T	T_{plus}		
1	2	8	12	16	72	0
2	3	6	10	14	81	0
3	5	5	9	13	84	0
4	6	18	22	26	55	21
5	2	19	23	27	51	21
6	2	18	22	26	51	16
7	4	10	14	18	63	1

Parameters were changed gradually to reach the set value at the end of the running time of each interval. Relative humidity and light regime were kept the same in all temperature treatments.

tion of 'temperature treatment × population' while all other terms were tested on the residuals. Besides direct responses to the environment, our measurements of plasticity might also be influenced by ontogenetic drift since the analyses did not account for plant size [72]. By including initial plant size (length of the longest leaf) as a covariate in the models we account for possible maternal effects [73].

We conducted phenotypic selection analyses on trait values in each temperature treatment to assess whether traits were under selection [58,74]. Selection gradients were calculated based on regressions of relative fitness, measured as total biomass (*T. montanum* and *R. bulbosus*) and above ground biomass (*B. media*), on standardized trait values of individual plants. A trait was under positive selection if trait values were positively correlated with genotype fitness. To test whether selection differed between treatments, we calculated ANCOVAs of relative fitness in which we included the trait as a covariate and its interaction with the temperature treatment [48].

Phenotypic selection analyses on trait plasticities were performed to test whether trait plasticity itself was under selection, i.e. if plasticity was positively correlated with overall fitness, indicating adaptive plasticity. Selection gradients on trait plasticity were determined by regressing average fitness of a population on population trait values averaged over both treatments (i.e. elevation of the reaction norm) and on population values of plasticity (i.e. steepness of the reaction norm) [75,76]. The elevation of the reaction norm was included to disentangle the fitness effect of the average value from the fitness effect of plasticity [77]. As proxies for fitness, we used number of inflorescences (*T. montanum* and *R. bulbosus*), total biomass (*T. montanum* and *R. bulbosus*) and above ground biomass (*B. media*). For both sets of selection gradient analyses, regression coefficients were standardized by expressing them in units of 1 SD to allow comparisons between traits and fitness measures.

Results

Environmental and Genetic Effects on Plant Variation

The model analyses allowed us to explain trait variation due to environmental effects of elevated temperature (E_{Temp}) and genetic differentiation (G_{Alt}), as well as their interactions ($G_{Alt} \times E_{Temp}$). The random term population nested within altitude of origin had essentially no influence on trait variation ($P>0.165$) except for a difference in the proportion of above-ground biomass in *T. montanum* and *B. media* ($P<0.001$), as well as a tendency for different budding starts in *R. bulbosus* ($P=0.056$; Tables 2 and 3). The covariate had a significant influence on the majority of traits (Tables 2 and 3). However, omitting the covariate did not change the results qualitatively (details not shown).

Plant responses to elevated temperature. *Trifolium montanum* responded to elevated temperature with an increase in total biomass (E_{Temp}: $P=0.002$; Tables 2 and S2; Fig. 1) and growth rate (E_{Temp}: $P=0.026$), but a slightly reduced proportion of above-ground biomass (E_{Temp}: $P=0.059$). SLA was unaffected by temperature (E_{Temp}: $P=0.668$). Flower bud initiation and anthesis were advanced under elevated temperature (E_{Temp}: $P<0.007$; Tables 3 and S3; Fig. 2) whereas number of flowers and reproductive allocation showed no temperature-induced plasticity (E_{Temp}: $P>0.9$).

In *R. bulbosus*, growth rate increased (E_{Temp}: $P=0.011$; Table 2; Fig. 1) and leaf length tended to increase (E_{Temp}: $P=0.053$) under elevated temperature whereas total biomass, SLA and proportion of above ground biomass were not affected by temperature (E_{Temp}: $P>0.9$). Flowering phenology was advanced under elevated temperature with buds and flowers appearing earlier (E_{Temp}: $P<0.001$; Table 3; Fig. 2) whereas number of flowers and reproductive allocation were unaffected by temperature (E_{Temp}: $P>0.139$).

In *B. media*, leaf length, above-ground biomass and growth rate increased under elevated temperature (E_{Temp}: $P<0.041$; Table 2; Fig. 1) and SLA was unaffected by temperature (E_{Temp}: $P>0.9$).

Genetic differentiation in traits and plasticity. In *T. montanum*, plants of low elevation origin produced almost twice as much biomass (G_{Alt}: $P<0.001$; Table 2, Fig. 1) and longer leaves (G_{Alt}: $P=0.005$) than high elevation plants whereas growth rate, SLA and the proportion of above-ground biomass did not differ between altitudes of origin ($P>0.9$). Budding and flowering started earlier in low as compared to high elevation plants (G_{Alt}: $P<0.043$; Table 3, Fig. 2). Number of flowers and reproductive allocation did not differ between low and high elevation plants (G_{Alt}: $P>0.9$). Furthermore, plasticity differed between altitudes of origin only in leaf length, where trait means almost doubled in low elevation plants but remained nearly unchanged in high elevation plants under elevated temperature ($G_{Alt} \times E_{Temp}$: $P<0.001$; Table 2, Fig. 1). Plasticity of total biomass, growth rate, SLA and proportion of above-ground biomass as well as reproductive and phenological traits did not differ between low and high elevation populations ($G_{Alt} \times E_{Temp}$: $P>0.1$; Tables 2 and 3; Figs. 1 and 2).

In *R. bulbosus*, low elevation plants produced longer leaves (G_{Alt}: $P=0.005$; Table 2; Fig. 1) and larger SLA (G_{Alt}: $P<0.001$) than high elevation plants whereas total biomass, growth rates and the proportion of above-ground biomass did not differ between low and high elevation plants (G_{Alt}: $P>0.158$). Low elevation plants tended to flower earlier than high elevation plants (G_{Alt}: $P=0.074$; Table 3; Fig. 2). Number of flowers, reproductive allocation and budding dates did not differ between low and high elevation plants (G_{Alt}: $P>0.139$). Furthermore, there were no significant differences in plasticity between low and high elevation populations neither in growth-related traits ($G_{Alt} \times E_{Temp}$: $P>0.07$; Table 2; Fig. 1) nor in reproductive and phenological traits ($G_{Alt} \times E_{Temp}$: $P>0.9$; Table 3; Fig. 2).

In *B. media*, low elevation plants produced more above ground biomass than high elevation plants (G_{Alt}: $P=0.043$; Table 2; Fig. 1). Leaf length, growth rate and SLA did not differ between low and high elevation plants (G_{Alt}: $P>0.534$). Furthermore, plasticity of leaf traits differed between altitudes of origin: leaf length and SLA increased in low elevation but decreased in high elevation plants under elevated temperature ($G_{Alt} \times E_{Temp}$: $P<0.001$; Table 2; Fig. 1). Plasticity of above ground biomass and growth rate did not differ between altitudes of origin ($G_{Alt} \times E_{Temp}$: $P>0.584$).

Selection Gradient Analyses on Traits and Plasticities

The selection gradient analyses on trait values in *T. montanum* revealed negative selection gradients for flower bud initiation and the appearance of the first flower averaged over both temperature treatments, which was reflected in a significant effect of these traits on fitness measured as total biomass ($P<0.008$; Table 4). In *R. bulbosus*, the selection analyses indicated a negative effect of SLA on fitness measured as total biomass ($P=0.003$; Table 4). In *B. media*, there was selection for higher growth rates averaged over both treatments ($P<0.001$). Selection gradient analyses on trait plasticities revealed no direct selection for plasticity in response to temperature in all traits and species ($P>0.148$; Table 5).

Table 2. Nested mixed-model ANCOVA for growth-related traits of *Trifolium montanum*, *Ranunculus bulbosus* and *Briza media* grown under ambient and elevated temperature.

		Biomass		Leaf length		Growth rate		SLA		Above-ground/total biomass	
	d.f.	MS	*F*	MS	*F*	MS	*F*	MS	*F*	MS	*F*
T. montanum											
Initial plant size	1	19.31	76.26***	167.18	16.77***	4.74	14.36**	105822.00	2.99	1.23	16.52***
Temp	1	7.07	27.91**	108.63	10.89**	4.16	12.59*	157334.00	4.45	0.73	9.83(*)
Orig	1	15.24	60.18***	198.74	19.93**	0.07	0.22	76416.00	2.16	0.49	6.57
Pop[Orig]	18	0.38	1.52	11.47	1.15	0.50	1.53	47855.00	1.35	0.28	3.75***
Temp × Orig	1	0.49	1.95	243.42	24.41***	0.33	0.99	38816.00	1.10	0.06	0.84
Temp × Pop	18	0.32	1.25	7.21	0.72	0.35	1.06	43784.00	1.24	0.08	1.04
Residuals	128	0.25		9.97		0.33		35383.00		0.07	
R. bulbosus											
Initial plant size	1	15.99	22.61***	0.26	0.21	394.70	12.96**	6230.20	5.16	0.00	0.13
Temp	1	1.06	1.50	10.56	8.49(*)	378.12	12.41*	2107.50	1.75	0.00	0.11
Orig	1	0.05	0.07	26.08	20.96**	504.64	16.56	30617.20	25.38***	0.03	0.77
Pop[Orig]	12	0.88	1.24	1.18	0.95	66.72	2.19	863.10	0.72	0.04	1.31
Temp × Orig	1	0.49	0.70	0.22	0.17	18.98	0.62	10155.70	8.42(*)	0.01	0.28
Temp × Pop	12	0.88	1.25	0.94	0.76	21.58	0.71	1267.10	1.05	0.02	0.46
Residuals	47–59	0.71		1.24		30.47		1206.40		0.03	
B. media											
Initial plant size	1	4.33	86.02***	54.40	14.45***	9.19	0.95	0.22	5.30(*)		
Temp	1	1.74	34.68***	14.84	3.94*	132.64	13.71*	0.04	0.87		
Orig	1	1.29	25.66*	0.05	0.01	24.62	2.54	0.09	2.14		
Pop[Orig]	24	0.17	3.36***	5.56	1.48	15.46	1.60	0.04	0.89		
Temp × Orig	1	0.11	2.13	333.25	88.50***	1.62	0.17	1.31	31.99***		
Temp × Pop	24	0.03	0.67	1.91	0.51	12.45	1.29	0.04	0.99		
Residuals	259	0.05		3.77		9.68		0.04			

d.f., degrees of freedom; MS, mean squares; *F*, *F*-values; Initial plant size, covariate; Temp, temperature treatment; Orig, altitude of population origin; Pop, population. Residual d.f. vary per trait due to missingness. Bonferroni-corrected significance levels within species are indicated by asterisks: ***$P_{Bonf} < 0.001$, **$P_{Bonf} < 0.01$, *$P_{Bonf} < 0.05$, (*)$P_{Bonf} < 0.08$.

Table 3. Nested mixed-model ANCOVA for reproductive and phenological traits of *Trifolium montanum* and *Ranunculus bulbosus* grown under ambient and elevated temperature.

		Number of flowers		Reproductive allocation		Budding start		Flowering start	
	d.f.	MS	*F*	MS	*F*	MS	*F*	MS	*F*
T. montanum									
Initial plant size	1	64.76	47.07***	1.96	42.38***	24721.90	30.61***	17.43	36.37***
Temp	1	0.25	0.18	0.01	0.17	5641.00	6.98**	11.24	23.46***
Orig	1	6.33	4.60	0.12	2.55	7049.00	8.73*	10.79	22.51**
Pop[Orig]	18	2.49	3	0.09	1.84	680.80	0.84	0.58	1.22
Temp × Orig	1	8.53	6.20	0.02	0.50	0.10	0.00	1.00	2.09
Temp × Pop	17–18	0.96	0.70	0.03	0.71	340.00	0.42	0.23	0.47
Residuals	77–128	1.38		0.05		807.80		0.48	
R. bulbosus									
Initial plant size	1	18.19	22.86***	0.12	8.82*	3312.60	23.61***	3925.90	26.58***
Temp	1	3.08	3.87	0.02	1.73	1188.00	8.47***	2238.00	15.15***
Orig	1	1.88	2.36	0.14	10.12	627.10	4.47	2197.50	14.88(*)
Pop[Orig]	12	0.97	1.23	0.02	1.27	391.80	2.79(*)	219.70	1.49
Temp × Orig	1	0.69	0.86	0.01	0.43	16.90	0.12	3.30	0.02
Temp × Pop	12	0.77	0.97	0.02	1.67	30.30	0.22	29.00	0.20
Residuals	45–59	0.80		0.01		140.30		147.70	

d.f., degrees of freedom; MS, mean squares; *F*, *F*-values; Initial plant size, covariate; Temp, temperature treatment; Orig, altitude of population origin; Pop, population. Residual d.f. vary per trait due to missingness. Bonferroni-corrected significance levels within species are indicated by asterisks: ***P_{Bonf} <0.001, **P_{Bonf} <0.01, *P_{Bonf} <0.05, (*)P_{Bonf}<0.08.

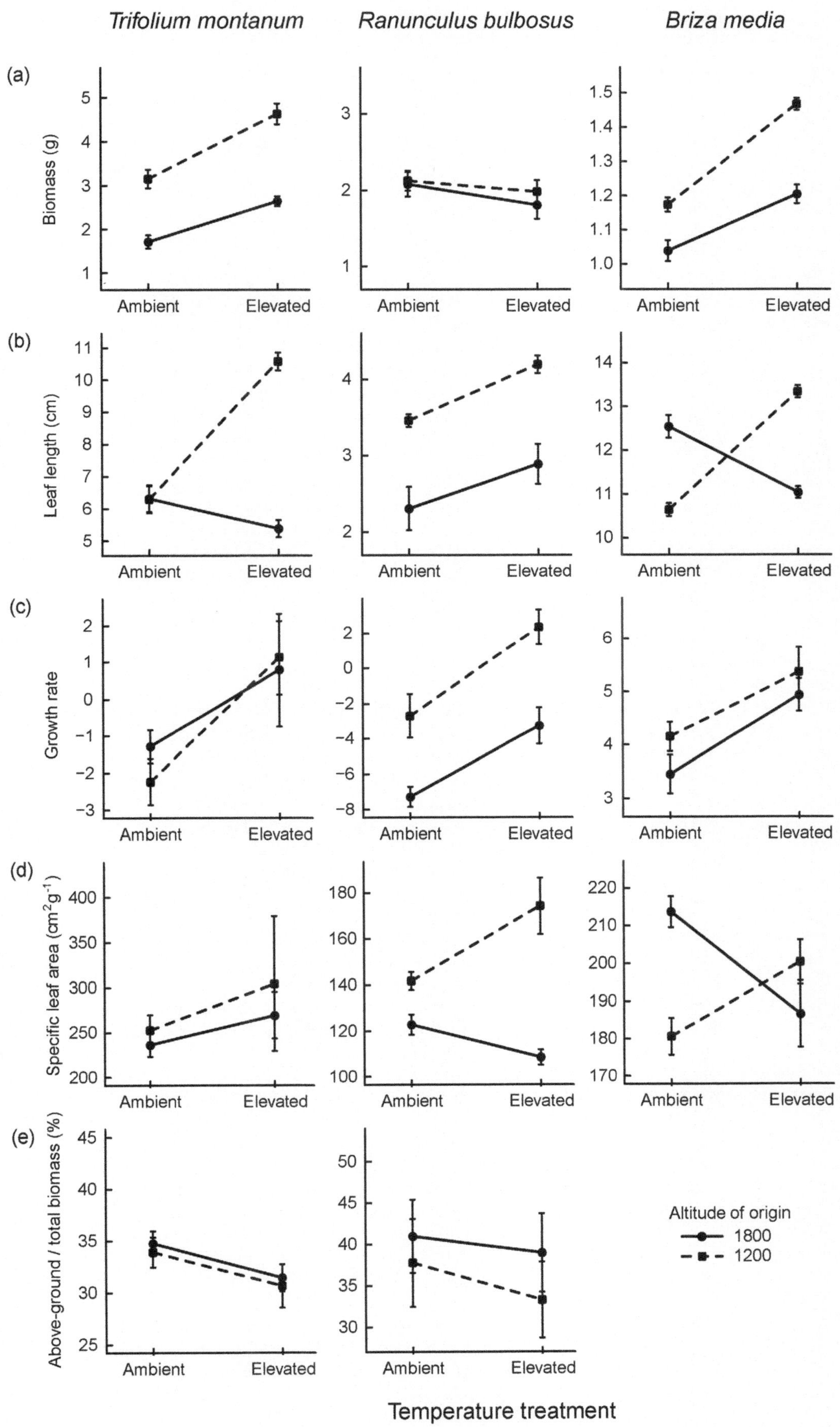
Trifolium montanum
Ranunculus bulbosus
Briza media
(a)
Biomass (g)
(b)
Leaf length (cm)
(c)
Growth rate
(d)
Specific leaf area ($cm^2 g^{-1}$)
(e)
Above-ground / total biomass (%)
Ambient
Elevated
Altitude of origin
1800
1200
Temperature treatment

Figure 1. Population average reaction norms for growth-related traits of *Trifolium montanum*, *Ranunculus bulbosus* and *Briza media* in response to ambient and elevated temperature in a climate chamber experiment. (a) total biomass (above-ground biomass for *Briza media*), (b) leaf length, (c) growth rate, (d) specific leaf area and (e) proportion of above ground biomass. Solid lines indicate high elevation plants (1800 m a.s.l.) and dashed lines low elevation plants (1200 m a.s.l.). Error bars indicate standard errors of trait means.

Discussion

In the present study, we evaluated the effects of elevated temperature on phenotypic variation of the three perennial grassland species *T. montanum*, *R. bulbosus* and *B. media*. We investigated if trait plasticity varies between low and high elevation populations and if trait values and plasticities were under selection.

Plastic Responses to Elevated Temperature

Positive warming effects on several growth-related traits in all three species (Table 2) indicate that plant growth is constrained by low temperature conditions, which is common in temperate species [6]. In arctic and temperate zones plant species spend the majority of their life at mean temperatures below the growth optimum [78]. Thus, a moderate warming could enhance plant performance. Meta-analyses of in-situ warming experiments with arctic and alpine tundra species revealed biome-wide trends of increased vegetative growth, albeit variable among species [79,80]. The variable responses to elevated temperature of the three study species might be associated with differences in characteristics of below-ground organs. Besides the positive responses in the other growth-related traits, the nitrogen-fixing *T. montanum* showed a slight decrease in the proportion of above ground biomass. The observed greater investment in roots under elevated temperature is contrary to the general trend that allocation to root biomass increases with decreasing temperature at higher elevation [54]. However, it is in line with the findings of Roughley and Dart [81] who described a positive effect of soil temperature on the formation of rhizobia in *Trifolium subterraneum*. The increased number of rhizobia enhances nitrogen availability, which stimulates root growth. *Fabaceae*, such as *T. montanum*, might therefore gain a competitive advantage over non-nitrogen-fixing species under warmer climate conditions. Competition experiments with elevated CO_2, which has a similar positive effect on nitrogen fixation as elevated temperature, provide evidence for such a competitive advantage, reviewed in [82]. The less pronounced warming effect on growth-related traits in *R. bulbosus* could be related to the species-specific life-form. The nutrient-storing corm might allow a temperature independent formation of new above ground tissue after the period of summer aestivation and buffer against short-term climatic fluctuations as shown for *R. nivalis* [83].

Preformation of flower buds might lead to a time lag in the reproductive response to warming potentially explaining why the experimental warming did not affect reproductive traits of *T. montanum* and *R. bulbosus* (Table 3). This phenomenon is relatively common among temperate herbaceous perennials [84] and has been documented for tundra species in in-situ warming experiments [79]. Thus, it might have required at least two growing seasons to provide a strong estimate of the effects of increased temperature on reproduction of the perennial study species. Although these species did not respond plastically to elevated temperature, they are likely to be affected by changes in other environmental factors (e.g. reduced precipitation and higher evapotranspiration) co-occurring in natural systems. Alternatively, the lack of reproductive responses could indicate that reproductive traits are genetically fixed suggesting that these species need to adjust to climate change through the relatively slow process of evolutionary adaptation [20,85]. In the context of rapid climate change, this process might be too slow to secure plant persistence [86–88]. Evidence for evolutionary adaptation to climate change is still scarce, and rapid evolution has been documented mainly in fast-growing annuals [27,89] but also in perennials [90].

Reproductive phenology of *T. montanum* and *R. bulbosus* was advanced under elevated temperature (Table 3) confirming the findings of other studies that these traits respond plastically to temperature, e.g. [2,5,79,91–95]. Moreover, our findings are in line with the advanced reproductive phenology of the same species grown in a common garden transplant experiment [57]. The advanced spring phenology facilitates setting flowers and seeds before the hot and dry midsummer period. It might therefore enhance plant fitness, but also bears the risk of damage by late frost events, especially in high elevation populations [54,96,97]. Furthermore, plant-pollinator-interaction models indicate that pollinating insects might not keep up with shifted flowering periods [98], which might impede insect pollination and thereby reduce reproductive success [99,100].

In summary, low and high elevation populations exhibited plasticity to temperature in reproductive phenology and several growth-related traits. The advanced flowering and the generally enhanced growth, albeit variable among species, indicate the ability of these species to cope with climate warming at least in the short term.

Genetic Variation in Traits and Differences in Plasticity between Low and High Elevation Populations

We found only little genetic variation among populations within each altitudinal origin (Tables 2 and 3) suggesting small selection differences and high genetic connectivity. Indeed, a previous study on similar populations of the same three species using neutral molecular markers revealed intermediate genetic diversity but only low genetic differentiation among populations and assigned this to extensive historic gene flow [101]. The genetic differentiation between low and high elevation populations in some growth-related and phenological traits (Tables 2 and 3) indicated that selection in the past has acted differently on low and high elevation plants. Alternatively, the observed genetic differences might be due to neutral genetic processes [24,102]. However, maternal effects were accounted for by including initial plant size as a covariate in the models and genetic drift is unlikely to have acted in the same direction on all sampled populations since seeds were sampled from multiple population pairs distributed over the Swiss Alps.

In contrast to the hypothesis of greater plasticity at more heterogeneous high elevation sites, plasticity of leaf length was reduced in high elevation populations of *T. montanum* (Table 2; Fig. 1). Similarly, a study with seedlings of European deciduous tree species found that high elevation provenances exhibited less temperature-induced plasticity in growth and leaf phenological traits [103,104]. The authors argued that low plasticity at high elevations was a result of different directional selection for reduced temperature sensitivity and a stronger influence of photoperiodism, which may reduce the risk of damage by unpredictable late spring frost events [105]. Moreover, the greater plasticity of leaf length in low elevation plants might have been induced by higher levels of competition at these sites [106]. In *B. media*, plastic responses of leaf length and SLA to elevated temperature were positive in plants from warmer low elevations, but negative in

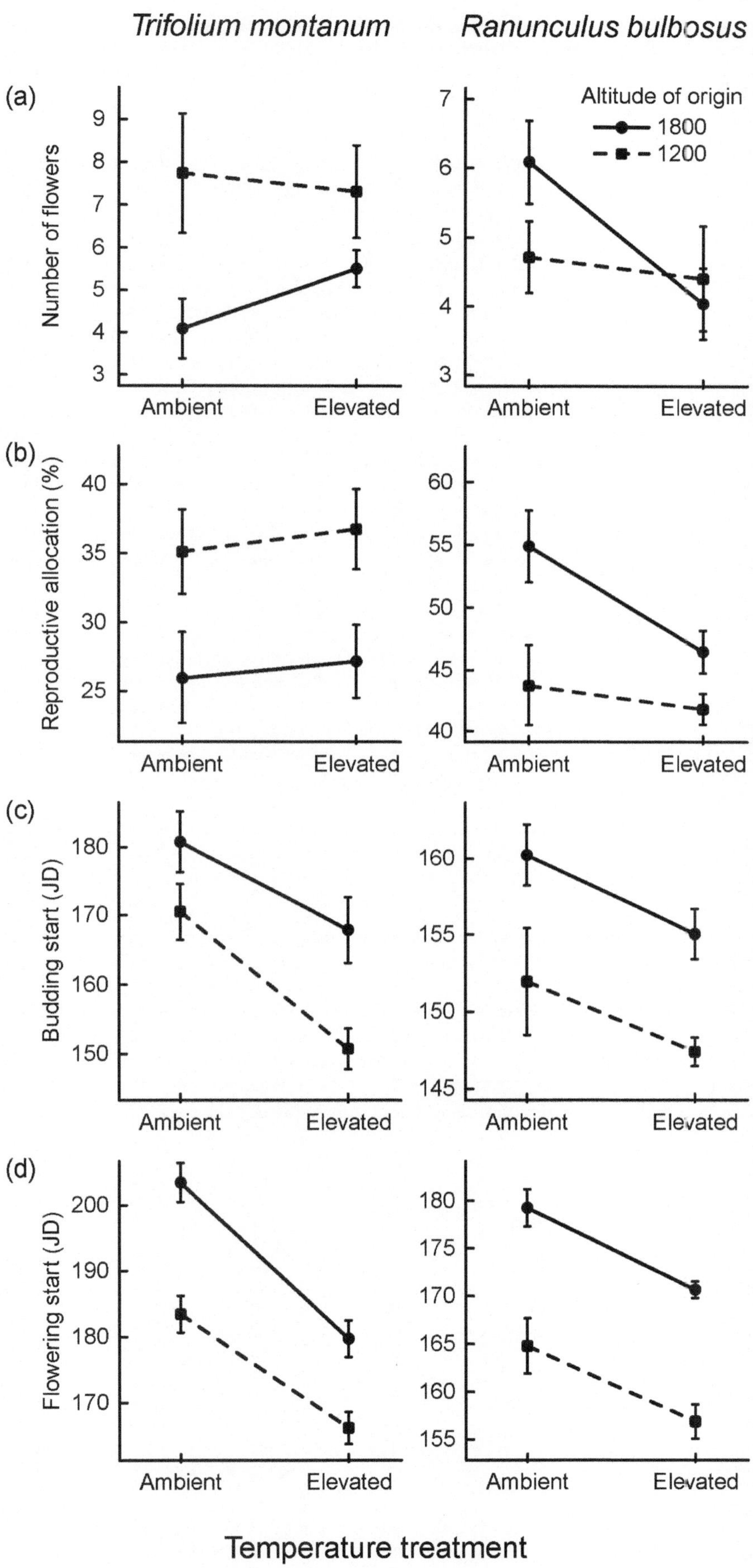

Trifolium montanum
Ranunculus bulbosus
(a)
Number of flowers
Altitude of origin
1800
1200
Ambient
Elevated
(b)
Reproductive allocation (%)
(c)
Budding start (JD)
(d)
Flowering start (JD)
Temperature treatment

Figure 2. Population average reaction norms for reproductive and phenological traits of *Trifolium montanum* and *Ranunculus bulbosus* in response to ambient and elevated temperature in a climate chamber experiment. (a) number of flowers, (b) reproductive allocation, (c) budding start and (d) flowering start. Solid lines indicate high elevation plants (1800 m a.s.l.) and dashed lines low elevation plants (1200 m a.s.l.). Error bars indicate standard errors of trait means.

plants from colder high elevations (Table 2; Fig. 1). In low elevation plants, elevated temperature induced longer leaves and greater SLA. Thus, the warming seems to have triggered a greater investment in vegetative growth. High elevation plants invested more in clonal propagation as can be concluded from their shorter and thicker leaves as indicated by the smaller leaf length and SLA in combination with an increase in number of ramets (details not shown). In-situ warming experiments showed similar response patterns, reviewed in [79]. The stronger growth response in low arctic plants was related to higher competition in these communities [107] whereas the greater reproductive response of high arctic plants was attributed to increased colonization efforts in these habitats where competition is less important [108].

In summary, plasticity differed between altitudes of origin in only a few traits and there was little genetic differentiation among populations within each altitudinal origin suggesting uniform responses to climate warming. Moreover, the high connectivity together with the moderate genetic differentiation between altitudes of origin might facilitate future adaptation to climate change.

Selection on Traits and their Plasticities

We found indication for selection on several growth-related and phenological traits (Table 4). In *T. montanum*, the two phenological traits budding and flowering start were under negative selection with respect to biomass: plants with earlier appearing flower buds and flowers ended up with higher biomass at the end of the growing season. Although we did not record leaf phenology, it is likely that the advanced flowering phenology was related to a similar advance in other spring phenophases [32] allowing these plants more time to accumulate biomass. In *B. media*, selection for higher growth rates can be explained by the fact that faster growing individuals acquired more above ground biomass [7].

Although plasticity to temperature was not adaptive in our experiment (Table 5), increased plant size under elevated temperature might eventually lead to higher fitness since flower bud preformation is common in perennials and sexual reproduction is often positively correlated with plant size [109]. Increased growth under warmer temperature might thus be beneficial for plant persistence in the longer term whereas the duration of our experiment might have been too short for observing fitness benefits of plasticity. The magnitude and even the direction of selection on a trait may differ for different components of fitness [74]. This might also have affected the results of our selection analyses and shows the need for more comprehensive fitness measures [110]. Furthermore, it is generally recommended to analyse phenotypic selection on trait means and their plasticities based on genotypic values [111] to avoid environmentally induced covariation [112]. However, our experimental design only allowed us to analyse selection based on individual values or population means, see [113]. This might have resulted in a loss of discriminative power because the observed variation is a combination of phenotypic plasticity and within population genetic variability. Thus, future selection studies at the genotype level are recommended to gain deeper insights into the mechanisms of plasticity in plant populations.

In summary, the absence of selection on phenotypic plasticity indicated that the observed trait plasticity was selectively neutral. Selection on several growth-related and phenological traits

Table 4. Selection gradients of growth-related and phenological traits of *Trifolium montanum*, *Ranunculus bulbosus* and *Briza media* under ambient and elevated temperature and F-values of ANCOVA testing for overall selection (Cov) and differences in selection between temperature treatments (Cov × Temp).

	Selection gradients		F-Values of ANCOVA	
	ambient	elevated	Cov	Cov × Temp
T. montanum				
Growth rate	−0.320	0.009	0.15	4.57
SLA	−0.332	0.009	0.85	0.17
Budding start	−0.543	−0.170	10.07**	0.80
Flowering start	−0.804	−0.297	19.58***	1.40
R. bulbosus				
Growth rate	−0.279	−0.071	4.88	1.35
SLA	−0.365	−0.309	12.68**	0.70
Budding start	−0.119	−0.145	0.94	0.04
Flowering start	−0.236	−0.309	4.50	0.23
B. media				
Growth rate	0.044	0.079	20.70***	0.02
SLA	−0.040	0.025	0.38	3.38

Selection gradients are expressed as standardized regression coefficients after regression of the fitness measures total biomass (*T. montanum*, *R. bulbosus*) and above ground biomass (*B. media*) on the respective trait. All calculations are based on individual plant values. Bonferroni-corrected significance levels are indicated by asterisks: ****P*<0.001, ***P*<0.01, **P*<0.05.

Table 5. Selection gradients of trait plasticities in response to temperature in *Trifolium montanum*, *Ranunculus bulbosus* and *Briza media*.

	T. montanum		*R. bulbosus*		*B. media*
	Total biomass	**Number of flowers**	**Total biomass**	**Number of flowers**	**Above ground biomass**
Growth rate	0.500	0.455	0.064	0.258	0.366
SLA	0.423	0.541	−1.380	−0.418	−0.577
Above-ground/total biomass	−0.497	−0.189	−1.334	−0.450	–
Budding start	0.289	0.480	−0.171	0.315	–
Flowering start	0.307	0.380	0.039	1.156	–

Selection gradients are expressed as standardized regression coefficients. All values are based on population means. Fitness measures are total biomass and number of flowers (*T. montanum, R. bulbosus*) and above ground biomass (*B. media*). Non of the selection gradients was significant.

suggested that there is potential for future evolution of mean trait values allowing the study species to adapt to elevated temperature provided there is sufficient genetic variation and heritability of the respective traits.

Conclusions

We investigated phenotypic variation in low and high elevation populations of nutrient-poor grassland species in response to elevated temperature. The three species exhibited trait plasticity with respect to temperature whereas genetic differentiation and differences in plasticity between low and high populations were less important. Plasticity in growth and flowering phenology determines the ability of the study species to respond to elevated temperature by buffering against detrimental effects of rapid climate change and allowing time for evolutionary adaptation. However, the capacity of species to respond to environmental changes through phenotypic plasticity is limited and plasticity alone might not be sufficient to cope with climate change. Thus, plants additionally need to adjust through the relatively slow process of evolutionary adaptation. Selection on several traits suggests that the three species have the potential for evolutionary changes, which might allow them to adapt to a future climate.

Supporting Information

Table S1 Sampling locations in the Swiss Alps of the source population pairs of *Trifolium montanum* (Tm), *Ranunculus bulbosus* (Rb) and *Briza media* (Bm).

Table S2 Means and standard errors (SE) grouped by altitude of population origin and temperature treatment for growth-related traits of *Trifolium montanum*, *Ranunculus bulbosus* and *Briza media*.

Table S3 Means and standard errors (SE) grouped by altitude of population origin and temperature treatment for reproductive and phenological traits of *Trifolium montanum* and *Ranunculus bulbosus*.

Acknowledgments

The authors thank M. Beck, M. Frei, S. Frischknecht, T. Hahn, M. Heggli, D. Kurth, J. Leuenberger, M. Leuzinger, M. Macsai and P. Matter who helped collecting seeds, cultivating and harvesting plants. They are grateful to F. Grassein, M. Kalisch, M. van Kleunen and I. Till-Bottraud for statistical advice as well as to E.S. Frei, M. Heggli and C.J. Kettle for valuable comments on the manuscript. The authors thank the editor and two anonymous reviewers for helpful comments.

Author Contributions

Conceived and designed the experiments: ERF ARP JG. Performed the experiments: ERF. Analyzed the data: ERF ARP. Contributed reagents/materials/analysis tools: ERF ARP. Wrote the paper: ERF JG ARP.

References

1. IPCC (2013) Climate Change 2013: The Physical Science Basis. Contribution of Working Group I to the Fifth Assessment Report of the Intergovernmental Panel on Climate Change; Stocker TF, Qin D, Plattner G-K, Tignor M, Allen SK et al., editors. Cambridge, United Kingdom and New York, NY, USA: Cambridge University Press. 1535 p.
2. Parmesan C, Yohe G (2003) A globally coherent fingerprint of climate change impacts across natural systems. Nature 421: 37–42.
3. Root TL, Price JT, Hall KR, Schneider SH, Rosenzweig C, et al. (2003) Fingerprints of global warming on wild animals and plants. Nature 421: 57–60.
4. Walker MD, Wahren CH, Hollister RD, Henry GHR, Ahlquist LE, et al. (2006) Plant community responses to experimental warming across the tundra biome. Proceedings of the National Academy of Sciences of the United States of America 103: 1342–1346.
5. Walther GR, Post E, Convey P, Menzel A, Parmesan C, et al. (2002) Ecological responses to recent climate change. Nature 416: 389–395.
6. Hopkins WG, Hüner NPA (2008) Introduction to plant physiology. Hoboken, New Jersey, United States: John Wiley & Sons.
7. Lambers H, Chapin FS III, Pons TL (2008) Plant physiological ecology. New York, United States: Springer.
8. Walther GR, Beissner S, Burga CA (2005) Trends in the upward shift of alpine plants. Journal of Vegetation Science 16: 541–548.
9. Frei E, Bodin J, Walther GR (2010) Plant species' range shifts in mountainous areas – all uphill from here? Botanica Helvetica 120: 117–128.
10. Gottfried M, Pauli H, Futschik A, Akhalkatsi M, Barancok P, et al. (2012) Continent-wide response of mountain vegetation to climate change. Nature Climate Change 2: 111–115.
11. Engler R, Randin CF, Thuiller W, Dullinger S, Zimmermann NE, et al. (2011) 21st century climate change threatens mountain flora unequally across Europe. Global Change Biology 17: 2330–2341.
12. Lenoir J, Gegout JC, Marquet PA, de Ruffray P, Brisse H (2008) A significant upward shift in plant species optimum elevation during the 20th century. Science 320: 1768–1771.
13. Huntley B (1991) How plants respond to climate change: migration rates, individualism and the consequences for plant communities. Annals of Botany 67: 15–22.
14. Davis MB, Shaw RG (2001) Range shifts and adaptive responses to quaternary climate change. Science 292: 673–679.
15. Collingham YC, Huntley B (2000) Impacts of habitat fragmentation and patch size upon migration rates. Ecological Applications 10: 131–144.
16. Honnay O, Verheyen K, Butaye J, Jacquemyn H, Bossuyt B, et al. (2002) Possible effects of habitat fragmentation and climate change on the range of forest plant species. Ecology Letters 5: 525–530.

17. Thomas CD, Cameron A, Green RE, Bakkenes M, Beaumont LJ, et al. (2004) Extinction risk from climate change. Nature 427: 145–148.
18. Lesica P, McCune B (2004) Decline of arctic-alpine plants at the southern margin of their range following a decade of climatic warming. Journal of Vegetation Science 15: 679–690.
19. Hoffmann AA, Sgro CM (2011) Climate change and evolutionary adaptation. Nature 470: 479–485.
20. Jump AS, Penuelas J (2005) Running to stand still: adaptation and the response of plants to rapid climate change. Ecology Letters 8: 1010–1020.
21. Nicotra AB, Atkin OK, Bonser SP, Davidson AM, Finnegan EJ, et al. (2010) Plant phenotypic plasticity in a changing climate. Trends in Plant Science 15: 684–692.
22. Gienapp P, Teplitsky C, Alho JS, Mills JA, Merila J (2008) Climate change and evolution: disentangling environmental and genetic responses. Molecular Ecology 17: 167–178.
23. Joshi J, Schmid B, Caldeira M, Dimitrakopoulos P, Good J, et al. (2001) Local adaptation enhances performance of common plant species. Ecology Letters 4: 536–544.
24. Linhart YB, Grant MC (1996) Evolutionary significance of local genetic differentiation in plants. Annual Review of Ecology and Systematics 27: 237–277.
25. Macel M, Lawson CS, Mortimer SR, Smilauerova M, Bischoff A, et al. (2007) Climate vs. soil factors in local adaptation of two common plant species. Ecology 88: 424–433.
26. Becker U, Colling G, Dostal P, Jakobsson A, Matthies D (2006) Local adaptation in the monocarpic perennial *Carlina vulgaris* at different spatial scales across Europe. Oecologia 150: 506–518.
27. Franks SJ, Sim S, Weis AE (2007) Rapid evolution of flowering time by an annual plant in response to a climate fluctuation. Proceedings of the National Academy of Sciences of the United States of America 104: 1278–1282.
28. Davis MB, Shaw RG, Etterson JR (2005) Evolutionary responses to changing climate. Ecology 86: 1704–1714.
29. Bradshaw AD (1965) Evolutionary significance of phenotypic plasticity in plants. Advances in Genetics 13: 115–155.
30. Valladares F, Sanchez-Gomez D, Zavala MA (2006) Quantitative estimation of phenotypic plasticity: bridging the gap between the evolutionary concept and its ecological applications. Journal of Ecology 94: 1103–1116.
31. Ghalambor CK, McKay JK, Carroll SP, Reznick DN (2007) Adaptive versus non-adaptive phenotypic plasticity and the potential for contemporary adaptation in new environments. Functional Ecology 21: 394–407.
32. Menzel A, Sparks TH, Estrella N, Koch E, Aasa A, et al. (2006) European phenological response to climate change matches the warming pattern. Global Change Biology 12: 1969–1976.
33. van Vliet AJH, de Groot RS, Bellens Y, Braun P, Bruegger R, et al. (2003) The European Phenology Network. International Journal of Biometeorology 47: 202–212.
34. Wolfe DW, Schwartz MD, Lakso AN, Otsuki Y, Pool RM, et al. (2005) Climate change and shifts in spring phenology of three horticultural woody perennials in northeastern USA. International Journal of Biometeorology 49: 303–309.
35. Atkin OK, Loveys BR, Atkinson LJ, Pons TL (2006) Phenotypic plasticity and growth temperature: understanding interspecific variability. Journal of Experimental Botany 57: 267–281.
36. Steadman KJ, Ellery AJ, Chapman R, Moore A, Turner NC (2004) Maturation temperature and rainfall influence seed dormancy characteristics of annual ryegrass (*Lolium rigidum*). Australian Journal of Agricultural Research 55: 1047–1057.
37. Qaderi MM, Cavers PB, Hamill AS, Downs MP, Bernards MA (2006) Maturation temperature regulates germinability and chemical constituents of Scotch thistle (*Onopordum acanthium*) cypselas. Canadian Journal of Botany-Revue Canadienne De Botanique 84: 28–38.
38. Kochanek J, Buckley YM, Probert RJ, Adkins SW, Steadman KJ (2010) Pre-zygotic parental environment modulates seed longevity. Austral Ecology 35: 837–848.
39. van Kleunen M, Fischer M (2005) Constraints on the evolution of adaptive phenotypic plasticity in plants. Research review. New Phytologist 166: 49–60.
40. Lande R (2009) Adaptation to an extraordinary environment by evolution of phenotypic plasticity and genetic assimilation. Journal of Evolutionary Biology 22: 1435–1446.
41. Chevin LM, Lande R, Mace GM (2010) Adaptation, plasticity, and extinction in a changing environment: towards a predictive theory. PLOS Biology 8: e1000357.
42. Draghi JA, Whitlock MC (2012) Phenotypic plasticity facilitates mutational variance, genetic variance, and evolvability along the major axis of environmental variation. Evolution 66: 2891–2902.
43. Price TD, Qvarnström A, Irwin DE (2003) The role of phenotypic plasticity in driving genetic evolution. Proceedings of the Royal Society of London Series B: Biological Sciences 270: 1433–1440.
44. Matesanz S, Gianoli E, Valladares F (2010) Global change and the evolution of phenotypic plasticity in plants. Annals of the New York Academy of Sciences 1206: 35–55.
45. Dudley SA, Schmitt J (1996) Testing the adaptive plasticity hypothesis: density-dependent selection on manipulated stem length in *Impatiens capensis*. American Naturalist 147: 445–465.
46. Gianoli E, Gonzalez-Teuber M (2005) Environmental heterogeneity and population differentiation in plasticity to drought in *Convolvulus chilensis* (Convolvulaceae). Evolutionary Ecology 19: 603–613.
47. Heschel MS, Sultan SE, Glover S, Sloan D (2004) Population differentiation and plastic responses to drought stress in the generalist annual *Polygonum persicaria*. International Journal of Plant Sciences 165: 817–824.
48. van Kleunen M, Lenssen JPM, Fischer M, de Kroon H (2007) Selection on phenotypic plasticity of morphological traits in response to flooding and competition in the clonal shore plant *Ranunculus reptans*. Journal of Evolutionary Biology 20: 2126–2137.
49. Springate DA, Kover PX (2014) Plant responses to elevated temperatures: a field study on phenological sensitivity and fitness responses to simulated climate warming. Global Change Biology 20: 456–465.
50. Pratt JD, Mooney KA (2013) Clinal adaptation and adaptive plasticity in *Artemisia californica*: implications for the response of a foundation species to predicted climate change. Global Change Biology 19: 2454–2466.
51. Etterson JR (2004) Evolutionary potential of *Chamaecrista fasciculata* in relation to climate change. 1. Clinal patterns of selection along an environmental gradient in the great plains. Evolution 58: 1446–1458.
52. Via S, Lande R (1985) Genotype-environment interaction and the evolution of phenotypic plasticity. Evolution 39: 505–522.
53. Sultan SE, Spencer HG (2002) Metapopulation structure favors plasticity over local adaptation. American Naturalist 160: 271–283.
54. Körner C (2003) Alpine plant life: functional plant ecology of high mountain ecosystems. Berlin, Germany: Springer.
55. Gonzalo-Turpin H, Hazard L (2009) Local adaptation occurs along altitudinal gradient despite the existence of gene flow in the alpine plant species *Festuca eskia*. Journal of Ecology 97: 742–751.
56. Byars SG, Papst W, Hoffmann AA (2007) Local adaptation and cogradient selection in the alpine plant, *Poa hiemata*, along a narrow altitudinal gradient. Evolution 61: 2925–2941.
57. Frei ER, Ghazoul J, Matter P, Heggli M, Pluess AR (2014) Plant population differentiation and climate change: responses of grassland species along an elevational gradient. Global Change Biology 20: 441–455.
58. Lande R, Arnold SJ (1983) The measurement of selection on correlated characters. Evolution 37: 1210–1226.
59. Aeschimann D, Lauber K, Moser DM, Theurillat JP (2004) Flora Alpina. Berne, Switzerland: Haupt.
60. Schleuning M, Niggemann M, Becker U, Matthies D (2009) Negative effects of habitat degradation and fragmentation on the declining grassland plant *Trifolium montanum*. Basic and Applied Ecology 10: 61–69.
61. Barling DM (1955) Some population studies in *Ranunculus bulbosus* L. Journal of Ecology 43: 207–218.
62. Steinbach K, Gottsberger G (1994) Phenology and pollination biology of 5 *Ranunculus* species in Giessen, Central Germany. Phyton 34: 203–218.
63. Coles SM (1973) *Ranunculus bulbosus* L. in Europe. Watsonia 9: 207–228.
64. Sarukhan J, Harper JL (1973) Studies on plant demography: *Ranunculus repens* L., *R. bulbosus* L. and *R. acris* L.: I. population flux and survivorship. Journal of Ecology 61: 675–716.
65. Dixon JM (2002) *Briza media* L. Journal of Ecology 90: 737–752.
66. Fischer SF, Poschlod P, Beinlich B (1996) Experimental studies on the dispersal of plants and animals on sheep in calcareous grasslands. Journal of Applied Ecology 33: 1206–1222.
67. Begert M, Seiz G, Schlegel T, Musa M, Baudraz G, et al. (2003) Homogenisierung von Klimamessreihen der Schweiz und Bestimmung der Normwerte 1961–1990: Schlussbericht des Projekts NORM90. Zürich, Switzerland: MeteoSchweiz. 170 p.
68. Bylesjoe M, Segura V, Soolanayakanahally RY, Rae AM, Trygg J, et al. (2008) LAMINA: a tool for rapid quantification of leaf size and shape parameters. BMC Plant Biology 8: 82–91.
69. Cornelissen JHC, Lavorel S, Garnier E, Díaz S, Buchmann N, et al. (2003) A handbook of protocols for standardised and easy measurement of plant functional traits worldwide. Australian Journal of Botany 51: 335–380.
70. R Development Core Team (2012) R: a language and environment for statistical computing. 2.15.0 ed. Vienna, Austria: R Foundation for Statistical Computing.
71. Faraway JJ (2005) Linear models with R. Boca Raton, United States: Chapman & Hall/CRC.
72. Coleman JS, McConnaughay KDM, Ackerly DD (1994) Interpreting phenotypic variation in plants. Trends in Ecology & Evolution 9: 187–191.
73. Roach DA, Wulff RD (1987) Maternal effects in plants. Annual Review of Ecology and Systematics 18: 209–235.
74. Kingsolver JG, Pfennig DW (2007) Patterns and power of phenotypic selection in nature. Bioscience 57: 561–572.
75. van Kleunen M, Fischer M (2001) Adaptive evolution of plastic foraging responses in a clonal plant. Ecology 82: 3309–3319.
76. Funk JL (2008) Differences in plasticity between invasive and native plants from a low resource environment. Journal of Ecology 96: 1162–1173.
77. Hahn MA, van Kleunen M, Muller-Scharer H (2012) Increased phenotypic plasticity to climate may have boosted the invasion success of polyploid *Centaurea stoebe*. PLOS ONE 7: e50284.
78. Pollock CJ (1990) The response of plants to temperature change. Journal of Agricultural Science 115: 1–5.

79. Arft AM, Walker MD, Gurevitch J, Alatalo JM, Bret-Harte MS, et al. (1999) Responses of tundra plants to experimental warming: meta-analysis of the international tundra experiment. Ecological Monographs 69: 491–511.
80. Elmendorf SC, Henry GHR, Hollister RD, Björk RG, Bjorkman AD, et al. (2012) Global assessment of experimental climate warming on tundra vegetation: heterogeneity over space and time. Ecology Letters 15: 164–175.
81. Roughley RJ, Dart PJ (1970) Growth of *Trifolium subterraneum* L. selected for sparse and abundant nodulation as affected by root temperature and *Rhizobium* strain. Journal of Experimental Botany 21: 776–786.
82. Warwick KR, Taylor G, Blum H (1998) Biomass and compositional changes occur in chalk grassland turves exposed to elevated CO^2 for two seasons in FACE. Global Change Biology 4: 375–385.
83. Molau U (1997) Responses to natural climatic variation and experimental warming in two tundra plant species with contrasting life forms: *Cassiope tetragona* and *Ranunculus glacialis*. Global Change Biology 3: 97–107.
84. Diggle PK (1997) Extreme preformation in alpine *Polygonum viviparum*: an architectural and developmental analysis. American Journal of Botany 84: 154–169.
85. Pulido F, Berthold P (2004) Microevolutionary response to climatic change. In: Moller AP, Fielder W, Berthold P, editors. Birds and Climate Change. Oxford, United Kingdom: Elsevier. 151–183.
86. Etterson JR, Shaw RG (2001) Constraint to adaptive evolution in response to global warming. Science 294: 151–154.
87. Visser ME (2008) Keeping up with a warming world; assessing the rate of adaptation to climate change. Proceedings of the Royal Society B-Biological Sciences 275: 649–659.
88. Bürger R, Lynch M (1995) Evolution and extinction in a changing environment - a quantitative-genetic analysis. Evolution 49: 151–163.
89. Whitney KD, Gabler CA (2008) Rapid evolution in introduced species, 'invasive traits' and recipient communities: challenges for predicting invasive potential. Diversity and Distributions 14: 569–580.
90. Didiano TJ, Turley NE, Everwand G, Schaefer H, Crawley MJ, et al. (2014) Experimental test of plant defense evolution in four species using long-term rabbit exclosures. Journal of Ecology 102: 584–594.
91. Menzel A (2003) Plant phenological anomalies in Germany and their relation to air temperature and NAO. Climatic Change 57: 243–263.
92. Parmesan C (2007) Influences of species, latitudes and methodologies on estimates of phenological response to global warming. Global Change Biology 13: 1860–1872.
93. Wilczek AM, Roe JL, Knapp MC, Cooper MD, Lopez-Gallego C, et al. (2009) Effects of genetic perturbation on seasonal life history plasticity. Science 323: 930–934.
94. De Frenne P, Brunet J, Shevtsova A, Kolb A, Graae BJ, et al. (2011) Temperature effects on forest herbs assessed by warming and transplant experiments along a latitudinal gradient. Global Change Biology 17: 3240–3253.
95. Donnelly A, Caffarra A, Kelleher CT, O'Neill BF, Diskin E, et al. (2012) Surviving in a warmer world: environmental and genetic responses. Climate Research 53: 245–262.
96. Inouye DW (2008) Effects of climate change on phenology, frost damage, and floral abundance of montane wildflowers. Ecology 89: 353–362.
97. De Valpine P, Harte J (2001) Plant responses to experimental warming in a montane meadow. Ecology 82: 637–648.
98. Thomson JD (2010) Flowering phenology, fruiting success and progressive deterioration of pollination in an early-flowering geophyte. Philosophical Transactions of the Royal Society B-Biological Sciences 365: 3187–3199.
99. Memmott J, Craze PG, Waser NM, Price MV (2007) Global warming and the disruption of plant-pollinator interactions. Ecology Letters 10: 710–717.
100. Hegland SJ, Nielsen A, Lazaro A, Bjerknes AL, Totland O (2009) How does climate warming affect plant-pollinator interactions? Ecology Letters 12: 184–195.
101. Hahn T, Kettle CJ, Ghazoul J, Frei ER, Matter P, et al. (2012) Patterns of genetic variation across altitude in three plant species of semi-dry grasslands. PLOS ONE 7: e41608.
102. Weiner J, Martinez S, Müller-Schärer H, Stoll P, Schmid B (1997) How important are environmental maternal effects in plants? A study with *Centaurea maculosa*. Journal of Ecology 85: 133–142.
103. Vitasse Y, Hoch G, Randin CF, Lenz A, Kollas C, et al. (2013) Elevational adaptation and plasticity in seedling phenology of temperate deciduous tree species. Oecologia 171: 663–678.
104. Vitasse Y, Lenz A, Kollas C, Randin CF, Hoch G, et al. (2014) Genetic vs. non-genetic responses of leaf morphology and growth to elevation in temperate tree species. Functional Ecology 28: 243–252.
105. Goldstein G, Rada F, Azocar A (1985) Cold hardiness and supercooling along in altitudinal gradient in Andean giant rosette species. Oecologia 68: 147–152.
106. Grime JP, Crick JC, Rincon JE (1986) The ecological significance of plasticity. In: Jennings DH, Trewavas AJ, editors. Plasticity in plants. Cambridge, United Kingdom: Company of Biologists. 5–29.
107. Parsons AN, Press MC, Wookey PA, Welker JM, Robinson CH, et al. (1995) Growth responses of *Calamagrostis lapponica* to simulated environmental change in the Sub Arctic. Oikos 72: 61–66.
108. Welker JM, Molau U, Parsons AN, Robinson CH, Wookey PA (1997) Responses of *Dryas octopetala* to ITEX environmental manipulations: a synthesis with circumpolar comparisons. Global Change Biology 3: 61–73.
109. Weiner J (1988) The influence of competition on plant reproduction. In: Lovett Doust J, Lovett Doust L, editors. Plant reproductive ecology: patterns and strategies. New York, United States: Oxford University Press. 228–245.
110. Kingsolver JG, Hoekstra HE, Hoekstra JM, Berrigan D, Vignieri SN, et al. (2001) The strength of phenotypic selection in natural populations. American Naturalist 157: 245–261.
111. Sultan SE (2000) Phenotypic plasticity for plant development, function and life history. Trends in Plant Science 5: 537–542.
112. Rausher MD (1992) The measurement of selection on quantitative traits - biases due to environmental covariances between traits and fitness. Evolution 46: 616–626.
113. Richards CL, Bossdorf O, Muth NZ, Gurevitch J, Pigliucci M (2006) Jack of all trades, master of some? On the role of phenotypic plasticity in plant invasions. Ecology Letters 9: 981–993.

Waterfowl Conservation in the US Prairie Pothole Region: Confronting the Complexities of Climate Change

Neal D. Niemuth[1]*, Kathleen K. Fleming[2], Ronald E. Reynolds¶[1]

1 United States Fish and Wildlife Service Habitat and Population Evaluation Team, Bismarck, North Dakota, United States of America, 2 United States Fish and Wildlife Service Division of Migratory Bird Management, Laurel, Maryland, United States of America

Abstract

The Prairie Pothole Region (PPR) is the most important waterfowl production area in North America. However, waterfowl populations there are predicted to decline because of climate-related drying of wetlands. Consequently, changes in the geographic focus of PPR waterfowl conservation have been recommended, which could have long-lasting and costly impacts. We used a 40-year dataset of pond counts collected in the PPR to test hypotheses about climate-related drying. We assessed May (1974–2013) and July (1974–2003) pond numbers in 20 waterfowl survey strata to determine if trends in pond numbers were consistent with predictions of drying. We also assessed trends in precipitation and temperature for the 20 strata and developed models describing May pond numbers from 1974 through 2010 as a function of precipitation, temperature, the previous year's pond numbers, and location. None of the 20 strata showed significant declines in May pond numbers, although seven strata showed increases over time. July pond numbers declined significantly in one stratum, and increased in seven strata. An index to hydroperiod showed significant increasing trends in three strata, and no strata had decreasing trends. Precipitation increased significantly in two strata and decreased in two from 1974 to 2010; no strata showed significant changes in temperature. The best linear model described pond numbers within all strata as a function of precipitation, temperature, the previous year's pond numbers, and the latitude and longitude of the stratum, and explained 62% of annual variation in pond numbers. We hypothesize that direct effects of climate change on prairie pothole wetlands and waterfowl may be overshadowed by indirect effects such as intensified land use and increased pressure to drain wetlands. We recommend that an adaptive, data-driven approach be used to resolve uncertainties regarding direct and indirect effects of climate change on prairie wetlands and waterfowl, and guide future conservation efforts.

Editor: Ben Bond-Lamberty, DOE Pacific Northwest National Laboratory, United States of America

Funding: The authors have no support or funding to report.

Competing Interests: The authors have declared that no competing interests exist.

* Email: Neal_Niemuth@fws.gov

¶ Retired.

Introduction

The millions of small wetlands and associated grassland nesting habitat of the Prairie Pothole Region (PPR; Figure 1) make it the most important area for waterfowl production in North America [1,2]. Consequently, the PPR is the focus of conservation programs that, in the United States alone, have permanently conserved >1.8 million ha of grasslands and wetlands (Table S1). Conservation efforts in the region are extensive and ongoing, as an additional 4.7 million ha of priority wetlands and grasslands must be protected to maintain waterfowl population goals in the US portion of the PPR as grasslands and wetlands continue to be converted to cropland [3,4].

Prairie pothole wetlands are extremely productive because their shallow waters warm early in spring and their dynamic nature facilitates nutrient cycling and regeneration of vegetation and associated macro-invertebrates [5–7]. Waterfowl population size (Figure S1), nesting propensity, clutch size, and brood presence are positively related to wetland numbers [8–11], which are highly variable among years [12,13]. Consequently, declines in the number and distribution of wetland basins containing water during the breeding season would reduce the ability of the PPR to attract and produce waterfowl.

Temperature and precipitation have generally increased in the PPR since the early to mid-1900s [14–16], although patterns differ between measures and among regions, time frames, and studies. Wetlands in the PPR may be particularly vulnerable to drying caused by increased temperatures associated with climate change because of their tenuous water balance and dynamic nature. Evaporation exceeds precipitation across much of the PPR, with conditions ranging from a positive water balance or lowest deficits in Iowa, where annual precipitation is highest, to highest deficits in northern Montana, southeastern Alberta, and southwestern Saskatchewan, although this gradient varies over space and time [17,18]. Atmosphere-ocean general circulation models project that future temperatures and precipitation in the PPR will be higher than historic levels [19]. Based on simulation models developed to assess potential effects of these climate projections on prairie wetlands, the PPR is forecast to experience "increased drought conditions…under nearly all global circulation model scenarios" ([20]:864), with consequences for waterfowl predicted to be negative due to drier conditions in the western and central PPR but positive in the eastern PPR, which is expected to become

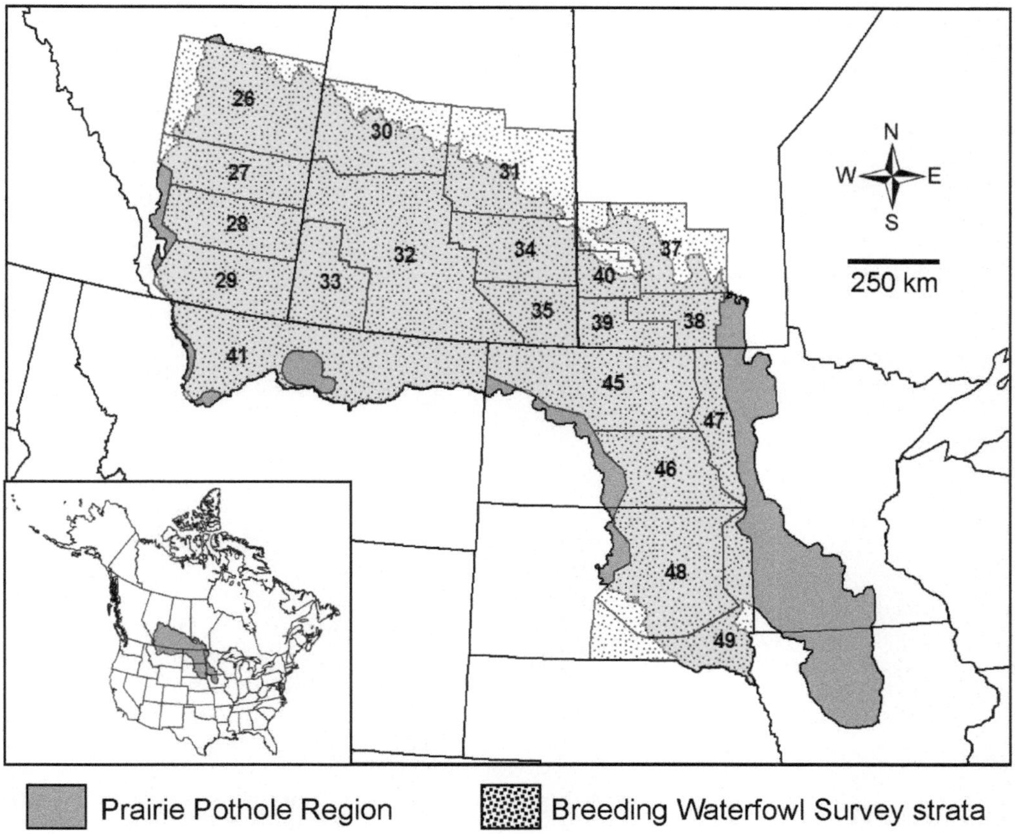

Figure 1. Location of North American Prairie Pothole Region. Twenty survey strata from the Waterfowl Breeding Population and Habitat Survey cover most of the land area, wetlands, and waterfowl resources of the Prairie Pothole Region. The survey strata extend from 96.5°W 42.5°N in the southeast to 114.8°W 54.0°N in the northwest.

wetter [20,21]. Consequently, a significant recommendation from studies that have addressed potential effects of climate change on wetlands and waterfowl conservation in the PPR is to shift conservation efforts from the central and western PPR (primarily Saskatchewan, North Dakota, and South Dakota) to the eastern portion of the PPR (primarily Minnesota and Iowa) [20,22]. However, in the short term, this shift may be costly: land protection and acquisition is more expensive in the eastern PPR, leading to less conservation "bang for the buck." Additionally, conservation efforts in the PPR will face stronger competition from intensification of agriculture, which has been identified as "the most dominant factor affecting the distribution, abundance, and reproductive success of the region's ducks" ([2]: 222; also [3,23]). Land use is important to waterfowl, as the nesting success and brood presence of upland-nesting species, which comprise the majority of waterfowl in the PPR, are positively related to the amount of grass in the landscape [11,24–27].

Given the magnitude of conservation efforts in the PPR (Table S1) and the potential impacts of climate change and other stressors on waterfowl and wetland-dependent species, recommendations to shift conservation efforts to the eastern PPR need to be carefully considered in terms of their short- and long-term consequences. Substantial uncertainty exists in our understanding of the complex interactions between climate, biological systems, socio-economic factors, and conservation costs. For example, how increased temperature and precipitation will affect wetland numbers and hydroperiod across the PPR is not easily predicted because of our limited understanding of hydrologic processes associated with widely varying precipitation/evapotranspiration ratios, geology, soil characteristics, and anthropogenic modifications [13,28,29].

We used 40 years of pond data collected during waterfowl surveys, along with broad-scale precipitation and temperature data, to investigate temporal and spatial trends in wetland numbers and hydroperiod and their relationship to precipitation and temperature trends in 20 waterfowl survey strata in the PPR. We sought to determine if trends in observed wetland numbers during this period were consistent with the decrease in wetland numbers predicted by published wetland models. In addition, we modeled wetland numbers as a function of precipitation and temperature to determine if broad-scale weather data could explain annual variation in wetland numbers. We discuss how the change in climate predicted for the PPR might interact with land use and other stressors to influence waterfowl ecology, and implications for the cost of effective conservation in the PPR. Because of considerable uncertainty and conflicting information related to climate change in the PPR, we also provide suggestions for reducing uncertainty relative to understanding the effects of climate change on waterfowl and conservation. We concentrate on waterfowl because they have been the primary focus for conservation and research related to climate change in the PPR, but our observations hold true for many other taxa. We focus our discussion on the US portion of the PPR, as conservation and agricultural programs and policies, as well as data availability, differ for Canada.

Methods

Trends in Wetland Numbers Over Time

Each May, biologists from the United States Fish and Wildlife Service and the Canadian Wildlife Service conduct aerial and ground surveys of wetlands and breeding waterfowl across major waterfowl-producing areas of the United States and Canada [30,31]. Surveys are conducted along multiple transects within geographically distinct, non-overlapping strata, which were identified by delineating areas of similar habitat and duck densities. Aerial crews count the number of artificial (e.g., stockdams, dugouts) and natural ponds, which are then corrected for visibility using data collected by crews on the ground. Only ponds large enough to support waterfowl are counted. Using data from the US Fish and Wildlife Service Migratory Bird Data Center (https://migbirdapps.fws.gov/mbdc/databases/db_selection.asp), we analyzed numbers of May ponds for waterfowl survey strata entirely or largely within the PPR, an area that spans >1,800 km from central Alberta to southeastern South Dakota ($n = 20$; Figure 1). Because long-term populations of breeding waterfowl in the PPR are strongly related to May pond numbers, we assessed temporal trends in numbers of May ponds for each stratum by regressing annual May pond numbers against year. Given the highly variable nature of climate in the PPR, we focused on long-term trends and did not assess short-term cyclical or non-linear patterns in wetland numbers. We used data from 1974 to 2013, which includes all May pond survey data that were corrected for visibility [30]. The same methods were used throughout the survey period and we treat estimates as an index, rather than a census, of wetland numbers.

Because the number of May ponds differed substantially among strata, we were concerned that slope estimates might not be comparable among strata, i.e., the change in ponds per unit of time would be proportional to the number of ponds in each stratum. Therefore, we compared slope estimates for untransformed, standardized, and log–log regression models. Slopes for standardized and log–log models were strongly ($r^2 = 0.64$ and 0.63, respectively) and linearly related to slopes estimated using untransformed data, so we present results for models developed using untransformed data for ease of interpretation.

We also analyzed trends in July pond data, which were collected during the July Production Survey along the same transects as the May survey until 2003, when the July survey was discontinued. Unlike the May pond count, the July survey was not visibility corrected, but because transects were sampled by the same aerial crews as in May, data from the survey provide a relative index to pond conditions during nesting and brood-rearing periods. We estimated regression slopes of pond estimates on year for each stratum ($n = 20$) using July pond data from 1974 to 2003. In order to analyze the May pond counts together with July counts, we converted May pond counts into uncorrected counts by dividing by the visibility correction factor [30]. We also used the percent change in ponds from May to July, (July ponds − May ponds) × 100/May ponds, as an index of hydroperiod length to test the hypothesis that, in a changing climate, increased evapotranspiration caused by higher summer temperatures would shorten wetland hydroperiods, reducing the number of July ponds available for duck broods. Early drying of ponds to which waterfowl have been attracted has been suggested to be an ecological trap [21]. We used the percent change in pond counts, rather than the difference, to minimize differences in the magnitude of counts among strata. We regressed the percent change in ponds on year for the period 1974–2003 for each stratum in the PPR ($n = 20$).

Climatic Trends and Factors Influencing Wetland Numbers

We acquired spatial data for monthly total rain gauge precipitation (mm) and monthly means of air temperature (degrees Celsius) interpolated from and cross-validated with weather station data [32,33] from http://climate.geog.udel.edu/~climate/html_pages/download.htm. The spatial resolution of these data was 0.5 degree (approximately 50 km). To assign data to individual waterfowl survey segments, we resampled the dataset to a higher resolution (0.005 degree, or 1/100 of the original pixel dimensions), and calculated mean temperature and precipitation of pixels falling within the surveyed area of each segment (200 m on each side of the segment) using Environmental Systems Research Institute's (ESRI) ArcGIS. For trend analysis and pond models the climate variables were the mean precipitation and temperature summarized over all segments within a stratum over the preceding year (May–April). As with the pond data, we used simple linear regression to assess trends in each stratum's mean precipitation and temperature. Because precipitation and temperature data were only available for 1974–2010, time periods for these variables did not coincide completely with pond counts. We also used multiple linear regression to model May pond numbers as a function of temperature and precipitation for all strata combined. In the full model we included the previous year's May pond numbers to account for temporal autocorrelation resulting from holdover in water conditions [34,35], as well as a temperature*precipitation interaction term. Because of concerns about lagged dependent variables biasing other variables toward zero, we compared coefficients of variables in models with and without the previous year's pond numbers. We standardized temperature and precipitation variables and the previous year's pond numbers to facilitate comparison of model coefficients, and included the latitude and longitude of each stratum centroid as predictors to account for spatial patterns in pond counts that were not explained by spatial variation in climate. We discriminated among a set of models (full and reduced) using Akaike's Information Criterion (AIC [36]) to identify the best model.

Because all data we used were collected remotely, no permits or guidelines for land access, handling of protected species, or animal husbandry were necessary. We conducted statistical analyses in the R environment and with Number Cruncher Statistical System 7.1 [37,38].

Results

Trends in Wetland Numbers Over Time

No strata showed statistically significant ($p<0.05$) declines in number of May ponds; however, seven strata showed significant increases (Figure 2). Inter-annual variation was high and cyclic patterns were sometimes evident; consequently, maximum model fit (coefficient of determination, r^2) for any stratum was 0.27 (Figure 3). The pattern in May pond numbers suggested a north–south gradient in trend, with strata in the southern portion of the PPR exhibiting significant increases in numbers of May ponds and strata in the northern portion of the PPR exhibiting non-significant increases or decreases (Figure 2).

Seven strata (38, 40, 45, 46, 47, 48, 49) showed significant increasing trends in July ponds over time, and 1 stratum (26) showed a significant decreasing trend (Figure 4). The maximum r^2 of these regression models was 0.39, again due to substantial inter-annual variation. A spatial pattern similar to that of the May pond counts existed in the trend in July pond counts: increasing trends were primarily found in strata in the eastern and southern portion

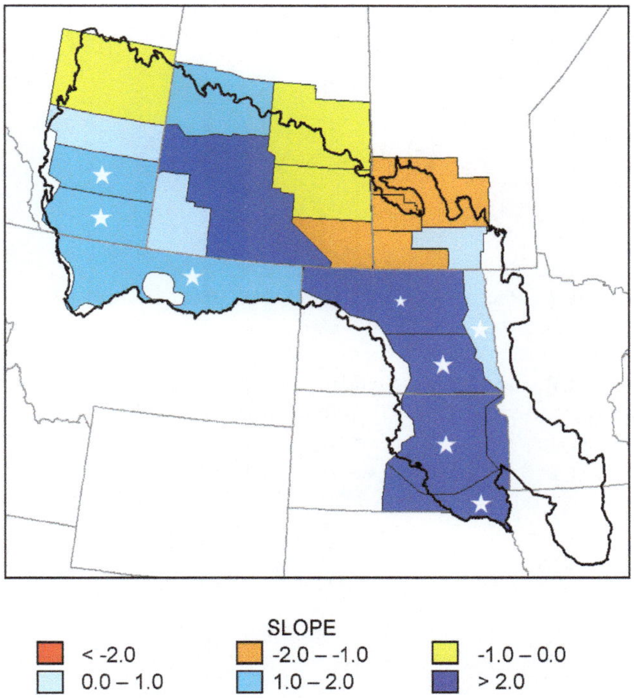

Figure 2. Trends in May pond numbers for 20 waterfowl survey strata, 1974–2013. Slope of regression models of May pond numbers (in thousands per year) in 20 strata of the Waterfowl Breeding Population and Habitat Survey as a function of year, 1974–2013. Large white stars indicate statistically significant trends ($p<0.05$); small white star indicates $0.05<p<0.1$.

of the PPR, while the stratum with a significant decreasing trend was located in the northwest corner of the region.

The percent change in ponds from May to July exhibited a significant increasing trend in 3 strata located in the northcentral portion of the PPR (31, 34, and 40); all other trends were not statistically significant (Figure 5). The maximum r^2 for these models was 0.30.

Climatic Trends and Factors Influencing Pond Numbers

Mean monthly total precipitation for the period 1974–2010 showed statistically significant increasing trends in 2 strata in the southeastern portion of the PPR and significant decreasing trends in 2 strata in the western portion of the PPR (Figure 6). None of the strata showed significant trends in temperature during the same time period, although fitted slopes for 18 of 20 strata were positive (Figure 7).

Using data from all strata combined, the best model predicting pond numbers contained all the predictor variables except the *ppt*temp* interaction term. All coefficients in this model were statistically significant ($P<0.0001$; adjusted $r^2=0.62$). However, the full model (with the interaction term) also ranked highly ($\Delta AIC=0.832$), with model weight 39.7%, which suggested that the full model also had a relatively high probability of being the best approximating model. Therefore, we calculated model-averaged parameter estimates [36], resulting in a final model with the form:

$$n=4344012+(134788*n_{t-1})+(62784*ppt)-(74513*temp)$$
$$-(10129*long)-(23660*lat)-(1932*temp*ppt)$$

where n is the number of ponds; n_{t-1} is the standardized previous year's number of ponds; *ppt* is the standardized mean monthly accumulated precipitation for the previous year, measured in mm; *temp* is the standardized previous year's mean air temperature, measured in degrees Celsius; *long* is the longitude of the stratum's centroid; and *lat* is the latitude of the stratum's centroid. All other models had ΔAIC values >137, so they were not considered competitive [36]. Although the lagged pond variable (previous year's pond numbers) was a stronger predictor in the highest ranking model than the climate variables we included, the coefficients of the precipitation (52744) and temperature (−59009) variables in the full model (but with the previous year's pond variable removed) were the same sign and slightly decreased in magnitude, indicating that these factors were contributing significant information in the current year's pond numbers not explained by a temporal trend in pond numbers and were not biased toward zero by including the previous year's pond numbers.

Discussion

Climate and Pond Trends

Increased numbers of May and July ponds over the 1974–2013 period (1974–2003 for July ponds) are not consistent with predictions of wetland drying [20,21,35,39]. Trends in the change in pond counts from May to July suggest that, in at least the central portion of the PPR, the predicted shortening of hydroperiods due to increased evapotranspiration has not been occurring during this period. We detected increasing trends in precipitation in the southeastern PPR, which is consistent with the location of strata showing greatest increases in May and July pond numbers; the southeastern PPR was the only area to show a decreasing trend in temperature, although this trend, like increases in the rest of the PPR, was not statistically significant. Our results are consistent with wetland simulation models that suggest that increased precipitation can offset some effects of temperature on PPR wetlands [20,34,35]. Although our analysis used only 40 years of wetland data, which is too short a period to support or refute predictions of climate change, the trends we observed are certainly occurring within a time frame relevant to conservation planning.

Our results suggest that the primary conservation strategy in the US PPR of protecting waterfowl breeding habitat in areas of high waterfowl density is not presently jeopardized by drying of wetlands. Pond and climate data both suggest a spatial pattern of more water in the central PPR from southern South Dakota through Saskatchewan, although patterns varied with metric and time period. Of course, effects of climate change may be manifested in ways other than linear trends in numbers of ponds, such as future shifts in migration and breeding phenology of wetland-dependent birds or changes in the variation in wetness that typifies prairie potholes and makes them so productive. Nevertheless, trends in pond numbers indicate that, during the past 40 years, increases in the amount of precipitation have been sufficient to offset predicted effects of climate change on numbers of May and July ponds in much of the PPR. However, it is possible that areas of high conservation priority in the Dakotas might become drier in the future as climatic conditions change.

Pond Models

Average daily temperature and total accumulated precipitation were statistically significant correlates of May pond numbers that, combined in a simple model with relatively coarse spatial and temporal resolution, explained 62% of variation in pond numbers. This suggests that, if current (i.e., past 37 years) observed trends in

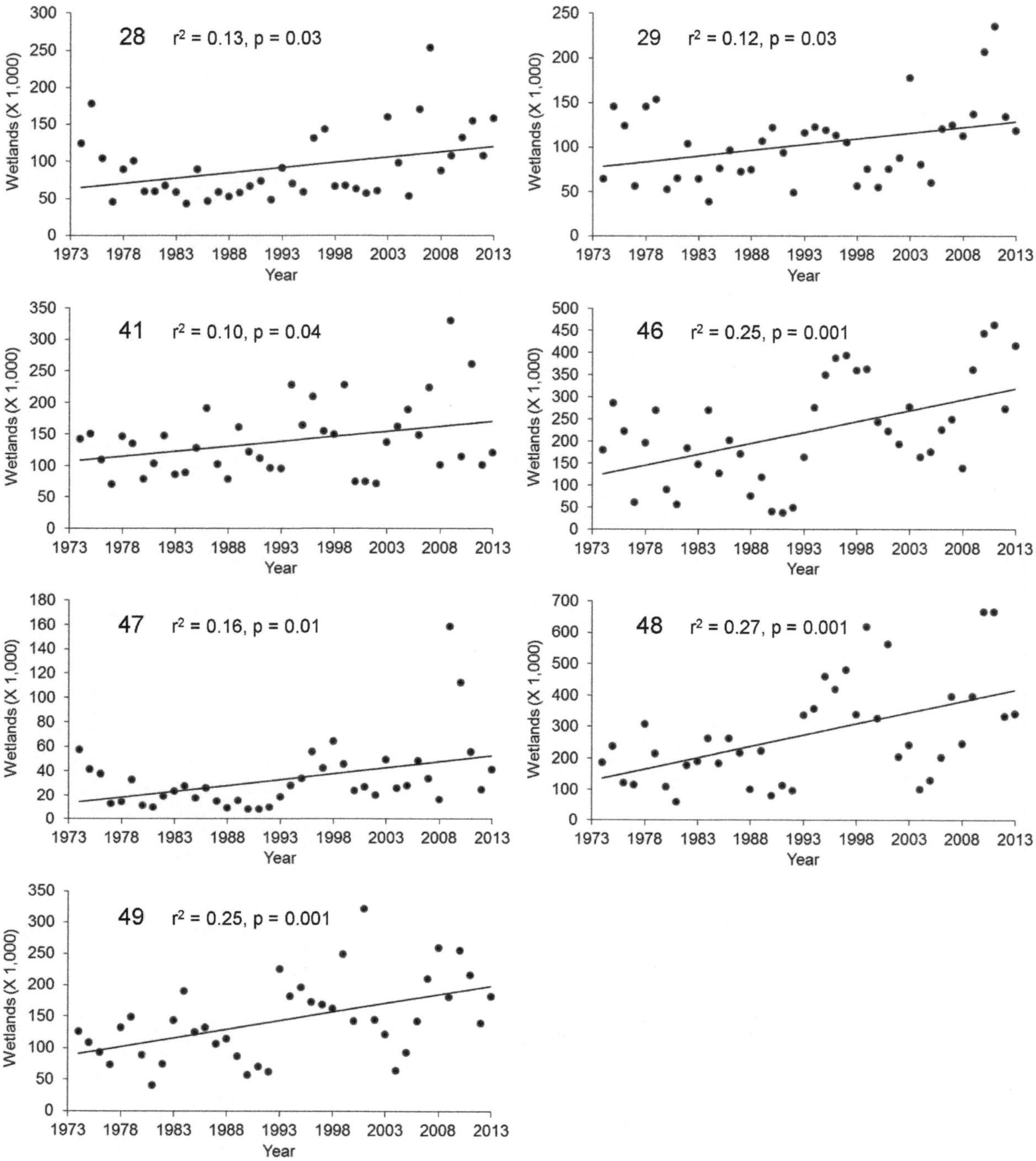

Figure 3. Trends in May pond numbers for seven waterfowl survey strata showing significant increases, 1974–2013. Number of May ponds (in thousands per year) increased significantly ($p<0.05$) in waterfowl survey strata 28, 29, 41, 46, 47, 48, and 49 of the Waterfowl Breeding Population and Habitat Survey, 1974–2013.

climate continue, pond numbers will respond in a similar manner in the near future. Of course, this does not imply that climate change is not happening in the PPR. Historical records suggest that the climate of the PPR is warming [14–16]; while we did not detect significant trends in temperature in the relatively short time period of our analysis, our pond model does predict that increasing temperature will result in fewer ponds. But in the short term, we propose that other stressors are having a much more pronounced impact on waterfowl populations in the PPR than climate-related drying of wetlands.

It is important to note that the short-term trends we report here, as well as the predicted long-term trends in climate change, do not capture the substantial cyclical and seasonal variability in wetland dynamics and climate in the PPR, which have a direct impact on waterfowl habitat quality. Coarse-scale climate data obscure much of the fine-scale variation in temperature and precipitation which influence wetland dynamics in the PPR. Variable hydrology may well be the defining characteristic of wetlands in the PPR because pond permanence and oscillating water levels are major drivers of ecological function of wetlands, influencing primary productivity,

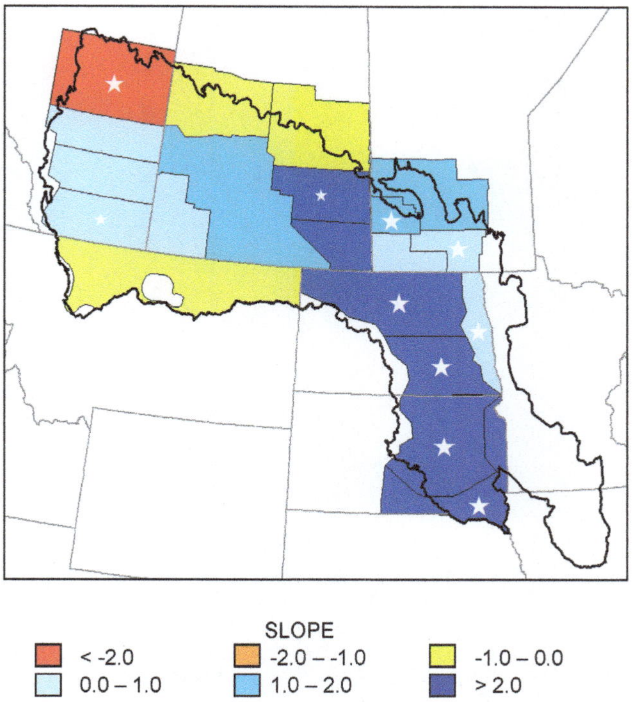

Figure 4. Trends in July pond numbers for 20 waterfowl survey strata, 1974–2003. Slope of regression models of July pond counts (in thousands per year) in 20 strata of the Waterfowl Breeding Population and Habitat Survey as a function of year, 1974–2003. Large white stars indicate statistically significant trends ($p<0.05$); small white stars indicate $0.05<p<0.1$.

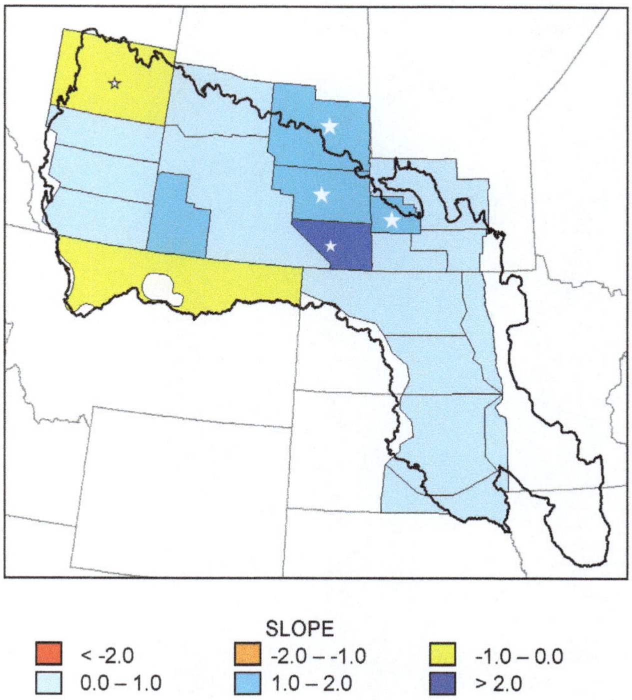

Figure 5. Trends in hydroperiod index for 20 waterfowl survey strata, 1974–2003. Slope of regression models of percent change in pond counts per year from May to July in 20 strata of the Waterfowl Breeding Population and Habitat Survey as a function of year (1974–2003). Large white stars indicate statistically significant trends ($p<0.05$); small white stars indicate $0.05<p<0.1$.

water salinity, nutrient cycling, invertebrate communities, composition and configuration of emergent vegetation, and wildlife [6–8,40–43]. At some point, wetland productivity may decrease if the wetter, eastern portion of the PPR becomes wetter still and wetland water levels stay at unusually high levels. Wetlands that do not dry out also are more likely to harbor fish, which, whether native or introduced, can influence availability of aquatic invertebrates and ultimately reduce growth rates and survival of ducklings [44,45].

Shifting the Geographic Focus: Implications to Waterfowl Conservation in the US PPR

The current conservation paradigm in the PPR calls for the establishment of a diverse portfolio of land protection and restoration to account for temporal dynamics and spatial configuration of multiple habitat components needed by waterfowl [3,46]. Changing the geographic focus of waterfowl conservation is more difficult than simply shifting the location of conservation efforts east to areas projected to have more water in the future. Waterfowl are necessarily associated with wetlands, but the majority of waterfowl in the PPR nest in adjacent uplands. Avian reproductive success is highly variable, but nesting success of many species of grassland-nesting birds, including waterfowl, typically increases with the amount of grass in the landscape due to reductions in predator densities associated with food crops and grassland fragmentation [26,47]. Waterfowl conservation benefits are known to be higher in areas with large, contiguous grasslands relative to areas dominated by row crops [48]. Therefore, efforts to conserve waterfowl in the eastern "corn belt" portion of the PPR, an area predicted to be warmer and wetter due to climate change, will likely be less productive because less grass is present in the landscape relative to other portions of the US PPR [48, Figure S2]. In addition, the amount of grassland in the PPR could decrease further if increased precipitation and warmer temperatures lead to intensification of farming in the region [4,15,49–51, Figure S3].

A second problem with shifting waterfowl conservation efforts to the eastern PPR is that large numbers of wetland basins in the eastern PPR would need to be restored [52, figures S4 and S5], which would cost several times that of conserving existing wetlands. For example, the cost of a simple restoration such as plugging a wetland drainage ditch exceeds USD $988 per ha, and removal of sediment that accumulated as a consequence of tillage and erosion is approximately $2,500 per ha (Scott McLeod, USFWS, personal communication). Using these values and average cost of cropland in each state from 2012 (http://www.nass.usda.gov/Charts_and_Maps/Land_Values_and_Cash_Rents/crop_value_map.asp), the cost of purchasing land in fee title and restoring wetlands would be approximately $21,500 per ha in Iowa and $13,500 in Minnesota. Based on the purchase of a total of 24,450 ha of easements from 1 October 2011 to 30 September 2012, the mean cost per ha of perpetual wetland easements and perpetual grassland easements was $2,372 and $714, respectively, in North Dakota (Tammy Fairbanks, USFWS, personal communication) and $1,305 and $1,080 in South Dakota (William Mulvaney, USFWS, personal communication). Consequently, given the same resources, substantially less land could be conserved in the eastern portion of the PPR than in the western portion [53].

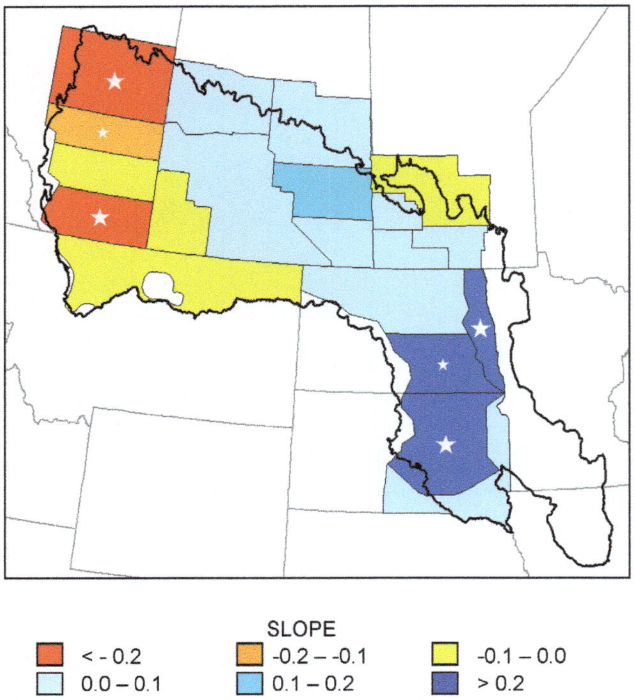

Figure 6. Trends in precipitation for 20 waterfowl survey strata, 1974–2010. Slope of trends in mean monthly total precipitation (mm per year) from 1974 to 2010 in 20 strata of the Waterfowl Breeding Population and Habitat Survey in the Prairie Pothole Region of the US and Canada. Data acquired from http://climate.geog.udel.edu/~climate/html_pages/download.htm [32]. White stars indicate statistically significant trends ($p<0.05$).

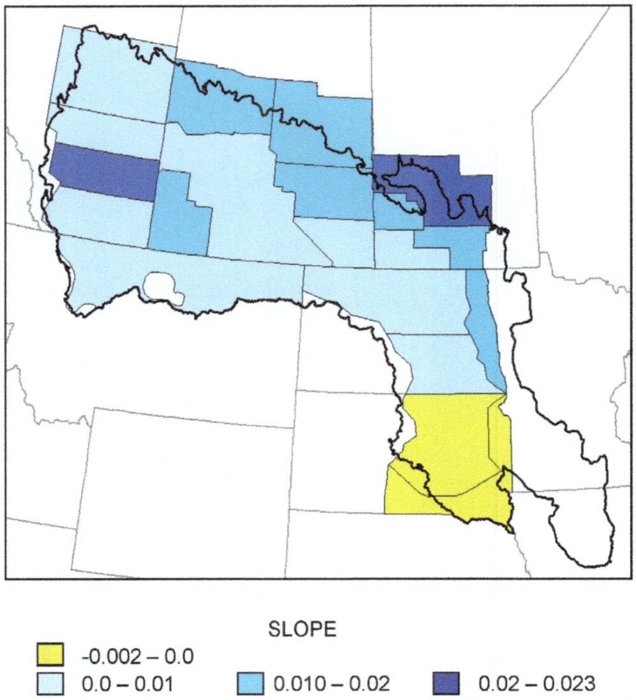

Figure 7. Trends in temperature for 20 waterfowl survey strata, 1974–2010. Slope of trends in mean monthly air temperature (degrees Celsius per year) from 1974 to 2010 in 20 strata of the Waterfowl Breeding Population and Habitat Survey in the Prairie Pothole Region of the US and Canada. Data acquired from http://climate.geog.udel.edu/~climate/html_pages/download.htm [33].

Reducing Uncertainty Relative to Climate Change in the PPR

Given the extent and diversity of the PPR, discrepancies between long-term projections and short-term observed trends in pond numbers, and the complexity of conservation in the region, we present several issues that should be addressed to reduce uncertainty relative to both direct and indirect effects of climate change in the PPR.

First, we need to better understand wetland dynamics and their relationship to climate, in both short and long-term temporal scales. Inferences about how wetlands respond to changes in temperature and precipitation, and their interaction, should be based on data from samples representing a broad range of wetland types and conditions. Most of the long-term wetland data from which PPR hydrologic models have been developed come from wetlands in managed wildlife areas that are rarely subject to anthropogenic disturbance, such as tillage agriculture. This limits the inferences that can be made from these data, illustrating the observation of Felton et al. ([54]:2244) that "the propensity of ecologists to work in essentially unmodified ecosystems may fundamentally hamper our ability to make useful recommendation in a world that is experiencing global change." An appropriate sampling framework is equally important to collecting data from which reliable inferences can be made. For example, 65% of the 3.25 million wetland basins identified by the National Wetlands Inventory in the PPR of North Dakota, South Dakota, Minnesota, and Iowa either touch or are surrounded by cultivated fields as identified through classified satellite imagery (Chuck Loesch and Rex Johnson, USFWS, unpublished data); data are needed to understand how these wetlands–which comprise such a large portion of the landscape–respond to future changes.

Second, collecting these data will require a new monitoring effort in the PPR, involving expansion of the extent or intensity of existing surveys, or concerted integration of multiple surveys and data sources. Open water is easily classified using remotely sensed, multi-spectral imagery, but the spatial resolution of remotely sensed data must be sufficiently fine to detect small wetlands, which dominate the PPR and are disproportionately important to waterfowl, as well as provide precise estimates of change in water area on small wetlands. In addition, correctly classifying water through emergent vegetation and collecting data during cloudy periods using remote sensing methods is problematic. Therefore, the sampling framework for future surveys must consider the full range of land use and physiographic variation across the PPR, with increased survey effort allocated to high-risk landscapes such as grasslands on fertile soils and wetlands in agricultural settings. Obtaining useful data may require new techniques or local collection, as classification error and thematic resolution of remotely sensed data will likely be insufficient to detect small changes typical of land surface change [55].

Finally, as conservationists working within a changing system, we need to embed this new monitoring effort within a context of adaptive decision making. Climate change will increase uncertainty in resource management [56]. Managers are faced with an optimization problem where the challenges of today must be confronted while considering the prospects of an uncertain tomorrow. Research needs to be focused on testing assumptions involved in our current decision making (in this case the allocation of conservation effort in regions of the PPR), investigating the interactions of climate and wetland dynamics in an anthropogenically modified system, and evaluating the relative threats of direct and indirect impacts of climate change.

Indirect effects of climate change that should be investigated include the relationship between increasing temperature and precipitation and land use intensification, specifically determining the rate of loss of native and tame grasslands, assessing changes in crop types, and determining the rate and type (e.g., tile versus ditch) of wetland loss. Other anthropogenic stressors that result from efforts to address climate change such as installation of wind turbines and planting of biofuel crops should also be monitored. Incorporating this information into our management framework allows us optimize the outcome of conservation decisions over both the short and long terms.

Conclusions

In an assessment of the most important determinants of changes in biodiversity at the global scale over the next 100 years, land-use change was expected to have a greater effect than climate change in both terrestrial and freshwater ecosystems [57]. Several lines of evidence suggest that this may also be the case in the PPR, where agriculture is the dominant land use and tillage agriculture, which has demonstrated negative impacts on waterfowl populations, is increasing. Climate change may be a new reality in portions of the PPR as temperature, precipitation, and pond numbers have changed from historic levels. However, the threat of drying PPR wetlands may be a conservation "red herring" in the short term, if resultant shifts in conservation efforts benefit fewer ducks at increased expense while habitat loss and other impacts–which also are influenced by climate change–are ignored. If current trends continue, agricultural intensification exacerbated by climate change will likely further increase in portions of the PPR that presently harbor the greatest numbers of waterfowl. Therefore, assessments of the effects of climate change on waterfowl conservation must fully consider ecological, economic, and social realities along with the potential for climate-induced changes to determine the most effective places for conservation.

More information is needed before substantial changes are made to conservation strategies in the US PPR, but this should not be construed to mean inaction; instead of reacting to scenario-based models we believe a better approach is to develop competing hypotheses about how climate change is impacting waterfowl in the PPR, monitor wetland change, land use, and waterfowl response using well-designed surveys, and respond adaptively to those effects supported by data.

Acknowledgments

We thank the many pilots and observers who have conducted May and July waterfowl surveys for many years; D. R. Hertel, R. R. Johnson, C. R. Loesch, and A. J. Ryba for providing or helping process spatial data; K. Hunter and S. Hutchcroft for providing data on protected lands; and E. Babij, B. Bond-Lamberty, G. Liu, R. R. Johnson, W. A. Meeks, K. D. Richkus, E. D. Silverman, C. L. Stemler, J. A. Walker, and one anonymous reviewer for helpful comments on earlier drafts of the manuscript. The findings and conclusions in this article are those of the authors and do not necessarily represent the views of the US Fish and Wildlife Service.

Supporting Information

Figure S1 Waterfowl populations are strongly related to wetland numbers. Annual population estimates for seven species of breeding dabbling ducks from the Waterfowl Breeding Population and Habitat Survey were strongly related to annual estimates of May pond numbers in the Canadian (upper) and US (lower) portions of the Prairie Pothole Region, 1974–2013. Data acquired from the US Fish and Wildlife Service Migratory Bird Data Center (https://migbirdapps.fws.gov/mbdc/databases/db_selection.asp).

Figure S2 Grassland in the US Prairie Pothole Region. The amount of grass in the landscape, which is positively associated with waterfowl nesting success, is generally correlated with a precipitation gradient that allows more intensive agriculture in the southeastern portion of the PPR. Grassland was defined as herbaceous cover, hay, and pasture cover classes identified by the 2006 National Landcover Database [58], available at www.mrlc.gov. We restricted the landscape portion of our analysis to the US because comparable data were not available for the PPR of Canada.

Figure S3 Changes in area of corn and soybeans harvested for grain, 1997–2007. Area of corn (Zea mays, upper) and soybeans (Glycine max, lower) harvested for grain increased substantially in the Prairie Pothole Region 1997–2007 relative to the rest of the conterminous United States, which showed decreases or small increases in area harvested per county. Data available at http://www.agcensus.usda.gov/index.php. We restricted the landscape portion of our analysis to the US because comparable data were not available for the PPR of Canada.

Figure S4 Wetland density in the US Prairie Pothole Region. Wetland density, which is positively associated with waterfowl settling and negatively influenced by agricultural development, is lowest in the southeastern portion of the US Prairie Pothole Region and is generally highest in central North Dakota and South Dakota. Wetland density derived from National Wetlands Inventory data [59] (available at http://www.fws.gov/wetlands/Data/DataDownload.html) processed to basins [60]. We restricted the landscape portion of our analysis to the US because comparable data were not available for the PPR of Canada.

Figure S5 Density and distribution of five priority duck species. Modeled density and distribution of five priority species of waterfowl (mallard [Anas platyrhynchos], gadwall [A. strepera], northern pintail [A. acuta], northern shoveler [A. clypeata], and blue-winged teal [A. discors]) in the US PPR are strongly related to wetland density. Figure derived following methodology of Reynolds et al. [27] and unpublished data courtesy of Sean P. Fields and Rex R. Johnson, USFWS.

Table S1 Hectares of land protected in fee title on National Wildlife Refuges (NWR) and Waterfowl Production Areas (WPA) or by perpetual wetland or grassland easement held by the US Fish and Wildlife Service in the Prairie Pothole Region of Montana, North Dakota, South Dakota, Minnesota, and Iowa.

Author Contributions

Conceived and designed the experiments: NN KF RR. Analyzed the data: NN KF. Wrote the paper: NN KF RR.

References

1. US Department of the Interior, Environment Canada (1986) North American Waterfowl Management Plan. Washington, DC: US Fish and Wildlife Service. 19 p.
2. Batt BDJ, Anderson MG, Anderson CD, Caswell FD (1989) Use of prairie potholes by North American ducks. In: van der Valk A, editor. Northern Prairie Wetlands. Ames: Iowa State University Press. 204–227.
3. Ringelman JK, editor (2005) Prairie Pothole Joint Venture 2005 Implementation Plan. Denver: Prairie Pothole Joint Venture.
4. Doherty KE, Ryba AJ, Stemler CL, Niemuth ND, Meeks WA (2013) Conservation planning in an era of change: state of the US Prairie Pothole Region. Wildlife Society Bulletin 37: 546–563.
5. Harris SW, Marshall WH (1963) Ecology of water-level manipulations on a northern marsh. Ecology 44: 331–343.
6. van der Valk AG, Davis CB (1978) The role of seed banks in the vegetation dynamics of prairie glacial marshes. Ecology 59: 322–335.
7. Murkin HR (1989) The basis for food chains in prairie wetlands. In van der Valk A, editor. Northern Prairie Wetlands. Ames: Iowa State University Press. 317–338.
8. Krapu GL, Klett AT, Jorde DG (1983) The effect of variable spring water conditions on mallard production. The Auk 100: 689–698.
9. Leitch WG, Kaminski RM (1985) Long-term wetland-waterfowl trends in Saskatchewan grassland. Journal of Wildlife Management 49: 212–222.
10. Pietz PJ, Krapu GL, Buhl DA, Brandt DA (2000) Effects of water conditions on clutch size, egg volume, and hatchling mass of mallards and gadwalls in the Prairie Pothole Region. The Condor 102: 936–940.
11. Walker J, Rotella JJ, Schmidt JH, Loesch CR, Reynolds RE et al. (2013) Distribution of duck broods relative to habitat characteristics in the Prairie Pothole Region. The Journal of Wildlife Management 77: 392–404.
12. Woodhouse CA, Overpeck JT (1998) 2000 years of drought variability in the central United States. Bulletin of the American Meteorological Society 79: 2693–2714.
13. Niemuth ND, Wangler B, Reynolds RE (2010) Spatial and temporal variation in wet area of wetlands in the Prairie Pothole Region of North Dakota and South Dakota. Wetlands 30: 1053–1064.
14. Zhang X, Vincent LA, Hogg WD, Niitsoo A (2000) Temperature and precipitation trends in Canada during the 20th Century. Atmosphere-Ocean 38: 395–429.
15. Karl TR, Melillo JM, Peterson TC, editors (2009) Global climate change impacts in the United States. Cambridge: Cambridge University Press. 192 p.
16. Millett B, Johnson WC, Guntenspergen G (2009) Climate trends of the North American prairie pothole region 1906–2000. Climatic Change 93: 243–267.
17. Winter TC (1989) Hydrologic studies of wetlands in the Northern Prairie. In van der Valk A, editor. Northern Prairie Wetlands. Ames: Iowa State University Press. 16–54.
18. Liu G, Schwartz FW (2012) Climate-driven variability in lake and wetland distribution across the Prairie Pothole Region: from modern observations to long-term reconstruction with space-for-time substitution. Water Resources Research 48. doi:10.1029/2011WR011539.
19. Meehl GA, Covey C, Delworth T, Latif M, McAvaney B et al. (2007) The WCRP CMIP3 multimodel dataset: a new era in climate change research. Bulletin of the American Meteorological Society 88: 1383–1394.
20. Johnson WC, Millett BV, Gilmanov T, Voldseth RA, Guntenspergen GR et al. (2005) Vulnerability of northern prairie wetlands to climate change. BioScience 55: 863–872.
21. Johnson WC, Werner B, Guntenspergen GR,Voldseth RA, Millett B et al. (2010) Prairie wetland complexes as landscape functional units in a changing climate. Bioscience 60: 128–140.
22. Ando AW, Mallory ML (2012) Optimal portfolio design to reduce climate-related conservation uncertainty in the Prairie Pothole Region. Proceedings of the National Academy of Sciences 109: 6484–6489.
23. Fargione JE, Cooper TR, Flaspohler DJ, Hill J, Lehman C et al. (2009) Bioenergy and wildlife: threats and opportunities for grassland conservation. BioScience 59: 767–777.
24. Klett A T, Shaffer TL, Johnson DH (1988) Duck nest success in the Prairie Pothole Region. Journal of Wildlife Management 52: 431–440.
25. Bethke RW, Nudds TD (1995) Effects of climate change and land use on duck abundance in Canadian prairie-parklands. Ecological Applications 5: 588–600.
26. Stephens SE, Rotella JJ, Lindberg MS, Taper ML, Ringelman JK (2005) Duck nest survival in the Missouri Coteau of North Dakota: landscape effects at multiple spatial scales. Ecological Applications 15: 2137–2149.
27. Reynolds RE, Shaffer TL, Loesch CR, Cox RR Jr. (2006) The Farm Bill and duck production in the Prairie Pothole Region: increasing the benefits. Wildlife Society Bulletin 34: 963–974.
28. Conly FM, van der Kamp G (2001) Monitoring the hydrology of Canadian prairie wetlands to detect the effect of climate change and land use changes. Environmental Monitoring and Assessment 67: 195–215.
29. Zhang H, Huang GH, Wang D, Zhang X (2011) Uncertainty assessment of climate change impacts on the hydrology of small prairie wetlands. Journal of Hydrology 396: 94–103.
30. Smith GW (1995) A critical review of the aerial and ground surveys of breeding waterfowl in North America. United States Department of Interior, National Biological Service Science Report 5. 252 p.
31. US Fish and Wildlife Service (2011) Waterfowl population status, 2011. Washington, DC: US Fish and Wildlife Service. 79 p.
32. Matsuura K, Willmot CJ (2012) Terrestrial precipitation: 1900–2010 gridded monthly time series version 3.02, Available: http://climate.geog.udel.edu/~climate/html_pages/Global2011/Precip_revised_3.02/README.GlobalTsP2011.html.
33. Matsuura K, Willmot CJ (2012) Terrestrial air temperature: 1900–2010 gridded monthly time series version 3.01, Available: http://climate.geog.udel.edu/~climate/html_pages/Global2011/README.GlobalTsT2011.html.
34. Larson DL (1995) Effects of climate on numbers of northern prairie wetlands. Climatic Change 30: 169–180.
35. Sorenson LG, Goldberg R, Root TL, Anderson MG (1998) Potential effects of global warming on waterfowl populations breeding in the northern Great Plains. Climatic Change 40: 343–369.
36. Burnham KP, Anderson DR. (2002) Model selection and multimodel inference: a practical information-theoretic approach. Springer-Verlag, New York, New York, USA.
37. R Core Team (2013) R: A language and environment for statistical computing. R Foundation for Statistical Computing, Vienna, Austria. Available: http://www.R-project.org/.
38. Hintze J. (2007) Number Cruncher Statistical System. NCSS, Kaysville, Utah.
39. Poiani KA, Johnson WC (1991) Global warming and prairie wetlands. BioScience 41: 611–618.
40. Kantrud HA, Krapu GL, Swanson GA (1989) Prairie basin wetlands of the Dakotas: a community profile. Washington, DC: US Fish and Wildlife Service Biological Report 85(7.28). 111 p.
41. LaBaugh JW (1989) Chemical characteristics of water in northern prairie wetlands. In: van der Valk A, editor. Northern Prairie Wetlands. Ames: Iowa State University Press. 56–90.
42. Murkin HR, Murkin EJ, Ball JP (1997) Avian habitat selection and prairie wetland dynamics: a 10-year experiment. Ecological Applications 7: 1144–1159.
43. van der Valk AG (2005) Water-level fluctuations in North American prairie wetlands. Hydrobiologia 539: 171–188.
44. Bouffard SH, Hanson MA (1997) Fish in waterfowl marshes: waterfowl managers' perspective. Wildlife Society Bulletin 25: 146–157.
45. Cox RR, Hanson MA, Roy CC, Euliss NH, Johnson DG et al. (1998) Mallard duckling growth and survival in relation to aquatic invertebrates. Journal of Wildlife Management 82: 124–133.
46. Prairie Habitat Joint Venture (2008) Prairie Habitat Joint Venture Implementation Plan 2007–2012. Edmonton, Albeta: Prairie Habitat Joint Venture.
47. Horn DJ, Phillips ML, Koford RR, Clark WR, Sovada MA et al. (2005) Landscape composition, patch size, and distance to edges: interactions affecting duck reproductive success. Ecological Applications 15: 1367–1376.
48. Reynolds RE, Shaffer TL, Renner RW, Newton WE, Batt BDJ (2001) Impact of the Conservation Reserve Program on duck recruitment in the U.S. Prairie Pothole Region. Journal of Wildlife Management 65: 765–780.
49. Krapu GL, Brandt DA, Cox RR Jr. (2004) Less waste corn, more land in soybeans, and the switch to genetically modified crops: trends with important implications for wildlife management. Wildlife Society Bulletin 32: 127–136.
50. Laingen C (2012) Delineating the 2007 Corn Belt region. Papers of the Applied Geography Conference 35: 173–181.
51. Rashford BS, Walker JA, Bastian CT (2011) Economics of grassland conversion to cropland in the Prairie Pothole Region. Conservation Biology 25: 276–284.
52. Johnson RR, Oslund FT, Hertel DR (2008) The past, present, and future of prairie potholes in the United States. Journal of Soil and Water Conservation 63(3): 84A–87A.
53. Loesch CR, Reynolds RE, Hansen LT (2012) An assessment of re-directing breeding waterfowl conservation relative to predictions of climate change. Journal of Fish and Wildlife Management 3: 1–22.
54. Felton A, Fischer J, Lindenmayer DB, Montague-Drake R, Lowe AR et al. (2009) Climate change, conservation, and management: an assessment of the peer-reviewed scientific journal literature. Biodiversity and Conservation 18: 2243–2253.
55. Sohl TL, Gallant AL, Loveland TR (2004) The characteristics and interpretability of land surface change and implications for project design. Photogrammetric Engineering and Remote Sensing 70: 439–448.
56. Nichols JD, Koneff MD, Heglund PJ, Knutson MG, Seamans ME et al. (2011) Climate change, uncertainty, and natural resource management. Journal of Wildlife Management 75: 6–18.
57. Sala OE, Chapin FS III, Armesto JJ, Berlow D, Bloomfield J, et al. (2000) Global biodiversity scenarios for the year 2100. Science 287: 1770–1774.
58. Fry J, Xian G, Jin S, Dewitz J, Homer C et al. (2011) Completion of the 2006 National Land Cover Database for the conterminous United States. Photogrammetric Engineering and Remote Sensing 77: 858–864.
59. Wilen BO, Bates MK (1995) The US Fish and Wildlife Service's National Wetlands Inventory project. Vegetatio 118: 153–169.
60. Cowardin LM, Shaffer TL, Arnold PM (1995) Evaluation of duck habitat and estimation of duck population sizes with a remote-sensing based system. Washington, DC: US Department of the Interior National Biological Service, Biological Science Report 2. 26 p.

18

Ocean Net Heat Flux Influences Seasonal to Interannual Patterns of Plankton Abundance

Tim J. Smyth*, Icarus Allen, Angus Atkinson, John T. Bruun, Rachel A. Harmer, Robin D. Pingree, Claire E. Widdicombe, Paul J. Somerfield

Plymouth Marine Laboratory, Plymouth, Devon, United Kingdom

Abstract

Changes in the net heat flux (NHF) into the ocean have profound impacts on global climate. We analyse a long-term plankton time-series and show that the NHF is a critical indicator of ecosystem dynamics. We show that phytoplankton abundance and diversity patterns are tightly bounded by the switches between negative and positive NHF over an annual cycle. Zooplankton increase before the transition to positive NHF in the spring but are constrained by the negative NHF switch in autumn. By contrast bacterial diversity is decoupled from either NHF switch, but is inversely correlated (r = −0.920) with the magnitude of the NHF. We show that the NHF is a robust mechanistic tool for predicting climate change indicators such as spring phytoplankton bloom timing and length of the growing season.

Editor: Hans G. Dam, University of Connecticut, United States of America

Funding: This work is supported by the Natural Environmental Research Council funded National Capability Western Channel Observatory. It also contributes to the projects Marinexus and Charm2 funded in part by the INTERREG IVA France (Channel) – England programme (FEDER), and to the EU FP7 projects Devotes (DEVelopment Of innovative Tools for understanding marine biodiversity and assessing good Environmental Status, Grant Agreement number 308392) and GreenSeas (Development of global plankton data base and model system for eco-climate early warning, Grant Agreement number 265294). The funders had no role in study design, data collection and analysis, decision to publish, or preparation of the manuscript.

Competing Interests: The authors have declared that no competing interests exist.

* E-mail: tjsm@pml.ac.uk

Introduction

The air-sea exchange of heat, critical for regulating the Earth's climate [1], is also related to the turbulence structure of the upper ocean [2] which in turn controls plankton dynamics [3]. As phytoplankton form the base of the marine food web, an understanding of their population dynamics, including species succession [4], is important for determining whole ecosystem trajectories. The spring bloom is manifested as a dramatic increase in the phytoplankton standing stock over a relatively short period of time [5]. This phenomenon is caused by a combination of increasing solar irradiance, abundant surface layer nutrients and a stabilizing water column [6–8] occurring in concert.

The idea of a critical depth above which the mixed layer shoals during the spring, marking the start of the bloom, has become synonymous with Sverdrup [9]. Recently, however, several authors have shown that the start of the spring bloom can precede the critical depth being reached by several weeks [3,10] leading to alternative theories being posited. These include: (i) the dilution-recoupling hypothesis [7] where the mixed layer depth deepening during winter causes a dilution between phytoplankton and their predators allowing an initial increase in phytoplankton biomass, before recoupling them as the mixed layer depth shoals; (ii) the critical turbulence hypothesis [3,6] where the spring bloom is triggered by a reduction in air-sea heat fluxes leading to a weakening of the turbulence in the mixed layer allowing phytoplankton sufficient residence time in a lit layer replete with nutrients to bloom and; (iii) the eddy driven hypothesis [11] where deep oceanic eddies cause initial stratification, rather than a springtime warming of the sea surface, favourable for the growth and multiplication of phytoplankton.

For an appraisal of the start of the spring bloom it is important to realise the conceptual difference between the mixed layer depth (MLD) and the depth of the layer which supports initial phytoplankton growth, which may be much shallower than this. The MLD requires a definition: this is often arbitrarily set to the depth at which there is a pre-determined (e.g. $-0.2°C$) change in temperature [12].

Due to winter mixing, the MLD in the Bay of Biscay can be > 500 m in March [13]. However, the spring bloom is initiated within a surface euphotic layer that may only be 10 m deep with temperature differences as little as 0.1°C. This is unrelated to, and certainly not discriminated by, a MLD definition of $-0.2°C$, which may still be >200 m at the height of the bloom [14].

Similarly, this observation is repeated on the NW European continental shelf [15], where the spring bloom is initiated in a mixed (to 0.01°C) water column, but with a stabilised surface layer of 0.1°C. From the two examples given above, using a MLD appraisal of conditions will lead to the conclusion that significant production takes place in the winter in temperate seas before any surface stabilisation.

To overcome the need for arbitrary MLD definitions, flux methods have previously been employed to determine the start of the productive season [16] with a balance (R) of potential energy (PE) and turbulent kinetic energy (TKE) production controlling near surface stabilisation:

$$R = \frac{PE}{TKE} = \frac{g\alpha Q_0}{2C_p\rho}\left[\frac{h}{C_D\overline{|u|}^3}\right] \quad (1)$$

where g is the acceleration due to gravity, α the volume coefficient of expansion, C_p the specific heat at constant pressure, Q_0 the surface heat flux, ρ the density of sea water, C_D the drag coefficient, h the water depth and |u| the magnitude of the instantaneous tidal stream velocity, u, with the overbar denoting the average value over one cycle. In this paper we use a limit of this approach, namely that the net heat flux (NHF) at the surface (Q_0) needs only to be a positive quantity for stabilisation.

The focus of this study on plankton bloom dynamics is centred on a tidally influenced, shelf seas time-series station (L4: 50° 15′N, 4° 13′W; depth = 54 m) situated in the western English Channel (Figure S1). We use long time-series, weekly resolved phytoplankton and zooplankton data to study relationships between plankton population dynamics, including biodiversity, and modelled NHF [17].

Materials and Methods

Ethics statement

No specific permissions were required for the plankton sampling at station L4 (50° 15′N, 4° 13′ W), and the field studies did not involve endangered or protected species.

Net Heat Flux

The air-sea flux of heat is governed by four processes: shortwave radiation from the sun (Q_{SW}), outgoing longwave radiation from the sea surface (Q_{LW}), sensible heat transfer resulting from air-sea temperature differences (Q_{SH}), and the latent heat transfer carried by evaporation of sea surface water (Q_{LH}). Direct observations (e.g. using ships and buoys) of the air-sea heat flux are rare because of the difficulty in taking such measurements. However, it is possible to determine the four components of the air-sea heat flux using bulk parameterizations which are a function of surface meteorological and oceanographic variables. We used the Woods Hole Oceanographic Institution air-sea exchange Matlab tools [18] to determine Q_{SW}, Q_{LW}, Q_{SH} and Q_{LH} [17], all in units of Wm^{-2}. The meteorological parameters used to run the heat flux model (HFM) were obtained from the European Centre for Medium Range Weather Forecasting (ECMWF) ERA-40 and Operational analyses and extracted for a gridpoint centred on 50° N, 4° W. These parameters were: air temperature (T_a in °C), dew point (T_d in °C), wind-speed at 10 m (U_{10} in m s^{-1}), cloud fraction (CF where 0: clear; 1: overcast) and atmospheric pressure (P in mb). Sea surface temperatures (T_s in °C) measured at stations L4 and E1 (50° 02′N, 4° 22′W) were used to run the HFM over different periods overlapping with the ECMWF data availability (1958–2011). Station L4 has available T_s between 1988 and 2011, whereas station E1 has records of T_s between 1903 and 2011. Q_{SW} was calculated as a function of date and position with correction for CF [19]; Q_{LW} as a function of T_a, T_s, T_d, CF using the Berliand bulk formula [20]. Both Q_{SH} and Q_{LH} were calculated as a function of T_a, T_s, T_d, CF, P and U_{10}. The sum of all four components results in the net heat flux (NHF) with the sign convention of positive NHF being into the water column used throughout.

The HFM was run between 1988 and 2010 for L4 and from 1958–2011 for E1, after the input T_s had been interpolated from the weekly sampling interval onto a daily time resolution grid. This was to enable the NHF to be calculated on the same time resolution as the available (daily) meteorological values. The day of each individual year where the NHF became positive, and remained positive for more than seven days, was determined. A period of seven days was chosen for two reasons. Firstly, studies of phytoplankton succession rate show that it requires between two and four generations of the ascendant species to alter the community structure. This equates to between 5 and 15 days [21], given the rate at which phytoplankton can double their numbers, which is likely to be temperature-dependent. Secondly, routine L4 sampling takes place on a weekly basis, so the granularity of the time-series is of order seven days before interpolation. The choice of a seven day period had the added benefit of ensuring that the sign of the NHF is properly established rather than being fleeting or ephemeral. On calm days, even in winter, if there are clear skies, solar insolation can be sufficient to ensure the NHF is positive. The day of NHF switching back to negative towards the end of the year was also determined and the number of days between the two NHF switching events (negative to positive and positive to negative) was calculated for each year.

Biological sampling

Key ecosystem parameters have been sampled in the western English Channel for over a century [22,23]. The weekly phytoplankton samples (since 1992: see Table S1 for species list) were taken at a depth of 10 m using a 10 L Niskin bottle at station L4. A 200 mL subsample was then removed from the bottle and immediately fixed with 2% (final concentration) Lugol's iodine solution [24]. A second 200 mL sub-sample was also taken and preserved with neutral formaldehyde for the enumeration of coccolithophores. The samples were stored in cool, dark conditions until taxonomic analysis at the Plymouth Marine Laboratory (PML) using light microscopy [25]. Samples for zooplankton have been collected on a weekly basis since 1988 (see Table S2 for species list), using a vertical net haul from 50 m depth to the surface using a WP2 net with a mesh-size of 200 μm and mouth area of 0.25 m^2 [26]. Two hauls are taken and the samples preserved and stored in 5% formalin. These are then sub-sampled, counted and identified using light microscopy at PML [27]. Bacterial abundance data (2003–2008) were generated by extracting nucleic acids from 5 L seawater samples collected from the surface and filtered immediately through a 0.22 μm Sterivex cartridge (Millipore), which was stored at −80°C. DNA was isolated from each sample and then stored at −20°C prior to 454-pyrosequencing [28].

Statistical analyses

The biological datasets were linearly interpolated onto a daily grid from the original weekly sampling intervals for the entire length of the time-series. This daily grid was then shifted in time with t = 0 representing the serial day number of the NHF becoming positive for each individual year (Figure S2). This puts the biological signal for each individual year into a common frame of reference. Once each year had been time-shifted to day zero, the biological signal for each day for all years was smoothed, using a median filter, to remove isolated high or low values. This reduces the multiple annual time-series to a single daily value (for each individual biological time-series) allowing the numerical analyses described below to be performed.

The bloom onset date was defined as the first day of rapid increase in total phytoplankton abundance which is then followed for more than seven days by values greater than the winter baseline i.e. a near step-wise change with a definite edge. This was numerically determined from the time-series data using an Auto-Regressive Integrated Moving Average (ARIMA) transfer function model [29]. The onset edge was estimated through use of a 'ragged-step' intervention function and uses the Akaike Information Criterion (AIC) to identify the relevant order of the ARIMA model to use in this setting. That specific model for the onset day is then estimated using maximum likelihood optimisation. This algorithm was coded in the R-statistical language.

The number of taxa (S) represented in the various samples was extracted from the phytoplankton, holozooplankton, merozooplankton and bacterial datasets. This was to give a measure of the biodiversity for the various planktonic functional groups and to examine plankton dynamics, as a function of heat flux, over a range of trophic levels.

Using the 54 complete year time-series of heat flux, positive NHF and negative NHF onset dates at station E1 (1958–2011) were determined. A spectral analysis approach was used to identify the regular repeating harmonic terms in the time-series signal [29]. The contribution of the harmonic and trend components to the signal were then estimated through the Box-Jenkins time-series model [30] approach.

Results

Figure 1 shows that when the NHF switches to positive, the total phytoplankton abundance increases rapidly, reaching peak abundance around 40 days later. The median NHF positive switch is serial day 71 (11 March, range: 19 February – 21 April) for L4. Using objective time-series statistical techniques [29], the first increase in the abundance of total phytoplankton was calculated to be day 70. The commencement of the rapid increase in phytoplankton abundance (Figure 1A.), precedes the maximum abundance by between 40 and 50 days (serial day 111–121) which is consistent with previously observed peak bloom timings in the western English Channel [31]. Total phytoplankton abundance (Figure 1A.) is initially dominated by species of phytoflagellates (Figure 1B. – e.g. *Phaeocystis*). This is followed by multiple peaks of diatoms (Figure 1C.) starting around 30 days after the NHF switch. Coccolithophores (Figure 1D.) and dinoflagellates (Figure 1E.) dominate later in the season. Zooplankton abundance (Figure 1F.) starts to increase (day 41) before either the NHF positive switch or the increase in total phytoplankton.

From Figure 1 it appears that the length of the growing season is constrained by the switch of sign from negative to positive NHF in the spring and the reverse in the autumn (dashed lines). The latter occurs between 185 (1999) and 258 (1997) days after the NHF positive switch at L4, for the period studied here (1988–2010). We therefore use the period between these switching events as a proxy for the length of the growing season.

Figure 2 shows the temporal evolution of plankton species richness (phytoplankton, holozooplankton, meroplankton and bacteria). The number of species (S) of phytoplankton (Figure 2A.) reaches a minimum of approximately 20 at the onset of the spring bloom, peaking at around 40 just before the NHF negative switch. Following this the number of species drops to around 25. The variability of phytoplankton species biodiversity with respect to the NHF differs markedly in comparison to other plankton types. For phytoplankton (Figure 2A.) the spring bloom is characterized by a smaller number of dominant species that are ideally suited to exploit the particular conditions of temperature, light and nutrients. Both holozooplankton (Figure 2B.) and merozooplankton (Figure 2C.) show an early season peak coincident with the NHF positive switch, gradually increasing later in the season before reducing markedly when the NHF switches to negative.

Station E1 provides a longer time-series of sea temperature to use in the calculations of NHF (Figure 3). This shows significant ($p = 0.002$) evidence of a 19 year repeating cycle in the timing of the change in sign of the NHF, with an envelope around this smooth oscillation of ± 10 days. Based on this 19 year harmonic of the time-series, the spring bloom timing changes on average by two days per year. This shift represents smooth background climatic variability; interannual variability is captured by the dashed line in Figure 3 (range $= 63$ days). There is also evidence of a longer term cycle ($p = 0.004$), represented by the dash-dot quadratic curve in Figure 3, which on average contributes to a changing start of spring bloom date by 0.4 days per year.

Discussion

Figure 1 clearly shows a strong link between the change to positive NHF and the start of the spring phytoplankton bloom indicating that stabilization of the water column gives phytoplankton [32] a window of opportunity to proliferate. The advantage of using the NHF is that it can be calculated using surface-only abiotic data: these are readily available from a combination of in situ observations, meteorological model analyses and satellite platforms. The NHF predictor can therefore be applied to real-time global ocean observations as well as to seasonal-decadal scale forecasts based on climate change projections.

Figure 1F, perhaps surprisingly, suggests that zooplankton have life strategies which allow them to pre-empt the spring bloom giving them a competitive edge at the start of the season. However, there are likely to be other mechanisms at work. The early spring increase in zooplankton is caused partly by merozooplankton such as barnacle larvae, whose spawning cues are likely different. The background levels of chlorophyll-a concentration in the western English Channel during the winter are around 0.5 mg m^{-3} [23] and support zooplankton reproduction at L4 year-round. During the narrow time (30 d) window between the increase in zooplankton abundance and NHF positive, grazing control by micro and mesozooplankton may be in operation. The reduction in turbulence, indicated by NHF becoming positive, will allow phytoplankton such as *Phaeocystis* (colonies) and diatoms (chains) to break through the grazing control "loophole" [33], due to increased residence time in the well-lit layer of the water column.

Towards the autumn the switch to negative NHF appears to be an important cut-off for phytoplankton species that require thermal stratification to thrive, such as coccolithophores [34] and dinoflagellates such as *Karenia mikimotoi* [25]. Dinoflagellates are likely to be favoured later in the season because, as the NHF becomes less positive, the exchange of nutrients across a weakening thermocline increases [35]. It is important to note that the mechanisms being invoked here are purely abiotic: it is likely other, as yet unquantified, loss mechanisms such as viral infection and parasitism, sinking and grazing play a part in the seasonal succession.

The NHF seems to be less of a factor for zooplankton (Figure 2B.), as they are more strongly governed by temperature and are able to easily migrate through the water column [36]. How prolific they are will depend on the amount and quality of available food.

Bacterial diversity (Figure 2D.) is disconnected completely from the switch from negative to positive NHF. Previous studies at L4 [28] concluded that bacterial diversity (Figure 2D.) is related to daylength (minimum/maximum at summer/winter solstices). Figure 4 shows that bacterial diversity is strongly negatively correlated with the magnitude of the NHF ($r = -0.920$). This implies that bacterial diversity is favoured in highly turbulent

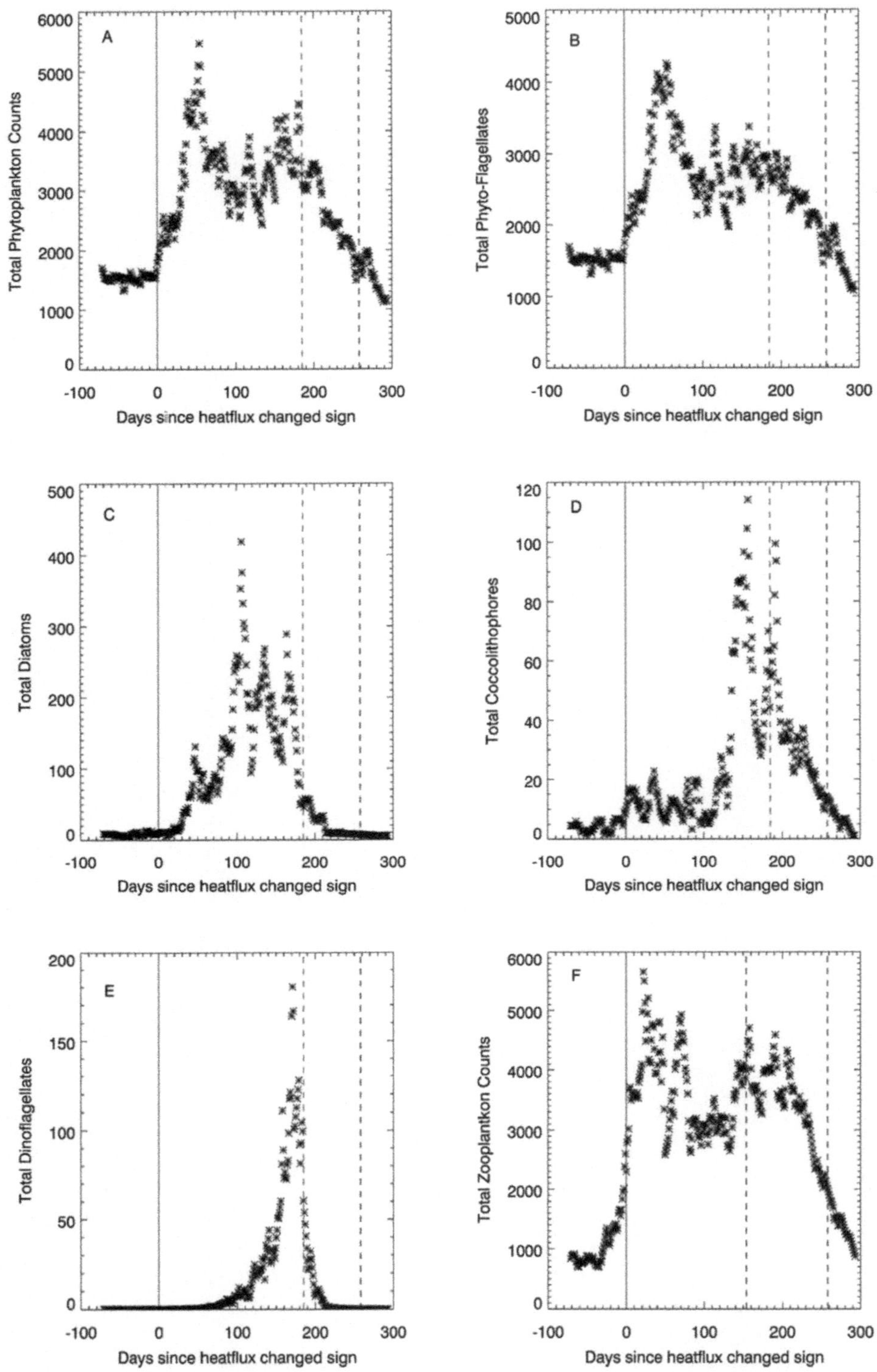

Figure 1. Entire time-normalized series at station L4 with a median filter applied to the phytoplankton (1992–2010) and zooplankton (1988–2010) abundance data (units: cells mL^{-1}, individuals m^{-3}, respectively). Phytoplankton species are aggregated as (A) total; (B) phytoflagellates; (C) diatoms; (D) coccolithophores and (E) dinoflagellates. Zooplankton are aggregated as (F) total zooplankton. The solid line indicates day zero, where the NHF switches from negative to positive and remains positive for more than 7 days. The two dashed vertical lines shown in all the panels indicate the minimum and maximum number of days after the positive NHF switch that the NHF returns to being negative (for more than 7 days).

conditions, providing a more plausible mechanistic explanation than a simple relationship with daylength. To test the wider applicability of this finding other high temporal (weekly) resolution bacterial time-series data would be required, from stations where the turbulent structure of the upper mixed layer is not necessarily in phase with the annual solar cycle. A weekly sampling routine should be adequate to describe seasonal variability in the bacterial diversity.

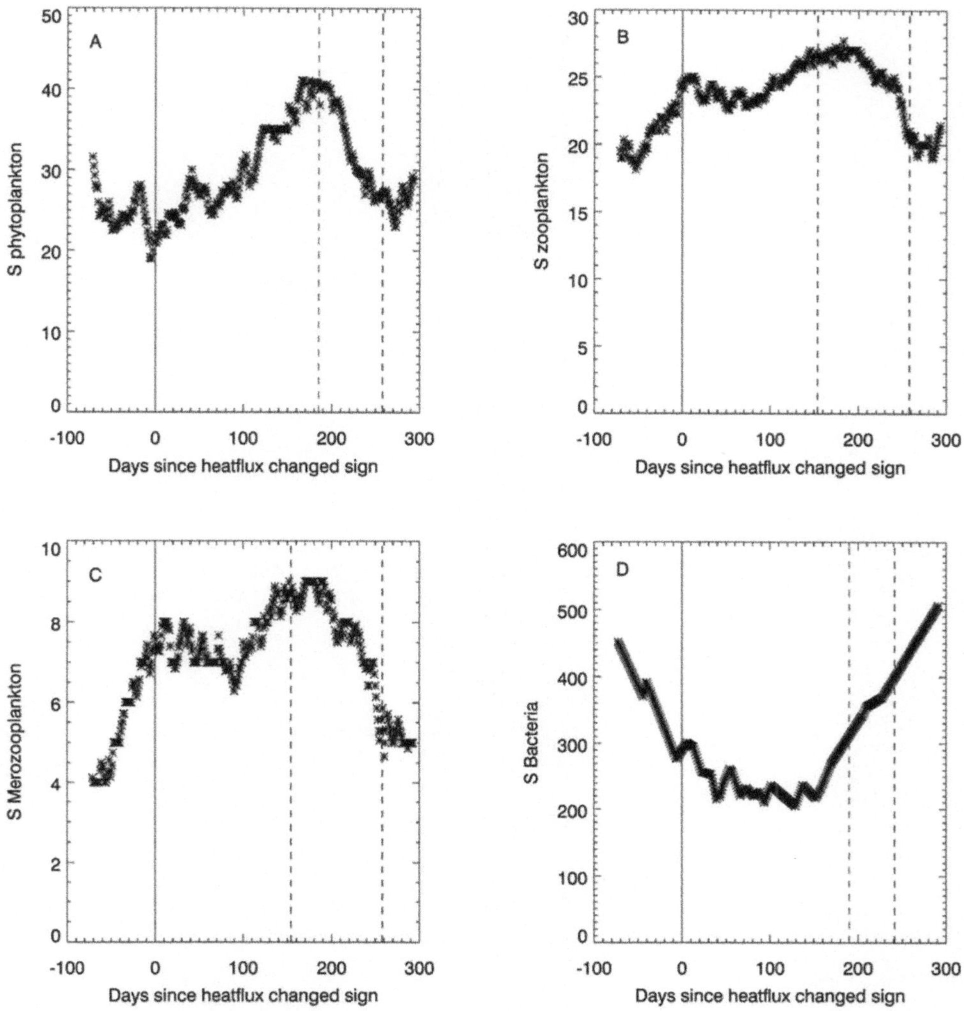

Figure 2. Median filtered daily interpolated biodiversity (species richness –S) data for (A) phytoplankton; (B) zooplankton; (C) merozooplankton and (D) bacteria relative to the change in sign for the NHF. The switch to negative heat flux (relative to the switch to positive and shown as the minimum and maximum range) is shown later in the season around 185–260 days afterwards, represented by the pair of dashed lines.

Figure S3 implies that the change from negative to positive (spring) and positive to negative (autumn) NHF are decoupled in regions of seasonal stratification i.e. an early onset of the spring bloom does not imply a late switch of NHF from positive to negative later in the year. Indeed there is no correlation ($r = -0.01$) between NHF positive and NHF negative date. Rather it seems that the length of the growing season is constrained by how early in the year positive NHF is established, i.e. there is more variability in spring bloom onset date (figure 3, dashed line) than the breakdown in stratification date in autumn (figure S4, dashed line). Mean NHF positive date (1958–2011) is 84.8 ± 16.0 whereas mean NHF negative date for this period is 287.3 ± 15.3. This is because the upper mixed layer is both wind mixed and convectively mixed as a result of surface heat flux, particularly at night [37]. The invariant factor in both onset and breakdown is daylight hours: these rapidly increase, and therefore promote positive NHF, in the spring and; rapidly decrease, and therefore promote negative NHF, in the autumn. Positive NHF is more difficult to establish in the spring than negative NHF in the autumn because of seasonal stratification at stations L4 and E1. In the spring, the water column is well mixed, so to promote NHF all the positive heat flux mechanisms (i.e. warmer atmosphere than sea-surface, reduction in wind, increasing levels of solar insolation) need to be acting in the same direction. A small perturbation to this, such as strong winds or a prolonged cold, cloudy period, will delay a switch to positive NHF and, as a consequence, thermal stratification. At the beginning of autumn, the top 10–15 m of the water column are thermally stratified, and because of the larger thermal heat capacity of water compared with air, as the nights get longer the tendency will be for all negative heat flux mechanisms to act in the same direction. This gives a more abrupt change in sign to negative NHF and hence a more distinct and invariant end point to the growing season (Figure S3).

One of the limitations of using a Eulerian time-series station such as L4, is that it does not allow the significant time-series noise to be resolved into local process and spatial patchiness components [38]. For such a partitioning to be made, fine scale spatial surveying would be required to smooth out the time-series. From the evidence of local, short duration Lagrangian studies, the effect of tides at station E1 tend to advect the water masses around in an ellipse, with the net result of little overall translational movement.

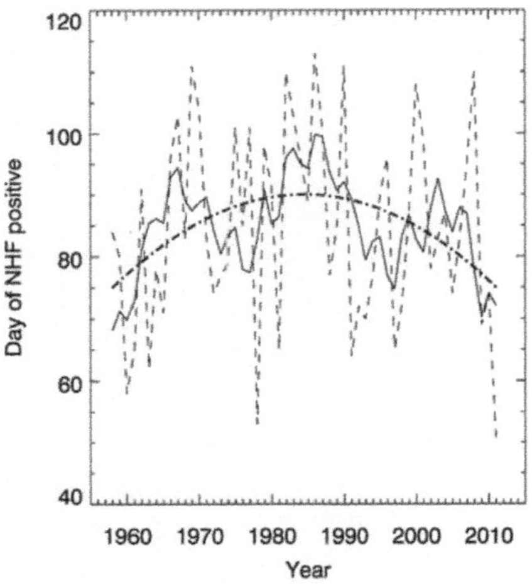

Figure 3. Onset of spring bloom day, determined using the NHF calculation (dashed line), between 1958 and 2011 at station E1. The solid line represents a spectral analysis of the NHF calculated, day of bloom onset, time-series using a Box-Jenkins [30] approach. The dash-dot line is a statistically significant ($p=0.004$; $r=0.294$) long-term cycle.

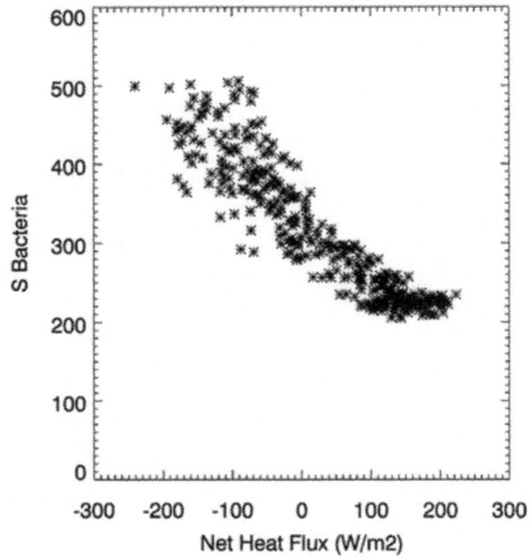

Figure 4. Bacterial diversity versus the magnitude of NHF for station L4 ($r=-0.920$). Each individual data point represents a daily average for the period 2003–2008.

At L4, there is a tendency for water masses to move in an east-west along-shore direction, with the added complication of episodic inputs from the River Tamar in terms of buoyancy, nutrient concentrations and material loading. However, the tides and the shallowness of L4 tend to damp variability [38].

The nineteen year repeating cycle in the timing of the change in sign of the NHF corresponds very closely to the 18.6 year tidal modulation period [39]. This shows the importance of the tides, probably in breaking down marginal stratification, for explaining seasonal changes in the abundance of phytoplankton [15].

The long-term quadratic temporal component could also be related to the Atlantic Multidecadal Oscillation (AMO), which has a periodicity of 65–70 years [40,41]; the AMO has a direct influence upon the North Atlantic Oscillation, which has previously been postulated to control timing of the spring bloom [42]. It is important to take account of both the tidal and large scale climatic oscillation harmonics when looking for evidence of background climate change. Indeed when both the tidal and long-term quadratic components are acting in phase, this can result in the date of the NHF advancing (or retreating) by 2.4 days per year. The end of the growing period (Figure S4) does not exhibit a long-term trend, but has an observable dual-frequency 20 year repeating pattern. With this repeating pattern, we note that over this 54 year period there are three events, of 5 years duration, which may indicate the emergence of a long-term increasing variability in the end day of the growing period. These statistical techniques therefore allow us to disentangle patterns of natural variability from any background climate change signal in the bloom onset and length of growing season.

Conclusions

We have demonstrated that the NHF provides a robust and powerful tool to determine essential climate change indicators such as the onset of the spring bloom and length of the growing season. The onset of the spring phytoplankton bloom has been shown to coincide with a switch from negative to positive NHF. At the end of the growing season, the switch back to negative NHF has been shown to control the succession of later blooming species of phytoplankton, effectively providing a cut-off to their proliferation. We have shown that zooplankton abundance pre-empts the NHF switch in spring, and that bacterial diversity patterns are structurally different to both phytoplankton and zooplankton. The latter suggests a tighter control of turbulence upon the diversity of the smallest organisms rather than a NHF switching mechanism which impacts upon phytoplankton diversity.

Using statistical tools on long-term ecological time-series data, we have been able to show the impact of a variety of physical and large scale climate forcing mechanisms, such as the 18.6 year tidal cycle and possibly the 65–70 year AMO, upon the length of the growing season.

As NHF can be calculated using surface only abiotic data, its predictive power can be applied to the global ocean using a combination of in situ observations, meteorological model analyses and satellite platforms. The NHF predictor can therefore be applied to real-time global ocean observations as well as to seasonal to decadal forecasts based on climate change projections.

Supporting Information

Figure S1 Map showing the locations of stations L4 (50° 15′N, 4° 13′W) and E1 (50° 02′N, 4° 22′W).

Figure S2 Daily interpolated phytoplankton abundance data for individual years 1992–2010 relative to the change in sign for the NHF. The figure shows the inherent difficulty in working with quantitative biological datasets as there is a large degree of inter- and intra-annual variability. However a general pattern of increasing phytoplankton abundance around the time of the NHF switch emerges. It could be argued that biomass should be used rather than numerical counts, as abundances of smaller phytoplankton (e.g. phytoflagellates), which contain less carbon than larger phytoplankton (e.g. diatoms), may be present in disproportionately large numbers. When the identical statistical time-series analysis (see text and Figure 1) was repeated for total phytoplankton biomass, the first increase was calculated to be on day 69.

Figure S3 Length of growing season, calculated using the period of time between NHF positive and switching back to negative, as a function of NHF positive day. Calculations are for the period 1958–2011 at station E1. Fitted line has r = −0.73.

Figure S4 Day of NHF negative (dashed line), between 1958 and 2011 at station E1, indicative of the end of the growing season. The solid line represents a spectral analysis of the calculated NHF time-series using a Box-Jenkins [30] approach.

Acknowledgments

The ECMWF operational and ERA-40 datasets were obtained from the British Atmospheric Data Centre. The authors also thank the crew of the RV Quest, Denise Cummings and James Fishwick for data acquisition and processing.

Author Contributions

Conceived and designed the experiments: TS PS. Performed the experiments: TS PS CW RH. Analyzed the data: TS PS CW RH AA. Contributed reagents/materials/analysis tools: TS PS CW RH AA RP IA JB. Wrote the paper: TS PS CW RH AA RP IA JB.

References

1. Yu L (2007) Sea surface exchanges of momentum, heat, and freshwater determined by satellite remote sensing. In: Steele J, Thorpe S, Turekian K, editors. Encyclopedia of Ocean Sciences, First Online Update. London, UK: Academic Press. p. 21.
2. Gregg MC, Peters H, Wesson JC, Oakey NS, Shay TJ (1985) Intensive measurements of turbulence and shear in the equatorial undercurrent. Nature 318: 140–144.
3. Huisman J, van Oostveen P, Weissing FJ (1999) Critical depth and critical turbulence: Two different mechanisms for the development of phytoplankton blooms. Limnology and Oceanography 44: 1781–1787.
4. Freund JA, Gruner N, Bruse S, Wiltshire KH (2012) Changes in the phytoplankton community at Helgoland, North Sea: lessons from single spot time series analyses. Marine Biology 159: 2561–2571.
5. Mann KH, Lazier JRN (2006) Dynamics of Marine Ecosystems: Biological-Physical Interactions in the Oceans Oxford: Blackwell Publishing Ltd. 512 p.
6. Taylor JR, Ferrari R (2011) Shutdown of turbulent convection as a new criterion for the onset of spring phytoplankton blooms. Limnology and Oceanography 56: 2293–2307.
7. Behrenfeld MJ (2010) Abandoning Sverdrup's Critical Depth Hypothesis on phytoplankton blooms. Ecology 91: 977–989.
8. Chiswell SM (2011) Annual cycles and spring blooms in phytoplankton: don't abandon Sverdrup completely. Marine Ecology Progress Series 443: 39–50.
9. Sverdrup HU (1953) On the conditions for the vernal blooming of phytoplankton. J Cons Perm Int Explor Mer 18: 287–295.
10. Townsend DW, Keller MD, Sieracki ME, Ackleson SG (1992) Spring phytoplankton blooms in the absence of vertical water column stratification. Nature 360: 59–62.
11. Mahadevan A, D'Asaro E, Lee C, Perry MJ (2012) Eddy-Driven Stratification Initiates North Atlantic Spring Phytoplankton Blooms. Science 337: 54–58.
12. Montegut CD, Madec G, Fischer AS, Lazar A, Iudicone D (2004) Mixed layer depth over the global ocean: An examination of profile data and a profile-based climatology. Journal of Geophysical Research-Oceans 109, C12, doi:10.1029/2004jc002378.
13. Pingree RD, New AL (1989) Downward propagation of internal tidal energy into the Bay of Biscay. Deep-Sea Research Part A-Oceanographic Research Papers 36: 735–758.
14. Garcia-Soto C, Pingree RD (1998) Late autumn distribution and seasonality of chlorophyll-a at the shelf-break/slope region of the Armorican and Celtic Shelf. Journal of the Marine Biological Association of the United Kingdom 78: 17–33.
15. Pingree RD, Holligan PM, Mardell GT, Head RN (1976) The influence of physical stability on spring, summer and autumn phytoplankton blooms in the Celtic Sea. J Mar Biol Ass UK 56: 845–873.
16. Pingree RD (1978) Mixing and stabilization of phytoplankton distributions on the northwest European continental shelf. In: Steele J, editor. Spatial Pattern in plankton communities. New York: Plenum Press: NATO Scientific Affairs Division.
17. Pawlowicz R, Beardsley R, Lentz S, Dever E, Anis A (2001) Software simplifies air-sea data estimates. EOS Trans AGU 82: 2–2.
18. Woods Hole Oceanographic Institution (2007) woodshole.er.usgs.gov/operations/sea-mat/air_sea-html.
19. Reed RK (1977) On estimating insolation over the ocean. J Phys Oceanogr 7: 482–485.
20. Fung IY, Harrison DE, Lacis AA (1984) On the variability of the net longwave radiation at the ocean surface. Reviews of Geophysics 22: 177–193.
21. Reynolds CS, Padisak J, Sommer U (1993) Intermediate disturbance in the ecology of phytoplankton and the maintenance of species-diversity – a synthesis. Hydrobiologia 249: 183–188.
22. Southward AJ, Langmead O, Hardman-Mountford NJ, Aiken J, Boalch GT, et al. (2005) Long-term oceanographic and ecological research in the western English Channel. Adv Mar Biol 47: 1–105.
23. Smyth TJ, Fishwick JR, Al-Moosawi L, Cummings DG, Harris C, et al. (2010) A broad spatio-temporal view of the Western English Channel observatory. Journal of Plankton Research 32: 585–601.
24. Throndsen J (1978) Productivity and abundance of ultra-plankton and nanoplankton in Oslofjorden. Sarsia 63: 273–284.
25. Widdicombe CE, Eloire D, Harbour D, Harris RP, Somerfield PJ (2010) Long-term phytoplankton community dynamics in the Western English Channel. Journal of Plankton Research 32: 643–655.
26. UNESCO (1968) Zooplankton Sampling. Paris, Unesco Press. 174 p.
27. Eloire D, Somerfield PJ, Conway DVP, Halsband-Lenk C, Harris R, et al. (2010) Temporal variability and community composition of zooplankton at station L4 in the Western Channel: 20 years of sampling. Journal of Plankton Research 32: 657–679.
28. Gilbert JA, Field D, Swift P, Newbold L, Oliver A, et al. (2009) The seasonal structure of microbial communities in the Western English Channel. Environmental Microbiology 11: 3132–3139.
29. Shumway RH, Stoffer DS (2010) Time Series Analysis and Its Applications: With R Examples. New York: Springer.
30. Box G, Jenkins G (1970) Time series analysis: Forecasting and control. San Francisco: Holden-Day.
31. Pingree RD (1975) The advance and retreat of the thermocline on the continental shelf. J Mar Biol Ass UK 55: 965–974.
32. Kiorboe T (2008) A mechanistic approach to plankton ecology. Princeton: Princeton University Press. 209 p.
33. Irigoien X, Flynn KJ, Harris RP (2005) Phytoplankton blooms: a 'loophole' in microzooplankton grazing impact? Journal of Plankton Research 27: 313–321.
34. Smyth TJ, Moore GF, Groom SB, Land PE, Tyrrell T (2002) Optical modeling and measurements of a coccolithophore bloom. Applied Optics 41: 7679–7688.
35. Pingree RD, Pennycuick L (1975) Transfer of heat, fresh water and nutrients through the seasonal thermocline. J Mar Biol Ass UK 55: 261–274.
36. Mackas DL, Greve W, Edwards M, Chiba S, Tadokoro K, et al. (2012) Changing zooplankton seasonality in a changing ocean: Comparing time series of zooplankton phenology. Progress in Oceanography 97: 31–62.
37. Pingree RD, Griffiths DK (1977) The bottom mixed layer on the Continental Shelf. Estuar Coast Mar Sci 5: 399–413.
38. Martin AP, Zubkov MV, Burkill PH, Holland RJ (2005) Extreme spatial variability in marine picoplankton and its consequences for interpreting Eulerian time-series. Biology Letters 1: 366–369.
39. Pingree RD, Griffiths DK (1987) Tidal friction for semidiurnal tides. Continental Shelf Res 7: 1181–1209.
40. Schlesinger ME, Ramankutty N (1994) An oscillation in the global climate system of period 65–70 years. Nature 367: 723–726.
41. Garcia-Soto C, Pingree RD (2012) Atlantic Multidecadal Oscillation (AMO) and sea surface temperature in the Bay of Biscay and adjacent regions. Journal of the Marine Biological Association of the United Kingdom 92: 213–234.
42. Sharples J, Ross ON, Scott BE, Greenstreet SPR, Fraser H (2006) Inter-annual variability in the timing of stratification and the spring bloom in the North-western North Sea. Continental Shelf Research 26: 733–751.

The Macroecology of Airborne Pollen in Australian and New Zealand Urban Areas

Simon G. Haberle[1]*, David M. J. S. Bowman[2], Rewi M. Newnham[3], Fay H. Johnston[4], Paul J. Beggs[5], Jeroen Buters[6], Bradley Campbell[7], Bircan Erbas[8], Ian Godwin[7], Brett J. Green[9], Alfredo Huete[10], Alison K. Jaggard[5], Danielle Medek[11], Frank Murray[12], Ed Newbigin[13], Michel Thibaudon[14], Don Vicendese[8], Grant J. Williamson[2], Janet M. Davies[15]

1 Department of Archaeology and Natural History, College of Asia and the Pacific, Australian National University, Canberra, Australian Capital Territory, Australia, **2** School of Plant Science, University of Tasmania, Hobart, Tasmania, Australia, **3** School of Geography, Environment and Earth Sciences, Victoria University of Wellington, Wellington, New Zealand, **4** Menzies Research Institute Tasmania, University of Tasmania, Hobart, Tasmania, Australia, **5** Department of Environment and Geography, Faculty of Science, Macquarie University, Sydney, New South Wales, Australia, **6** Center for Allergy and Environment, Technical University of Munich, Munich, Germany, **7** School of Agriculture and Food Science, The University of Queensland, St Lucia, Queensland, Australia, **8** School of Public Health and Human Biosciences, La Trobe University, Bundoora, Victoria, Australia, **9** Allergy and Clinical Immunology Branch, Health Effects Laboratory Division, National Institute for Occupational Safety and Health, Centre for Disease Control and Prevention, Morgantown, West Virginia, United States of America, **10** Plant Functional Biology and Climate Change, University of Technology, Sydney, New South Wales, Australia, **11** School of Medicine, Australian National University, Canberra, Australian Capital Territory, Australia, **12** School of Environmental Science, Murdoch University, Murdoch, Western Australia, Australia, **13** School of Botany, University of Melbourne, Melbourne, Victoria, Australia, **14** European Aerobiology Society, Réseau National de Surveillance Aérobiologique, Lyon, Rhône-Alpes, France, **15** Lung and Allergy Research Centre, School of Medicine, and Translational Research Institute, The University of Queensland, Woolloongabba, Queensland, Australia

Abstract

The composition and relative abundance of airborne pollen in urban areas of Australia and New Zealand are strongly influenced by geographical location, climate and land use. There is mounting evidence that the diversity and quality of airborne pollen is substantially modified by climate change and land-use yet there are insufficient data to project the future nature of these changes. Our study highlights the need for long-term aerobiological monitoring in Australian and New Zealand urban areas in a systematic, standardised, and sustained way, and provides a framework for targeting the most clinically significant taxa in terms of abundance, allergenic effects and public health burden.

Editor: Gil Bohrer, The Ohio State University, United States of America

Funding: Funding support for the Working Group came from the Australian Centre for Ecological Analysis and Synthesis (ACEAS). Terrestrial Ecosystem Research Network (TERN). Merck Sharp and Dohme provided additional independent untied co-sponsorship for the Working Group. The funders had no role in study design, data collection and analysis, decision to publish, or preparation of the manuscript.

Competing Interests: The authors have declared that no competing interests exist.

* E-mail: simon.haberle@anu.edu.au

Introduction

Pollen is the microscopic haploid stage of the plant life cycle. Pollen grains from different plant species have a remarkable diversity of shapes, sizes, and biochemical compositions. The identification of fossilised pollen in sedimentary sequences (palynology) has been pivotal in illuminating past environmental changes caused by, or associated with, climate change, human impacts and natural disturbances such as fires and tectonic activity. This technique is dependent on the assumption that deposited pollen provides a faithful representation of the vegetation patterns surrounding sedimentary traps. Palynologists test this assumption by collecting pollen from the ground surface within or between typical vegetation types or by sampling pollen in the atmosphere (e.g., D'Costa and Kershaw [1]; Wilmshurst and McGlone [2]; Fletcher and Thomas [3]; Tng *et al.* [4]). These investigations of 'modern pollen rain' have focused on wilderness or rural areas, to provide a basis for reconstructing historic biogeographic patterns. How pollen rain varies geographically within urban environments in Australia and New Zealand, where over 85% of the population live in Australia and New Zealand, remains poorly characterised (see Tng *et al.* [4] for an exception).

Parallel studies have produced substantial time series of airborne pollen data for urban locations to underpin investigations of seasonal allergic reaction such as rhinitis or asthma in humans. These large data sets have enabled the development of predictive models between meteorological variables and the concentration of pollen from specific allergenic pollen taxa such as grass and birch (e.g., Schäppi *et al.* [5]; Emberlin *et al.* [6]; Rodríguez-Rajo *et al.* [7]; Sofiev *et al.* [8]). In addition, in some urban areas, particularly in Europe, the aerobiological datasets now span several decades and are beginning to provide insights into biogeographic variation in landscape phenological patterns, and how these patterns respond to current climate change [9–12].

In contrast to the well-studied Northern Hemisphere, aerobiological studies of the Southern Hemisphere have been conducted in isolation and limited to monitoring specific cities for narrow time periods with few attempts made to discern generalised patterns across a country or region (see [13] for an exception). Indeed, many of the existing aerobiological studies located in

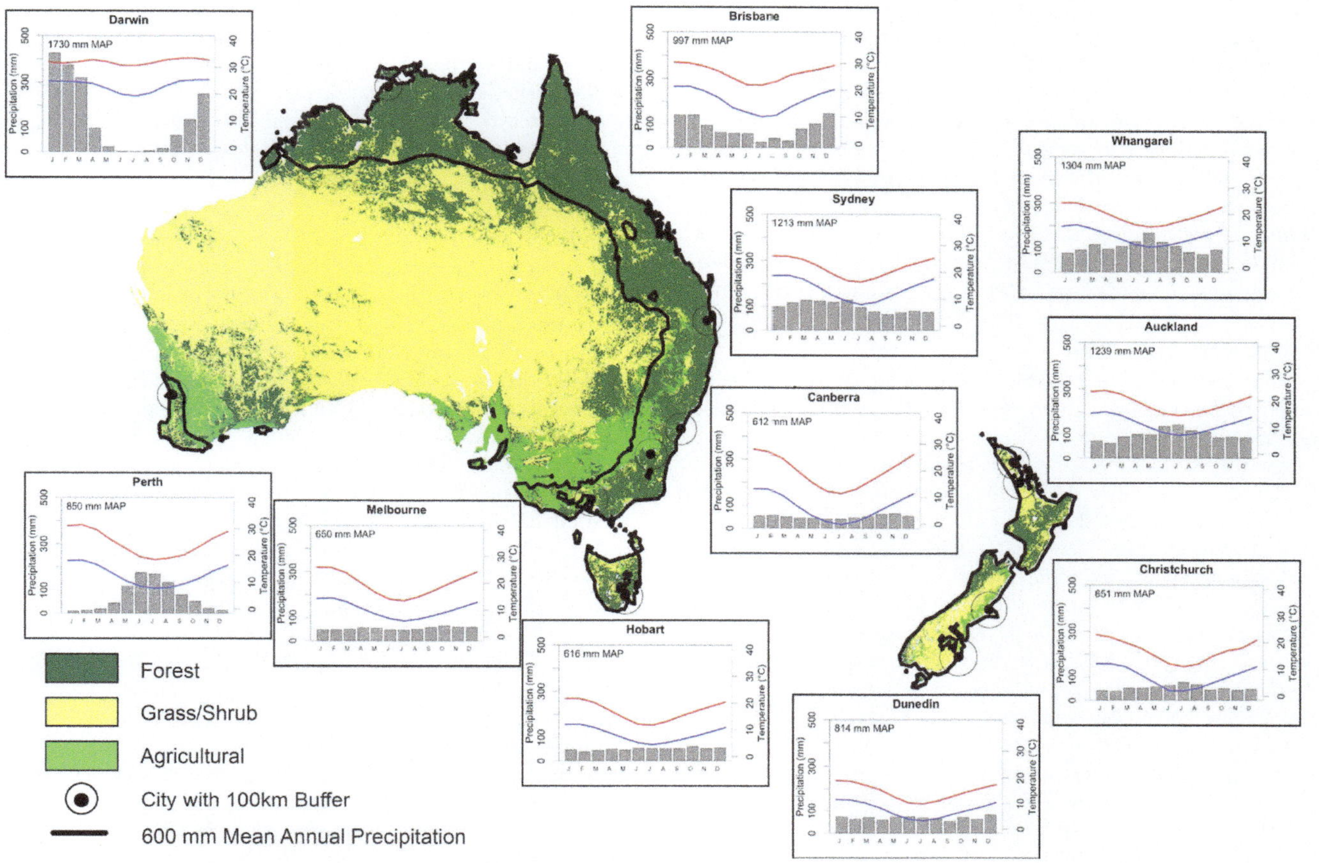

Figure 1. Land cover attribute for Australia and New Zealand with climatological summaries for each major urban centre associated with an aerobiology study. Climate summaries include average monthly precipitation and minimum and maximum temperatures for each urban area (see Table 3 for data sources). Data shown for Whangarei represents the Kaikohe pollen count site.

Australia and New Zealand have been motivated by an interest in the public health burden of pollen sensitization and allergic asthma at major population centres. Aerobiology in the Southern Hemisphere is as complex as that in the Northern Hemisphere, given the variability of abiotic factors and environmental gradients that span the tropics to the temperate zones. The coexistence of very distinct indigenous vegetation, introduced Northern Hemisphere ornamentals, as well as exotic invasive species are additional variables that contribute to the aerobiology of each Southern Hemisphere regional center. Here we compile and analyse an historic atmospheric pollen dataset from 11 cities across Australia and New Zealand and examine the regional variations in pollen content and relative abundance. The geographical distribution of the 11 Australasian sites was broad enough to enable further examination of the role that biogeography plays in regional and urban variations in pollen composition. These data can potentially provide insights into how pollen production and dispersal derived from both native and exotic taxa may respond to changes in climate and urban and peri-urban land use. The influence of these variables and how that may affect allergic diseases is also discussed.

We emphasise that these results were achieved by collating data sets that were originally obtained for other purposes and were often of limited duration and collected using a variety of methods. The combined data set provides impetus for further work and suggests the potential of a more systematic and coordinated pollen monitoring network in Australasia for long-term studies into environmental change and the production of more accurate pollen forecasting systems to help allergy sufferers and their health care providers better manage their conditions. We conclude by considering some important issues that such an initiative could address.

Materials and Methods

Sampling Sites

Aerobiological data have been collected across Australia and New Zealand using a range of methods and counting periods (see Table 1). This paper describes the re-analysis of approved university studies of aerobiological pollen in Australia and New Zealand. All studies have been published (Table 1) with the exception of pollen data from Perth and Canberra, where unpublished data were provided by the investigators at these sites (also authors in this paper FM and SGH). In cities where more than one station is operating the secondary station is not included in the numerical analysis as they overlap and are shorter records than the nearby stations in the same cities. Land cover attributes for Australia and New Zealand with climate summaries for each major urban centre associated with aerobiology studies are shown in Figure 1. Table 1 provides the location details of each aerobiological station including the collection periods of 14 aerobiology stations across 11 urban centres, which have been in operation for at least one season over the last 25 years. In aerobiology, pollen traps are situated well above ground level, both to avoid the complicating influence of local vegetation and anthropogenic disturbance as well as to better sample the regional pollen flux that are dominated by the major anemophilous (wind-dispersed) taxa [19]. This group of wind-pollinated species comprise many of the important allergenic sources [20]. The data presented here were obtained at pollen sampling sites ranging in elevation above ground level that, along with other differences in collection methods described below, are likely to affect the total volume of pollen sampled. Nevertheless the methods used at each location provide estimates of airborne pollen volume (in grains/ cm^3) that can be compared between regions and through time.

Pollen Data Collection

The most common method of daily pollen counts in Australia is based on the deployment of a seven-day Hirst-type volumetric pollen and spore trap [21] located on structures (mostly rooftops) ranging from between 2–14 m above the ground. In New Zealand, an alternative device, the Intermittent Cycling Rotorod sampler [22] was deployed at all sites, but provides similar volumetric data. The Hirst-type sampler uses a range of adhesive surface compounds including vaseline and 10% paraffin wax in toluene on Melinex™ tape (Burkard Manufacturing Co. Ltd., Rickmansworth, Hertfordshire, UK), silicon-based adhesive (Lanzoni s.r.l., Bologna, Italy) or a 50% saturated Dow Corning high vacuum grease in solvent [23]. The seven-day tapes are then cut into 24-hr segments and mounted on glass slides with a stain such as fuchsine stained Gelvatol [24], Calberla's stain, or 2% Saffranin O in glycerol jelly. Alternatively daily pollen monitoring can be done using a glass microscope slide that is coated and stained as described. Analysis of each 24-hr period is conducted by counting between one and four transects at 400 magnification [25], though at one site the entire slide surface was counted (Brisbane, [26]). Pollen and spore counts were then converted to grains/m^3 of air and expressed as a daily mean value [21]. Hirst-type pollen and spore traps are known to show an instrumental variation of about 25% [27]. In New Zealand the Intermittent Cycling Rotorod sampler is an impaction collector with a retracting collector rod sampling head [22]. Particles are collected on the leading, greased, edge of two 1.59×32 mm clear polystyrene collector rods spun intermittently by an electric motor at 2400 rev./min. The samplers were set-up to operate for 6 min every hour. Sampling rods were replaced every 24 hours, stained with Calberla's solution, and examined under a transmitted light microscope. Raw pollen counts for each of the pollen/spore types identified were converted to the volumetric index (grains/m^3 of air) using a standard formula taking into account the sampling period and volume of air sampled [22]. The daily mean concentration of fungal spores, particularly *Alternaria*, was also recorded at some stations (Darwin, Brisbane, Sydney, Canberra, Melbourne and Hobart), though detailed analysis of these data is not presented here. The raw pollen counts for each site used in this paper have been archived in excel files at the Australian Centre for Ecological Analysis and Synthesis data portal (http://aceas-data.science.uq.edu.au/portal/). Links to specific station datasets are provided in Table 1.

Pollen Identification

Pollen identification was aided by the existing reference collections held by individual analysts or collated into digital or online reference collections (e.g., the Department of Archaeology and Natural History at the Australian National University, http://apsa.anu.edu.au, the CD published by Hjelmroos *et al.* [28]). In a number of these aerobiological surveys, it was not possible to place identifications at a species or genus taxonomic level. In these cases the taxon name reverted to the higher level (i.e., all genus-level identifications were placed into the relevant family) such as the Poaceae and Myrtaceae. The introduced tree taxa are referred to by their generic name (*Alnus, Betula, Pinus, Quercus, Salix and Ulmus*) as all are represented in Australia and New Zealand by more than one species in the genus and these are not distinguished in the available datasets.

Pollen Count Analysis

In order to reduce variation in pollen abundance records due to variations in equipment or handling protocols, the datasets were transformed into percentage values based on a total pollen sum (total pollen abundance of a single taxon recorded over the time of study divided by the total pollen abundance of all taxa recorded over the time of study) (Table 2). This data transformation mitigated other sources of variation in abundance between monitoring sites such as location, as coastal cities are exposed to a smaller proportion of land area and hence vegetation sources than inland cities (Figure 1).

The pollen season has been defined in many ways by different authors [29]. To date, no standardized approach to identifying the pollen season has been defined in Australia or New Zealand [13,15–17]. In general, the pollen season is said to have four parameters, the start, duration, peak and the end. In this study we adopted the Nilsson and Persson [30] approach, where the 'start day' is when the sum of daily pollen concentrations reaches 5% of the total yearly count and the 'end day' when the sum reaches 95% of the yearly count. The 'year' for all subtropical and temperate sites began on July 1 and ends on June 30. The exception was the tropical site (Darwin), where the 'year' begins on January 1 and ends on December 31. The season is thus the period when 90% of the total annual pollen count is collected on the trapping surface (see example given in Figure 2). Here we also point out that the years of sampling varied between certain sites, ranging from 1988 to 2012, with some sites sampled for a single season, a single year, or a small number of years. This factor must be considered when comparing phenological observations such as the timing of pollen season parameters between sites, as pollen season for the same species may vary substantially between years [31].

Similarities between the airborne pollen assemblages recorded within the 11 urban areas were calculated using non-metric multidimensional scaling ordination (nMDS, [32]), a technique that reduces the dimensionality of multivariate data, and graphically represented using the relative dissimilarities amongst (a) urban areas and (b) dominant taxa. The robust nature of the ordination is measured through the numerical stress value. A stress less than two corresponds to a good ordination and useful two-dimensional picture of sample similarity [32]. The nMDS outputs show vector positions for each variable within the environmental space. The length of the vectors correspond to the square-root of the r^2 values, so weak predictors have a shorter length than strong predictors. High r^2 values indicate a vector that is strongly associated with site/species variation in the ordination space. Low r^2 values would indicate a vector that doesn't really explain much in the ordination space. Environmental vectors representing three meteorological variables (Mean annual maximum and minimum temperature and mean annual precipitation) and the proportion of four broad land cover types in a 100 km radius around each urban area were fitted to the species ordination using the envfit function of the vegan package in R [33], with random permutations of environmental variables performed in order to assess their significance. We make the assumption that the proportion of land cover types found in a 100 km radius around pollen sampling

Table 1. Sites and methodologies employed in aerobiology recording stations across Australia and New Zealand.

Location name	Suburb (site)	Country (State/ Territory)	Latitude (S)	Longitude (E)	Elevation (m)	Monitor Height (m)	Start Date	End Date	Duration (days)	Time step	Time Start/End	Monitor type	Magnification	Counting method
Darwin	Casuarina (Charles Darwin University)	A (NT)	12°22′	130°52′	13	14	18/03/2004	15/11/2005	607	24 hrs	0:00	Burkard	400×	4 transects
Darwin	Palmerston	A (NT)	12°28′	130°58′	24	14	1/04/2004	31/03/2005	364	24 hrs	0:00	Burkard	400×	4 transects
Brisbane	Rocklea	A (QLD)	27°29′	153°08′	13	2	7/06/1994	17/05/1999	1805	7 day	unknown	Burkard	250×	entire slide
Perth	Murdoch (Murdoch University)	A (WA)	32°04′	115°50′	30	unknown	1/09/2006	31/12/2006	121	24 hrs*	0:00	Burkard	×400	3 transects
Sydney	Campbelltown (University of Western Sydney)	A (NSW)	34°04′	150°47′	74	10	1/01/1993	31/12/1995	1094	24 hrs	unknown	Burkard	400×	3 transects
Canberra	Acton (Australian National University)	A (ACT)	35°16′	149°7′	569	8	26/09/2007	13/12/2009	809	24 hrs	0:00	Burkard	400×	4 transects
Canberra	Holder	A (ACT)	35°19′	149°2′	570	14	27/8/09, 29/9/10	31/1/10, 26/11/10	210	24 hrs	10am (2009), 4pm (2010)	Burkard	×40 (2009), ×20 (2010)	1 transect
Melbourne	Parkville (Melbourne University)	A (VIC)	37°48′	144°17′		14	1/09/2009	31/12/2011	852	24 hrs	16:00	Burkard	20× with a field of view of 1 mm	1 transect
Hobart	Sandy Bay (University of Tasmania)	A (TAS)	42°54′	147°19′	58	12	6/09/2007	31/12/2010	1212	24 hrs	0:00	Burkard	400×	4 transects
Kaikohe	Kaikohe	NZ	35°24′	173°48′	198	2	15/11/1988	11/02/1989	88	24 hrs	12:00	Rotorod	400×	n/a
Auckland	Grafton (Museum)	NZ	36°51′	174°46′	61	8	28/10/1989	27/04/1990	181	24 hrs	9:00	Rotorod	400×	n/a
Auckland	Onehunga	NZ	36°51′	174°46′	15	3	27/10/1989	30/04/1990	185	24 hrs	9:00	Rotorod	400×	n/a
Christchurch	Christchurch (Canterbury University)	NZ	43°31′	172°35′	14	20	17/11/1988	13/02/1989	88	24 hrs	12:00	Rotorod	400×	n/a
Dunedin	Dunedin	NZ	45°51′	170°30′	0	5	1/10/1992	31/01/1993	122	24 hrs	9:00	Rotorod	400×	n/a

Data derived from unpublished and published data (Newnham *et al.* [13]; Bass and Morgan [14]; Ong *et al.* [15]; Green *et al.* [16]; Stevenson *et al.* [17]; Tng *et al.* [4]; Medek *et al.* [18]). Note that data from secondary stations in Darwin (Palmerston), Canberra (Holder) and Auckland (Grafton Museum) are not included in the numerical analysis in this paper. The raw pollen counts for each site used in this paper have been archived in excel files at the Australian Centre for Ecological Analysis and Synthesis data portal (http://aceas-data.science.uq.edu.au/portal/). Click on the following hyperlinked text to download data from each aerobiology recording station. Darwin, Brisbane, Perth, Sydney, Canberra, Melbourne, Hobart, Kaikohe, Auckland, Christchurch, Dunedin.

Assumptions: All air sampled at 10 l/minute (except NZ rotorod: 1.59×32 mm rods spun at 2400 rev/min for 6 min/hr), and all express concentration as grains/m3 of air.

*Not Saturdays and Sundays.

A; Australia, NZ; New Zealand, NT; Northern Territory, QLD; Queensland, WA; Western Australia, NSW; New South Wales, ACT; Australian Capital Territory, VIC; Victoria, TAS; Tasmania.

Table 2. Aerobiologically significant pollen taxa contributing >80% to the total annual atmospheric pollen in urban areas across Australia and New Zealand.

Taxa	Darwin	Brisbane	Perth	Sydney	Canberra	Melbourne	Hobart	Kiakohe	Auckland	Christchurch	Dunedin	Ave (rank)
Arecaceae	26.0											*26.0
Casuarina	3.1	6.5		6.2	1.8	0.4	4.3					3.7 (8)
Urticaceae	1.6	1.8		0.7	1.9	0.9			0.1			1.2 (16)
Acacia	3.1	0.2		0.3	0.3	0.3	0.5		<0.1			0.7 (18)
Myrtaceae	31.0	3.1	6.0	11.0	5.7	5.6	8.2	<0.1	5.2	3		7.9 (4)
Chenopodiaceae	0.3	0.1			0.8	0.2	0.3	<0.1	11.0	6.8		2.4 (12)
Cyperaceae	5.5	1.3		0.3	1.2		0.5			0.03		1.5 (13)
Asteraceae	0.1	0.7	2.5	1.0	0.8	0.4	0.5	0.3	2.1	2		1.0 (17)
Cupressaceae	5.2	9.0	27.5	23.0	22.5	58.0	13.3	0.4	1.0	1.7	28.5	17.3 (2)
Pinus	0.1	4.5	31.0	3.0	17.0	1.2	3.3	0.3	0.5	2.4	11.4	6.8 (5)
Poaceae	18.0	71.0	23.0	17.5	15.0	10.8	10.4	84.0	43.7	51.6	25.8	33.7 (1)
Plantago		0.2		5.1	3.2	1.8	3.4	11.1	16.0	8.6		6.2 (7)
Ulmus				<0.1	3.8	0.4	1.6					1.5 (14)
Oleaceae				20.0	1.8	5.2	5.2		<0.1		7.2	6.6 (6)
Salix				0.2	1.7		5.0		<0.1	0.2		1.4 (15)
Rumex					4.3	0.1	2.3	3.6	3.5	6.5		3.4 (9)
Quercus				0.1	3.1	4.9	1.6	<0.1	0.2	11.2		3.0 (10)
Alnus					3.9	1.8	2.0					2.6 (11)
Betula					1.7	4.5	16.7		0.6	0.35	27.0	8.5 (3)
Coprosma							0.4	<0.1	0.1	1.7		0.6 (19)
Sub-total	93.9	98.3	90.0	88.3	90.5	96.3	79.4	99.8	84.0	96.1	99.9	92.4
Other	6.1	1.7	10.0	11.7	9.5	3.7	20.6	0.2	16	3.9	0.1	7.6

Pollen taxa values are expressed as a percentage of the total pollen count for the record period available in each urban area (*excluded from ranking as only occurs in one urban area). The average and rank of percentage values across all urban areas is given in the right hand column. The category "Other" includes all pollen taxa counted in the total pollen sum and not identified to taxa here.

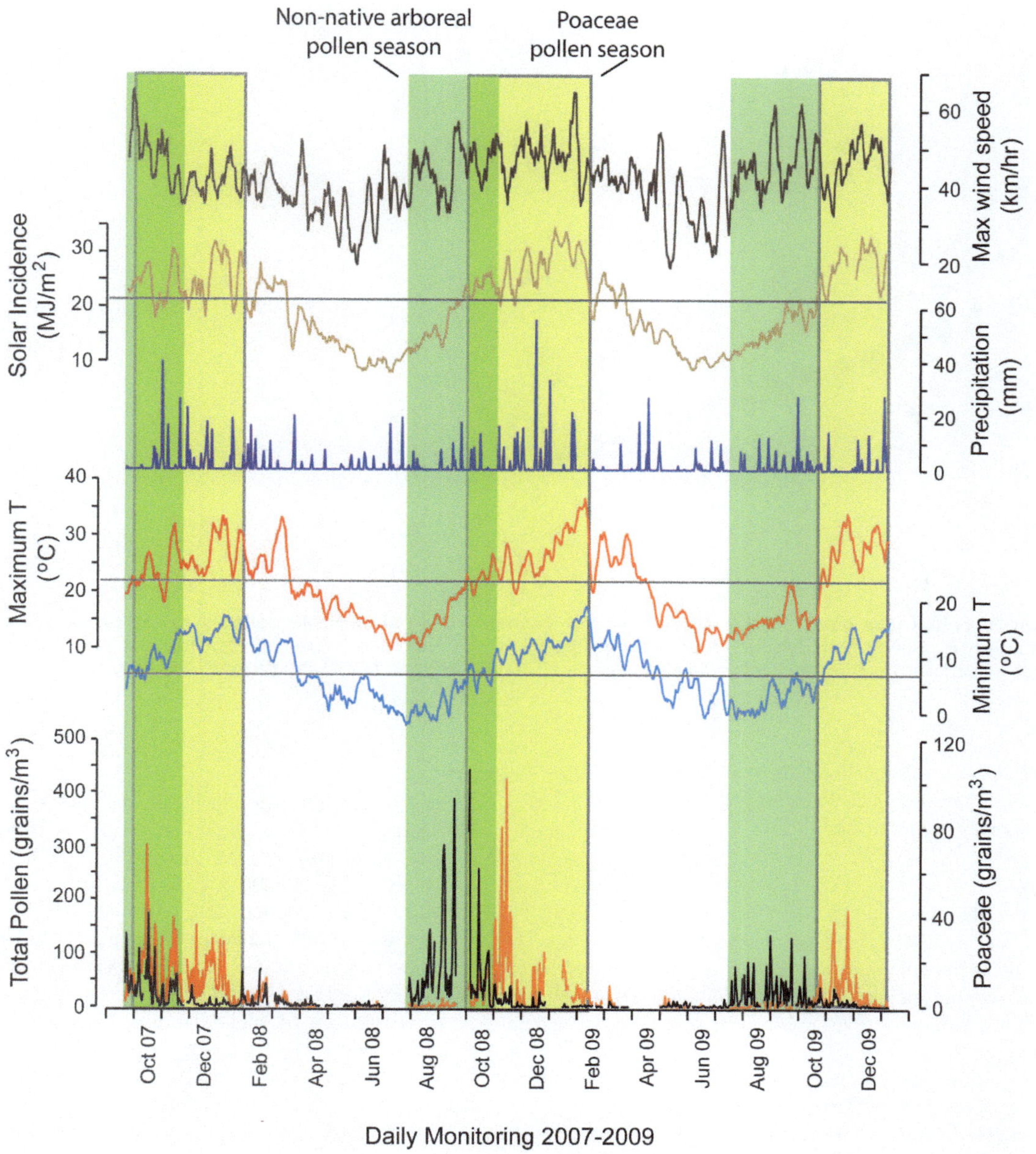

Figure 2. Climate summary and daily airborne pollen (Poaceae and Non-native arboreal taxa) for Canberra (26 Sept 2007–31 Dec 2009). Pollen season is depicted by the shaded columns and defined by 90% of the airborne pollen falling in this time for each year (July 1 to June 30). Climate data from the Australian Bureau of Meteorology.

stations is representative of the total potential pollen dispersal area. However, it must be acknowledged that little is known of the pollen dispersal characteristics of Australasian pollen types and that further research is required to understand the influence that meteorological factors such as wind strength and distance from source vegetation might have on pollen concentrations in the atmosphere.

Results

Pollen Ranking Among Urban Areas

The airborne pollen data were ranked by percentage representation, with the top 20 pollen taxa making up >80% of the total pollen collected across 11 cities in Australia and New Zealand (Table 2 and Figure 3). The dataset incorporates taxa of high relative percentage representation that commonly occur in two or more urban areas (Arecaceae is an exception because of the high values recorded in Darwin). The most significant taxa across all sites are Poaceae and Cupressaceae, making up over 50% of the total airborne pollen in urban environments throughout the year. These are followed by *Betula*, the trees and shrubs in Myrtaceae, *Pinus*, Oleaceae, *Casuarina*, and the important herbaceous taxa such as *Plantago* and *Rumex*. Differences between the pollen taxa rankings (highest to lowest percentage representation) across the broad biogeographic regions of Australia (tropics to temperate: Australian Tropical/Subtropical = Darwin, Brisbane; Australian Temperate = Perth, Sydney, Canberra, Melbourne, Hobart) and New Zealand (North and South Island: NZ North = Kaikohe, Auckland; NZ South = Christchurch, Dunedin), reflect the strong climate controls on plant distributions, particularly those associ-

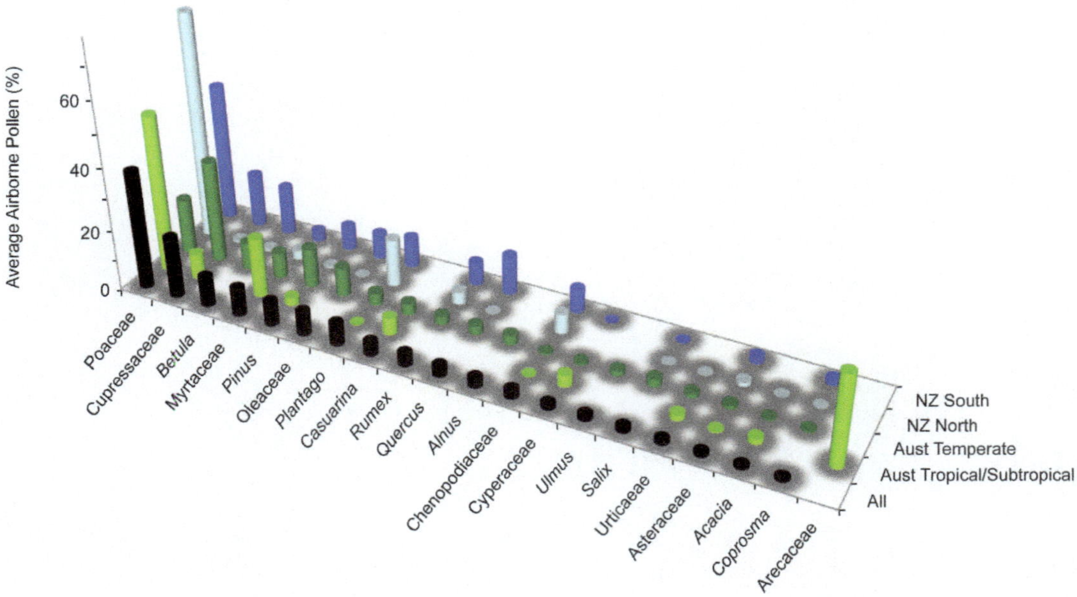

Figure 3. Ranking for top 20 pollen taxa based on average percentage representation of airborne pollen areas across all urban areas (black bars). These urban areas are then grouped into regional biogeographic zones (blank space = taxa not recorded). Australian Tropical/ Subtropical = Darwin, Brisbane; Australian Temperate = Perth, Sydney, Canberra, Melbourne, Hobart; NZ North = Kaikohe, Auckland; NZ South = - Christchurch, Dunedin.

ated with Northern Hemisphere introductions such as *Betula*, *Quercus*, *Alnus* and *Ulmus* that contribute high levels of pollen into the atmosphere of the southern temperate cities.

Biogeographic Patterns

The two dimensional ordination had a low stress level (stress = 0.11) indicating it was a robust graphical representation of the similarity matrix among the cities and the dominant taxa. The ordination displays a gradation in vegetation formation (and land cover) across the set of urban areas (Figure 4A). The x-axis separates the tree dominant urban landscapes (negative values associated with pie charts dominated by tree taxa) from the grassland dominated urban landscapes (positive values associated with pie charts dominated by herbaceous taxa). The y-axis is associated with latitudinal position of Australian and New Zealand sites with positive values corresponding to lower latitude urban areas and negative values corresponding with higher latitude urban areas. The most significant environmental variable explaining the difference between airborne pollen in each urban area was minimum annual temperature (MinT, $r^2 = 0.77$, $P = 0.002$), which is closely aligned to the latitudinal transect along which the urban areas lie (Table 3). The second environmental factor that showed a significant correlation with the airborne pollen was mean annual precipitation (MAP, $r^2 = 0.66$, $P = 0.01$). Increased grass pollen dominance is clearly apparent in Figure 4, particularly in the savanna dominated tropics (Darwin), dry sclerophyll subtropical forests (Brisbane) and pastoral dominated temperate landscapes of New Zealand, where associated ruderal taxa such as *Rumex*, *Plantago*, Asteraceae and Chenopodiaceae are also prominent. The contribution of grass pollen was reduced in the southern temperate cities, where other woody taxa such as Cupressaceae (including both native and introduced species) and the introduced Northern Hemisphere trees (*Betula*, *Alnus*, *Ulmus*, and *Quercus*) become significant contributors.

Pollen Seasons

Figure 5 illustrates the Australasian pollen calendar for the top 20 taxa across 11 urban areas. The matrix is organised according to geographical distribution from tropical to temperate for both urban areas and pollen taxa. The clear shifts from long pollen seasons in the tropics to shorter periods in the temperate regions reflects the strong control of solar radiation incidence on pollen production during spring and summer months in the southern urban areas [36]. There is also a discernible shift in the initiation and length of pollen season towards higher latitude. For example, Poaceae has a long flowering season in the northern cities of Darwin and Brisbane (across the dry season and into the wet season as different grass species flower through the year), compared to the progressively shorter and later initiation of the pollen season in the southern cities.

The effect of temperature and solar incidence on the beginning and end of the pollen season in the temperate zone is illustrated with an example from Canberra in Figure 2, where the relationship between airborne pollen and key climate variables for Poaceae and total non-native tree species (Northern Hemisphere) is shown. While this is simply a visual comparison, previous studies from Melbourne [15,37], Brisbane [26] and New Zealand [38] have also documented the significant influence of temperature and precipitation on the commencement and duration of the grass pollen season. The lack of consistent pollen season in some taxa such as Oleaceae may be a function of the multiple species from different genera being incorporated into this taxon (including *Olea* and *Fraxinus*), reflecting a significant limitation imposed by the inability to resolve pollen to the species level.

Discussion

The history of aerobiology research in Australia and New Zealand, like the rest of the world, can be characterised by its focus on local site (urban area) issues. Unlike other continents however,

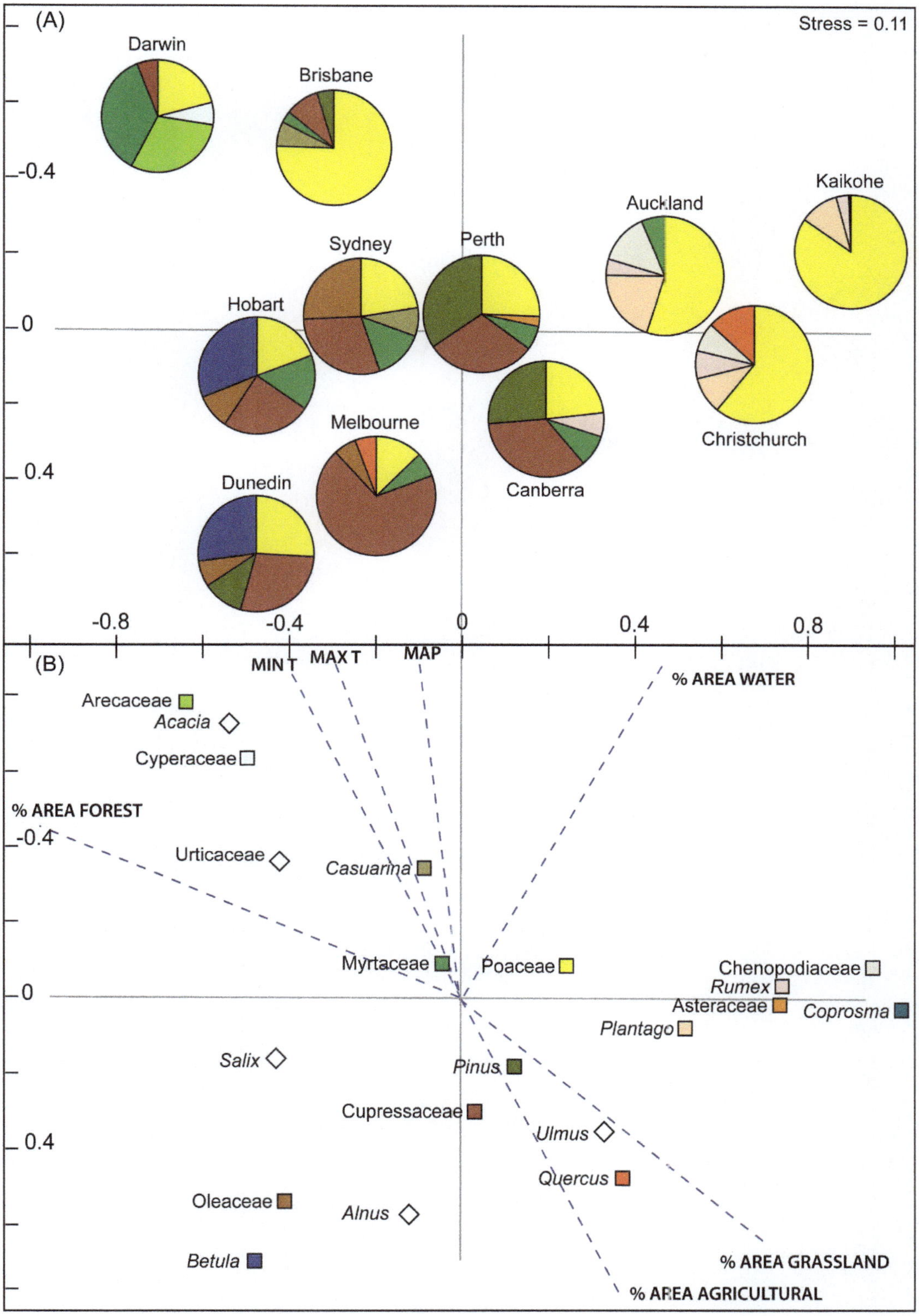

Figure 4. Non-metric multidimensional scaling ordination (nMDS) of the major pollen taxa in Australia and New Zealand. (**A**) Distribution of urban areas using a matrix of percentage representation of major pollen taxa (see Table 2 and using top eight pollen taxa at each site). The pie charts depict the relative contribution of the most abundant pollen taxa in each urban area. (**B**) Distribution of pollen taxa contributing to the differentiation of aerobiology of each urban area Note: coloured squares associated with each taxa depicted in (B) match the pie chart colours shown in (A), with diamonds showing taxa with a low percentage representation. The vectors (dotted lines) for the environmental variables show positions for each variable within the environmental space. Longer vectors (higher r^2 values, see Table 3) indicate a stronger association of the environmental variable with site/species variation in the ordination space.

Table 3. Land cover attributes within a 100 km radius associated with each aerobiology recording station, nearest weather station summary data and nMDS axis 1 and axis 2 scores.

Urban Area	%Area Water	%Area Agricultural	%Area Grass	%Area Forest	MaxT (°C)	MinT (°C)	MAP (mm)
Darwin	45.5	0.1	0.2	54.3	33.3	19.3	1730
Brisbane	34.9	1.5	0.1	10.1	30.3	10.0	997
Perth	38.1	31.6	3.5	26.8	31.4	7.7	850
Sydney	51.1	4.2	0.3	44.3	25.9	8.0	1213
Canberra	2.2	30.2	5.1	62.5	28.0	−0.1	612
Melbourne	19.0	27.6	14.3	39.1	25.9	6.0	650
Hobart	40.2	6.3	7.3	46.2	21.6	4.5	616
Kaikohe	65.1	0.6	22.2	12.2	24.3	7.8	1304
Auckland	62.9	2.2	24.8	10.1	23.7	7.1	1239
Christchurch	57.9	20.5	18	3.6	23.0	2.0	651
Dunedin	62.4	6.1	31.1	0.4	18.9	3.2	814
nMDS Axis 1	0.48	0.36	0.72	−0.89	−0.29	−0.40	−0.11
nMDS Axis 2	−0.88	0.93	0.70	−0.45	−0.96	−0.92	−0.99
r^2	0.14	0.39	0.33	0.34	0.38	**0.77**	**0.66**
Pr(>r)	0.568	0.164	0.205	0.195	0.154	**0.002**	**0.013**

The significant relationships ($p<0.05$) between environmental variables and airborne pollen are indicated in bold (T = temperature, MAP = Mean Annual Precipitation). Meteorological data derived from Worldclim [34] and land cover data from Giri *et al.* [35].

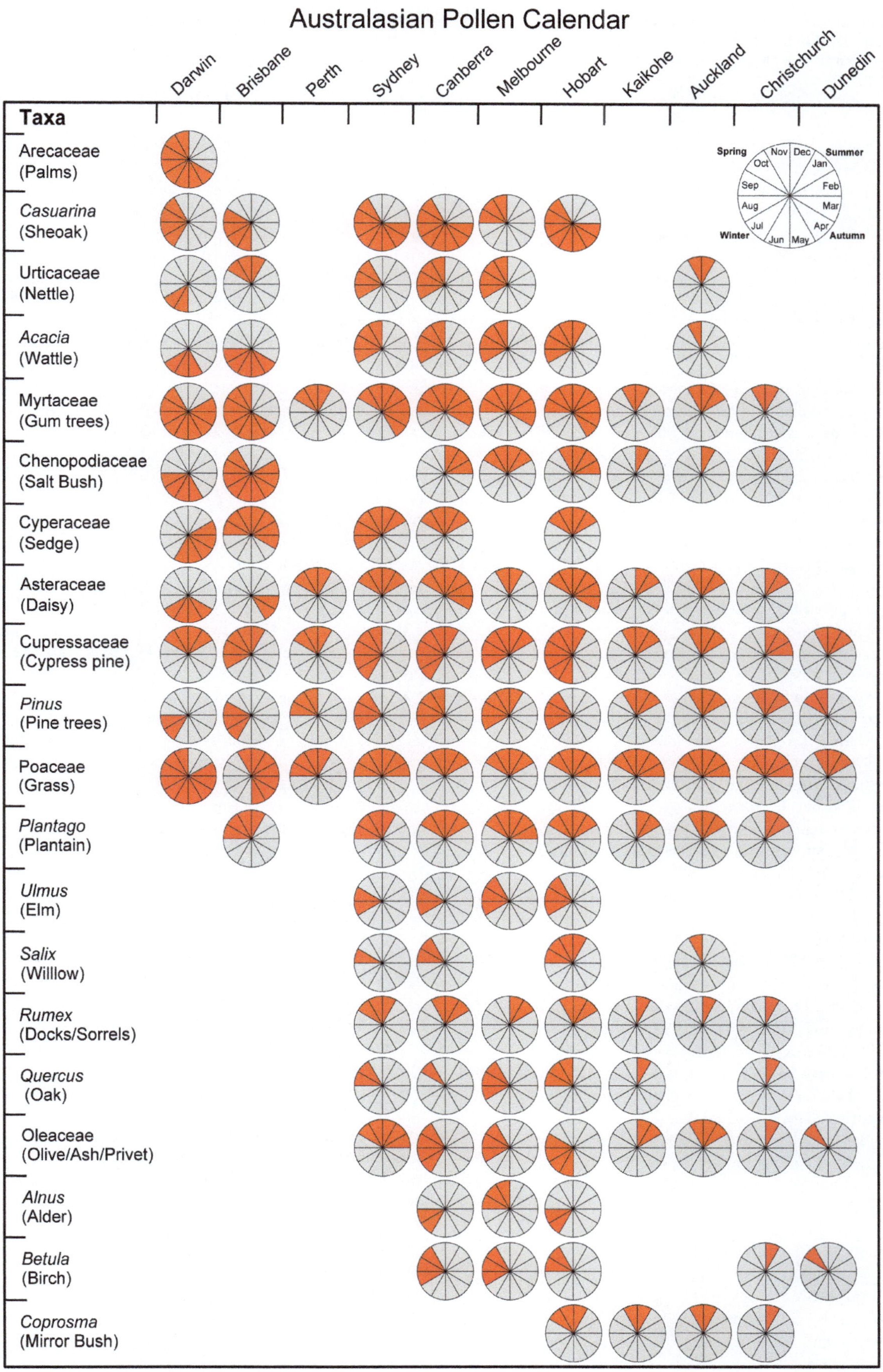
Australasian Pollen Calendar
Darwin
Brisbane
Perth
Sydney
Canberra
Melbourne
Hobart
Kaikohe
Auckland
Christchurch
Dunedin
Taxa
Arecaceae (Palms)
Casuarina (Sheoak)
Urticaceae (Nettle)
Acacia (Wattle)
Myrtaceae (Gum trees)
Chenopodiaceae (Salt Bush)
Cyperaceae (Sedge)
Asteraceae (Daisy)
Cupressaceae (Cypress pine)
Pinus (Pine trees)
Poaceae (Grass)
Plantago (Plantain)
Ulmus (Elm)
Salix (Willow)
Rumex (Docks/Sorrels)
Quercus (Oak)
Oleaceae (Olive/Ash/Privet)
Alnus (Alder)
Betula (Birch)
Coprosma (Mirror Bush)
Spring
Summer
Winter
Autumn
Nov
Dec
Oct
Jan
Sep
Feb
Aug
Mar
Jul
Apr
Jun
May

Figure 5. Pollen calendar for aerobiologically significant pollen taxa (contributing 80%+ to the annual atmospheric pollen) in Australian and New Zealand urban areas. The pie charts are divided into monthly segments with the red shade depicting the pollen season for each taxa to the nearest month in a given urban area. The pollen season for each taxon is determined using the period encompassing 90% of the annual pollen rain (see Figure 2).

where considerable effort has been applied to standardise methodologies, teams working in Australia and New Zealand have tended to use a variety of collection and counting methodologies with little or no co-ordination between aerobiology stations in different urban areas. Despite the idiosyncrasies of the data, this study supports a basic principle that underpins palynology – that airborne pollen is a sensitive proxy of the current climate and flora of a region. Our results indicate that this principle applies as much to the introduced flora as to indigenous vegetation, despite concerted human effort to manipulate the former.

We found that Australian and New Zealand urban areas from a similar climate zones have similar pollen spectra but important differences occur due to surrounding land use and the establishment of non-native plants. Likewise, in similar climate zones, pollen season of each taxon is similar but not the same. For example, urban areas are often surrounded by agricultural landscapes that have a diversity of pollen predominantly represented by grass pollen and characterized by short seasons (Figure 5) compared to urban areas where there is more surrounding forest cover (e.g. Sydney and Hobart). Changes in land use on urban boundaries also have the potential to affect the types of airborne pollen inside the urban area [39]. The predominance of exotic and invasive tree and shrub species (and certainly non-native grasses, albeit these cannot be taxonomically resolved) highlights the profound changes that have occurred following European colonisation of Australia and New Zealand. For instance, in the urban areas of southern Australia and New Zealand, the widely planted birches are particularly dominant.

Temperature and rainfall are known to be primary controls on the daily distribution of airborne pollen in Australia and New Zealand, however, other climate variables that may influence the dissemination of these allergenic pollen types in the atmosphere are not well understood. Other variables such as the El Niño-Southern Oscillation (ENSO) have not been investigated, even though this large-scale ocean-atmosphere anomaly has been shown to modify climatic patterns, leading to droughts and floods, which have local and regional implications on the biosphere as well as vector borne diseases. A similar climate oscillation in western Europe, the North Atlantic Oscillation, has been linked with seasonal variation in grass pollen (e.g. [40]). ENSO and other short term climate oscillations such as the Southern Annular Mode and Indian Ocean Dipole may be important variables that could account for interseasonal differences observed in airborne pollen counts and seasonal starting dates in Australasia. The broad-scale patterns in the distribution, abundance and season of pollen amongst Australasian urban areas provide some clues as to potential changes in aerobiology due to climate change. Evidence from Europe and North America demonstrates that climate change has already increased the abundance and seasonal duration of allergenic pollens such as birch and ragweed and possibly increased concentrations of allergenic compounds [9,12,41]. Changes in the burden of allergenic disease related to changing climate have also been demonstrated [41,42].

In marked contrast to developed countries of the northern hemisphere [43], the potential allergenic impacts of endemic southern hemisphere plants are poorly characterized despite many being identified as important allergens. Given the widespread population exposure to known allergenic plants we endorse the recommendation of the 2007 study of the economic impact of allergic diseases in Australia by Access Economics that *'studies of the aerobiology and clinical significance of potential native Australian triggers of respiratory allergic disease should be made a priority'* [44]. Possible triggers include both native and non-native taxa. Native genera include common woody plants (*Eucalyptus*, *Melaleuca*, *Callistemon* and *Acacia* [45–51]) and grasses (*Sorghum*, *Sarga* and *Andropogon* [52–54]). Non-native plants include a number of taxa known to be allergenic. For example, birch is an important allergenic pollen in Scandinavia and a notable allergen throughout north-central Europe [6,55] and it has been identified as an important allergenic pollen type in southern Australian urban areas [56]. Similarly, the non-native gamba grass (*Andropogon gayensis*) is thought to have lengthened the pollen season and increased the community burden of allergic rhinitis in Darwin [57].

Exotic allergenic plants have the potential to cause profound public health impacts if their ranges were to expand and their population to increase. For example, pollen produced by olive trees (*Olea europaea*) is the leading cause of seasonal allergic diseases in some regions of southern Europe [58], and this species is expanding its range across mediterranean climate zones in Australia [59] adding a new allergic pollen in areas where there are already allergic pollen in the atmosphere. If Ragweed species (*Ambrosia* spp.) were to expand their range and abundance they would add a new source of allergenic into autumn [60–62], a season currently with very few types of pollen (see Figure 5). Because allergic sensitisation to multiple plant allergens is common [63], our results show that the juxtaposition of non-native tree pollen such as birch in early-mid spring with Poaceae pollen in mid-late spring and summer could result in a lengthened period of risk for people allergic to pollens (cf. Figure 2).

Conclusions

Our study highlights the need to monitor changes to the aerobiology and provides a framework for targeting the most important taxa in terms of abundance and allergenic effects for each urban area. Establishing systematic regionally-based monitoring of airborne pollen will enable Australian and New Zealanders to better understand the high taxonomic diversity and seasonal variability of allergenic pollens. This will redress the paucity of research on the clinical and public health impacts and treatments for common endemic allergenic species such as *Eucalyptus* and *Sorghum*. Monitoring the increasing abundance of allergenic exotic species populations that can move from a dormant 'sleeper' populations to aggressive expansion phases, Ragweed (*Ambrosia* spp.) being a example, will also be a priority. Understanding the impacts that climate change will have on the phenological cycles and range of allergenic species into the future will be a critical step in the advancement of aerobiology studies in the Australasian region [9].

Acknowledgments

The authors wish to thank the staff of Australian Centre for Ecological Analysis and Synthesis, Terrestrial Ecosystem Research Network for assistance in organising the Workshop 1 of the Working Group "Understanding Australian aerobiology to monitor environmental change and human allergenic exposure", North Stradbroke Island, Australia (11–15 March, 2013). Alison Specht, Kaitao Lai and Siddeswara Guru assisted with the construction and implementation of the online aerobiology data set. We thank Doctors Diana Bass and Geoffrey Morgan for provision of their published pollen count data from Sydney. Alison Jaggard has been assisted by the Environmental Trust (project reference number 2011/RD/0049), New South Wales Government. The authors also acknowledge the contributions a diverse group of people made to this study through counting pollen and maintaining the pollen records that we consolidated and analysed.

Author Contributions

Conceived and designed the experiments: SGH DMJSB RMN FHJ PJB JB BC BE IG BJG AJ DM EN MT AH DV JMD. Analyzed the data: SGH GJW. Wrote the paper: SGH DMJSB RMN FHJ. Contributed data: EN BE SGH DM DMJSB FM BJG FHJ RMN. The application for funding support was led by the principal investigator: JMD.

References

1. D'Costa D, Kershaw AP (1997) An expanded recent pollen database from south-eastern Australia and its potential for refinement of palaeoclimatic estimates. Aust J Bot 45: 583–605.
2. Wilmshurst JM, McGlone MS (2005) Origin of pollen and spores in surface lake sediments: comparison of modern palynomorph assemblages in moss cushions, surface soils and surface lake sediments. Rev Palaeobot Palynol 136: 1–15.
3. Fletcher M-S, Thomas I (2007) Modern pollen–vegetation relationships in western Tasmania, Australia. Rev Palaeobot Palynol 146: 146–168.
4. Tng DYP, Hopf F, Haberle SG, Bowman DMJS (2010) Seasonal pollen distribution in the atmosphere of Hobart, Tasmania: preliminary observations and congruence with flowering phenology. Aust J Bot 58: 440–452.
5. Schäppi GF, Taylor PE, Kenrick J, Staff IA, Suphioglu C (1998) Predicting the grass pollen count from meteorological data with regard to estimating the severity of hayfever symptoms in Melbourne (Australia). Aerobiologia 14: 29–37.
6. Emberlin J, Detandt M, Gehrig R, Jaeger S, Nolard N, et al. (2002) Responses in the start of *Betula* (birch) pollen seasons to recent changes in spring temperatures across Europe. Int J Biometeorol 46: 159–170. [Erratum published (2003) 47: 113–115.].
7. Rodríguez-Rajo FJ, Frenguelli G, Jato MV (2003) Effect of air temperature on forecasting the start of the Betula pollen season at two contrasting sites in the south of Europe (1995–2001). Int J Biometeorol 47: 117–125.
8. Sofiev M, Siljamo P, Ranta H, Linkosalo T, Jaeger S, et al. (2013) A numerical model of birch pollen emission and dispersion in the atmosphere. Description of the emission module. Int J Biometeorol 57: 45–58.
9. Ziska L, Knowlton K, Rogers C, Dalan D, Tierney N, et al. (2011) Recent warming by latitude associated with increased length of ragweed pollen season in central North America. Proc Natl Acad Sci U S A 108: 4248–4251.
10. Ziello C, Sparks TH, Estrella N, Belmonte J, Bergmann KC, et al. (2012) Changes to airborne pollen counts across Europe. PLoS ONE 7: e34076.
11. Ziska LH, Beggs PJ (2012) Anthropogenic climate change and allergen exposure: The role of plant biology. J Allergy Clin Immunol 129: 27–32.
12. Newnham RM, Sparks TH, Skjøth CA, Head K, Adams-Groom B, et al. (2013) Pollen season and climate: Is the timing of birch pollen release in the UK approaching its limit? Int J Biometeorol 57: 391–400.
13. Newnham RM, Fountain DW, Cornford CC, Forde MB (1995) A national survey of airborne pollen and grass flowering in New Zealand, with implications for respiratory disorder. Aerobiologia 11: 239–252.
14. Bass D, Morgan G (1997) A three year (1993–1995) calendar of pollen and *Alternaria* mould in the atmosphere of south western Sydney. Grana 36: 293–300.
15. Ong EK, Singh MB, Knox RB (1995) Seasonal distribution of pollen in the atmosphere of Melbourne: an airborne pollen calendar. Aerobiologia 11: 51–55.
16. Green BJ, Dettmann ME, Yli-Panula E, Rutherford S, Simpson R (2004) Aeropalynology of Australian native arboreal species in Brisbane, Australia. Aerobiologia 20: 43–52.
17. Stevenson J, Haberle SG, Johnston FH, Bowman DMJS (2007) Seasonal distribution of pollen in the atmosphere of Darwin, tropical Australia: Preliminary results. Grana 46: 34–42.
18. Medek DE, Kljakovic M, Fox I, Pretty DG, Prebble M (2012) Hay fever in a changing climate: linking an internet-based diary with environmental data. Ecohealth 9: 440–447.
19. European Aerobiology Society (2011) Minimum requirements to manage aerobiological monitoring stations included in a national network involved in the EAN. International Aerobiology Newsletter 72: 1.
20. Knox RB (1979) Pollen and Allergy. Southampton, Edward Arnold, 60pp.
21. Hirst JM (1952) An automatic volumetric spore trap. Ann Appl Biol 39: 257–265.
22. Chapman JA (1982) The enhancement of the practice of clinical allergy with daily pollen and spore counts. Immunol Allergy Pract 4: 13–18.
23. Comtois P, Mandrioli P (1997) Pollen capture media: a comparative study. Aerobiologia 13: 149–154.
24. Lacey ME, West JS (2006) The air spora. A manual for catching and identifying airborne biological particles. Dordrecht: Springer.
25. Comtois P, Alcazar P, Néron D (1999) Pollen counts statistics and its relevance to precision. Aerobiologia 15: 19–28.
26. Green BJ, Dettmann ME, Rutherford S, Simpson RW (2002) Airborne pollen of Brisbane, Australia: a five-year record, 1994–1999. Grana 41: 242–250.
27. Buters JTM, Thibaudon M, Smith M, Kennedy R, Rantio-Lehtimäki A, et al. (2012) Release of Bet v 1 from birch pollen from 5 European countries. Results from the HIALINE study. Atmos Environ 55: 496–505.
28. Hjelmroos M, Beyon F, Culliver S, Jones AS, Tovey E (1999) Airborne allergens: Interactive identification of allergenic pollen and fungal spores. Compact Disc, Institute of Respiratory Medicine Limited, Sydney.
29. Jato V, Rodríguez-Rajo FJ, Alcázar P, De Nuntiis P, Galán C, et al. (2006) May the definition of pollen season influence aerobiological results? Aerobiologia 22: 13–25.
30. Nilsson S, Persson S (1981) Tree pollen spectra in the Stockholm region (Sweden), 1973–1980. Grana 20: 179–182.
31. Spieksma FThM, Emberlin JC, Hjelmroos M, Jäger S, Leuschner RM (1995) Atmospheric birch (*Betula*) pollen in Europe: Trends and fluctuations in annual quantities and the starting dates of the seasons. Grana 34: 51–57.
32. Clarke KR (1993) Non-parametric multivariate analyses of changes in community structure. Aust J Ecol 18: 117–143.
33. Oksanen J, Blanchet FG, Kindt R, Legendre P, Minchin PR, et al. (2013) vegan: Community Ecology Package. R package version 2.0-8. http://CRAN.R-project.org/package = vegan
34. Hijmans RJ, Cameron SE, Parra JL, Jones PG, Jarvis A (2005) Very high resolution interpolated climate surfaces for global land areas. Int J Climatol 25: 1965–1978.
35. Giri C, Zhu Z, Reed B (2005) A comparative analysis of the Global Land Cover 2000 and MODIS land cover data sets. Remote Sens Environ 94: 123–132.
36. Hutchinson MF, McIntyre S, Hobbs RJ, Stein JL, Garnett S, et al. (2005) Integrating a global agro-climatic classification with bioregional boundaries in Australia. Global Ecol Biogeogr 14: 197–212.
37. Ong EK, Singh MB, Knox RB (1995) Grass pollen in the atmosphere of Melbourne: seasonal distribution over nine years. Grana 34: 58–63.
38. Newnham RM (1999) Monitoring biogeographical response to climate change: The potential role of aeropalynology. Aerobiologia 15: 87–94.
39. Emberlin J, Mullins J, Corden J, Jones S, Millington W, et al. (1999) Regional variations in grass pollen seasons in the UK, long-term trends and forecast models. Clin Exp Allergy 29: 347–356.
40. Smith M, Emberlin J, Stach A, Rantio-Lehtimaki A, Caulton E, et al. (2009) Influence of the North Atlantic Oscillation on grass pollen counts in Europe. Aerobiologia 25 (4): 321–332.
41. Beggs PJ (2004) Impacts of climate change on aeroallergens: past and future. Clin Exp Allergy 34: 1507–1513.
42. Beggs PJ, Bambrick HJ (2005) Is the global rise of asthma an early impact of anthropogenic climate change? Environ Health Perspect 113: 915–919.
43. Patel NJ, Bush RK (2000) The role of environmental allergens in rhinitis. Immunol Allergy Clin North Am 20: 323–353.
44. Access Economics PL (2007) The economic impact of allergic disease in Australia: not to be sneezed at. Australasian Society of Clinical Immunology and Allergy (ASCIA). 111 p.
45. Howlett BJ, Hill DJ, Knox RB (1982) Cross-reactivity between *Acacia* (wattle) and rye grass pollen allergens: Detection of allergens in *Acacia* (wattle) pollen. Clin Allergy 12: 259–268.
46. Sweeney M, Hosseiny S, Hunter S, Klotz SD, Gennaro RN, et al. (1994) Immunodetection and comparison of melaleuca, bottlebrush, and bahia pollens. Int Arch Allergy Immunol 105: 289–296.
47. Liam C-K, Loo K-L, Wong CM-M, Lim K-H, Lee T-C (2002) Skin prick test reactivity to common aeroallergens in asthmatic patients with and without rhinitis. Respirology 7: 345–350.
48. Altintaş DU, Karakoç GB, Yilmaz M, Pinar M, Kendirli SG, et al. (2004) Relationship between pollen counts and weather variables in East-Mediterranean Coast of Turkey: Does it affect allergic symptoms in pollen allergic children? Clin Dev Immunol 11: 87–96.
49. Sritipsukho P (2004) Aeroallergen sensitivity among Thai children with allergic respiratory diseases: a hospital-based study. Asian Pac J Allergy Immunol 22: 91–95.
50. Rutherford S (2001) Air pollution and asthma severity in South East Queensland: exposure and effects [PhD]. Griffith University.
51. Hanigan IC, Johnston FH (2007) Respiratory hospital admissions were associated with ambient airborne pollen in Darwin, Australia, 2004–2005. Clin Exp Allergy 37: 1556–1565.
52. Davies JM, Li H, Green M, Towers M, Upham JW (2012) Subtropical grass pollen allergens are important for allergic respiratory diseases in subtropical regions. Clin Transl Allergy 2:4.

53. Lazarides M, Hacker JB, Andrew MH (1991) Taxonomy, cytology and ecology of indigenous Australian sorghums (*Sorghum* Moench: Andropogoneae: Poaceae). Aust Syst Bot 4: 591–635.
54. Spangler RE (2003) Taxonomy of *Sarga*, *Sorghum* and *Vacoparis* (Poaceae: Andropogoneae). Aust Syst Bot 16: 279–299.
55. Burbach GJ, Heinzerling LM, Edenharter G, Bachert C, Bindslev-Jensen C, et al. (2009) GA^{2}LEN skin test study II: Clinical relevance of inhalant allergen sensitizations in Europe. Allergy 64: 1507–1515.
56. Baldo BA, Wrigley CW (1984) Allergy in Australia: Symptoms, diagnosis and treatment. Special Supplement. Med J Aust 141: S12–S18.
57. Johnston FH, Hanigan IC, Bowman DMJS (2009) Pollen loads and allergic rhinitis in Darwin, Australia: a potential health outcome of the grass-fire cycle. Ecohealth 6: 99–108.
58. D'Amato G, Liccardi G (2002) The increasing trend of seasonal respiratory allergy in urban areas. Allergy 57: 35–36.
59. Cuneo P, Leishman MR (2006) African olive (*Olea europaea* subsp. *cuspidata*) as an environmental weed in eastern Australia: a review. Cunninghamia 9: 545–77.
60. White JF, Bernstein DI (2003) Key pollen allergens in North America. Ann Allergy Asthma Immunol 91: 425–435; quiz 35–6, 92.
61. Bass DJ, Delpech V, Beard J, Bass P, Walls RS (2000) Ragweed in Australia. Aerobiologia 16: 107–111.
62. Bass DJ, Delpech V, Beard J, Bass P, Walls RS (2000) Late summer and fall (March–May) pollen allergy and respiratory disease in Northern New South Wales, Australia. Ann Allergy Asthma Immunol 85: 374–381.
63. Greiner AN, Hellings PW, Rotiroti G, Scadding GK (2011) Allergic rhinitis. Lancet 378: 2112–2122.

20

Fire Activity and Severity in the Western US Vary along Proxy Gradients Representing Fuel Amount and Fuel Moisture

Sean A. Parks[1]*, Marc-André Parisien[2], Carol Miller[1], Solomon Z. Dobrowski[3]

1 Aldo Leopold Wilderness Research Institute, Rocky Mountain Research Station, USDA Forest Service, Missoula, Montana, United States of America, **2** Northern Forestry Centre, Canadian Forest Service, Natural Resources Canada, Edmonton, Alberta, Canada, **3** Department of Forest Management, College of Forestry and Conservation, University of Montana, Missoula, Montana, United States of America

Abstract

Numerous theoretical and empirical studies have shown that wildfire activity (e.g., area burned) at regional to global scales may be limited at the extremes of environmental gradients such as productivity or moisture. Fire activity, however, represents only one component of the fire regime, and no studies to date have characterized *fire severity* along such gradients. Given the importance of fire severity in dictating ecological response to fire, this is a considerable knowledge gap. For the western US, we quantify relationships between climate and the fire regime by empirically describing both fire activity and severity along two climatic water balance gradients, actual evapotranspiration (AET) and water deficit (WD), that can be considered proxies for fuel amount and fuel moisture, respectively. We also concurrently summarize fire activity and severity among ecoregions, providing an empirically based description of the geographic distribution of fire regimes. Our results show that fire activity in the western US increases with fuel amount (represented by AET) but has a unimodal (i.e., humped) relationship with fuel moisture (represented by WD); fire severity increases with fuel amount and fuel moisture. The explicit links between fire regime components and physical environmental gradients suggest that multivariable statistical models can be generated to produce an empirically based fire regime map for the western US. Such models will potentially enable researchers to anticipate climate-mediated changes in fire recurrence and its impacts based on gridded spatial data representing future climate scenarios.

Editor: Matthew Germino, US Geological Survey, United States of America

Funding: This research was conducted with National Fire Plan funds provided by the Rocky Mountain Research Station, USDA Forest Service. The funders had no role in study design, data collection and analysis, decision to publish, or preparation of the manuscript.

Competing Interests: The authors have declared that no competing interests exist.

* E-mail: sean_parks@fs.fed.us

Introduction

Fire is a ubiquitous ecosystem process across the globe. The concept of the fire regime has been used to describe the role of fire in an ecosystem in terms of its spatial-temporal patterns and ecosystem impacts [1]. As such, maps of fire regimes, in terms of fire frequency and severity, have been produced [2]. These maps not only allow geographic comparisons of fire regime components [3], but are also a necessary first step for describing shifts in fire regimes resulting from factors such as climate change [4], fire suppression [5], and invasive species [6]. Though the importance of mapping fire regimes is long acknowledged [7,8], efforts to date have been largely qualitative and thus have been unable to make direct linkages to factors driving both fire activity and severity (but see [9]).

Recently, however, linkages between environmental gradients and fire activity have been identified. Several studies have shown that wildland fires across the globe tend to avoid environmental extremes of productivity and moisture [10–13]. For example, in very dry areas such as deserts, fire activity (i.e., fire occurrence and area burned) is limited by a lack of biomass [14]. Conversely, in the wettest places on Earth (e.g., rainforests), there is ample biomass but fire activity is limited because climate conditions promoting combustion are rare [15]. To date, studies of the geographic distribution of fire, or pyrogeography, have focused more on patterns of *fire activity* (e.g., [16–18]) than on explaining patterns of *fire severity* (but see [19,20]). Given that fire severity – a measure of ecosystem change – can strongly dictate the response of biological communities to fire [21–23], an understanding of its environmental controls is a prerequisite for understanding the role of fire in ecosystems. Furthermore, without an understanding of these controls, modeling and prediction of climate-mediated changes in fire regimes is tenuous.

The seminal biome classification of Whittaker [24] exemplifies the utility of representing large scale ecological features along an energy-moisture biplot (temperature and precipitation in this case). As a refinement of Whittaker's framework, Stephenson [25] developed a bivariate representation of North American's main vegetation types using actual evapotranspiration (AET) and climatic water deficit (WD; potential evapotranspiration minus AET), as measures of productivity and drought, respectively. Specifically, AET represents a measure of available moisture to plants, whereas WD is a measure of unmet atmospheric demand for water (i.e., how much water could be evaporated and transpired were it available) [25]. Though related to precipitation and temperature, AET and WD take into account the concurrent availability and demand for moisture and therefore better represent environmental controls on ecosystems. As such, these metrics have been shown to be strong predictors of plant physiognomic types [25] and tree species distributions [26].

Recent studies have shown that water balance metrics such as AET and WD are predictive of annual to decadal fire activity [27,28]. AET is strongly correlated with biomass production [29,30], and as such, it is a suitable proxy for fuel amount in many ecosystems. WD is a measure of absolute drought and should reflect the moisture status and flammability of live [31] and large-diameter dead fuels [28]. Previous research has suggested a unimodal (i.e., humped) response of fire activity to fuel amount (e.g., [11]) and fuel moisture [32]; that is, less fire occurs at the extremes of these gradients than at intermediate levels. Less clearly understood are the relationships between severity and these gradients. Some studies have suggested that fire severity depends upon amount of biomass [33] and therefore would be expected to increase with AET; the relationship between fire severity and fuel moisture, however, is unknown.

We seek to better understand how contemporary fire regimes across the western US vary along macroscale environmental gradients. Our first objective is to examine how fire activity and fire severity each vary as a function of proxy gradients for fuel amount and fuel moisture (represented by AET and WD, respectively). Our second objective is to present an empirically based characterization of fire regimes that includes both fire activity and fire severity. Specifically, we characterize fire activity and severity by ecoregion, thus quantitatively describing ecoregions within a pyrogeographic context that includes more than one fire regime component. Because anthropogenic activities can obscure fire-climate relationships [34], we focus our analysis on protected areas. This study explicitly links fire regime components with physical environmental gradients, and therefore is an important first step toward forecasting and mitigation of projected changes in fire regimes.

Methods

Data

Fire data for the 1984–2010 period were obtained from the Monitoring Trends in Burn Severity (MTBS) project [35], which has mapped fire perimeters and fire severity for all fires ≥400 ha in the western US. We define fire severity as the degree of fire-induced change to vegetation and soils as measured by the delta normalized burn ratio (dNBR), a satellite-inferred index that differences pre- and post-fire Landsat images [36]. Raw dNBR values were adjusted to account for differences due to phenology or precipitation between the pre- and post-fire images by subtracting the average dNBR of pixels outside the burn perimeter; this adjustment can be important when comparing dNBR among fires [37]. As dNBR values increase, there is generally an increase in char and scorched/blackened vegetation and a decrease in moisture content and vegetative cover [36]. The dNBR is strongly associated with field-assessed measures of fire severity over a broad range of ecosystem types [38–41]. In very limited cases, fire perimeters had no associated dNBR; this was generally the case for fires that occurred in 1984 and pre-fire Landsat TM data were unavailable.

Gridded annual AET and WD data (30 arc-second resolution; ~800 m) were obtained from Dobrowski et al. [42] and were derived from a modified Thornthwaite-Mather soil water balance model. The model was run on a monthly time step and accounts for atmospheric demand (potential evapotranspiration – calculated using the Penman-Monteith equation [43]), soil water storage, and includes the effect of temperature and radiation on snow hydrology via a snowmelt model. Annual values for AET and WD were averaged for each pixel over the 1984–2010 time period, corresponding to the fire data used in this study. Low values of AET represent areas with low potential productivity and thus limited fuel amount, whereas high values indicate the potential for high fuel amounts. Low values of WD indicate that water is less limiting (i.e., high fuel moisture), whereas high values indicate severe water limitation. Given that these water balance metrics are correlated in this study (Pearsons's r>0.75), it may be difficult to disentangle the unique contribution of each term.

Analysis

The western US was partitioned into 50,000-ha (500 km^2) hexels in which fire activity and severity were summarized. Although the spatial resolution of fire-related analyses has been shown to influence outcomes [44,45], this hexel size was deemed a reasonable compromise for capturing variability among sample units while maintaining an adequate sample size for analysis, especially given the inherent spatial variability associated with the relatively short fire record we analyzed (1984–2010). Fire activity was calculated as the proportion of each hexel (excluding nonfuel, such as barren, ice/snow, etc. [3]) that burned from 1984 to 2010; this proportion was subsequently square root transformed to homogenize variance in model residuals. Fire severity was calculated by averaging the dNBR of all pixels of all fires that intersected each hexel; pixels classified as nonfuel were excluded in the calculation of the mean. AET and WD were also aggregated to the hexel scale (Fig. 1).

To characterize how fire activity varies along gradients in fuel amount and fuel moisture, we built bivariate statistical models of area burned as a function of AET and WD; similar models were produced with dNBR to characterize how fire severity varies along these gradients. The two models of area burned were generated using generalized linear models (family = quasibinomial) with a quadratic term (i.e., 2nd order polynomial) for AET and WD. dNBR was assessed as a function of AET using an ordinary least squares (ols) model, whereas dNBR was related to WD using a nonlinear exponential function. All model fits (except the dNBR vs. AET model) were evaluated using the coefficient of determination (denoted here as R^2), which is the r^2 of a linear regression between observed and predicted values (cf. [39,46]); the dNBR vs. AET model was evaluated using ols r^2. In our initial data exploration, we also evaluated dNBR as a function of AET using an exponential function, but this resulted in a decreased model strength compared to the linear fit. To reduce the effect of human infrastructure that can disrupt the spread of fires (e.g., roads) and vegetation management activities that can alter fuel structures (e.g., silviculture), we limited this analysis to hexels comprising ≥ 80% designated wilderness and national park; a total of 153 hexels met this criterion for the fire activity models. An additional criterion was added for the evaluation of fire severity to limit variability in mean dNBR values that are associated with small sample sizes (cf. [47]): at least 400 ha had to have burned within a hexel from 1984–2010. A total of 99 hexels met this additional criterion for the fire severity models.

To place our data and findings into a broader biogeographic context, we summarized the hexels (using the previously described subset based on minimum of 80% wilderness and national park) in each 'ecological section' (hereafter ecoregion) as defined by ECOMAP [48] (Fig. 1). We plotted each ecoregion along axes of AET and WD by averaging the values among hexels in each ecoregion. We similarly plotted each ecoregion along axes of area

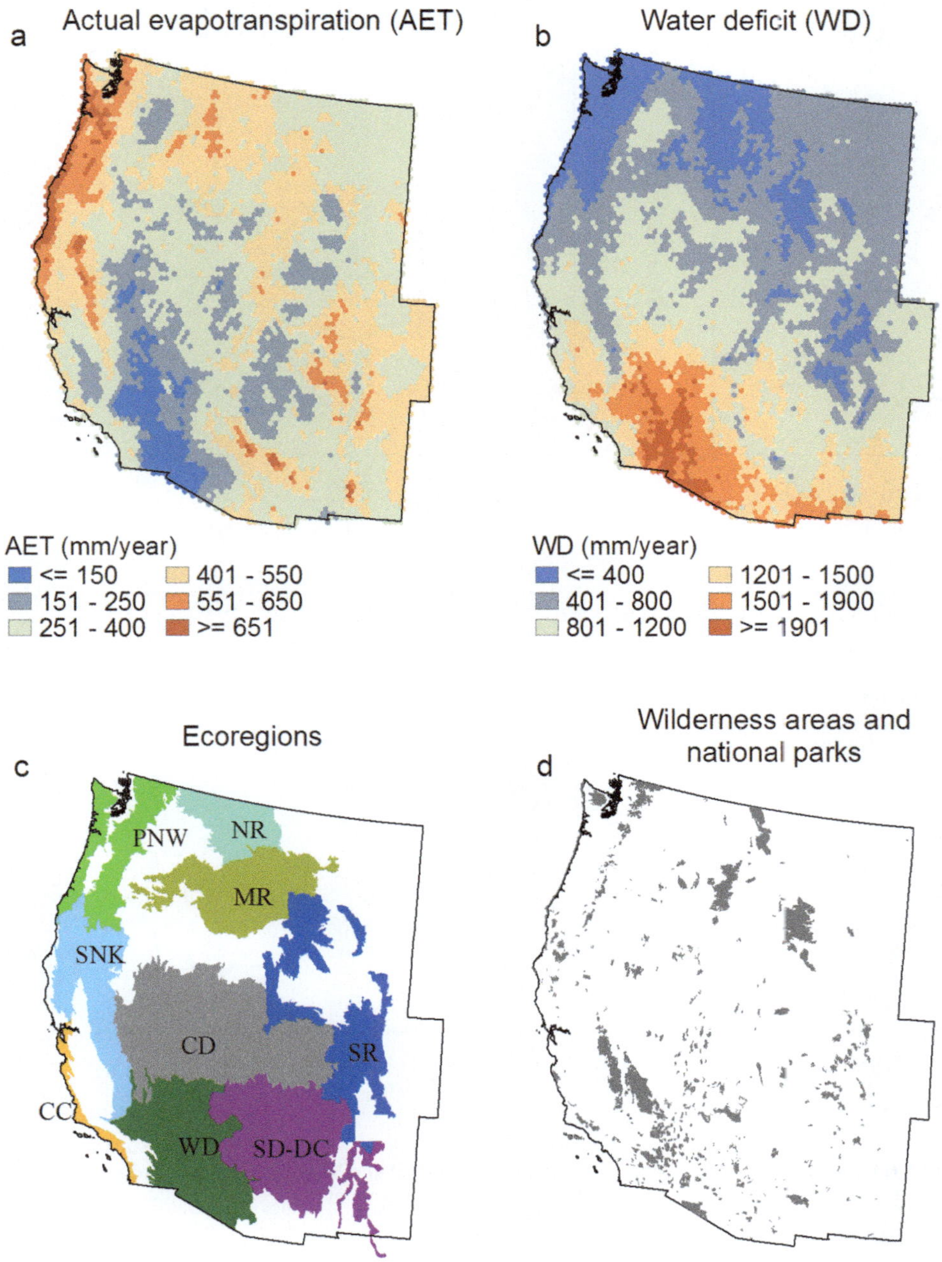

Figure 1. Mean actual evapotranspiration (AET) (a) and water deficit (WD) (b) per 50,000 ha hexel for the western US from 1984–2010. Also shown are the ecoregions (USDA Forest Service, 2007) we analyzed (c): California chaparral (CC), cold desert (CD), middle Rocky Mountains (MR), northern Rocky Mountains (NR), Pacific Northwest (PNW), semi desert – dry conifer (SD-DC), Sierra Nevada and Klamath (SNK), southern Rocky Mountains (SR), warm desert (WD). Cold desert and semi desert – dry conifer are both the result of merging two ecoregions; all ecoregions were renamed for easier interpretation. Areas in white represent ecoregions that did not contain enough fire and/or wilderness and national park area to be included in this study (see Methods). Locations of wilderness areas and national parks in the western US (d).

burned and dNBR, thereby providing a broad-scale concurrent characterization of two components of the fire regime. All ecoregions were renamed and two pairs of two ecoregions were merged to 'cold desert' and 'semi desert – dry conifer' for easier interpretation.

Although not a primary objective of our study, we were able to evaluate how changing the sampling criteria for inclusion of hexels into the models affected the strength of the relationships between the fire characteristics (area burned and dNBR) and the climatic water balance metrics (AET and WD). For each of the four relationships, we built 20 models with different minimum requirements for amount of within-hexel designated wilderness and national park, ranging from zero to 95% (5% increments).

Results

The relationships of area burned and dNBR to AET and WD are moderately strong. Area burned increases with AET ($R^2 = 0.40$), as does dNBR ($r^2 = 0.21$) (Fig. 2a, 2b). Area burned and WD have a unimodal relationship ($R^2 = 0.34$), whereas dNBR decreases with WD ($R^2 = 0.43$) (Fig. 2c, 2d).

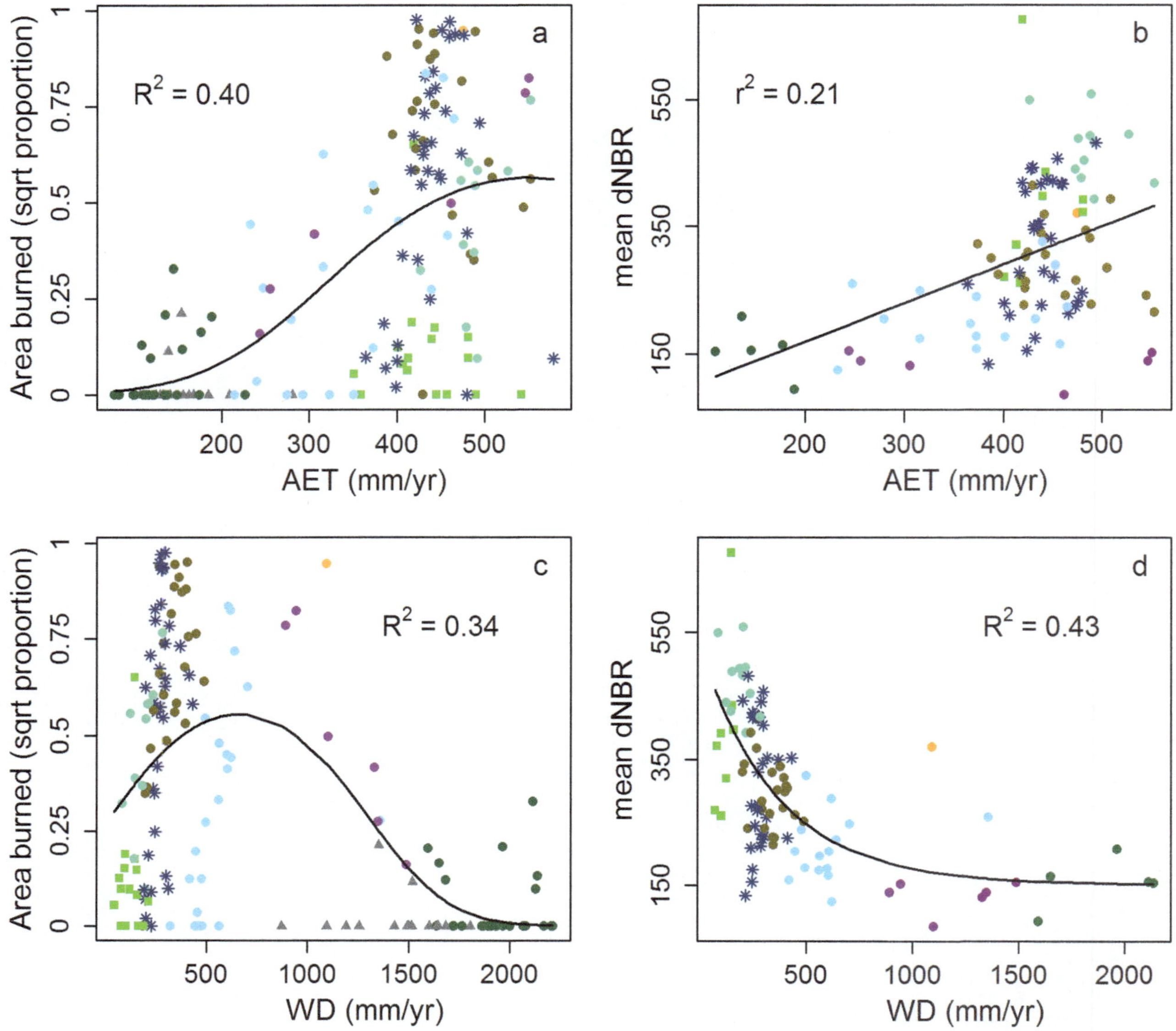

Figure 2. Area burned plotted against AET (a) and WD (b). Mean dNBR plotted against AET (c) and WD (d). Fitted lines are generated by models of each relationship (see Methods); the strength of each model is reported (R^2 or r^2). Symbol colors correspond to the ecoregion in which it is located (Fig. 1).

Broad ecoregion-level biogeographic patterns are revealed in the biplot of AET and WD (Fig. 3a); extreme differences in both AET and WD among contrasted ecoregions, such as the warm desert (WD) and Pacific Northwest (PNW), are evident. In general, ecoregions occupy distinct portions of the AET and WD bi-dimensional space, but the Pacific Northwest and all Rocky Mountain ecoregions are relatively tightly grouped (Fig. 3a). Ecoregional differences in fire regime characteristics are also apparent in the biplot of area burned and dNBR (Fig. 3b). The tight grouping of some ecoregions seen in the biplot of AET and WD (Fig. 3a) is not seen in the biplot of fire regime characteristics (Fig. 3b).

The relationships generally strengthened as we increased the requisite percent of wilderness and national park for inclusion in the models (Fig. 4). In one of the four cases (area burned vs. AET), the R^2 increased from 0.03 to 0.53 when the minimum within-hexel percent wilderness and national park increased from 0% to 95%, respectively.

Discussion

Fire activity increased with AET, the proxy we used for fuel amount; the unimodal relationship that previous theoretical and empirical studies have reported was not seen here [11–13]. This is likely because our study was limited to the western US, and therefore, we evaluated a truncated range of AET (the highest AET values in the US are located in the Southeast). Had we been able to include ecosystems with higher AET values in our analysis, we expect we would have seen a limitation on fire activity at the extreme upper values of AET. Fire activity had a unimodal relationship with fuel moisture, represented by WD, concurring with the theoretical model described by Pausas and Bradstock [32]. Consistent with the "varying constraints hypothesis" posited by Krawchuk and Moritz [11], areas in ecoregions with the highest fuel moisture (lowest WD) (i.e., Pacific Northwest) experienced relatively low fire activity because, despite being biomass rich, they are rarely dry enough to burn. At the other end

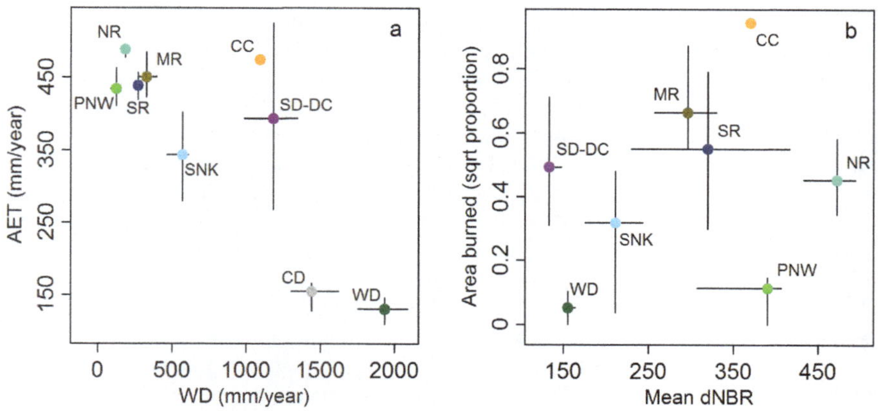

Figure 3. AET and WD averaged across hexels in each ecoregion [48] **(a).** Area burned and dNBR averaged across hexels in each ecoregion (b). Although the cold desert ecoregion did experience some fire within the hexels we analyzed (see Fig. 2), there were no dNBR data for this ecoregion (see Methods); therefore, this ecoregion is not included in the plot on the right (b). Vertical and horizontal lines represent the middle 50 percent (25th to 75th percentile) of the hexel values in each ecoregion.

of the gradient, fire activity in ecoregions with low fuel moisture (i.e., warm desert) was limited because low productivity limits the amount of biomass that can burn.

Fire severity increased with fuel amount, providing support to the idea that fire severity is limited by available biomass [33,49]. The relationship between fire severity and fuel moisture distinctly showed an exponential decrease in fire severity with increasing WD (see Fig. 2d). This is a somewhat counterintuitive and unexpected result because lower fuel moisture has been found to result in higher severity fires [20,50], as drought-stressed trees may be more susceptible to damage by fire and low fuel moisture increases the flammability of fuels. The apparent discrepancy likely involves key differences and interpretations between analyses that evaluate the influence of short-term (e.g., annual) variability (e.g., [20,50]) and those that evaluate long-term averages (our study). For example, Parisien et al. [51] found that the factors controlling fire activity were strikingly different when a statistical model was built incorporating annual variability compared to one using long-term (~30 years) averages. In our study, we evaluated fairly long-term fuel moisture (i.e., averaged over 26 years), as measured by WD, which doesn't necessarily reflect the short-term moisture conditions under which fires actually occur. Consequently, the influence of fuel moisture on fire severity seen in this study may be due to its long-term climatic influence on the establishment and perpetuation of forest types that are adapted to particular moisture conditions; those forest types in turn influence fire severity. For example, on those occasions when moist forest types experience fire (i.e., unusually dry years), they tend to burn at high severity due to their high tree density and abundance of ladder fuels [52]. Nevertheless, the aforementioned correlation between our proxies for fuel amount and moisture (AET and WD, respectively) cannot be ignored and, consequently, completely disentangling their influence is challenging.

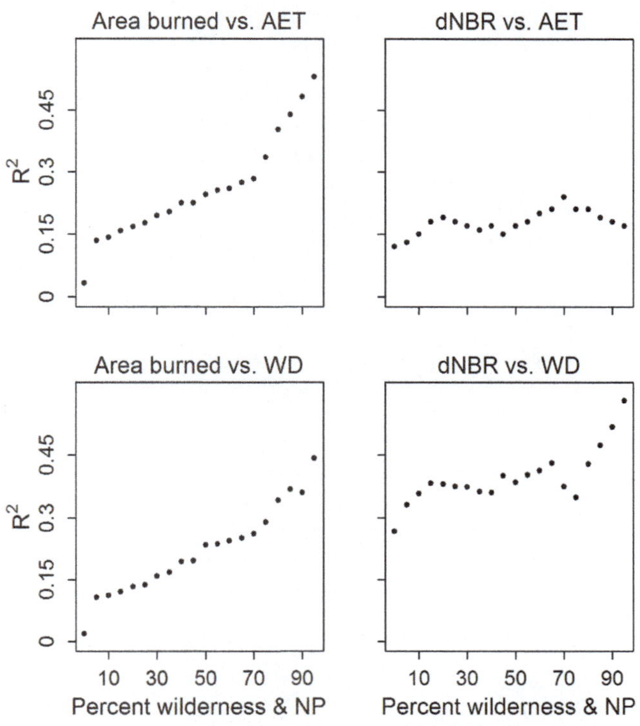

Figure 4. Strength of the relationship (R^2 for all models except the dNBR vs. AET model, which is evaluated with the r^2; see Methods) for all models as a function of increasing percentage of wilderness and national park in each hexel.

Broad biogeographic differences among ecoregions were evident when they were partitioned in terms of AET (fuel amount) and WD (fuel moisture). The concurrent characterization of fire activity and severity also clearly distinguished ecoregions. For example, the warm desert ecoregion experienced low fire activity and severity, whereas the Pacific Northwest ecoregion experienced low fire activity but high fire severity (Fig. 3b). However, ecoregions that were tightly grouped in terms of AET and WD (Pacific Northwest and all Rocky Mountain ecoregions) (Fig. 3a) did not exhibit a similar grouping of fire regime characteristics (Fig. 3b). This discrepancy could be due to difficulties associated with describing the fire regime using only AET and WD or the relatively short fire record that we analyzed, which potentially over- or under-emphasizes fire activity for certain regions (e.g., 1988 Yellowstone fires) [53]; other factors such as macroscale spatial variability in ignition sources [54] and fire suppression effectiveness could also be responsible. Ultimately, however, aggregating hexels by ecoregion may oversimplify complex fire regime dynamics and their effects on ecosystems.

Changes in fire activity and severity are certain to occur as water balance metrics shift in concert with a changing climate, perhaps in complex and unexpected ways [55]. The temporal covariance between energy and moisture availability will play a key role in determining how fuel amount and fuel moisture are geographically distributed in the future, with clear implications for pyrogeography. The complex interplay between climate and the seasonal timing of water and energy availability illustrates the need for mechanistic bioclimatic predictors that can account for these contingencies [25]. The framework developed in this study is an excellent starting point for predicting fire regime shifts under a changing climate though we acknowledge that climate-induced changes in fire regimes will likely have strong feedbacks with the dominant vegetation and ecosystem processes in any given area [56].

We focused this study on wilderness areas and national parks in the western US, where anthropogenic influences (e.g., forest management) are minimal relative to unprotected lands. One consequence of our focus on protected areas is that the fairly low number of hexels available to use in our models of area burned and fire severity represent only 2.5% and 1.6% of the land area in the western US, respectively. As such, the relationships we derived from these relatively unmanaged, natural lands may not well represent the majority of the western US; indeed, the relationships between fire and climate clearly weaken as the human footprint increases (see Fig. 4). Archibald et al. [34] noted a similar finding in the African savannas, in that annual climatic variability was strongly associated with large fire occurrence in areas of low human impact but not in human-dominated areas. Although the hexels we used admittedly represent a small proportion of the western US, the data we analyzed provide valuable insight regarding the "natural" relationships between fire and climate. Consequently, our approach can potentially be used to identify areas with disrupted fire regimes and in need of restoration treatments [57].

Although we used a single metric (mean dNBR) as a convenient measure of ecosystem change, fire severity is the result of many complex physical and ecological factors that are difficult to represent with simplistic measures such as dNBR. For example, lodgepole pine forests experience high-severity fire regimes that favor the regeneration and perpetuation of lodgepole pine on a site, because, as fire often kills most or all of the trees, it also opens their serotinous cones [58,59]. In this example, although dNBR would indicate high severity fire, the change to the ecosystem would be considered far less drastic because lodgepole pine will remain the dominant tree species. As such, our analyses do not address if fire severity is within its historical range of variability [60] for any given vegetation community. Furthermore, dNBR in some ecosystems such as semi-arid shrublands (low AET and high WD) may be relatively low even though the degree of fire-induced change may be very high due to factors such as high rates of post-fire erosion [61] and conversion of perennial woodlands to annual grasslands [62]. Consequently, the relationships reported in this study could be different were other aspects of fire severity (those that cannot be measured with satellites) evaluated. It has also been argued that *relative* metrics of burn severity (i.e., those that measure change relative to the amount of pre-fire vegetation) are more appropriate than absolute metrics (i.e., dNBR) on sites where pre-fire vegetation is low [63,64]. The reasoning for this argument is that dNBR values will be lower on sparsely vegetated sites such as deserts and shrublands, regardless of the degree of fire-induced vegetation mortality, compared to relative metrics such as the relativized delta normalized burn ratio (RdNBR) [63] or relativized burn ratio (RBR) [37]. Indeed, we explored how RBR [37] varied along gradients of AET and WD; shapes of the relationships were nearly identical to those we reported for dNBR, but the strengths of the relationships were weaker. In summary, although fire severity has ecological significance beyond what can be inferred from dNBR, we used dNBR as a convenient and standardized way to assess fire severity.

Several additional factors should be considered when interpreting our results. First, although we focused our study on wilderness and national parks, fire regimes are not entirely natural in those areas. Fire exclusion does occur, which reduces fire activity [65] and increases the potential for high severity fire in some forest types [5]. This factor may blur the relationship between fire and climate. Second, the spatial resolution of our analysis (i.e., size of hexels) does not allow us to adequately represent the substantial variability in climate and fire regime characteristics within each hexel [52]. Third, though we used AET and WD as proxies for fuel amount and fuel moisture, respectively, there may be some ecosystems where these proxies are less suitable. For example, AET may poorly represent fuel amount in ecosystems with high rates of decomposition (tropics), or those that experience frequent fire. Finally, because wilderness areas and national parks are often located at high elevations (e.g., at the tops of mountain ranges), the climatic conditions within hexels included in this study are not necessarily representative of the ecoregions as a whole [66]; for example, many of the hexels in the Pacific Northwest and Sierra Nevada and Klamath ecoregions have a relatively high proportion of unproductive alpine environments, and they hence have lower AET than the ecoregion on average.

Conclusion

A growing body of literature on pyrogeography points to relatively simple energetic controls on fire occurrence [58]: fires require fuel and that fuel needs to be dry enough to burn. To the extent that AET and WD are suitable proxies for fuel amount and fuel moisture, respectively, we found that fire activity across the western US is clearly limited by lack of available fuel and fuel moisture conditions. Furthermore, we found that fire severity is positively related to both fuel amount and fuel moisture. To our knowledge, this is the first broad-scale characterization of fire activity *and* severity along physical environmental gradients. A pyrogeographic perspective that includes both fire activity and severity provides an enhanced view of contemporary fire regimes that complements existing classifications of fire activity. As such, the concurrent characterization of fire activity and severity was effective in distinguishing contemporary fire regime properties of most ecoregions in the western US. The explicit links between fire activity and severity with physical environmental gradients provides a necessary first step for generating multivariable statistical models to produce an empirically based fire regime map for the western US. Such models will potentially enable researchers to predict the geographic distribution of future fire regime characteristics based on gridded spatial data representing future climate scenarios. Consequently, our framework should advance the study of climate-mediated impacts on fire regimes, because, although numerous studies have predicted changes in fire activity due to climate change (e.g., [67]), none have yet examined how fire severity is predicted to change.

Author Contributions

Conceived and designed the experiments: SP MAP CM SD. Performed the experiments: SP. Analyzed the data: SP. Contributed reagents/materials/analysis tools: SD. Wrote the paper: SP MAP CM SD.

References

1. Agee JK (1993) Fire Ecology of Pacific Northwest Forests. Washington, D.C.: Island Press. 493 p.
2. Schmidt KM, Menakis JP, Hardy CC, Hann WJ, Bunnell DL (2002) Development of Coarse-Scale Spatial Data for Wildland Fire and Fuel Management. USDA Forest Service, Rocky Mountain Research Station, Fort Collins, CO, GTR-RMRS-087. 42.
3. Rollins MG (2009) LANDFIRE: a nationally consistent vegetation, wildland fire, and fuel assessment. International Journal of Wildland Fire 18: 235–249.
4. Westerling AL, Hidalgo HG, Cayan DR, Swetnam TW (2006) Warming and earlier spring increase western US forest wildfire activity. Science 313: 940–943.
5. Naficy C, Sala A, Keeling EG, Graham J, DeLuca TH (2010) Interactive effects of historical logging and fire exclusion on ponderosa pine forest structure in the northern Rockies. Ecological Applications 20: 1851–1864.
6. Mack MC, D'Antonio CM (1998) Impacts of biological invasions on disturbance regimes. Trends in Ecology & Evolution 13: 195–198.
7. Morgan P, Hardy CC, Swetnam TW, Rollins MG, Long DG (2001) Mapping fire regimes across time and space: Understanding coarse and fine-scale fire patterns. International Journal of Wildland Fire 10: 329–342.
8. Murphy BP, Williamson GJ, Bowman D (2011) Fire regimes: moving from a fuzzy concept to geographic entity. New Phytologist 192: 316–318.
9. Archibald S, Lehmann CER, Gomez-Dans JL, Bradstock RA (2013) Defining pyromes and global syndromes of fire regimes. Proceedings of the National Academy of Sciences of The United States of America 110: 6442–6447.
10. Whitlock C, Higuera PE, McWethy DB, Briles CE (2010) Paleoecological perspectives on fire ecology: revisiting the fire-regime concept. The Open Ecology Journal 3: 6–23.
11. Krawchuk MA, Moritz MA (2011) Constraints on global fire activity vary across a resource gradient. Ecology 92: 121–132.
12. Pausas JG, Ribeiro E (2013) The global fire productivity relationship. Global Ecology and Biogeography 22: 728–736.
13. Ryan KC (1991) Vegetation and wildland fire - implications of global climate change. Environment International 17: 169–178.
14. Westerling AL, Gershunov A, Brown TJ, Cayan DR, Dettinger MD (2003) Climate and wildfire in the western United States. Bulletin of the American Meteorological Society 84: 595–604.
15. Littell JS, McKenzie D, Peterson DL, Westerling AL (2009) Climate and wildfire area burned in western U.S. ecoprovinces, 1916–2003. Ecological Applications 19: 1003–1021.
16. Parisien M-A, Snetsinger S, Greenberg JA, Nelson CR, Schoennagel T, et al. (2012) Spatial variability in wildfire probability across the western United States. International Journal of Wildland Fire 21: 313–327.
17. Krawchuk MA, Moritz MA, Parisien M-A, Van Dorn J, Hayhoe K (2009) Global pyrogeography: the current and future distribution of wildfire. PLoS ONE 4: e5102.
18. Guyette RP, Stambaugh MC, Dey DC, Muzika RM (2012) Predicting fire frequency with chemistry and climate. Ecosystems 15: 322–335.
19. Dillon GK, Holden ZA, Morgan P, Crimmins MA, Heyerdahl EK, et al. (2011) Both topography and climate affected forest and woodland burn severity in two regions of the western US, 1984 to 2006. Ecosphere 2: art130.
20. van Mantgem PJ, Nesmith JCB, Keifer M, Knapp EE, Flint A, et al. (2013) Climatic stress increases forest fire severity across the western United States. Ecology Letters 16: 1151–1156.
21. Turner MG, Romme WH, Gardner RH (1999) Prefire heterogeneity, fire severity, and early postfire plant reestablishment in subalpine forests of Yellowstone National Park, Wyoming. International Journal of Wildland Fire 9: 21–36.
22. Smucker KM, Hutto RL, Steele BM (2005) Changes in bird abundance after wildfire: Importance of fire severity and time since fire. Ecological Applications 15: 1535–1549.
23. McHugh CW, Kolb TE (2003) Ponderosa pine mortality following fire in northern Arizona. International Journal of Wildland Fire 12: 7–22.
24. Whittaker RH (1975) Communities and Ecosystems. New York: Macmillan.
25. Stephenson NL (1990) Climatic control of vegetation distribution: the role of the water balance. The American Naturalist 135: 649–670.
26. Lutz JA, van Wagtendonk JW, Franklin JF (2010) Climatic water deficit, tree species ranges, and climate change in Yosemite National Park. Journal of Biogeography 37: 936–950.
27. Littell JS, Gwozdz RB (2011) Climatic water balance in regional fire years in the Pacific Northwest, USA: Linking regional climate and fire and at landscape scales. In: McKenzie D, Miller C, Falk DA, editors. The Landscape Ecology of Fire: Springer Netherlands. 177–139.
28. Abatzoglou JT, Kolden CA (2013) Relationships between climate and macroscale area burned in the western United States. International Journal of Wildland Fire 22: 1003–1020.
29. Cleveland CC, Townsend AR, Schimel DS, Fisher H, Howarth RW, et al. (1999) Global patterns of terrestrial biological nitrogen (N-2) fixation in natural ecosystems. Global Biogeochemical Cycles 13: 623–645.
30. Mu Q, Heinsch FA, Zhao M, Running SW (2007) Development of a global evapotranspiration algorithm based on MODIS and global meteorology data. Remote Sensing of Environment 111: 519–536.
31. Dennison PE, Roberts DA, Thorgusen SR, Regelbrugge JC, Weise D, et al. (2003) Modeling seasonal changes in live fuel moisture and equivalent water thickness using a cumulative water balance index. Remote Sensing of Environment 88: 442–452.
32. Pausas JG, Bradstock RA (2007) Fire persistence traits of plants along a productivity and disturbance gradient in mediterranean shrublands of south-east Australia. Global Ecology and Biogeography 16: 330–340.
33. Mitchell SR, Harmon ME, O'Connell KEB (2009) Forest fuel reduction alters fire severity and long-term carbon storage in three Pacific Northwest ecosystems. Ecological Applications 19: 643–655.
34. Archibald S, Nickless A, Govender N, Scholes RJ, Lehsten V (2010) Climate and the inter-annual variability of fire in southern Africa: a meta-analysis using long-term field data and satellite-derived burnt area data. Global Ecology and Biogeography 19: 794–809.
35. Eidenshink J, Schwind B, Brewer K, Zhu ZL, Quayle B, et al. (2007) A project for monitoring trends in burn severity. Fire Ecology 3: 3–21.
36. Key CH, Benson NC (2006) Landscape assessment (LA). In: Lutes D, Keane RE, Caratti JF, Key CH, Benson NC et al., editors. FIREMON: Fire Effects Monitoring and Inventory System. Fort Collins, CO: RMRS-GTR-164. 55.
37. Parks SA, Dillon GK, Miller C (2014) A new metric for quantifying burn severity: the Relativized Burn Ratio. Remote Sensing 6: 1827–1844.
38. Wimberly MC, Reilly MJ (2007) Assessment of fire severity and species diversity in the southern Appalachians using Landsat TM and ETM plus imagery. Remote Sensing of Environment 108: 189–197.
39. Soverel NO, Perrakis DDB, Coops NC (2010) Estimating burn severity from Landsat dNBR and RdNBR indices across western Canada. Remote Sensing of Environment 114: 1896–1909.
40. Cocke AE, Fule PZ, Crouse JE (2005) Comparison of burn severity assessments using Differenced Normalized Burn Ratio and ground data. International Journal of Wildland Fire 14: 189–198.
41. Arkle RS, Pilliod DS, Welty JL (2012) Pattern and process of prescribed fires influence effectiveness at reducing wildfire severity in dry coniferous forests. Forest Ecology and Management 276: 174–184.
42. Dobrowski SZ, Abatzoglou J, Swanson AK, Greenberg JA, Mynsberge AR, et al. (2013) The climate velocity of the contiguous United States during the 20th century. Global Change Biology 19: 241–251.
43. Allen RG, Pereira LS, Raes D, Smith M (1998) Crop Evapotranspiration-Guidelines for Computing Crop Water Requirements-FAO IRRIGATION and Drainage. Food and Agriculture Organization of the United Nations, Rome, Italy.
44. Parks SA, Parisien MA, Miller C (2011) Multi-scale evaluation of the environmental controls on burn probability in a southern Sierra Nevada landscape. International Journal of Wildland Fire 20: 815–828.
45. Parisien MA, Parks SA, Krawchuk MA, Flannigan MD, Bowman LM, et al. (2011) Scale-dependent controls on the area burned in the boreal forest of Canada, 1980–2005. Ecological Applications 21: 789–805.
46. Girardin MP, Ali AA, Carcaillet C, Mudelsee M, Drobyshev I, et al. (2009) Heterogeneous response of circumboreal wildfire risk to climate change since the early 1900s. Global Change Biology 15: 2751–2769.
47. Wisz MS, Hijmans RJ, Li J, Peterson AT, Graham CH, et al. (2008) Effects of sample size on the performance of species distribution models. Diversity and Distributions 14: 763–773.
48. USDA Forest Service (2007) Ecological Subregions: Sections and Subsections for the Conterminous United States.
49. Parks SA, Miller C, Nelson CR, Holden ZA (2014) Previous fires moderate burn severity of subsequent wildland fires in two large western US wilderness areas. Ecosystems 17: 29–42.
50. Renkin RA, Despain DG (1992) Fuel moisture, forest type, and lightning-caused fire in Yellowstone National Park. Canadian Journal of Forest Research 22: 37–45.
51. Parisien MA, Parks SA, Krawchuk MA, Little JM, Flannigan MD, et al. (2014) Comparing the influence of temporally averaged versus annually varying controls on fire probabilities across the boreal forest of Canada. Ecological Applications in press.
52. Schoennagel T, Veblen TT, Romme WH (2004) The interaction of fire, fuels, and climate across rocky mountain forests. Bioscience 54: 661–676.
53. Romme WH, Boyce MS, Gresswell R, Merrill EH, Minshall GW, et al. (2011) Twenty years after the 1988 Yellowstone fires: lessons about disturbance and ecosystems. Ecosystems 14: 1196–1215.
54. McWethy DB, Higuera PE, Whitlock C, Veblen TT, Bowman D, et al. (2013) A conceptual framework for predicting temperate ecosystem sensitivity to human impacts on fire regimes. Global Ecology and Biogeography 22: 900–912.
55. Batllori E, Parisien MA, Krawchuk MA, Moritz MA (2013) Climate change-induced shifts in fire for Mediterranean ecosystems. Global Ecology and Biogeography 22: 1118–1129.
56. Turner MG (2010) Disturbance and landscape dynamics in a changing world. Ecology 91: 2833–2849.
57. Schoennagel T, Nelson CR (2010) Restoration relevance of recent National Fire Plan treatments in forests of the western United States. Frontiers in Ecology and the Environment 9: 271–277.

58. McKenzie D, Miller C, Falk DA (2011) Toward a Theory of Landscape Fire. In: McKenzie D, Miller C, Falk DA, editors. The Landscape Ecology of Fire: Springer Netherlands. 3–25.
59. Baker WL (2009) Fire ecology in Rocky Mountain Landscapes. Washington DC: Island Press. 605 p.
60. Landres PB, Morgan P, Swanson FJ (1999) Overview of the use of natural variability concepts in managing ecological systems. Ecological Applications 9: 1179–1188.
61. Wagenbrenner NS, Germino MJ, Lamb BK, Robichaud PR, Foltz RB (2013) Wind erosion from a sagebrush steppe burned by wildfire: Measurements of PM10 and total horizontal sediment flux. Aeolian Research 10: 25–36.
62. D'Antonio CM, Vitousek PM (1992) Biological invasions by exotic grasses, the grass fire cycle, and global change. Annual Review of Ecology and Systematics 23: 63–87.
63. Miller JD, Thode AE (2007) Quantifying burn severity in a heterogeneous landscape with a relative version of the delta Normalized Burn Ratio (dNBR). Remote Sensing of Environment 109: 66–80.
64. Norton J, Glenn N, Germino M, Weber K, Seefeldt S (2009) Relative suitability of indices derived from Landsat ETM+ and SPOT 5 for detecting fire severity in sagebrush steppe. International Journal of Applied Earth Observation and Geoinformation 11: 360–367.
65. Heyerdahl EK, Brubaker LB, Agee JK (2001) Spatial controls of historical fire regimes: A multiscale example from the interior west, USA. Ecology 82: 660–678.
66. Batllori E, Miller C, Parisien M-A, Parks SA, Moritz MA (2014). Is US climatic diversity well represented within the existing federal protection network? Ecological Applications in press.
67. Littell JS, Oneil EE, McKenzie D, Hicke JA, Lutz JA, et al. (2010) Forest ecosystems, disturbance, and climatic change in Washington State, USA. Climatic Change 102: 129–158.

21

Future Oceanic Warming and Acidification Alter Immune Response and Disease Status in a Commercial Shellfish Species, *Mytilus edulis* L.

Clara L. Mackenzie[1]*, Sharon A. Lynch[2], Sarah C. Culloty[2], Shelagh K. Malham[3]

1 Centre for Marine Biodiversity and Biotechnology, School of Life Sciences, Heriot-Watt University, Edinburgh, United Kingdom, **2** Aquaculture and Fisheries Development Centre, School of Biological, Earth and Environmental Sciences, University College Cork, Cork, Ireland, **3** Centre for Applied Marine Sciences, Bangor University, Menai Bridge, Anglesey, United Kingdom

Abstract

Increases in atmospheric carbon dioxide are leading to physical changes in marine environments including parallel decreases in ocean pH and increases in seawater temperature. This study examined the impacts of a six month exposure to combined decreased pH and increased temperature on the immune response and disease status in the blue mussel, *Mytilus edulis* L. Results provide the first confirmation that exposure to future acidification and warming conditions via aquarium-based simulation may have parallel implications for bivalve health. Collectively, the data suggests that temperature more than pH may be the key driver affecting immune response in *M. edulis*. Data also suggests that both increases in temperature and/or lowered pH conditions may lead to changes in parasite abundance and diversity, pathological conditions, and bacterial incidence in *M. edulis*. These results have implications for future management of shellfish under a predicted climate change scenario and future sustainability of shellfisheries. Examination of the combined effects of two stressors over an extended exposure period provides key preliminary data and thus, this work represents a unique and vital contribution to current research efforts towards a collective understanding of expected near-future impacts of climate change on marine environments.

Editor: Pikul Jiravanichpaisal, Fish Vet Group, Thailand

Funding: This work was funded by the project entitled: Shellfish Productivity in the Irish Sea (SUSFISH). SUSFISH is partially funded by the European Regional Development Fund (ERDF) through the Ireland Wales Territorial Co-operation (INTERREG 4A) Programme 2007-2013. INTERREG 4A is managed in Ireland by the Southern and Eastern Regional Assembly on behalf of the Irish Government, the Welsh Government and the European Commission (http://www.irelandwales.ie). The funders had no role in study design, data collection and analysis, decision to publish,or preparation of the manuscript.

Competing Interests: The authors have declared that no competing interests exist.

* E-mail: clm32@hw.ac.uk

Introduction

Increases in atmospheric carbon dioxide are leading to physical changes in marine environments. Under the IPCC IS92a CO_2 emission scenario, ocean pH is expected to decrease by 0.2–0.4 units by the year 2100 [1,2]. In addition to changes in ocean pH, organisms will also have to contend with simultaneous increases in seawater temperature with the current average sea surface temperature of 19.7°C predicted to rise to 22.7°C by the end of the century [2,3].

The ability of marine organisms to adjust to future climate change conditions will be critical to their health and ultimate survival. Bivalves are unable to thermo-regulate and are considered poor regulators of haemolymph acid-base balance [4,5]. Consequently, pH and temperature changes in the environment will likely have direct influence on bivalve haemolymph pH and temperature. Additionally, any physiological mechanism associated with the bivalve circulatory system is liable to be affected under hypercapnic conditions. The bivalve immune system, for example, exists as an integrative part of the circulation system [6,7]. The circulating cells, or haemocytes, represent the major cellular component of the animal's immune response and are responsible for a number of defence activities including phagocytosis or encapsulation of foreign or diseased cells, release of reactive oxygen metabolites and enzymes, and secretion of cytotoxic molecules [6,8,9,10].

While prior investigations of the consequences of near-future changes in ocean pH and temperature have demonstrated a number of effects on marine organism physiology including impacts on growth, calcification and acid-base status [4,11–18], less confirmed is how coinciding stressors might influence immune response in organisms [19]. Changes to single environmental factors, particularly temperature [20–24] but also salinity [25], pH [26] and dissolved oxygen [27] have demonstrated impacts on various immunological parameters including haemocyte numbers, phagocytosis and oxidative burst response.

The immune system is a major physiological mechanism ensuring host survival in the battle with pathogenic or parasitic organisms [28]. As stress impacts immune function and such mechanisms are the primary line of defence against pathogens, there is likely a strong link between immune response and the outbreak of disease in shellfish culture [29,30]. Thus, climate change impacts on immunological aspects of physiology may be paralleled by associated changes in bivalve disease status.

Global climate change has increased pathogen development and survival, disease transmission and host susceptibility in the last thirty years [31–40]. Increased disease development in certain ecosystems can contribute to species decline and even extinction. Disease dynamics in the marine environment are being influenced by physical, chemical and biological alterations driven by climate change and such impacts will affect the sustainability of the shellfish industry. Multiple climate drivers such as ocean acidification and warming ocean temperatures have the potential to promote pathogen range expansion, infected host invasions and native host decline [41,42] and tropical pathogen range contractions may also occur [43]. Additionally, it is possible that certain pathogens will be detrimentally affected by climate change conditions via changes to their required conditions thus allowing a relatively small number of emergent pathogens to infect new hosts with little or no tolerance [31].

Previous studies have shown associations between parasites and changing ocean temperatures, with a warmer environment being more favourable to the parasite [44–49]. Higher temperatures resulting in lower oxygen levels may stress organisms thus increasing their susceptibility to disease [50,51,52]. In the temperate zone, shorter milder winters are expected to increase the frequency and intensity of transmission of diseases [31]. Certain parasites also have higher growth rates, higher reproductive ability and decreased generation time with higher temperatures [53,54].

Prior investigation also indicates that ocean acidification and associated stressors will be negative for calcifying organisms with more energy being required to build shells, which will reduce the amount of energy available to find food, reproduce or resist parasites and disease [26,55,56,57]. It is possible that ocean acidification will also have a regulatory effect on certain parasite populations especially those with free-living developmental stages [58,59]. However it is also possible that increased pathogenicity will occur if certain parasites are less affected by an acidified environment than their respective hosts [60].

This research aimed to examine the impacts of exposure to future ocean warming and acidification on the immune response and disease status of a commercially important bivalve species, *Mytilus edulis* (Linnaeus). The impacts of six months exposure to lowered pH and increased temperature on immunological responses, including haemocyte counts, phagocytosis, and nitroblue tetrazolium (NBT) reduction, and the health status of the mussels by histological and molecular examination were investigated.

Materials and Methods

Ethics statement

No specific permits were required for the study, which complied with all relevant regulations. The species collected in this study is not endangered or protected.

Study organisms

Mytilus edulis (mean shell length = 50.51 mm, SD = 3.68 mm) were collected from a sub-tidal population in the Menai Strait, North Wales, UK on May 3rd, 2011 and acclimatised in large (200 L) flow-through holding tanks (12.5°C±SD 0.26°C, pH 8.01±SD 0.08, Salinity 34± SD 1 psu, 12L:12D light regime) at the School of Ocean Sciences, Bangor for a 3 week time period. During this time, mussels were drip-fed concentrated algal feed (Instant Algae Shellfish Diet 1800, Reed Mariculture, Campbell, CA, USA; 40% *Isochrysis* sp., 15% *Pavlova* sp., 25% *Tetraselmis* sp. and 20% *Thalassiosira weissglogii* (Grunow); 52% protein, 16.1% lipid, 22.0% carbohydrate and 9.9% ash) at a ration of 27 mg dry mass $mussel^{-1}$ day^{-1}.

Experimental set-up

Experimental set-up and conditions for this study are described in detail in Mackenzie *et al.* [61]. Briefly, *M. edulis* were exposed to current (~400 µatm) and future (~1000 µatm) pCO_2 levels in an aquarium-based CO_2 system. Current and future seawater conditions were simulated with seawater held at two temperatures (ambient, ambient +4°C) and two pH levels (ambient, ambient - 0.4 pH units) as predicted under the IPCC IS92a CO_2 emission scenario [62]. This set-up allowed for examination of the interactive effects of varying pH and temperature conditions and included four treatments: ambient pH at ambient temperature (ambient), ambient pH at ambient temperature +4°C (warming), reduced pH at ambient temperature (acidified) and reduced pH at ambient temperature +4°C (acidified+warming).

Following input of seawater from the Menai Strait to an internal settling tank, sea water was filtered and UV treated before delivery to each of four header tanks (150 L) representative of the four experimental treatments. In-line heaters (Elecro Titanium Digital Heater) and cooling units (Aqua Medic TITAN 200) were used to maintain experimental temperatures. A pH controller (Walchem dual input pH controller) regulated addition of CO_2 (g) to achieve acidified treatments.

Experimental header tanks gravity-fed (flow rate ~5 ml $second^{-1}$) seawater to 20 replicate 1.5 L tanks per treatment with overflows running to waste. Eight individual mussels were assigned to each tank per treatment. Mussels were acclimated to temperature treatments at an increase of 0.5°C day^{-1}. Mussel tanks were cleaned three times per week and mussels were fed according to amounts and rates described for the acclimatisation period.

Seawater parameters

Collection, storage and analyses of seawater carbonate samples are described in detail in Mackenzie *et al.* [61]. Briefly, 60 mL total alkalinity and dissolved inorganic carbon (DIC) seawater samples were collected fortnightly and sent to the Carbonate System Facility (LIMS) at the National Oceanography Centre, Southampton for total alkalinity and DIC analysis. Temperature (Mettler Toledo SG2 SevenGO, MT Ltd., Leicester) and salinity (TMC V2 ATC) of samples were also recorded at time of collection. Additionally, 30 mL nutrient samples were collected, filtered and frozen for analysis of phosphate and silicate concentrations at the Scottish Marine Institute. All values were entered into the CO2SYS model [63] to determine seawater carbonate parameters (i.e. pH, pCO_2, HCO_3, CO_3^{-2}, ΩAr, ΩCa) using the thermodynamic constants of Mehrbach *et al.* [64] refitted by Dickson and Millero [65]. Carbonate chemistry values for the four experimental treatments are provided in Table 1. Values represent the mean ±SD of bimonthly measures taken over the six months exposure period (n = 12 for each treatment).

Immunological response

Mussels. An initial group of adult *M. edulis* (n = 5) were sampled to determine baseline immune status. Following six months of exposure, five mussels were randomly selected from five experimental tanks (i.e. replicates) within each treatment (n = 5). Haemolymph samples of 0.3–0.6 mL were withdrawn from the sinus of the posterior adductor muscle of each mussel using 2 mL syringes and 25 G ½ needles. Five mussels were bled consecutively so that total haemocyte counts, phagocytosis counts, and NBT analysis could be carried out concurrently.

Total haemocyte counts. A 20 µL haemolymph subsample from each mussel was used to determine total haemocytes ml^{-1}, counted using an improved Neubauer haemocytometer [7].

Table 1. Seawater carbonate chemistry values for experimental treatments.

Treatment	T	Sal	TA	DIC	pH_T	pCO_2	HCO_3^-	CO_3^{-2}	Ωarag	Ωcalc
	(°C)	(‰)	(μmol/kg)	(μmol/kg)		(μatm)	(μmol/kg)	(μmol/kg)		
Ambient	12.14	33.58	2263.57	2093.78	7.99	465.82	1959.02	125.46	1.91	3.00
	(±0.48)	(±0.72)	(±15.70)	(±12.69)	(±0.05)	(±63.15)	(±30.99)	(±12.67)	(±0.20)	(±0.31)
Warming	15.84	33.71	2263.75	2087.47	7.95	516.77	1942.52	131.45	2.02	3.15
	(±0.27)	(±0.69)	(±15.99)	(±13.67)	(±0.03)	(±40.94)	(±18.28)	(±8.00)	(±0.12)	(±0.19)
Acidified	12.18	33.63	2264.69	2213.62	7.65	1087.69	2113.12	63.35	0.97	1.52
	(±0.48)	(±0.71)	(±16.97)	(±19.08)	(±0.06)	(±169.87)	(±21.27)	(±9.57)	(±0.14)	(±0.23)
Acidifed+Warming	16.11	33.54	2265.32	2205.95	7.63	1161.35	2101.03	69.07	1.06	1.66
	(±0.28)	(±0.72)	(±16.50)	(±15.69)	(±0.05)	(±136.86)	(±17.23)	(±7.37)	(±0.11)	(±0.17)

Standard deviations shown in parentheses (±SD) below corresponding mean value (n = 12). Measured values: temperature (T), salinity (Sal), total alkalinity (TA), and dissolved inorganic carbon (DIC). Modelled values: pH total scale (pH_T), CO_2 partial pressure (pCO_2), bicarbonate (HCO_3^-), carbonate (CO_3^{-2}), aragonite saturation state (Ω_{arag}) and calcite saturation state (Ω_{calc}).

Phagocytosis. Phagocytosis assay was performed as previously described [29,66] with some modifications. Briefly, a 20 μL subsample of haemolymph was placed onto a glass slide and allowed to adhere for 20 min in a moist incubation chamber before the addition of 20 μL of fluorescein 5-isothiocyanate Isomer 1 (FITC) labelled *E. coli* (Sigma). Glass slides were then incubated for a further 20 min after which all were rinsed with Tris-HCl buffer solution containing 2% NaCl (pH 7.6). Unphagocytosed bacteria were counterstained with ethidium bromide (Sigma) and then rinsed clear with Tris-HCL. Triplicate counts of 200 cells were immediately carried out using a 488 nm emission filter on a Zeiss microscope.

Nitroblue tetrazolium (NBT) reduction. A modified version of the nitroblue tetrazolium (NBT) assay, as described by Pipe [67] and Bussell *et al.* [25] was applied. Twenty μL of haemocyte suspension were added to 96 well plates (Fisherbrand) in triplicate for each sample. Negative and positive controls were incorporated. Following incubation for 20 min, 20 μL of nitroblue tetrazolium (NBT) (Sigma) (2 mg/ml in Tris–HCl buffer containing 2% NaCl (pH 7.6)) were added to one set of triplicate wells for each sample. Twenty μL Tris buffer were added to corresponding negative control wells. The well plates were then incubated for a further 20 min before centrifugation (1000 g at 10°C for 10 min). All wells were then washed with Tris-HCl buffer solution before another centrifugation (x2) (1000 g) after which 20 uL of methanol (50%) were added before a final centrifugation. The plates were air dried before the addition of 240 uL potassium hydroxide (KOH) and 280 uL dimethylsulphoxide (DMSO) to all plate wells. Optical density (OD) values were measured on a Dynex MRX-II spectrophotometer (λ = 620 nm).

Disease status

Mussels. An initial sample of adult *M. edulis* (n = 30) was screened from the Menai Strait in May 2011 to determine the health status of the mussels in the field. Following six months of exposure, twelve mussels were randomly selected from twelve experimental tanks (i.e. replicates) within each treatment for determination of health status by histological and molecular examination (n = 12).

DNA extraction. DNA was extracted from the initial mussel sample (n = 30) (gill tissue stored in 95% ethanol) using 10% Chelex 100 resin [68,69]. DNA was extracted from the corresponding paraffin-embedded tissue of each mussel from the laboratory trial consisting of twelve mussels per treatment (n = 12). Deparaffinization of the tissue was carried out [70] and DNA extractions were undertaken using protein precipitate and cell lysis using the QIAamp DNA Mini Kit (Qiagen).

Standard polymerase chain reaction (PCR). Several PCRs were carried out to determine (1) which Mytilus species was being screened, (2) general bacterial screening, (3) general microbial screening, (4) presence or absence of Marteilia refringens (Grizel et al) and general haplosporidian spp. screening.

(1) PCR for species identification: Initially a PCR was carried out to detect the nuclear DNA markers Me15/Me16 [71] to confirm that *M. edulis* was being screened as hybrid zones consisting of *M. edulis*, *Mytilus galloprovincialis* (Lamarck) and hybrids of both parent species are known to occur on the southwest coast of the UK. The PCR mastermix was modified slightly to include 5× green buffer. Amplification was conducted in 25 μl of the reaction mixture containing 14 μl ddH_2O, 5 μl 5× green buffer (Promega), 2.5 μl of each of the four deoxyribonucleotide triphosphates dNTPs (dATP, dCTP, dGTP, dTTP) (0.2 mM), 1.5 μl $MgCl_2$ (25 mM stock), 0.5 μl of the primer Me15 (5′-CCAGTATACAAACCTGTGAAGA-3′), 0.5 μl of the

primer Me16 (5′-TGTTGTCTTAATGGTTTGTAAGA-3′) (100 pmol/μL stock), 1 μl of Go Taq DNA polymerase (0.03 U) and 1 μl of total DNA. Negative controls consisting of ddH_20 were used. The following conditions were used for the PCR in a thermocycler: 94°C for 30 s, 55°C for 45 s and 70°C for 90 s (40 cycles). The expected product size for *M. galloprovincialis* is 126 bp while in *M. edulis* it is 180 bp. In hybrids both bands occur simultaneously at 126 bp and 180 bp.

(2) Bacterial screening: PCR was carried out to screen for general bacterial species that might be present using the universal primers EUBB and EUBA [72], which amplify the entire 16srRNA region. The PCR mastermix was modified to include 5 μl 5× green buffer, 5 μl dNTP (0.2 mM), 0.5 μl of $MgCl_2$ (25 mM stock), 0.25 μl of each primer (EUBB and EUBA) (100 pmol/μL stock) and 0.1 μl GoTaq DNA polymerase (0.03 U) per PCR reaction. The PCR reaction mix was made up to a volume of 20 μl using ddH20 and 5 μl of undiluted genomic DNA. The following conditions were used for the PCR in a thermocycler: 95°C for 5 min, 35 cycles of 95°C for 30 s, 55°C for 30 s and 72°C for 30 s with a final extension at 72°C for 10 min. The expected product size is 1.5 Kbp. A second pair of primers, UNIV16s EUB f933/UNIV16s EUB r 1387 [73] which are specific for universally conserved bacterial 16s rDNA region were used for the bacterial screening. The PCR mastermix and thermocycling conditions were modified. The PCR mastermix was made up to a volume of 49 μl for each PCR reaction: 37.75 μl ddH_20, 10 μl 5× green buffer, 1 μl dNTPs (0.2 mM), 3 μl $MgCl_2$ (25 mM stock), 1 μl of each primer (100 pmol/μL stock) and 0.25 μl GoTaq DNA polymerase (0.03 U). 1 μl of undiluted genomic DNA was screened. The PCR was carried out in a thermocycler as follows: 95°C for 1 min, 35 cycles at 94°C for 20 s, 56°C for 30 s, 72°C for 30 s and a final extension at 72°C for 7 min.

(3) Microbial screening: Universal primers 18ScomF1/18ScomR1 [74], which amplify the 18Scom region, were used for a general health screen to assess if any other foreign DNA in mussel tissue could be detected.

(4) Screening for potential mussel parasites: A PCR protocol targeting the ITS1 was carried out to detect *M. refringens* [75]. Several generic haplosporidian PCRs were carried out using the HAP-F1 and HAP-R3 primers [76] and the ssu980 and HAP-R1 primers [76,77]. Electrophoresis of the amplification products from all PCRs was conducted in a 2% agarose gel. Fifteen μL of ethidium bromide, EtBr, (10 mg/ml stock) was added to the agarose gel and the gel was run at 110 V for 60 min.

Sequencing. Direct sequencing was carried out on products amplified by eurofins MWG using ABI 3730xl 96-capillary DNA Analyzers. DNA from pooled PCR products (n = 3/n = 4) from multiple individuals in each sample were isolated and purified using the Qiaquick gel extraction kit (Qiagen) prior to direct sequencing Both the forward and reverse strands of DNA samples were sequenced commercially (MWG eurofins). Each sequence was matched against the National Centre for Biotechnology Information (NCBI) nucleotide database with BLASTn (Basic Local Alignment Search Tool) to identify the species present.

Histology. A transverse section of mussel tissue (~1 cm^2) containing mantle, gill, digestive gland and gonad tissue was excised and fixed in 96% ethanol initially and tissue sections were subsequently transferred to Davidson's fixative at 4°C for 48 hrs [78] before being processed. The fixed tissue was dehydrated through an ascending ethanol series and embedded in paraffin. 5 μm tissue sections were stained using haematoxylin and eosin. Tissue slides were scanned for any pathological or morphological observations and macro- or microparasites at a magnification of 40× and under oil at 100x. The location and intensity of each type of parasite were recorded and the prevalence and intensity of infection were calculated where possible.

Statistical analyses

All statistical analyses were carried out in SPSS 14.0 for Windows (2005). Two-way ANOVAs were applied to determine the effect of pH, temperature and their interaction on immunological parameters ($p<0.05$). Data was tested for normality and homogeneity of variances (Levene's Test). One-way ANOVA were used to directly compare treatments. If necessary, data was log-transformed to meet assumptions. Significant difference in the diversity and prevalence of each parasite group and pathological condition, with varying pH and temperature, was calculated using chi-square (χ^2) analysis with significance determined at $p<0.05$.

Results

Immunological response

Total haemocyte counts. Two-way ANOVA identified a significant negative effect of increased temperature on haemocyte counts but neither an influence of pH nor interaction between pH and temperature (Table 2). Haemocyte counts varied significantly between groups ($F = 24.336$; $p<0.001$) with *post-hoc* tests revealing that baseline animals had significantly lower haemocyte counts than any experimental treatments (ambient ($p<0.001$), warming ($p = 0.001$), acidified ($p<0.001$), acidified+warming ($p = 0.001$)). Animals held at either ambient temperature treatment (ambient, acidified) had significantly higher haemocyte counts than those under increased temperature treatments at either pH (warming ($p = 0.005$, $p<0.001$, respectively), acidified+warming ($p = 0.004$, $p<0.001$, respectively)) (Fig. 1).

Phagocytosis. Two-way ANOVA showed a significant positive effect of increased temperature on percentage of phagocytosed cells as well as a significant interaction between pH and temperature but no effect of pH (Table 2). The percentage of phagocytosed cells varied significantly between all groups ($F = 3.846$; $p = 0.020$) but failed assumptions of homogeneity of variances even after transformed. *Post-hoc* tests identified a significantly higher percentage of phagocytosed cells in haemolymph of mussels held at decreased pH and warming conditions (acidified+warming) than in baseline animals ($p = 0.047$) or at ambient temperature treatments at either pH (ambient ($p = 0.016$), acidified ($p = 0.001$)). Additionally, animals held at increased temperature and ambient pH (warming) had a significantly higher percentage of phagocytosed cell than those under ambient temperature and decreased pH (acidified) ($p = 0.044$) (Fig. 2).

Nitroblue tetrazolium (NBT) reduction. Two-way ANOVA identified a significant positive effect of both decreased pH and increased temperature on NBT reduction but no interaction between the two factors (Table 2). NBT reduction varied significantly between treatments ($F = 6.349$; $p = 0.002$) with *post-hoc* tests revealing that mussels reared under ambient temperature and pH conditions (ambient) had significantly lowered reduction than all other experimental treatments (warming ($p = 0.004$), acidified ($p = 0.022$), acidified+warming ($p<0.001$)) Animals held under warming conditions at either pH (warming, acidified+warming) had significantly increased reduction than baseline animals ($p = 0.012$ and $p = 0.002$, respectively) (Fig. 3).

Disease status

PCR and direct sequencing. (1) A product of 180 bp was amplified in the Me15/Me16 PCR confirming the mussel species

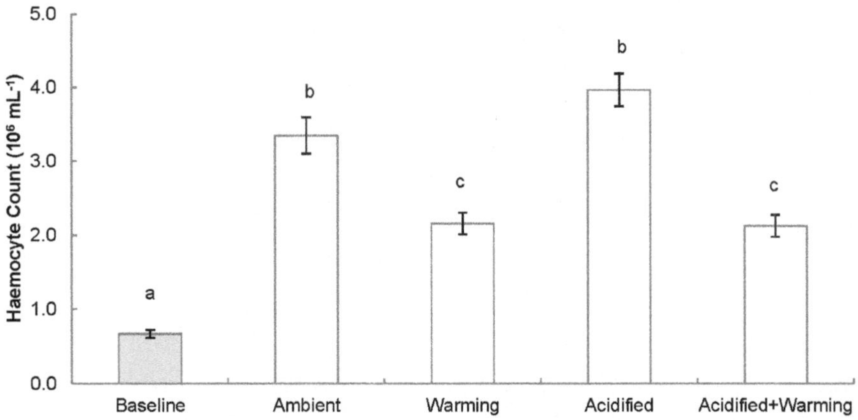

Figure 1. Effects of temperature and pH on haemocyte count (10^6 mL^{-1}) (±1 SE) in *Mytilus edulis* (n = 5) following a six month exposure period to varying temperature and pH conditions. Baseline value shown in grey scale. Lowercase letters indicate significant differences.

to be *M. edulis* (Table 3). Direct sequencing confirmed the mussel species to be *M. edulis*.

(2) In the PCR using the universal primers for bacteria screening, EUBB and EUBA [72], multiple products were amplified at 800 bp, 900 bp and 1000 bp. The second PCR used to screen for bacteria using the Univ16s EUBf and Univ16s EUBr primers amplified a product at ~500 bp (Table 3). Several bacterial species were detected in the initial sample and in the mussels from all four treatments. The highest prevalence was observed in mussels from the initial sample (92%), ambient treatment (92%) and acidified+warming treatment (92%) followed by the acidified treatment (83%) and warming treatment (67%). In the bacterial screening, no DNA sequences were recovered for the products amplified in the EUBA and EUBB PCR. Five forward and five reverse sequences were obtained from the products amplified in the Univ16sEUBf and Univ16s EUBr PCR from a subsample of infected individuals (n = 6). Bacterial species detected included a Gammaproteobacterium sp. (99% Query coverage, 98% Maximum identity), *Pseudomonas* sp. (98% Query coverage, 95% Maximum identity), *Pectobacterium carotovorium* (Jones) (100% Query coverage, 97% Maximum identity), *Xanthomonadaceae* sp. (99% Query coverage, 98% Maximum identity) and *Serratia* sp. (99% Query coverage, 97% Maximum identity).

(3) A product (~350 bp) was amplified in the 18Scom region using the Zhang *et al.* [74] modified PCR. This product was amplified in mussels from the initial sample, warming and ambient treatments but was absent in mussels from the acidified and acidified+warming treatments. The free-living amoebae *Hartmannella vermiformis* (Page) (85% Query coverage, 99% Maximum identity) was identified using the 18Scom PCR.

(4) No products were amplified in the ITS region using the *M. refringens* PCR (Table 3). A product was amplified (350 bp) using the generic haplosporidian primers [76] in eighteen individuals in the initial sample (60% prevalence), however, no products were amplified in mussels from any of the experimental treatments. Direct sequencing of the products to confirm haplosporidian identification in the initial sample using the generic haplosporidian primers was unsuccessful. Using the second pair of haplosporidian primers (ssu980 and HAP-R1), a product was amplified at 430 bp in a single individual in the initial sample, which also amplified a

Table 2. Two-way ANOVA results comparing the effects of pH and temperature on immunology (a. haemocyte counts, b. phagocytosis, c. NBT) in *Mytilus edulis* following a 6 month exposure period.

Source of variation	df	SS	MS	*F*-ratio	*P*-value
a. Haemocyte Counts					
pH	1	0.435	0.435	0.923	0.351
Temperature	1	11.476	11.476	24.343	**<0.001**
pH*Temperature	1	0.528	0.528	1.120	0.306
b. Phagocytosis					
pH	1	30.854	30.854	0.271	0.610
Temperature	1	1720.224	1720.224	15.133	**0.001**
pH*Temperature	1	649.686	649.686	5.715	**0.030**
c. NBT					
pH	1	0.186	0.186	4.740	**0.045**
Temperature	1	0.419	0.419	10.668	**0.005**
pH*Temperature	1	0.045	0.045	1.137	0.302

Significant values shown in bold ($p<0.05$).

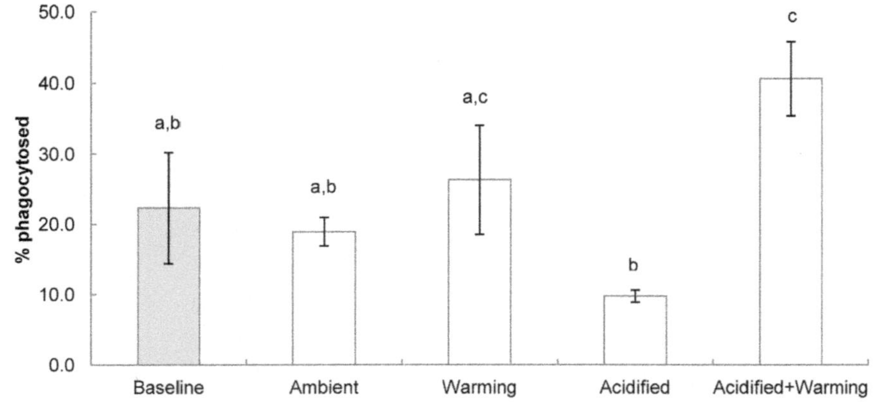

Figure 2. Effects of temperature and pH on phagocytosis (% phagocytosed haemocytes) (±1 SE) in haemocytes from *Mytilus edulis* (n = 5) following a six month exposure period to varying temperature and pH conditions. Baseline value shown in grey scale. Lowercase letters indicate significant differences.

product using the HAP-F1 and HAP-R3 primers. No products were amplified in the mussels from the experimental treatments (Table 3). In the direct sequencing, two forward and two reverse DNA sequences were obtained from the replicate PCR products amplified by the single mussel in the ssu980 and Hap-R1 PCR [76,77]. Blastn analysis of the DNA isolated from *M. edulis* confirmed the DNA to be that from a previously undescribed haplosporidian (Accession no. KC852876 and KC852877) most similar to *Minchinia chitonis* (Lankester) (Accession no. AY449711.1, 95–100% Query coverage, 92–93% Maximum identity) and a haplosporidian of the Florida marsh clam *Cyrenoida floridana* (Dall) (Accession no. AY449712.1, 90–98% Query coverage, 93–95% Maximum identity). A phylogenetic tree was generated based on the majority rule jackknife consensus parsimony analysis with 1000 pseudo replicates with 75% character deletion using a heuristic search with 1000 random sequence additions each.

Histology. In total, seven parasite groups were detected in *M. edulis*. Haplosporidian like organisms (HLOs) were observed in the tissues of ten mussels (33% prevalence) from the initial sample (Table 4) but were not observed or detected in the final sample of mussels screened from any treatment group of the laboratory trial. The HLOs consisted of a cluster of cells of varying sizes (approximately 10 to 30 microns) each with an eccentric nucleus.

No single parasite species was detected across the initial sample and all four mussel treatment groups. However, the copepod *Mytilicola intestinalis* (Steuer) and the amoeba *Hartmannella sp.* were observed in the initial sample and three of the experimental treatment groups, being absent only in mussels from the warming treatment. The highest parasite diversity was observed in the initial sample (6 species) followed by the acidified treatment (4 species), the ambient and acidified+warming treatments (3 species) and warming treatment (2 species). A significant decrease in parasite diversity from the initial sample was observed in three of the experimental treatments (ambient ($p = 0.046$), warming ($p = 0.014$), acidified+warming ($p = 0.046$). All other diversity comparisons were not significant.

Most prevalence comparisons were significantly different for each parasite and treatment, with the parasite being either absent or the prevalence increasing or decreasing compared to the initial sample. However, some comparisons did have a similar prevalence. Comparing parasite prevalence within each treatment, in the initial sample most comparisons were significantly different except for ciliates and *Nematopsis* sp. ($p = 1$) and *M. intestinalis* and trematodes ($p = 0.411$). In the warming treatment, a similar prevalence was observed for ciliates and trematodes, the only parasite groups present ($p = 1$). Of the three parasite groups

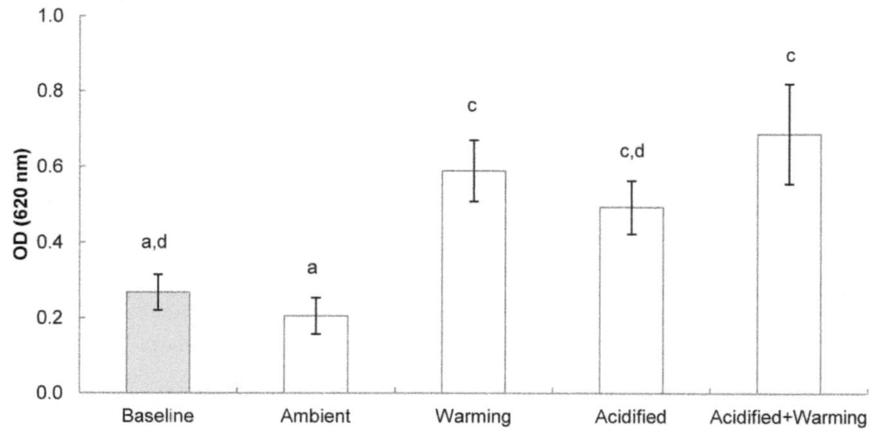

Figure 3. Effect of temperature and pH on the reduction of Nitroblue tetrazolium (NBT) (±1 SE) in haemocytes from *Mytilus edulis* (n = 5) following a six month exposure period to varying temperature and pH conditions (OD = optical density at $\lambda = 620$ nm). Baseline value shown in grey scale. Lowercase letters indicate significant differences.

Table 3. PCR screening results of the Menai Strait *Mytilus edulis* using different primer pairs.

Treatment	Me15	EUBA	Univ16s EUBf	18ScomF1	ITs	HAP-F1	ssu980
	Me16	EUBB	Univ16s EUBr	18ScomR1		HAP-R3	HAP-R1
Initial Sample	*Mytilus edulis*	0%	92%	7%	0%	60%	3%
			(28/30)	(2/30)		(18/30)	(1/30)
Ambient	*Mytilus edulis*	0%	92%	67%	0%	0%	0%
			(11/12)	(8/12)			
Warming	*Mytilus edulis*	33%	67%	25%	0%	0%	0%
		(4/12)	(8/12)	(3/12)			
Acidified	*Mytilus edulis*	0%	83%	0%	0%	0%	0%
			(10/12)				
Acidified+Warming	*Mytilus edulis*	0%	92%	0%	0%	0%	0%
			(11/12)				

Table 4. Prevalence of parasites in *Mytilus edulis* by histology.

Treatment	Ciliate	*Mytilicola intestinalis*	Trematode	*Hartmannella* sp.	*Nematopsis* sp.	Prokaryote inclusions	*Marteilia refringens*	Haplosporidian sp.
Initial sample	6%)	27%	22%	50%	6%	0%	0%	33%
	(2/30	(8/30)	(6/30)	(15/30)	(2/30)			(10/30)
Ambient	0%	33%	0%	42%	0%	58%	0%	0%
		(4/12)		(5/12)		(7/12)		
Warming	25%	0%	25%	0%	0%	0%	0%	0%
	(3/12)		(3/12)					
Acidified	0%	33%	0%	33%	8%	83%	0%	0%
		(4/12)		(4/12)	(1/12)	(10/12)		
Acidified+Warming	0%	42%	8%	8%	0%	0%	0%	0%
		(5/12)	(1/12)	(1/12)				

Initial prevalence and prevalence following 6 month exposure to ambient, warming, acidified or acidified+warming seawater conditions are presented.

present in the ambient treatment, a significant difference in prevalence was observed between *M. intestinalis* and prokaryote inclusions ($p = 0.0003$) and *Hartmannella* sp. and prokaryote inclusions ($p = 0.024$). No significant difference was observed between *M. intestinalis* and *Hartmannella* sp. ($p = 0.189$). In the four parasite groups present in the acidified treatment mussels, most prevalence comparisons were significantly different ($p<0.0001$) except for *M. intestinalis* and *Hartmannella* sp. ($p = 1$). In the acidified+warming treatment, no significant difference ($p = 1$) was observed between trematodes and *Hartmannella* sp.

Ciliates were only present in mussels from the initial sample and the warming treatment, with prevalence quadrupled in the latter showing a significant difference in prevalence between both samples ($p = 0.0002$). An increase in the prevalence of *M. intestinalis* was observed in three of the four treatments (ambient, acidified, acidified+warming,) compared to the initial sample, with the prevalence increasing significantly ($p = 0.026$) in the acidified+ warming treatment but was not significantly different in the ambient and acidified treatments ($p = 0.3$). A slight increase in the prevalence of trematodes from the initial sample was observed in the warming treatment, which was not significant ($p = 0.6$), whereas a significant decrease ($p = 0.0056$) in trematode prevalence was observed in the only other treatment (acidified+ warming) in which trematodes were observed. Across all treatments, *Nematopsis* sp. was only present in one individual of the acidified treatment. Prokaryote inclusions were observed in a high proportion of mussels from the ambient and acidified treatments (58% and 83%, respectively), with a significant difference being observed between both treatments ($p = 0.0001$) (Table 4, Fig. 4).

Pathological conditions. In total, three pathological conditions were observed in the *M. edulis*. These conditions included melanin aggregates in the mantle tissue, lipofuscin within the kidney epithelium [79] and focal haemocyte infiltration in the vascular connective tissue (Table 5, Fig. 5). All three conditions were observed in the four treatments however, melanin aggregates were not observed in the initial sample nor haemocyte infiltration in the acidified treatment. Mussels with the highest prevalence of melanin aggregates and lipofuscin, within the kidney epithelium, were observed in the acidified+warming treatment while the highest prevalence of haemocyte infiltration was observed in the warming treatment. No significant difference was calculated for the number of pathological conditions observed between samples.

Discussion

This is the first confirmation that coinciding exposure to future acidification and warming conditions has parallel implications for immune and disease status in *M. edulis*. Collectively, the data suggests that temperature more than pH may be the key driver affecting immune response in *M. edulis*. Haemocyte counts were lower under 16°C temperature conditions and phagocytosis activity increased under such conditions. Conversely, NBT reduction increased under 16°C conditions and/or pH 7.68 conditions, and therefore the effect of pH to immune function should not be discounted. Likewise, data suggests that both increases in temperature and/or lowered pH conditions may lead to changes in parasite diversity and prevalence, pathological conditions, and bacterial incidence in *M. edulis*. These changes may result in the removal or emergence of a parasite and an increase or decrease in parasite prevalence. No significant parasites such as *M. refringens* for example, were detected in this study, however, the detection of a previously undescribed HLO in the initial sample and its impact on that mussel stock, which has yet to be determined, is significant.

Previous research has shown that increased stress increases total haemocyte counts (THC) in bivalves [23,80,81]. However, investigation into the specific effect of temperature stress on THC has demonstrated both decreases [82,83,84] and increases [23] in marine bivalves. Our results suggest that temperature negatively influences THC as warming resulted in a decrease in circulating haemocytes.

Any metabolic adaptations to warming and/or acidification stressors could result in a transfer of energy resources to other physiological systems [85], potentially at the expense of maintaining haemocyte abundance. Therefore, the observed decline in THC may be indicative of an overall reduced energy state of *M. edulis* under stress conditions as been seen previously in marine bivalves under stress conditions [86]. Similarly, such declines may be representative of an overall depletion of energy reserves due to the extended exposure period (which included the species' natural spawning windows). Any immunological response mounted by the species to deal with stressors may also incur energetic costs including altered energy metabolism, altered fatty acid composition and reduced reproductive investment [86] and thus, an animal that has to sustain such a response over time will inevitably incur reduced energy stores. These in turn may lead to a decreased ability to mount an effective immunological response particularly if parallel shifts in pathology and parasite loads coincide. Consistent

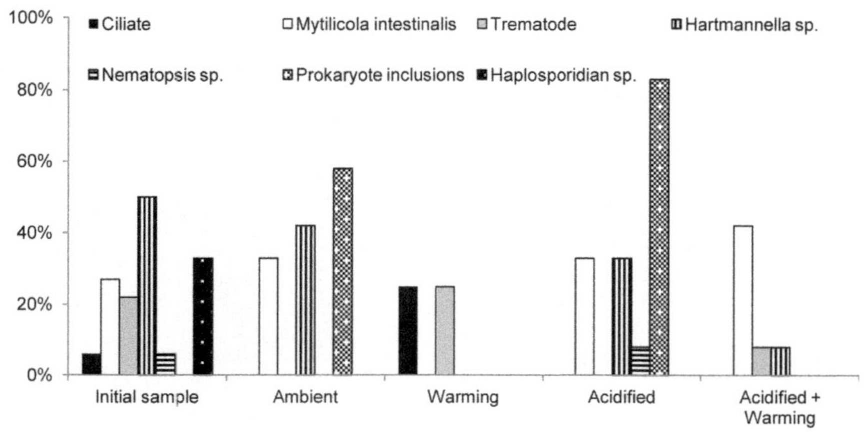

Figure 4. Prevalence of parasites in *Mytilus edulis*.

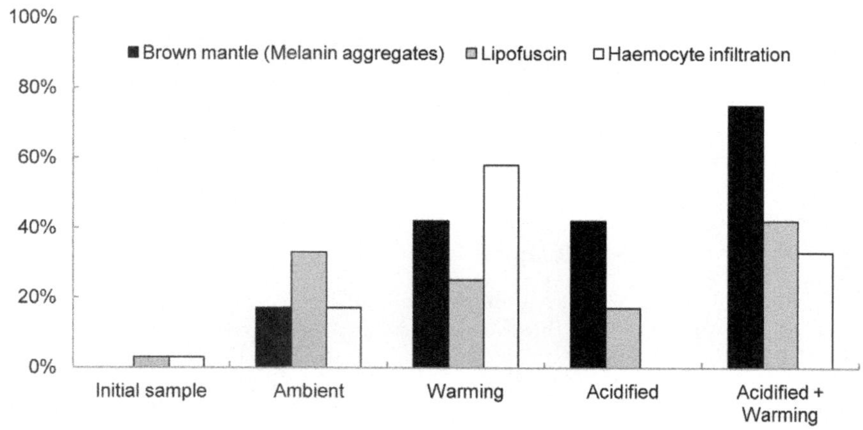

Figure 5. Prevalence of morphological changes in *Mytilus edulis*.

increases in THC in all treatments from baseline values could be a result of either proliferation of cells, or transfer of cells from tissues into circulation [8] in response to stress conditions. Acidification conditions did not appear to affect THC. Examination of the sole effect of acidified seawater conditions by Bibby *et al.* [26] also found no effect on THC.

Phagocytosis is considered a good measure of immune competency in bivalves [87]. Previous studies have demonstrated that phagocytic activity of haemocytes is affected by changes in both temperature [28] and acidification [26]. Stress conditions predominantly cause decreases in phagocytic activity [23,26,30,88], Conversely, the current study observed significant increased phagocytosis in *M. edulis* under increased temperature with potentially some effect of lowered pH. Such changes were paralleled by an increase in oxidative burst (NBT). Reduced oxidative burst has been shown to decline in other bivalve species under conditions of temperature stress [20]. Increases in both immunological parameters may be indications of the general decline in condition of the animal and could also represent a higher "immunological load" as presented by changing pathogen and parasite condition induced by temperature and/or pH stressors.

A reduction in parasite diversity in the treatments compared to the initial sample may have been due to the artificial holding conditions in the laboratory or it may have been due to the altered pH, temperature and ecology of the parasite, such as the removal of other hosts. The HLOs observed in the initial sample were not observed in the mussels from the different treatments at the end of the trial. This is likely due to the removal of mortalities during the trial, which may have been been infected with the HLOs, while the survivors were uninfected giving a negative result in the final sample at the end of the trial.

Prokaryote inclusion bodies, which are bacterial cellular reserve material used during periods of special growth phases or under special environmental conditions [89], were not detected in the initial sample but were observed in both treatments held at 12°C and were most prevalent in the reduced pH treatment. A high prevalence of bacteria detected using PCR in all treatment samples would be expected due to the laboratory holding conditions.

The prevalence of *M. intestinalis*, which causes 'red worm disease', almost doubled in the acidified+warming treatment, when compared to the initial sample. Increasing temperature has been shown to be a factor influencing infestation rates [90], however, *M. intestinalis* was not detected in mussels from the

Table 5. Prevalence of morphological changes in *Mytilus edulis* by histology.

Treatment	Brown mantle (Melanin aggregates)	Lipofuscin	Haemocyte infiltration
Initial sample	0%	3%	3%
		(1/30)	(1/30)
Ambient	17%	33%	17%
	(2/12)	(4/12)	(2/12)
Warming	42%	25%	58%
	(5/12)	(3/12)	(7/12)
Acidified	42%	17%	0%
	(5/12)	(2/12)	
Acidified+Warming	75%	42%	33%
	(9/12)	(5/12)	(4/12)

Initial prevalence and prevalence following 6 month exposure to ambient, warming, acidified or acidified+warming seawater conditions are presented.

warming treatment in this study but was present in all other three treatments which might indicate that this copepod prefers an environment with a reduced pH when seawater temperatures are elevated. *Hartmannella* sp., a free-living amoeba, was observed in most samples but was absent (warming) or significantly reduced (acidified+warming) in warming treatments indicating that this species may be sensitive to elevated seawater temperatures. Ciliates were only detected in the warming treatment but were absent in the acidified+warming treatment, which might indicate that they are sensitive to a reduced pH environment.

Trematodes were only observed in the two treatments held at 16°C, however, a prevalence reduction of almost a third was observed in the acidified-warming treatment. These results may indicate that the trematodes are sensitive to a more acidic and cooler environment, due to their absence in treatments held at 12°C.

In this study, all pathological conditions observed in the initial sample increased significantly in the four treatments, with an additional condition (melanin aggregates), which is part of a general inflammatory-like response in invertebrates [91], being observed in the laboratory trial mussels. Lipofuscin, lipid-containing residues of lysosomal digestion, is considered one of the ageing or "wear and tear" pigments and may be symptomatic of membrane damage or damage to mitochondria and is thus associated with the break down and absorption of damaged blood cells [92,93]. In this study, all mussel samples had lipofuscin within the kidney epithelium with the acidified+warming treatment having the highest prevalence. Haemocyte infiltration had a significantly higher prevalence in mussels in the warming treatment but was absent in mussels in the acidified treatment. A positive correlation between haemocyte number and increasing seawater temperature has been recorded in the Mediterranean mussel, *M. galloprovincialis* [94] and in clams, *Ruditapes philippinarum* (Adams & Reeve) [95]. Monari *et al.* [23] suggested that the increased number of haemocytes found in the clam *Chamelea gallina* (Linnaeus) under elevated seawater temperatures, might have been a consequence of a mobilisation of cells from tissues to haemolymph, in order to respond to bacteria.

Intrinsic factors such as host immunity and pathogen virulence will be more difficult to predict in climate change studies [96]. Knowledge of the rate of pathogen evolution and host evolutionary response is critical in predicting disease spread and subsequent effects on biodiversity [31,97]. Certain climate change driven stressors may also have a more negative impact on the parasite than on hosts [98]. Consequently, it is apparent that climate change-disease interactions are complex [38,43,96,99,100] and that an interdisciplinary approach should be taken [101].

Conclusions

Ocean warming and acidification pose substantial threat to marine bivalves and similar thermo-conforming species with limited ability to regulate haemolymph pH. The immune system of marine bivalves such as *M. edulis* may become comprised under such conditions. Furthermore, the reduced immune capacity of *M. edulis* to cope with climate-induced stressors such as ocean warming and acidification while dealing with co-occurring changes in parasite loads and pathogen incidence poses advanced threat to the health of the species. As such, parallel impacts to the immune status and pathology of *M. edulis* could have serious implications for global fisheries that subsist on the production and export of this commercially valuable species [102]. In addition, any ecosystem services provided by *M. edulis* (e.g. uptake and recycling of energy and nutrients, bioturbation and bioirrigation of marine sediments, sediment/shoreline stabilization and habitat formation [103]) could also be threatened. We therefore highlight the need for further investigation into the long-term effects of collective climate-induced stressors to combined aspects of bivalve health and physiology so as to provide a comprehensive illustration of impacts, which could inform management and protection of *M. edulis* as well as other marine species of high global ecological and economic value.

Author Contributions

Conceived and designed the experiments: CLM SAL SCC SKM. Performed the experiments: CLM SAL SCC SKM. Analyzed the data: CLM SAL. Contributed reagents/materials/analysis tools: SAL SCC SKM. Wrote the paper: CLM SAL SCC SKM. Maintenance of animals: CLM.

References

1. Bindoff NL, Willebrand J, Artale A, Cazenave A, Gregory J, et al. (2007) Observations: Oceanic Climate Change and Sea Level. In: Solomon S, Qin D, Manning M, Chen Z, Marquis M, Averyt KB, Tignor M, Miller HL, editors. Climate Change 2007: The Physical Science Basis. Contribution of Working Group I to the Fourth Assessment Report of the Intergovernmental Panel on Climate Change. Cambridge: Cambridge University Press. pp 385–482.
2. Nicholls RJ, Wong PP, Burkett VR, Codignotto JO, Hay JE, et al. (2007) Coastal systems and low-lying areas. In: Parry ML, Canziani OF, Palutikof JP, van der Linden PJ, Hanson CE, editors. Climate Change 2007: Impacts, Adaptation and Vulnerability.Contribution of Working Group II to the Fourth Assessment Report of the Intergovernmental Panel on Climate Change. Cambridge: Cambridge University Press. pp. 315–356.
3. Sabine CL, Feely RA, Gruber N, Key RM, Lee K, et al. (2004) The oceanic sink for anthropogenic CO_2. Science (New York, NY) 305(367): 367–71.
4. Michaelidis B, Ouzounis C, Paleras A, Pörtner H (2005) Effects of long-term moderate hypercapnia on acid-base balance and growth rate in marine mussels *Mytilus galloprovincialis*. Mar Ecol Prog Ser 293: 109–118.
5. Whiteley NM (2011) Physiological and ecological responses of crustaceans to ocean acidification. Mar Ecol Prog Ser 430: 257–271.
6. Wootton EC, Dyrynda E, Ratcliffe N (2003) Bivalve immunity: comparisons between the marine mussel (*Mytilus edulis*), the edible cockle (*Cerastoderma edule*) and the razor-shell (*Ensis siliqua*). Fish Shellfish Immunol 15(3): 195–210.
7. Wootton EC, Pipe RK (2003) Structural and functional characterisation of the blood cells of the bivalve mollusc, *Scrobicularia plana*. Fish Shellfish Immunol 15(3): 249–262.
8. Pipe RK, Coles JA (1995) Environmental contaminants influencing immune function in marine bivalve molluscs. Fish Shellfish Immunol 5(8): 581–595.
9. Pipe RK, Farley SR, Coles JA (1997) The separation and characterisation of haemocytes from the mussel *Mytilus edulis*. Cell Tissue Res 289: 537–545.
10. Cao A, Novás A, Ramos-Martínez JI, Barcia R (2007) Seasonal variations in haemocyte response in the mussel *Mytilus galloprovincialis* Lmk. Aquaculture 263(1–4): 310–319.
11. Riebesell U, Zondervan I, Rost B, Tortell PD, Zeebe R, et al. (2000) Reduced calcification of marine plankton in response to increased atmospheric CO_2. Nature 407: 364–367.
12. Portner HO, Langenbuch M, Reipschlger A (2004) Biological Impact of Elevated Ocean CO_2 Concentrations: Lessons from Animal Physiology and Earth History. J Oceanogr 60(4): 705–718.
13. Berge JA, Bjerkeng B, Pettersen O, Schaanning MT, Øxnevad S (2006) Effects of increased sea water concentrations of CO_2 on growth of the bivalve *Mytilus edulis* L. Chemosphere 62(4): 681–7.
14. Gazeau F, Quiblier C, Jansen JM, Gattuso J-P, Middelburg JJ, et al. (2007) Impact of elevated CO_2 on shellfish calcification. Geophys Res Lett 34: 1–5.
15. Miles H, Widdicombe S, Spicer JI, Hall-Spencer J (2007) Effects of anthropogenic seawater acidification on acid-base balance in the sea urchin *Psammechinus miliaris*. Mar Pollut Bull 54(1): 89–96.
16. Spicer JI, Raffo A, Widdicombe S (2007) Influence of CO_2-related seawater acidification on extracellular acid-base balance in the velvet swimming crab, *Necora puber*. Mar Biol 151: 1117–25.
17. Wood HL, Spicer JI, Widdicombe S (2008) Ocean acidification may increase calcification rates, but at a cost. P Roy Soc B-Biol Sci 275(1644): 1767–73.
18. Gooding RA, Harley CDG, Tang E (2009) Elevated water temperature and carbon dioxide concentration increase the growth of a keystone echinoderm. Proc Natl Acad Sci USA 106: 9316–9321.

19. Callaway R, Shinn AP, Grenfell SE, Bron JE, Burnell G, et al. (2012) Review of climate change impacts on marine aquaculture in the UK and Ireland. Aquat Conserv 22, 389–421.
20. Cheng W, Hsiao I-S, Hsu C-H, Chen J-C (2004) Change in water temperature on the immune response of Taiwan abalone *Haliotis diversicolor supertexta* and its susceptibility to *Vibrio parahaemolyticus*. Fish Shellfish Immunol 17(3): 235–43.
21. Parry HE, Pipe RK (2004) Interactive effects of temperature and copper on immunocompetence and disease susceptibility in mussels (*Mytilus edulis*). Aquat toxicol (Amsterdam, Netherlands) 69(4): 311–25.
22. Malagoli D, Casarini L, Sacchi S, Ottaviani E (2007) Stress and immune response in the mussel *Mytilus galloprovincialis*. Fish Shellfish Immunol 23(1): 171–7.
23. Monari M, Matozzo V, Foschi J, Cattani O, Serrazanetti GP, et al. (2007) Effects of high temperatures on functional responses of haemocytes in the clam *Chamelea gallina*. Fish Shellfish Immunol 22: 98–114.
24. Matozzo V, Marin MG (2011) Bivalve immune responses and climate changes: is there a relationship? Invertebrate Surviv J 8: 70–77.
25. Bussell JA, Gidman EA, Causton DR, Gwynn-Jones D, Malham SK, et al. (2008) Changes in the immune response and metabolic fingerprint of the mussel, *Mytilus edulis* (Linnaeus) in response to lowered salinity and physical stress. J Exp Mar Bio Ecol 358(1): 78–85.
26. Bibby R, Widdicombe S, Parry H, Spicer J, Pipe R (2008) Effects of ocean acidification on the immune response of the blue mussel *Mytilus edulis*. Aquat Biol 2: 67–74.
27. Cheng W, Li C-H, Chen J-C (2004) Effect of dissolved oxygen on the immune response of *Haliotis diversicolor supertexta* and its susceptibility to *Vibrio parahaemolyticus*. Aquaculture 232(1–4): 103–115.
28. Ellis RP, Parry H, Spicer JI, Hutchinson TH, Pipe RK, et al. (2011) Immunological function in marine invertebrates: responses to environmental perturbation. Fish Shellfish Immunol 30(6): 1209–1222.
29. Lacoste A, Malham SK, Gélébart F, Cueff A, Poulet SA (2002) Stress-induced immune changes in the oyster *Crassostrea gigas*. Dev Comp Immunol 26(1): 1–9.
30. Malham SK, Lacoste A, Gélébart F, Cueff A, Poulet SA (2003) Evidence for a direct link between stress and immunity in the mollusc *Haliotis tuberculata*. J Exp Zool 295A: 136–144.
31. Harvell CD, Mitchell CE, Ward JR, Altizer S, Dobson AP, et al. (2002) Climate warming and disease risks for terrestrial and marine biota. Science 296: 2158–2162.
32. Walther GR, Post E, Convey P, Menzel A, Permesan C, et al. (2002) Ecological responses to recent climate change. Nature 416: 398–395. ISSN: 0028-0836.
33. Parmesan C, Yohe G (2003) A globally coherent fingerprint of climate change impacts across natural systems. Nature 421: 37–42. ISSN: 0028-0836.
34. Lafferty KD, Porter JW, Ford SE (2004) Are diseases increasing in the ocean? Ann Rev Ecol Evol S 35: 31–54.
35. Ward JR, Lafferty KD (2004) The elusive baseline of marine disease: are diseases in ocean ecosystems increasing? PLoS Biol 2: 542–547.
36. Thomas CD, Williams SE, Cameron A, Green RE, Bakkenes M, et al. (2004) Biodiversity conservation: Uncertainty in predictions of extinction risk/Effects of changes in climate and land use/Climate change and extinction risk (reply). Nature 430: 6995.
37. Epstein PR, Diaz HF, Elias S (1998) Biological and physical signs of climate change: focus on mosquito-borne disease. B Am Meteorol Soc 78: 409–417.
38. Patz JA, Campbell-Lendrum D, Holloway T, Foley JA (2005) Impact of regional climate change on human health. Nature 438: 310–317.
39. Jones KE, Patel NG, Levy MA, Storeygard A, Balk D (2008). Global Trends in Emerging Infectious Diseases. Nature 451: 990–93.
40. Pilling P, Hoffman I (2011) Climate change and animal genetic resources for food and agriculture - State of knowledge, risks and opportunities. FAO CGRFA Background Study Paper No. 53, p. 28.
41. Ford SE (1996) Range extension by the oyster parasite *Perkinsus marinus* into the northeastern United States: response to climate change? J Shellfish Res 15: 45–56.
42. Cook T, Folli M, Klinck J, Ford S, Miller J (1998) The relationship between increasing sea surface temperature and the northward spread of *Perkinsus marinus* (Dermo) disease epizootics in oysters. Estuar Coast Shelf Sc 40: 587–97.
43. Lafferty KD (2009) The ecology of climate change and infectious diseases. Ecology 90: 888–900.
44. Colwell RR (1996) Global climate and infectious disease: the cholera paradigm. Science 274: 2031–2025.
45. Harvell CD, Kim K, Burkholder JM, Colwell RR, Epstein PR, et al. (1999) Review: Marine ecology - Emerging marine diseases - Climate links and anthropogenic factors. Science 285: 1505–1510.
46. Pascual M, Rodo X, Ellner SP, Colwell R, Bouma MJ (2000) Cholera dynamics and the El Nino-Southern oscillation. Science 289: 1766–1769.
47. Harvell CD, Kim K, Quirolo C, Smith G, Weir J (2001) Mass mortality of *Briareum asbestinum* associated with the 1998 Caribbean coral bleaching. Hydrobiologia 460: 97–104.
48. Porter JW, Dustan P, Jaap WC, Patterson KL, Kosmynin V (2001) Patterns of spread of coral disease in the Florida Keys. Hydrobiologia 460: 1–24.
49. Harvell CD (2004) The rising tide of ocean diseases: unsolved problems and research priorities. Front Ecol Environ 2: 375–382.
50. Holmes JC (1996) Parasites as threats to biodiversity in shrinking ecosystems. Biodivers Conserv 5: 975–83.
51. Scott ME (1988) The impact of infection and disease on animal populations: implications for conservation biology. Conserv Biol 2: 40–56.
52. Kim K, Harvell CD, Kim PD, Smith GW, Merkel SM (2000) Fungal disease resistance of Caribbean sea fan corals (*Gorgonia* spp.). Mar Biol 136: 259–267.
53. Chubb JC (1979) Seasonal occurrence of helminths in freshwater fish. Part II. Trematoda. Adv Parasitol 17: 141–313.
54. Pojmanska T, Grabda-Kazubska B, Kazubski SL, Machalska J, Niewiadomska K (1980) Parasite fauna of five fish species from the Konin lakes complex, artificially heated with thermal effluents, and from Gopło Lake. Acta Parasitol Polonica 27: 319–357.
55. McDonald MR, McClintock JB, Amsler CD, Rittschof D, Angus RA, et al. (2009) Effects of ocean acidification over the life history of the barnacle *Amphibalanus amphitrite*. Mar Ecol Prog Ser 385: 179–187.
56. Kroeker KJ, Kordas RL, Crim RN, Singh GG (2010) Meta-analysis reveals negative yet variable effects of ocean acidification on marine organisms. Ecol Lett 13: 1419–1434.
57. Amaral V, Cabral HN, Bishop MJ (2011) Resistance among wild invertebrate populations to recurrent estuarine acidification. Estuar Coast Shelf Sci 93: 460–467.
58. Pietrock M, Marcogliese DJ (2003) Free-living endohelminth stages: at the mercy of environmental conditions. Trends Parasitol 19: 293–299.
59. Al Kurashi M, Eastick FA, Kuchipudi SV, Rauch C, Madouasse A, et al. (2011) Influence of culture medium pH on internalization, growth and phenotypic plasticity of *Neospora caninum*. Vet Parasitol 177: 267–274.
60. MacLeod CD, Poulin R (2012) Host-parasite interactions: a litmus test for ocean acidification? Trends Parasitol 28: 365–9.
61. Mackenzie CL, Ormondroyd GA, Curling SF, Ball RJ, Whiteley NM, et al. (2014) Ocean warming, more than acidification, reduces shell strength in a commercial shellfish species during food limitation. PLOS ONE 9(1): e86764.
62. Houghton JT, Ding Y, Griggs DJ, Nogouer M, van der Linden PJ, et al. (editors) (2001) Climate change 2001: the scientific basis. Contribution of Working Group I to the Third Assessment Report of the International Panel on Climate Change. Cambridge: Cambridge University Press. 881 p.
63. Lewis E, Wallace D (1998) Program developed for CO_2 System calculations. Oakridge National Lab, US Department of Energy, Oakridge. 38 p.
64. Mehrbach C, Culberson CH, Hawley JE, Pytkowicz RM (1973) Measurements of the apparent dissociation constants of carbonic acid in seawater at atmospheric pressure. Limnol Oceanogr 18: 897–907.
65. Dickson AG, Millero FJ (1987) A comparison of the equilibrium constants for the dissociation of carbonic acid in seawater media. Deep Sea Res 34: 1733–1743.
66. Mortensen SH, Glette J (1996) Phagocytic activity of scallop (*Pecten maximus*) haemocytes maintained in vitro. Fish Shellfish Immunol 6: 111–121.
67. Pipe RK (1992) Generation of reactive oxygen metabolites by the haemocytes of the mussel *Mytilus edulis*. Dev Comp Immunol 16: 111–122.
68. Walsh PS, Metzger DA, Higuchi R (1991) Chelex 100 as a medium for simple extraction of DNA for PCR-based typing from forensic material. Biotechniques 10: 506–513.
69. Lynch SA, Mulcahy MF, Culloty SC (2008) Efficiency of diagnostic techniques for the parasite, *Bonamia ostreae*, in the flat oyster, *Ostrea edulis*. Aquaculture 281: 17–21.
70. Shi SR, Cote RJ, Wu L, Liu C, Datar R et al. (2002) DNA extraction from archival formalin-fixed, paraffin-embedded tissue sections based on the antigen retrieval principle: heating under the influence of pH. J Histochem Cytochem 50(8): 1005–1011.
71. Inoue K, Waite JH, Matsuoka M, Odo S, Harayama S (1995) Interspecific variation in adhesive protein sequences of *Mytilus edulis*, *M. galloprovincialis* and *M. trossulus*. Biol Bull 189: 370–375.
72. Weisberg WG, Barns SM, Pelletier DA, Land DJ (1991) 16S ribosomal DNA amplification for phylogenetic study. J Bacteriol 173: 697–703.
73. Kawai M, Matsutera E, Kanda H, Yamaguchi N, Tani K, et al. (2002) 16S ribosomal DNA-based analysis of bacterial diversity in purified water used in pharmaceutical manufacturing processes by PCR and denaturing gradient gel electrophoresis. Appl Environ Microbiol 68: 699–704.
74. Zhang H, Bhattacharya D, Lin S (2005) Phylogeny of dinoflagellates based on mitochondrial cytochrome *b* and nuclear small subunit rDNA sequence comparisons. J Phycol 41: 411–420.
75. Le Roux F, Lorenzo G, Peyret P, Audemard C, Figueras A, et al. (2001) Molecular evidence for the existence of two species of *Marteilia* in Europe. J Eukaryot Microbiol 48: 449–454.
76. Renault T, Stokes NA, Chollet B, Cochennec N, Berthe F, et al. (2000) Haplosporidiosis in the Pacific oyster *Crassostrea gigas* from the French Atlantic coast. Dis Aquat Organ 42: 207–214.
77. Molloy DP, Giambérini L, Stokes NA, Burreson EM, Ovcharenko MA (2012) *Haplosporidium raabei* n. sp (Haplosporidia): a parasite of zebra mussels, *Dressina polymorpha* (Pallas, 1771). Parasitology 139: 463–477.
78. Shaw BL, Battle HI (1957) The gross and microscopic anatomy of the digestive tract of the oyster *Crassostrea virginica* (Gmelin). Can J Zoolog 35: 325–347.
79. Stevens A, Chalk BT (1996) Pigments and minerals. In: Bancroft JD, Stevens A, editors. Theory and practice of histological techniques, Fourth edition. London: Churchill Livingstone. pp. 243–267.
80. Mayrand E, St-Jean SD, Courtenay SC (2005) Haemocyte responses of blue mussels (*Mytilus edulis* L.) transferred from a contaminated site to a reference site: can the immune system recuperate? Aquac Res 36(10): 962–971.

81. Renwrantz L, Siegmund E, Woldmann M (2013) Variations in hemocyte counts in the mussel, *Mytilus edulis*: Similar reaction patterns occur in disappearance and return of molluscan hemocytes and vertebrate leukocytes. Comp Biochem Physiol A Mol Integr Physiol 164(4): 629–37.
82. Smith VJ, Chisholm JRS (1992) Non-cellular immunity in crustaceans. Fish Shellfish Immunol 2: 1–31.
83. Cheng W, Chen J-C (2000) Effects of pH, temperature and salinity on immune parameters of the freshwater prawn *Macrobrachium rosenbergii*. Fish Shellfish Immunol 10(4): 387–91.
84. Malham SK, Cotter E, O'Keeffe S, Lynch SA, Culloty SC, et al. (2009) Summer mortality of the Pacific oyster, *Crassostrea gigas*, in the Irish Sea: The influence of temperature and nutrients on health and survival. Aquaculture 287(1–2): 128–138.
85. Sokolova IM, Frederich M, Bagwe R, Lannig G, Sukhotin AA (2012) Energy homeostasis as an integrative tool for assessing limits of environmental stress tolerance in aquatic invertebrates. Mar Environ Res 79: 1–15.
86. Ellis RP (2013) The impact of ocean acidification, increased seawater temperature and a bacterial challenge on the immune response and physiology of the blue mussel, *Mytilus edulis*. PhD Thesis. University of Plymouth, UK.
87. Hooper C, Day R, Slocombe R, Handlinger J, Benkendorff K (2007) Stress and immune responses in abalone: limitations in current knowledge and investigative methods based on other models. Fish Shellfish Immunol 22: 363–379.
88. Fisher SW, Auffret M, Balouet G (1987) Response of European flat oyster (*Ostrea edulis*) hemocytes to acute salinity and temperature changes. Aquaculture 67: 179–190.
89. Lengeler J, Drews G, Schlegel H (2009) Biology of the prokaryotes. Google ebook ISBN: 978-1-4443-1330-7. Hoboken: John Wiley & Sons. 984 p.
90. Robledo JAF, Santarém MM, Figueras A (1994) Parasite loads of rafted blue mussels (*Mytilus galloprovincialis*) in Spain with special reference to the copepod, *Mytilicola intestinalis*. Aquaculture 127: 287–302.
91. Nappi AJ, Christensen BM (2005) Melanogenesis and associated cytoxic reactions: applications to insect innate immunity. Insect Biochem Molec 35: 443–59.
92. Katz ML, Robison WG Jr (2002) What is lipofuscin? Defining characteristics and differentiation from other autofluorescent lysosomal storage bodies. Arch Gerontol Geriatr 34: 169–84.
93. Jung T, Hohn A, Grune T (2010) Lipofuscin: detection and quantification by microscopic techniques. Methods Mol Biol 594: 173–193.
94. Carballal MJ, Villalba A, Lopez C (1998) Seasonal variation and effects of age, food availability, size, gonadal development, and parasitism on the hemogram of *Mytilus galloprovincialis*. J Invertebr Pathol 72: 304–312.
95. Paillard C, Allam B, Oubella R (2004) Effect of temperature on defence parameters in Manila clam *Ruditapes philippinarum* challenged with *Vibrio tapetis*. Dis Aquat Organ 59: 249–262.
96. Rohr JR, Schotthoefer AM, Raffel TR, Carrick HJ, Halstead N, et al. (2008) Agrochemicals increase trematode infections in a declining amphibian species. Nature 455: 1235–1250.
97. Palumbi S (2001) Humans as the world's greatest evolutionary force. Science 293: 1786–1790.
98. Lafferty KD (1997) The ecology of parasites in a salt marsh ecosystem. In: Beckage NE, Zuk M, editors. Parasites and Pathogens: Effects on Host Hormones and Behavior. New York: Chapman and Hall. pp. 316–332.
99. Hay SI (2002) Hot topic or hot air? Climate change and malaria resurgence in East African highlands. Trends Parasitol 18: 530–534.
100. Rogers DJ, Randolph SE (2006) Climate change and vector-borne diseases. Adv Parasitol 62: 345–381.
101. Rohr JR, Halstead NT, Raffel TR (2011) Modelling the future distribution of the amphibian chytrid fungus: the influence of climate and human-associated factors. J Appl Ecol 48: 174–176.
102. Food and Agriculture Organization (FAO) (2012) The State of World Fisheries and Aquaculture (SOFIA). FAO, Rome.
103. Shumway SE, Davis C, Downey R, Karney R, Kraeuter J, et al. (2003) Shellfish aquaculture - In praise of sustainable economies and environments. World Aquac 34: 15–19.

22

Surface Sediments in the Marsh-Sandy Land Transitional Area: Sandification in the Western Songnen Plain, China

Xiaofei Yu[1], Michael Grace[2], Yuanchun Zou[1], Xuefeng Yu[3], Xianguo Lu[1], Guoping Wang[1]*

1 Key Lab of Wetland Ecology and Environment, Northeast Institute of Geography and Agroecology, Chinese Academy of Sciences, Changchun, China, **2** Water Studies Centre and School of Chemistry, Monash University, Clayton, VIC, Australia, **3** State Key Laboratory of Loess and Quaternary Geology, Institute of Earth Environment, Chinese Academy of Sciences, Xi'an, China

Abstract

The development of sandification process was studied, by monitoring the changes of sediment characteristics, at marsh-sandy land intersections in China's Songnen region. A series of sediment collection plates were deployed in the region; after one year, sediments in these plates were analyzed for changes of mass and chemical characteristics. The sediment flux and the sand content of the sediments decreased with the increasing longitudinal distance between the sampling site and the centre line of a sand dune. The mean sediment flux was 29±14 kg m^{-2} yr^{-1} and 0.6±0.3 kg m^{-2} yr^{-1} in the sandy land and marsh, respectively. Strong, positive correlations were found between the concentrations of organic matter, total nitrogen, P, Fe, Ti, V and Zr, all of which were also negatively correlated with the sand content. The concentrations of organic matter, total nitrogen, P, Fe, Ti, V and Zr in the marsh sediment samples were all significantly greater than the corresponding concentrations of the sandy land ($p<0.001$). Sand content and Ti, V and Zr concentrations all proved to be valid indicators of sandification intensity, and they showed that the marsh could be divided into three distinct zones. Sand expansion extended about 88 m into the marsh. The mean sand content in the sediments of the sandy land was 91% and then 64% in the marsh, which in turn was higher than that of marshes outside the influence of sandification, suggesting that the marsh in the marsh-sandy land transitional area has already undergone extensive sandification in the past. The study results provide information on the wetland's function of indicating and buffering the sandification process.

Editor: Vanesa Magar, Centro de Investigacion Cientifica y Educacion Superior de Ensenada, Mexico

Funding: The National Basic Research Program (No. 2012CB956100), the National Natural Science Foundation of China (No. 41101195) and the Science Foundation of State Key Laboratory of Loess and Quaternary Geology (No. SKLLQG0910) supplied the financial supports of this paper. The funders had no role in study design, data collection and analysis, decision to publish, or preparation of the manuscript.

Competing Interests: The authors have declared that no competing interests exist.

* E-mail: wangguoping@iga.ac.cn

Introduction

Sandification refers to the coarsening process of the land surface after fine particles are lost to aeolian erosion. Sandification forms a stage in the development of desertification [1]. The Songnen Plain, as the largest plain of 18.3×10^4 km^2 in Northeast China and one of the commodity grain production bases in China, has been identified as a key area of sandification control [2]. Its western part is a wide semi-arid sandy land on the eastern extension of the Gobi desert and Inner Mongolian steppe [3]. Historically, the Songnen Plain is an alluvial–proluvial and lacustrine plain during the Mesozoic. During the Pleistocene, strong sand driving winds and sandstorms caused by Baikal and Mongolian high-pressure systems, through which sandification formed and expanded into the middle and east of Songnen Plain [4]. Since the mid-20th century, desertification has increased rapidly under the threat of improper land use and climate change. The sandification area covers 7.25% of the total area at present [5]. Because of its geological background, considerable wetlands distribute in the Songnen Plain with the area of 2.96×10^4 km^2 [6]. Zhalong, Xianghai and Momoge National Nature Reserves and other local wetland reserves protect the endangered and rare waterfowl (i.e. *Grus japonensis*, *Grus leucogeranus*) with international importance. The domain wetland plant is *Phragmites australis*.

The Plain has alternating landscapes of sandy lands and wetlands (marshes and lakes) with frequent changeovers between strong and weak impacts of sandification. Little attention has been focused on the relationship between sandification and marshes in the wetland-sandy land transitional areas. Numerous studies have focused on the relationships between loess sediments and sandification in the loess-desert transitional areas [7,8,9]. The sediments of the loess and the wetlands both contain historical archives of surrounding landscapes, leading to the hypothesis that the sediments of a wetland should record sandification in the wetland-sandy land transitional areas.

Wetlands are strongly influenced by input of matter from surrounding landscapes and play an important role in recording changes in climate and environment-altering human activities [10,11,12]. Extensive studies of wetland sediment cores reveal large amounts of information on the development history of both the wetland and the surrounding landscapes [13,14]. Although the surface sediments of wetlands should reflect current environment events, there has been very little research done on this [15].

Sandy land contains mostly sand-size particles (coarser than 63 μm in diameter), while wetlands contain mostly silt and clay-size particles (finer than 63 μm) [16]. The sand-sized particles are usually carried by the wind and trapped in the wetlands directly next to the sandy land because of the dense vegetation cover and

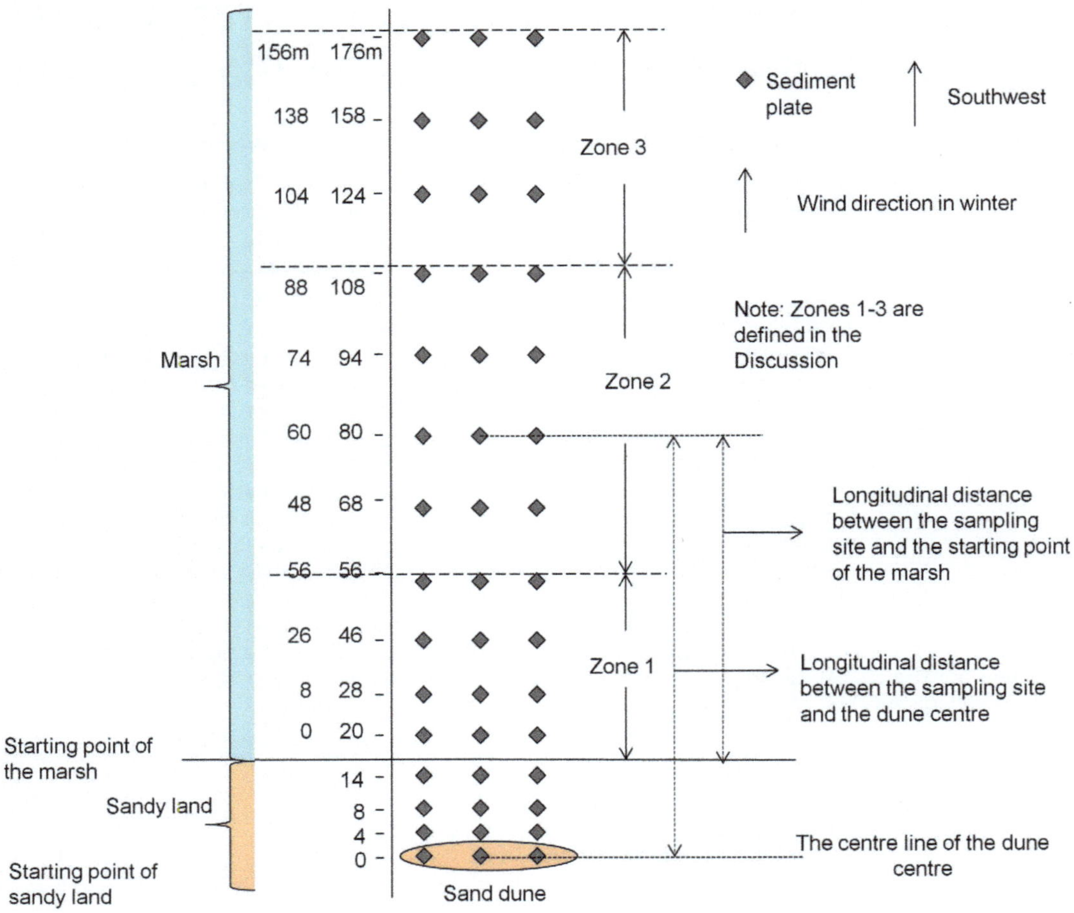

Figure 1. Arrangement of sampling sites in the field experiment. The sediment plate matrix (15 rows ×3 columns) was set in a marsh-sandy land transitional area. The length of the column was 176 m, while that of the row was 20 m. Four rows were set in the sandy land, and eleven rows were set in the marsh. The column length in the marsh was 156 m. The sediment plates on a column were at longitudinal distances of 0, 4, 8..., 176 m to the centre line of the dune. The plates in each row were replicates (n = 3) with the same longitudinal distance to the centre line of the dune, with a lateral interval of 10 m between each plate.

wet ground. Whether the sand contents in wetland sediments can be used to determine the intensity of sandification has not been proved. Moreover, in a wetland ecosystem, sediment deposits act as a sink for a wide range of environmental chemicals [17]. We hypothesize that the concentrations of some inert elements can be used as indicators of sandification intensity.

Some wetlands are located at the margin of the sandy lands and may act as barriers against sand expansion, which is qualitatively easy to observe and understand, as the wetland's dense vegetation cover slows down the wind speed [18,19] and traps the sands carried by the wind, which then decreases the transport distance of the sand. However, to verify and quantify these barrier effects requires specific field investigations. We hypothesize that there will be a gradient in sand content in general and also concentrations of some chemical elements, in the surface sediments from the sandy land to the downwind wetland. We therefore predict that these data will allow us to determine the degree to which the wetland is defending against sand expansion.

This study was conducted in a marsh-sandy land transitional area on the western Songnen Plain, China. There were three major, specific objectives to this work: (1) to quantify the sediment flux, the sand content and geochemical characteristics in surface sediments in the marsh-sandy land transitional area; (2) to link these characteristics with sandification and assess the indicators of the degree of sandification; and (3) to test the assumption that the marsh is blocking sand expansion and to determine the distance of the barrier to sand expansion. The results will facilitate the protection of wetlands and the governance of the sandy land in the wetland-sandy land transitional areas.

Materials and Methods

Location of the Study

The Niuxintaobao Reed Field (123°21′E, 45°14′N) is situated in the western Songnen Plain, China. At an altitude of 140 m, this area of more than 40 km^2 is the floodplain of the Huolin River [20], in which wetlands, sandy land, farmland and forest land are distributed. This area is located in a semi-arid climatic zone, with an annual average temperature of 4.3°C (varying from −18°C in January to 23°C in July) and an average annual precipitation of 413 mm, more than 70% of which occurs from June to August [2]. The climate is classified as temperate continental monsoon [20]. The prevailing wind direction is southwest and northeast in summer and winter, respectively. Seasonal droughts in winter, spring and autumn mainly lead to the sandification. The wetlands are being degraded due to seasonal drought, sandification and high-intensity agricultural activities [2]. This increasing sandifica-

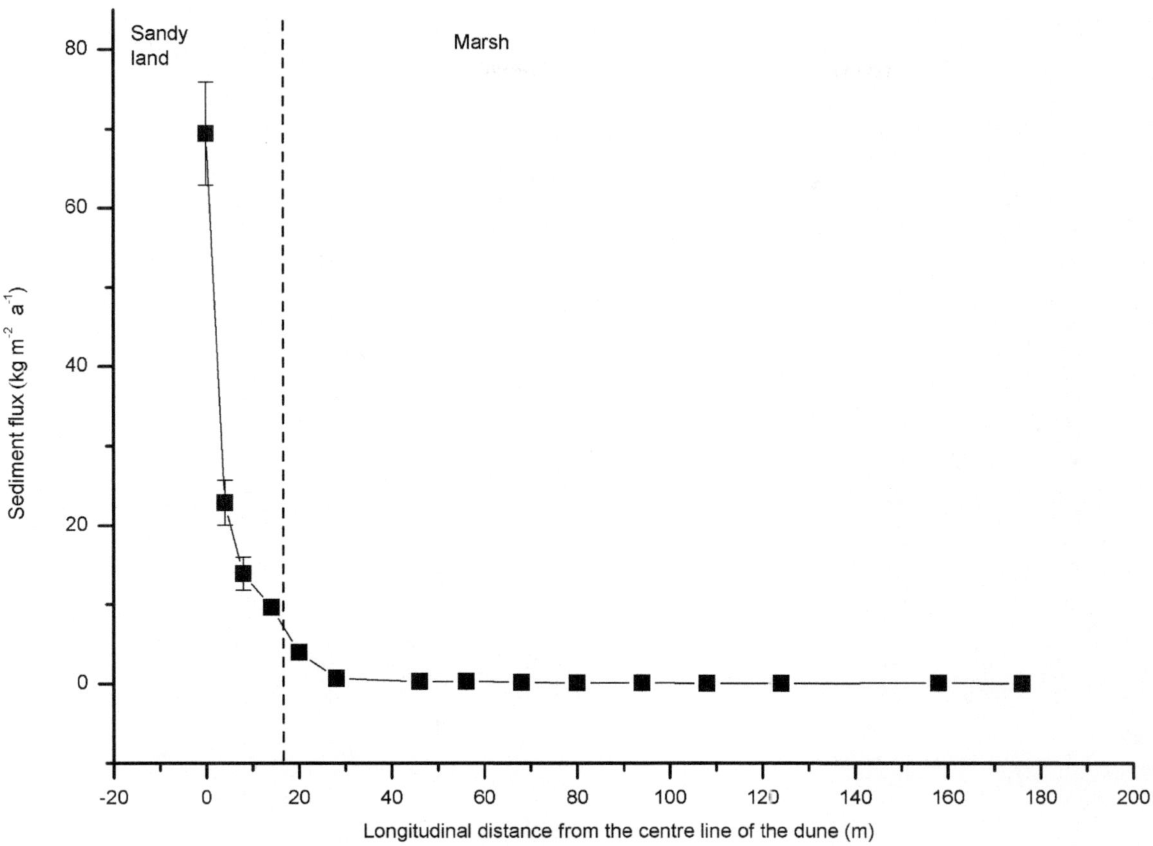

Figure 2. Sediment flux in the marsh-sandy land transitional area. Error bars represent the standard error of the mean of three parallel samples.

tion and the degradation of the wetlands are the main environmental and ecological problems of this area.

Arrangement of sampling sites

A marsh-sandy land transitional area was chosen for the field experiment. The dominant marsh plant is *Phragmites australis*, with an average height of 150 cm and an average density of 80–100 stem m^{-2}. The start point of the marsh is defined as the edge of the *Phragmites australis* and is about 17 m from the centre line of the nearest sand dune. The sediment plate array (15 rows ×3 columns) was installed in the marsh-sandy land transitional area on April 19th, 2010, as depicted in Figure 1. The prevailing northeast wind blew towards the marsh from the sandy land in winter, which also represents the direction of expanding sandification and the column direction in the experimental array. During summer the prevailing wind is a southwesterly. Under these conditions the water and the dense vegetation in the marsh restrict any sand blowing from the marsh toward the sandy land. The length of the plate column was 176 m. Four rows were set in the sandy land, and eleven rows in the marsh. The column length in the marsh was 156 m. The longitudinal distance of 0, 4, 8..., 176 m of each row is to the centre line of the dune. The three plates in each row had the same longitudinal distance to the centre line of the dune, with a lateral separation of 10 m between each plate. Each sediment plate consisted of a collecting plate and a fixed rod. The square (30×30 cm) collecting plate was made of fired clay with a coarse surface for the immobilization of sediments. The 40 cm fixed steel rod penetrated through the centre of the collecting plate and was fixed by a bolt and nut at the upper end. The lower, sharpened end was inserted into the soil for fixation.

Sampling method

After field deployment for one year, the fixed rods were pulled out, and the net sediments on the plates were collected carefully and completely, and then stored in polyethylene bags. It is noted that some material (especially finer material) was lost from the plates during rainfall events without any plant coverage on the plates, which is the same as the sediments in field. Compared with the sediments formed in many years, the sediments we collected just deposited in one year and we call them surface sediments. All the field studies got the permission from administrative authority of Niuxintaobao National Wetland Park and did not involve any endangered or protected species (123°21′E, 45°14′N).

Sample Analyses

Any vegetation litter was removed from the sediment samples before analysis. Each sample was weighed (as 'fresh weight') and then air-dried for 1 month. These dried samples were then reweighed. Grain-size distributions in the air-dried samples were determined using a Mastersizer2000 Laser Grainsizer (Malvern Instruments Ltd. UK; measuring range: 0.02–2000 mm). The dried samples were then crushed and sieved through a 0.18-mm nylon sieve. The organic matter (OM) was measured by the potassium dichromate volumetric method [21], which required heating at 180°C for 5 minutes. Total nitrogen (TN) was determined by the Kjeldahl digestion–constant distillation–titration method [21]. Concentrations of P, Fe, Al, Ti, V and Zr were

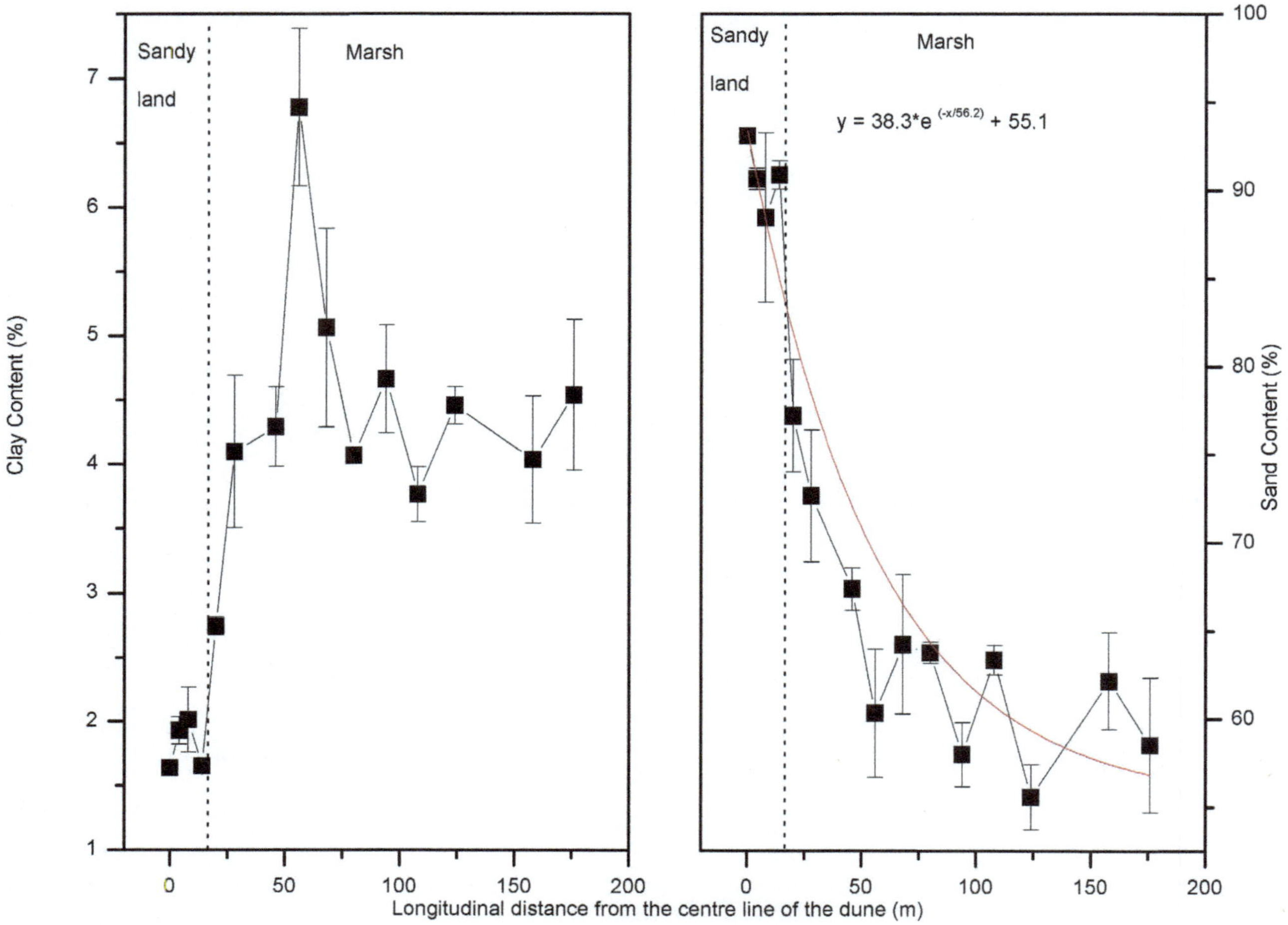

Figure 3. Sand (>63 μm) and clay (<2 μm) composition of the sediments. Error bars represent the standard error of the mean of three parallel samples.

determined by ICP-MS-7500 (Shimadzu) after sample digestion using a HNO_3–HF–$HClO_4$ solution for 0.5 h [21].

Data analyses

The sediment flux was calculated according to:

$$SedimentFlux = DryWeight/(PlateArea \times 1yr) \quad (1)$$

The plate area was 0.09 m^2 as described above.

Statistical analysis was performed with IBM SPSS 21.0. The t-test analyses were conducted to compare sediments from the marsh and the sandy land. The following parameters were examined in this analysis: the sediment flux, sand content, and the concentrations of OM, TN, P, Fe, Al, Ti, V and Zr in the sediments. Pearson correlation analyses were conducted to test relationships between the sand content, clay content, and concentrations of OM, TN, P, Fe, Al, Ti, V and Zr in the sediments. A Hierarchical Cluster Analysis was conducted to classify different zones in the marsh with different sand content, and Ti, V and Zr in the sediments. A one-way ANOVA was performed to document the differences in the sand content, and concentrations of Ti, V and Zr in the sediments of the three zones in the marsh.

Results

Sediment flux

The trends in sediment flux are presented in Figure 2. The sediment flux decreased with increasing longitudinal distance from the centre line of the dune. The sediment fluxes in the sandy land were significant higher than that in the marsh ($p = 0.003$). The mean sediment flux was 29±14 kg m^{-2} yr^{-1} in the sandy land, while that was 0.6±0.3 kg m^{-2} yr^{-1} in the marsh.

Sediment grain-size distributions

The trends in the relative content (by mass) of sand (>63 μm) and clay (<2 μm) in the sediment are presented in Figure 3. The sand content in the sediment of the sandy land was much greater than that of the marsh (Figure 3). The sand content in the marsh sediment decreased as the longitudinal distance between the sampling site and the centre line of the dune increased (Figure 3). The sand content at the sampling site 20 m from the centre line of the dune was 77.19%, while it is 58.51% at the site 176 m from the centre line of the sand dune. The mean sand content in the sediments of sandy land was 91±1%, while it was 64±6% of marsh. The mean clay content in the sediments of the sandy land was 1.8±0.1%, while it was 4±1% in the marsh. The clay content of the sandy land was significant lower than that of the marsh ($p < 0.001$). The sand content was negatively correlated at a level of 0.1% to the clay content ($r = -0.878$).

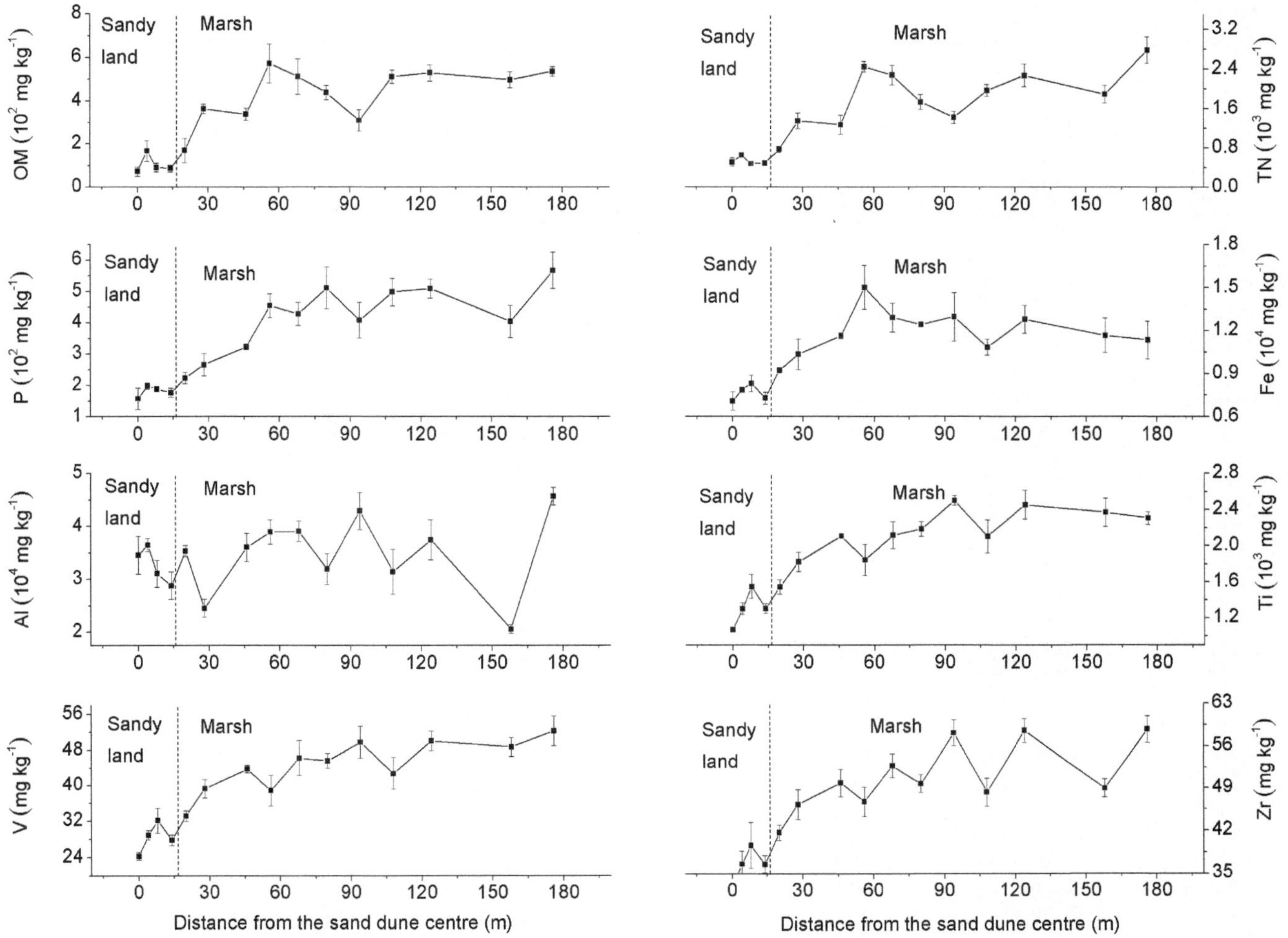

Figure 4. Concentrations of OM, TN, P, Fe, Al, Ti, V and Zr in the sediments. Error bars represent the standard error of the mean of three parallel samples. OM and TN are organic matter and total nitrogen, respectively.

Sediment geochemical characteristics

The concentrations of OM, TN, P, Fe, Al, Ti, V and Zr in sediments at increasing longitudinal distances from the centre line of the dune are presented in Figure 4. When sample concentrations from different distances were pooled together, the concentrations of OM, TN, P, Fe, Ti, V and Zr in the marsh sediments were all significant greater than in the sandy land ($p=0.001$ for TN and P, $p<0.001$ for others). There was no significant difference in Al concentration between the two sediment types ($p=0.459$). The results of Pearson correlation analyses between concentrations of the various sediment constituents: OM, TN, P, Fe, Al, Ti, V and Zr, are presented in Table 1. There were significant positive correlations between the concentrations of OM, TN, P, Fe, Ti, V and Zr. The Al concentration was not significantly correlated to any other sediment constituent.

Discussion

Indicators of sandification intensity

The sediment flux and the sand content in the sediments both decrease with increasing longitudinal distance between the sampling site and the centre line of the dune (Figures 2, 3), indicating that sand deposition decreases further away from the centre line of the dune. Approximately 90% of the sediments from sandy land are sand (Figure 3), which reflects the mean kinetic energy of eolian transport and represent the typical eolian environment [5]; however, because of the existence of reed marsh, the sandy land does not extent and the sand content of marsh sediments decrease dramatically. These findings support our hypothesis that the sand content in the sediment is an indicator of the sandification intensity in the marsh-sandy land transitional area. Ding *et al.* [22] used sediment cores in a desert-loess area to reveal sandification history. Our results show that sand content trapped in the surface sediment can indicate current sandification progress. Coupling these findings suggests that marsh sediment cores may also record the history of sandification.

The chemical analysis showed that Al and Fe concentrations are much higher than the rest elements (Figure 4), which suggests that the main components of these sediments were silicon, aluminum and iron oxides [3]. There are positive correlations between the concentrations of OM, TN, P, Fe, Ti, V and Zr in the sediments (Table 1). However, these components were all negatively correlated (0.1% significance level) with the sand composition of the sediments, indicating that the sandy land is not the source of these elements. The much lower concentrations of these elements in sandy land sediments may have originated from transfer by water and wind from the surrounding marsh, meadow or farmland. OM, TN and P may also come from the decomposition

Table 1. Pearson correlation results for relationships between OM, TN, P, Fe, Al, Ti, V and Zr concentrations in the sediments.

		OM	TN	P	Fe	Al	Ti	V	Zr
OM	*r*	1.000	0.970**	0.922**	0.867**	0.172	0.802**	0.834**	0.788**
	p		<0.001	<0.001	<0.001	0.539	<0.001	<0.001	<0.001
TN	*r*		1.000	0.936**	0.836**	0.338	0.776**	0.832**	0.817**
	p			<0.001	<0.001	0.218	0.001	<0.001	<0.001
P	*r*			1.000	0.818**	0.346	0.856**	0.882**	0.862**
	p				<0.001	0.207	<0.001	<0.001	0.001
Fe	*r*				1.000	0.317	0.809**	0.800**	0.802**
	p					0.250	<0.001	<0.001	<0.001
Al	*r*					1.000	0.196	0.253	0.419
	p						0.484	0.363	0.120
Ti	*r*						1.000	0.986**	0.955**
	p							<0.001	<0.001
V	*r*							1.000	0.970**
	p								<0.001
Zr	*r*								1.000
	p								

**Correlation is significant at the 0.01 level; OM and TN are organic matter and total nitrogen, respectively.

of detritus from dead plants and animals in the marsh. Sites in the marsh with greater sand deposition have lower concentrations of these elements. As a result, the sand content and the concentrations of Ti, V and Zr in the marsh sediments can be used as the indicators of sandification intensity in this area. The applicability of these indicators to other marsh-sandy land transitional areas requires validation.

The OM concentration is strongly correlated to Fe, Ti, V and Zr (Table 1). Cheshire *et al.* [23] found that Fe, Ti and V were extracted from the soil OM, suggesting that these elements in our sediment samples may also be associated with organic matter. There is no correlation between Al and other elements (Table 1). There is no significant difference in the Al concentrations between the marsh and the sandy land (Figure 4(e)), as Al exists in both the clay and the sand [24].

Variations of Sand Content in the Sediments

The sand content in the marsh sediment samples are significantly lower than those of sandy land ($p<0.001$), demonstrating that the sand deposition rate is slowed down in the marsh. There are two contributing factors to this finding. Firstly, the longitudinal distance from the centre line of the dune is bigger in the marsh than in the sandy land and the deposition rate of sand decreases with the increasing distance. Secondly, the dense plant cover blocks some sand from penetrating into the marsh.

The sand content in the marsh sediment exponentially decays with increasing longitudinal distance between the sampling site and the start point of the marsh (Equation 2, Figure 3b, $p=0.79$):

$$y=38.3\times e^{-x/56.2}+55.1 \qquad (2)$$

where y is the % sand content and x is the longitudinal distance (m) between the sampling site and the start point of the marsh. The equation and Figure 3b show that the sand content from these marsh sediment samples is at least 55%; this is much higher than the sand content (20.2%) in the surface sediment of Boluo Pao wetland (124°48′E, 44°22′N), situated outside any sandification areas on the Songnen Plain [25]. The sand content of surface sediment sampled from the Xianghai Wetlands Nature Reserve (122°05–31′E, 44°55–09′N) situated in the western border of the Songnen Plain is about 10~20%, which is also lower than our data [11]. These reveal that the sand content in the sediments of this marsh should be greater than other marshes outside the influence of sandification, for the combination of wet ground and dense plant cover allows the marsh to act as an efficient sand trap [26].

The higher sand content in our study also demonstrates that the marsh in this marsh-sandy land transitional area has a one-year record of sand addition and is facing the threat of sandification. Sandification may negatively affect ecological functions in the marsh, including primary production and carbon sequestration

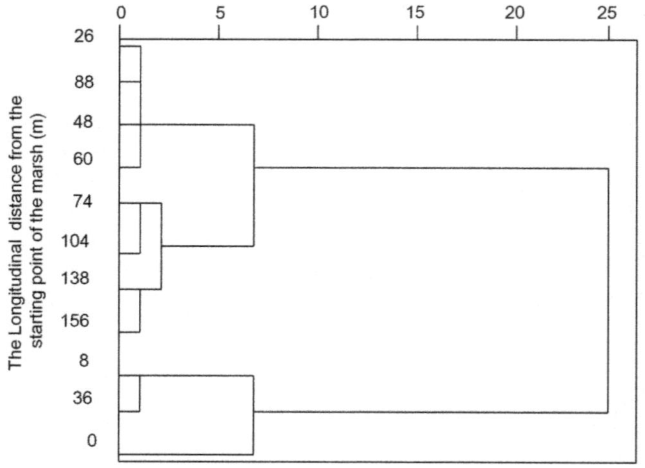

Figure 5. Hierarchical Cluster Analyses on sand content, concentrations of Ti, V and Zr in marsh sediments.

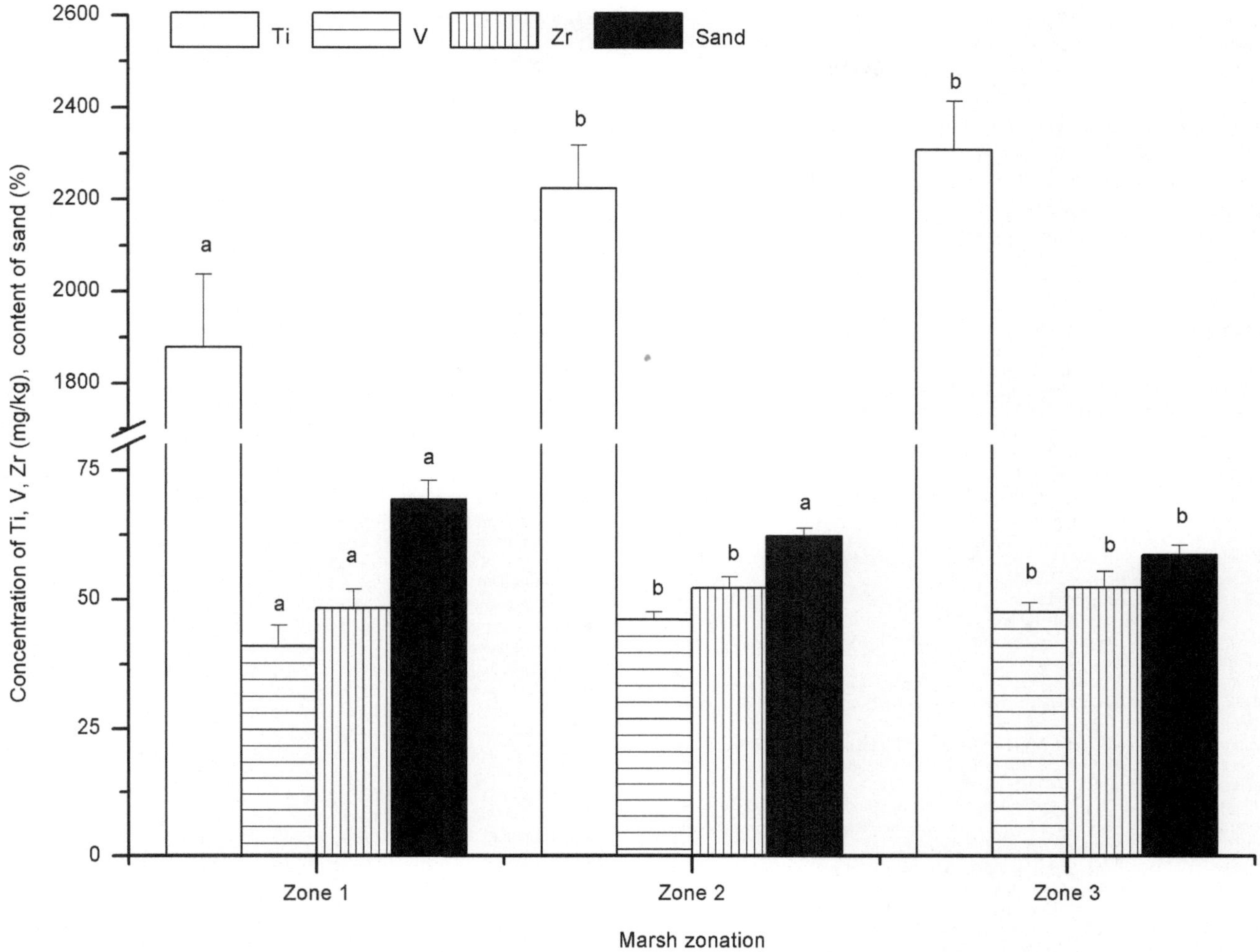

Figure 6. Mean Ti, V and Zr concentrations and sand content in marsh sediments of different zones. Error bars represent the standard error of the mean of twelve, twelve and nine parallel samples in Zone 1, Zone 2 and Zone 3, respectively. The letters 'a' and 'b' indicate mean concentrations of Ti, V and Zr and % sand content which are significantly different at $p = 0.05$.

[27,28]. During the sandification process, the nutrient conditions of wetland soils deteriorate, characterized as lower concentrations of organic matter, nitrogen and phosphorus, leading to reduced biomass and biodiversity. As a result, the primary production and carbon sequestration of wetland are weakened. Moreover, a decline in biomass, biodiversity, soil nutrients and soil structure makes the wetland system less resilient, with greater vulnerability and sensitivity to disturbance from human activities and climate change.

Buffering Zones of Sandification

As the farming-pastoral ecotone, the Songnen Plain has suffered from severe land use/cover changes [29]. The croplands reclaimed from grasslands or wetlands (wet meadow) in wet years were unfortunately abandoned in dry years when farming becomes no longer practical, which had caused land degradation locally and accelerated the extension of sand dunes under the action of predominant wind direction without vegetation coverage of topsoil meanwhile [30]. Soil water conservation and vegetation coverage were recognized as the important countermeasures of sand desertification [31].

A hierarchical cluster analysis was conducted to classify different zones in the marsh according to the different sand content and concentrations of Ti, V and Zr in the sediments. According to the results shown in Figure 5, the marsh can be divided into three distinct zones (Zone 1, Zone 2 and Zone 3; Figure 2). One-Way ANOVA analysis shows that the sand contents in the sediments of Zones 1 and 2 are significantly higher than that of Zone 3 (Figure 6). The concentrations of Ti, V and Zr in the sediments of Zone 1 are significantly higher than that of Zones 2 and 3 (Figure 6), while those concentrations in Zone 2 are higher than Zone 3, but this difference is not statistically significant (p>0.01). As a result, Zone 1 and Zone 2 are acting as a buffer zone, limiting further sand expansion during our field experiment. The buffer length can therefore be estimated as about 88 m (the total length of Zones 1 and 2) into the marsh. It is clear from this data that the marsh provides a means of controlling the expansion of sand, and suggests that an alternating distribution of marshes and sandy lands may both impede sandification and assist in the protection of wetlands in wetland-sandy land transitional areas.

Conclusions

The study presents sediment flux measurements, sand and clay content and the geochemical characteristics of sediments in a marsh-sandy land transitional area on the western Songnen Plain, China. The sediment flux and the sand content of the sediments decreased with the increasing longitudinal distance between the sampling site and the centre line of a sand dune. The sand content is negatively correlated with the concentrations of organic matter, total nitrogen, P, Fe, Ti, V and Zr. The concentrations of organic matter, total nitrogen, P, Fe, Ti, V and Zr in the sediments of the marsh were all significantly greater than those of the sandy land. Sand content and Ti, V and Zr concentrations all proved to be valid indicators of sandification intensity, and they showed that the marsh could be divided into three distinct zones.

Acknowledgments

We gratefully acknowledge Guo Jiawei and Li Hongnan for plate installation and sediment sampling, and Zhang Yuxia, Zhao Haiyang, and Lou Bokun for sample analyses.

Author Contributions

Conceived and designed the experiments: Xiaofei Yu GPW. Performed the experiments: Xiaofei Yu YCZ GPW. Analyzed the data: Xiaofei Yu MG YCZ. Contributed reagents/materials/analysis tools: Xuefeng Yu XGL GPW. Wrote the paper: Xiaofei Yu MG GPW.

References

1. Zha Y, Gao J (1997) Characteristics of desertification and its rehabilitation in China. Journal of Arid Environments: 419–432.
2. Wang GP, Zhai ZL, Liu JS, Wang JD (2008) Forms and profile distribution of soil phosphorus in four wetlands across gradients of sand desertification in Northeast China. Geoderma 145: 50–59.
3. Chen B, Kitagawa H, Hu K, Jie DM, Yang JP, et al. (2009) Element and mineral characterization of dust emission from the saline land at Songnen Plain, Northeastern China. Journal of Environmental Sciences 21: 1363–1370.
4. Bian JM, Zu LY, Dong ZY (2003) Study of the influence of the resources development to the wetland eco-environment in the Songnen Plain. Areal research and development 22: 68–70.
5. Qiu SW (2007) The pattern and evolution of the sandy land in Western Northeast China. In: Liu JQ editor.The historical evolution of natural environment and the effects of human activities in Northeast China.Beijing, China: Science press. pp 86–153.
6. Bian JM, Tang J, Lin NF (2008) Relationship between saline–alkali soil formation and neotectonic movement in Songnen Plain, China. Environmental Geology 55: 1421–1429.
7. Zhou WJ, Dodson J, Head MJ, Li BS, Hou YJ, et al. (2002) Environmental variability within the Chinese desert-loess transition zone over the last 20000 years. The Holocene 12: 107–112.
8. Maher BA, Mutch TJ, Cunningham D (2009) Magnetic and geochemical characteristics of Gobi Desert surface sediments: Implications for provenance of the Chinese Loess Plateau. Geology 37: 279–282.
9. Guan QY, Pan BT, Li N, Li Q, Zhang JD, et al. (2010) Loess record of the evolution history of severe sandstorms in the Tengger Desert during the Last Interglacial Period (MIS5). Geosciences Journal 14: 155–162.
10. Kim JG, Rejmankova E (2002) Recent history of sediment deposition in marland sand-based marshes of Belize, Central America. Catena 48: 267–291.
11. Wang GP, Liu JS, Tang J (2004) Historical variation of heavy metals with respect to different chemical forms in recent sediments from Xianghai wetlands, Northeast China. Wetlands 24: 608–619.
12. Pigati JS, Latorre C, Rech JA, Betancourt JL, Martínez KE, et al. (2012) Accumulation of impact markers in desert wetlands and implications for the Younger Dryas impact hypothesis. Proceedings of the National Academy of Sciences 109: 7208–7212.
13. Kokfelt U, Reuss N, Struyf E, Sonesson M, Rundgren M, et al. (2010) Wetland development, permafrost history and nutrient cycling inferred from late Holocene peat and lake sediment records in subarctic Sweden. Journal of Paleolimnology 44: 327–342.
14. Lombardo U, Szabo K, Capriles JM, May J, Amelung W, et al. (2013) Early and middle Holocene hunter-gatherer occupations in Western Amazonia: The hidden shell middens. PLOS ONE 8: e72746.
15. Tweel AW, Turner RE (2012) Landscape-scale analysis of wetland sediment deposition from four tropical cyclone events. PLOS ONE 7: e50528.
16. Azmon E, Offer ZY (1994) Mineralogy and granulometry of settled dust in a dust-dune-loess-liquid waste system in the northern Negev Desert, Israel. The Science of the Total Environment 149: 155–166.
17. Keddy PA (2000) Wetland Ecology Principles and Conservation. Cambridge University Press. 189–212p.
18. Soons MB (2006) Wind dispersal in freshwater wetlands: Knowledge for conservation and restoration. Applied Vegetation Science 9: 271–278.
19. White BL, Nepf HM (2008) A vortex-based model of velocity and shear stress in a partially vegetated shallow channel. Water Resources Research 44: W01412.
20. Wen BL, Liu XT, Li XJ, Yang FY, Li XY (2012) Restoration and rational use of degraded saline reed wetlands: A case study in western Songnen Plain, China. Chinese Geographical Science 22: 167–177.
21. Bao SD (2000) Soil Agrochemical Analysis (in Chinese). Third Edition. China Agriculture Press. 30–226p.
22. Ding ZL, Sun JM, Liu DS (1999) The deposition indicators of desert-losses coupling evolution. Science in China (Series D) 29: 82–87.
23. Cheshire MV, Berrow ML, Goodman BA, Mundie CM (1977) Metal distribution and nature of some Cu, Mn and V complexes in humic and fulvic acid fractions of soil organic matter. Geochimica et Cosmochimica Acta 41: 1131–1138.
24. Pan MH, Li YC, Sumner ME (2011) Handbook of Soil Sciences Properties and Processes. Second Edition. CRC Press. 20-3–20-4p.
25. Zhai ZL (2006) Desertification information records derived from the marsh cores in the West of Songnen Plain. M.Sc. Dissertation. Graduate University of Chinese Academy of Sciences. China.
26. Prospero JM, Ginoux P, Torres O, Nicholson S, Gill T (2002) Environmental characterization of global sources of atmospheric soil dust identified with the NIMBUS7 Total Ozone Mapping Spectrometer (TOMS) Absorbing Aerosol Product. Review of Geophysics 40: 2-1–2-31.
27. De Groot AV, Veeneklaas BM, Bakker JP (2011) Sand in the salt marsh: Contribution of high-energy conditions to salt-marsh accretion. Marine Geology 282: 240–254.
28. Li Y, Zhang HP, Liu L, Wang Y, Sun YH, et al. (2005) Influence of sandification on ecological function of Xianghai wetland. Bulletin of Soil and Water Conservation 25: 83–86.
29. Liu DW, Wang ZM, Song KS, Zhang B, Hu LJ, et al. (2009) Land use/cover changes and environmental consequences in Songnen Plain, Northeast China. Chinese Geography Science 19: 299–305.
30. Liu YS, Wang DW, Gao J, Deng W (2005) Land use/cover changes, the environment and water resources in Northeast China. Environment Management, 36: 691–701.
31. Tang HP, Zhang XS (2003) Establishment of Optimized Eco-productive Paradigm in the Farming-Pastoral Zone of Northern China. Acta Botanica Sinica 45: 1166–1173.

23

Increasing Minimum Daily Temperatures Are Associated with Enhanced Pesticide Use in Cultivated Soybean along a Latitudinal Gradient in the Mid-Western United States

Lewis H. Ziska*

Crop Systems and Global Change Laboratory, United States Department of Agriculture, Agricultural Research Service, Beltsville, Maryland, United States of America

Abstract

Assessments of climate change and food security often do not consider changes to crop production as a function of altered pest pressures. Evaluation of potential changes may be difficult, in part, because management practices are routinely utilized *in situ* to minimize pest injury. If so, then such practices, should, in theory, also change with climate, although this has never been quantified. Chemical (pesticide) applications remain the primary means of managing pests in industrialized countries. While a wide range of climate variables can influence chemical use, minimum daily temperature (lowest 24 h recorded temperature in a given year) can be associated with the distribution and thermal survival of many agricultural pests in temperate regions. The current study quantifies average pesticide applications since 1999 for commercial soybean grown over a 2100 km North-South latitudinal transect for seven states that varied in minimum daily temperature (1999–2013) from −28.6°C (Minnesota) to −5.1°C (Louisiana). Although soybean yields (per hectare) did not vary by state, total pesticide applications (kg of active ingredient, ai, per hectare) increased from 4.3 to 6.5 over this temperature range. Significant correlations were observed between minimum daily temperatures and kg of ai for all pesticide classes. This suggested that minimum daily temperature could serve as a proxy for pesticide application. Longer term temperature data (1977–2013) indicated greater relative increases in minimum daily temperatures for northern relative to southern states. Using these longer-term trends to determine short-term projections of pesticide use (to 2023) showed a greater comparative increase in herbicide use for soybean in northern; but a greater increase in insecticide and fungicide use for southern states in a warmer climate. Overall, these data suggest that increases in pesticide application rates may be a means to maintain soybean production in response to rising minimum daily temperatures and potential increases in pest pressures.

Editor: Matthew Germino, US Geological Survey, United States of America

Funding: This author has no support or funding to report.

Competing Interests: The authors have declared that no competing interests exist.

* E-mail: l.ziska@ars.usda.gov

Introduction

Considerable research effort has focused on determining the impact of anthropogenic climate change on global agriculture [1–4]. Of merited interest in this regard are the physical aspects of climate (e.g. carbon dioxide, temperature, precipitation, extreme weather events) that directly alter crop biology (e.g. growth, phenology, sterility, yields) and the resulting consequences for food security [5–9].

However, research efforts related to assessing the agricultural impacts of rising CO_2 and climate change do not always consider trophic interactions. Overall, changes to the biology and competitive abilities of agricultural pests (insects, pathogens, weeds) relative to potential crop yield losses has not been well quantified [10–11]. This is an important omission as the role of pests on constraining crop production is significant and well recognized. For example, weed competition can result in potential crop losses of ~34% globally, with insect pests and pathogens resulting in additional losses of ~18 and 16%, respectively [12].

Such omissions may reflect the complex challenges in relating atmospheric CO_2 and climate variables to potential reductions in crop production related to increased pest pressures. For example, weed growth and fecundity can be directly affected by increasing atmospheric CO_2 as well as rising temperature; insects and pathogens can also be directly affected by temperature, but indirectly by CO_2 and/or climate induced changes to their weed hosts [12–13]. Overall, while a number of pest studies have been conducted, empirical evidence has been eclectic, although it has been suggested that pest pressures will probably increase with climate change (e.g., [14]).

Yet, even if pest pressures regarding crop production were unequivocal and well-characterized in regard to climate change, it would still be difficult to quantify yield reductions *in situ*. This is because there are strong economic incentives at the field level to manage agro-ecosystems to prevent or minimize pest damage. While management methods vary greatly and may include cultural, mechanical, chemical and biological options, among developed countries, such as Australia and the United States, application of chemicals, usually as pesticides, represents the most widely used method for pest control.

But could quantification of pesticide usage in turn, provide an alternative means to gauge changes in pest pressures associated with changes in climate? Among climate variables, it is generally recognized that in temperate regions, the distribution and survival of agricultural pests is often limited by low winter temperatures; i.e., minimum thermal thresholds) [15]. Rising minimum temperatures associated with anthropogenic climate change could extend the potential geographic range of pest species and/or alter their demographics, although long-term changes in species diversity are unclear [16–17]. Climate change assessments have also emphasized that the current and projected increases in global warming are not uniform, and enhanced land-surface temperatures (relative to the global average) are more probable for minimum (Winter) than maximum (Summer) temperatures [18–19].

To ascertain if minimum temperatures reflect pesticide usage, regression analysis of the interrelationship between insecticide, fungicide and herbicide application rates and the minimum daily (24 h) observed temperature was conducted on a commercial crop, soybean, grown over a latitudinal transect of seven mid-western states within a humid region. This analysis was performed using a multi-year (1999–2013), multi-state (Minnesota, Wisconsin, Iowa, Missouri, Arkansas, Mississippi and Louisiana) data series obtained from USDA-NASS pesticide usage surveys in conjunction with state based minimum temperature. If significant correlations were observed between minimum temperature and usage for a given class of pesticide, longer trends (1977–2013) in minimum temperature were analyzed and then used to project near-term (decadal, to 2023) changes in pesticide rates by state.

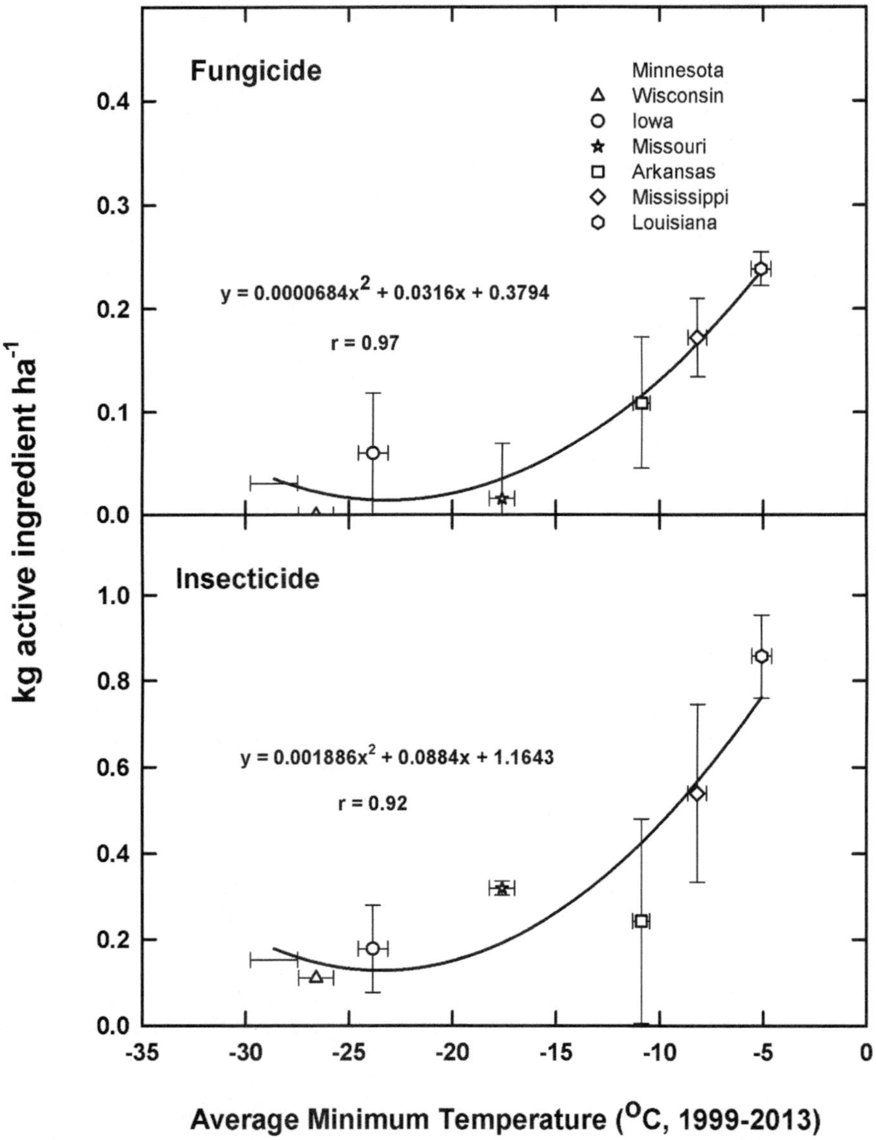

Figure 1. Minimum temperature by state averaged for the period 1999–2013 as a function of insecticide and fungicide usage (kg active ingredient, ai, per hectare) as determined from the National Agricultural Statistical Service survey for the years 1999, 2000, 2001, 2002, 2004, 2006, and 2012) for a north-south transect of seven Midwestern states where commercial soybean is grown. Minimum temperatures were determined as the lowest recorded temperature for a 24 h period during a given calendar year for four locations for a soybean growing area within a state. Lines represent the "best fit" second order polynomial. Bars are ±SE. See Methods for additional details.

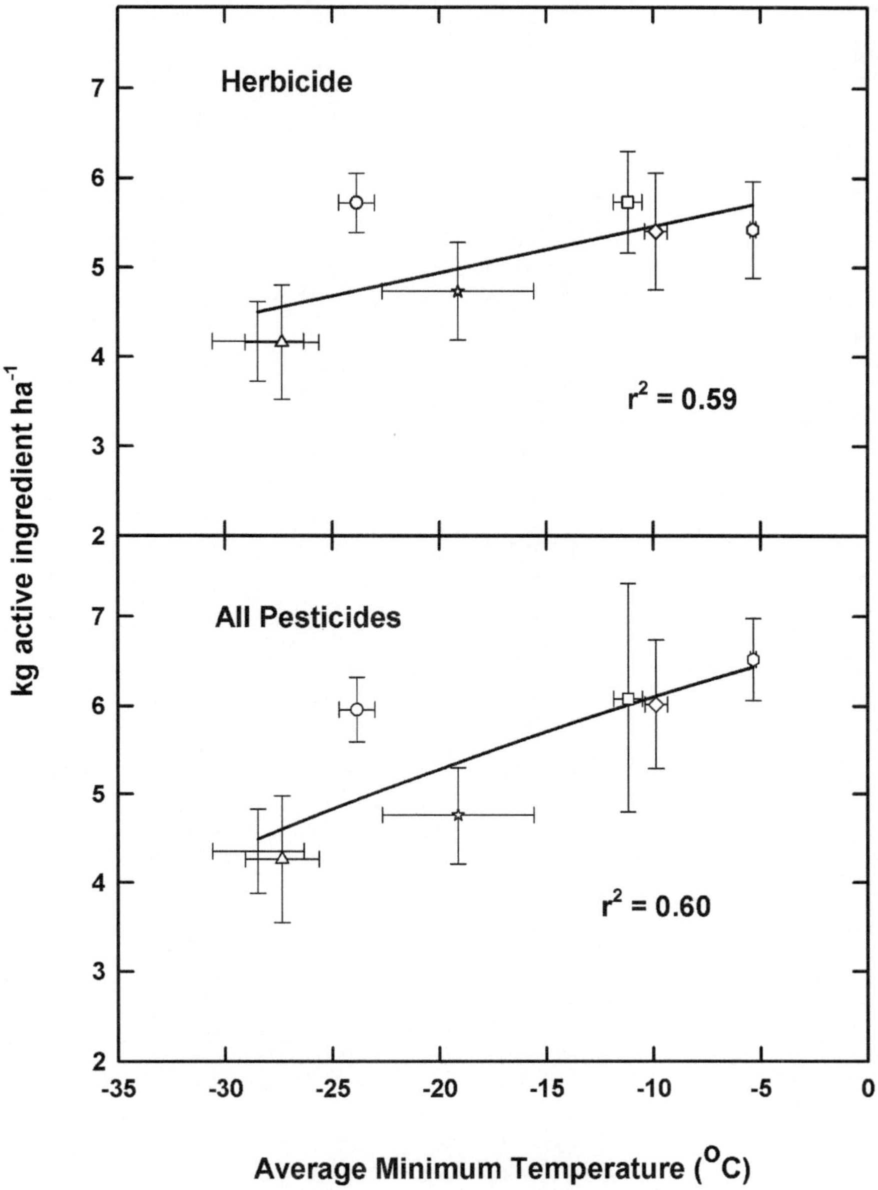

Figure 2. Symbols and regression determined as for figure 1, but for herbicide use and total pesticide (insecticide, fungicide and herbicide) usage for soybean.

Results

For two classes of pesticides, fungicides and insecticides, a second order quadratic function provided the 'best-fit' for minimum temperature and amounts of active ingredient applied for the period 1999–2012. There was a significant correlation (r values of 0.97 and 0.92, **,P<0.01) for fungicide and insecticide soybean applications, respectively over the north-south transect (**Figure 1**). A linear function between minimum daily temperature and herbicide application rates was also observed to be positively correlated and significant (r value of 0.84: *,P<0.05) for herbicide use (**Figure 2**). As most pesticide applications are herbicides, a similar relationship (r = 0.84) was observed for total pesticide use (**Figure 2**).

As the correlation between minimum daily temperature and pesticide usage was significant in all cases, states within the transect were examined to quantify longer term temperature changes (1977 through 2013) in order to assess the increase in minimum winter temperature per decade. Overall, these data indicated that the rise in minimum temperatures was a function of latitude, with states such as Minnesota showing a more rapid increase in minimum temperatures than southern states (e.g., Louisiana) (**Table 1, Figure 3**). However, average soybean yields (2009–2013, MT Ha^{-1}) did not significantly vary as a function of the north-south transect (**Table 1**).

The average decadenal increase in minimum daily temperature was used to estimate potential near term (to 2023) changes in pesticides by category. Because a first order regression was the best-fit for herbicide use, increases in minimum temperature result in a relatively greater increase in herbicide applied for northern compared to southern states. In contrast, because the relationship between insecticide/fungicides and minimum temperatures is best described by a second degree polynomial, temperature changes above a critical threshold, (ca −20°C, **Figure 2**) resulted in

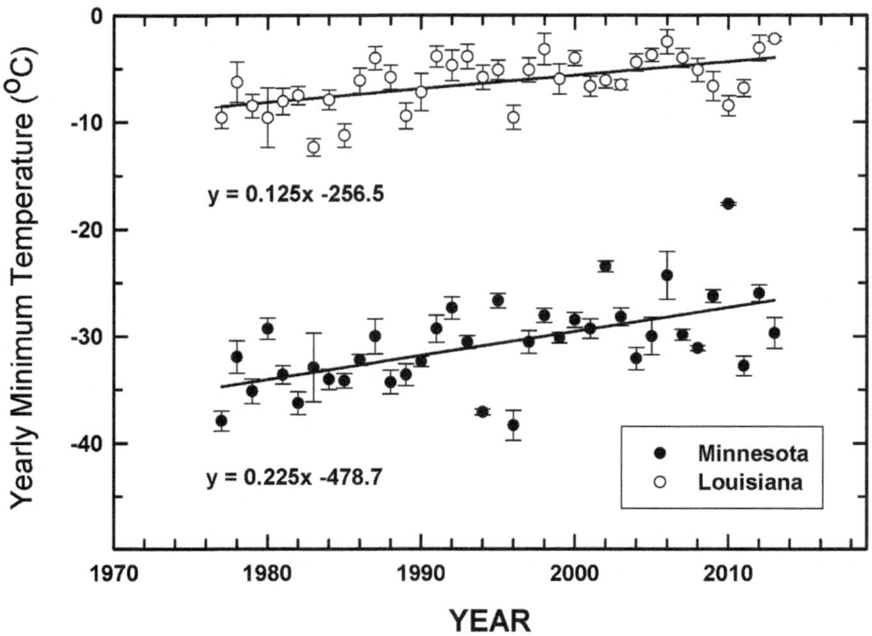

Figure 3. Change in average (±SE) minimum temperature (lowest recorded temperature for a 24 h period during a given calendar year for four locations) for the northernmost and southernmost states (Minnesota, closed circles) and Louisiana (open circles) used in the Midwestern transect. Slope of the regression line indicates the average increase in minimum temperature (°C) per year for each location.

proportionally larger changes in insecticide/fungicide usage. Hence, for those states with minimum temperatures above this threshold, further increases in temperature result in disproportionally larger projected changes in insecticide and fungicide usage (**Table 1**).

Discussion

The need to assess the relative impact of pests on agricultural production in a changing climate has long been recognized [20]; however, few quantitative *in situ* assessments are available. An earlier economic analysis [21] indicated increased pesticide costs as a function of increased rainfall and hotter weather during the 1990s. Application of this analysis to projected climate scenarios to 2090 also indicated general increases in average pesticide cost for a range of crop species [21]. However, this study, while useful, did not distinguish between classes of pesticide, or quantify changes in application amounts as a function of a specific climate variable.

Although agriculture is widespread, there are, in fact, only a handful of crops that are grown over a wide range of climates. For the U.S., soybean, corn and wheat fall into this category; however wheat is grown temporally over a latitudinal gradient as spring (e.g. North Dakota) or winter wheat (e.g. Kansas) with significant differences in pest management. In addition, wheat is not grown widely in the southern states. In contrast, corn is often grown in conjunction with soybean, but data on pesticide use for corn in the southern states (e.g. Mississippi, Arkansas, and Louisiana) is currently not available through NASS. Alternatively, the distribution of soybean in this study encompassed a broad range of minimum daily temperatures (i.e., a ~23°C difference from Minnesota to Louisiana). Such a range would include projections of future surface temperatures associated with anthropogenic climate change, e.g. if minimum daily temperatures in Minnesota became like those of Louisiana [22]; hence soybean was selected for analysis.

There is, as shown here, a significant positive correlation between the range of minimum daily temperatures and pesticide usage for soybean, particularly in regard to fungicides and insecticides. The second order functions for soybean are consistent with insect and fungal biology; i.e., once temperature has reached a critical thermal threshold, it is a significant driver of shifts in insect and pathogen demography [13,23]. For example, in the UK increasing winter temperatures above a biological minimum have been associated with increased northward migration of aphids in Scotland and increased genotype variation among aphid populations [24]. Similar correlations have also been observed between minimum winter temperatures and the southern pine beetle [25].

Patterson et al. [14] provide an extensive list of temperature thresholds in this regard including associated phenological responses and potential shifts in the expansion of insect ranges. These data indicate that for temperate regions, warming would result in increased winter survival as well as an increase in growth and insect fecundity. Similarly, it has been recognized that pathogen development is affected by warm conditions, with mild winters and warmer weather associated with increased outbreaks of powdery mildew; leaf spot disease, leaf rust and rizomania disease [14]. Treharne [26] in turn, has also suggested that increases in minimum daily temperatures would shift the occurrence of plant disease into cooler regions. The relationship between insecticide and fungicide use observed here for soybean in response to rising minimal temperatures is, overall, consistent with the biology of thermal limits and expected shifts in insect and fungal populations in a warmer climate (e.g., [14,23,27]).

For the current study, herbicide applications were a linear function of rising minimum daily temperatures in soybean. In contrast to insect and fungal populations, thermal limits for the occurrence of weeds in soybean were not evident. However, if climate is suitable for crop production, it will also be, de facto, suitable for weed growth.

Although herbicide tolerance in agriculture is increasing [28], for the current study the observed increase in herbicide

Table 1. States included in the latitudinal transect.

State	Soybean (Ha *1000)	Soybean (MT Ha^{-1})	T_{min}/decade	r value	Insecticide (2023)	Fungicide (2023)	Herbicide (2023)
Minnesota	2,513	2.81	+2.25	0.58,**	0	0	+160.2
Wisconsin	628	2.91	+1.99	0.57,***	0	0	+140.0
Iowa	3,741	3.20	+1.58	0.51,**	+20.3	0	+85.1
Missouri	2,247	2.52	+1.56	0.51,**	+35.6	+12.0	+73.5
Arkansas	1,308	2.62	+1.31	0.54,***	+110.0	+36.5	+70.0
Mississippi	806	2.77	+1.31	0.52,***	+117.7	+39.3	+70.3
Louisiana	449	2.81	+1.25	0.55,***	+170.0	+41.1	+58.6

Soybean area is given as hectares (*1000), average yield was determined from 2009–2013 as metric tons (MT) per hectare, the T_{min} is the increase in minimal daily (24 h) temperature (°C) per decade (determined from 1977–2013). The r value is the correlation coefficient of T_{min} over time, with $P<0.05$, *; $P<0.01$, ** and $P<.001$, ***. The values for insecticide, fungicide and herbicide are determined for 2023 (10 year projection) using the functions for figures 1–3 and are shown as changes in grams of active ingredient (ai) per hectare, with a value of zero indicating no change in application rates over 2013 levels.

applications with minimum temperatures did not appear to be related to a greater number of herbicide resistant weeds along the north-south transect (**Table 2**). Nor was there any apparent relationship with photosynthetic pathway (i.e. the number of C_4 weeds is greater in Minnesota than Louisiana). This suggests no difference in recent CO_2 increases on latitudinal weed selection between C_3 and C_4 weedy species *per se*. However the number of perennial or facultative perennial weed species was significantly greater for the southern soybean locations (e.g. Louisiana, **Table 2**). In weed management, it is generally recognized that perennial weeds can be more competitive and more difficult to control. More competitive because of their ability to quickly regenerate from below-ground structures in the spring and more difficult to control chemically because the elimination of perennial weeds requires the killing of all plant parts (including belowground stems, rhizomes, tubers, etc.) that are capable of producing new shoots [29]. Potential increases in perennial weed establishment with increasing minimum temperatures could, in turn, result in concomitant increases in herbicide usage.

The quantitative differences in decadenal increases in minimum temperature reported here (e.g., 2.25 vs. 1.25°C per decade for Minnesota and Louisiana, respectively since 1977) are consistent with the Intergovernmental Panel on Climate Change (IPCC) projections regarding enhanced warming as a function of latitude [19]. Because of the variation in pesticide category by minimal temperature for soybean, near-term projections (2023) suggest a greater relative increase in herbicide use for the upper Midwest, whereas greater insecticide and fungicide use is projected for the southern states for this same period. This may be due, in part, to the southern states having already surpassed the minimum winter temperatures needed to support insect and pathogen populations for soybean in the southern region. However, if minimum temperatures continue to rise, then similar increases in use for all pesticide categories could occur for the upper Midwest.

At present, soybean in the mid-western United States is grown over a ~2100 km north-south transect that provides a wide range of yearly temperatures. Although such a range of temperatures should, ostensibly, increase pest pressures and subsequent crop loss, no production losses are evident in average yield per hectare (e.g., average yields in Minnesota and Wisconsin do not differ from those of Mississippi and Louisiana). However, insecticide, fungicide and herbicide use increase significantly in conjunction with minimum daily temperatures along the latitudinal transect. Assuming that prophylactic use is not widespread in any one region, this increase should reflect increased pest pressures *per se*.

It should be emphasized however, that the biological basis for rising minimum temperatures and changes in pesticide usage are likely to be complex. While there are empirical correlations between population demographics and temperature, temperature by itself does not reflect the complexity of pest population causations in agriculture. For example, pest-pest interactions, insects as disease vectors, temporal asynchrony between insects and host plants, soil management and herbicide efficacy (e.g. precipitation, windspeed), *inter alia* [10,23,30] will also need to be considered. Rising levels of carbon dioxide are also likely to directly alter secondary plant chemistry and alter plant-arthropod interactions depending on plant species and insect group [31]. In addition, the current analysis assumes future monotonic poleward shifts in minimum temperatures influencing pesticide usage; whereas an increase in extreme climatic events may be the norm [32].

However, the ability to include realistic impacts of agricultural pests in future assessments of climate change and food security is imperative. While there are numerous challenges remaining before

Table 2. Top ten lists of most troublesome weed species in soybean for three states along the North-South transect.

State	Common Name	Scientific Name	Photosynthetic Pathway	Growth Habit
Louisiana	Morning glory spp.	*Ipomea spp.*	C_3	Annual or Perennial
	Pigweed spp.	*Amaranthus spp.*	C_4	Annual or Perennial*
	Browntop millet	*Urochloa ramose*	C_4	Perennial
	Nutsedge spp.	*Cyperus spp.*	C_4	Perennial
	Redvine	*Brunnichia ovate*	C_3	Perennial
	Henbit	*Lamium amplexicaule*	C_3	Annual*
	Dayflower	*Commelina spp.*	C_3	Annual or Perennial
	Johnson grass	*Sorghum halapense*	C_4	Perennial
	Prickly sida	*Sida spinosa*	C_3	Annual
	Red rice	*Oryza sativa*	C_3	Annual
Missouri	Waterhemp	*Amaranthus rudis*	C_4	Annual*
	Morning glory	*Ipomea spp.*	C_3	Annual or Perennial
	Palmer amaranth	*Amaranthus palmeri*	C_4	Annual*
	Giant ragweed	*Ambrosia trifida*	C_3	Annual*
	Johnson grass	*Sorghum halapense*	C_4	Perennial
	Asiatic dayflower	*Commelina communis*	C_3	Annual
	Horseweed	*Conyza canadensis*	C_3	Biennial*
	Prickly sida	*Sida spinosa*	C_3	Annual
	Common ragweed	*Ambrosia artemisiifolia*	C_3	Annual*
	Eastern black nightshade	*Solanum ptycanthum*	C_3	Annual or Perennial
Minnesota	Lambsquarters	*Chenopodium album*	C_3	Annual
	Giant foxtail	*Setaria faberii*	C_4	Annual
	Waterhemp	*Amaranthus rudis*	C_4	Annual*
	Wooly cupgrass	*Eriochloa villosa*	C_4	Annual
	Giant ragweed	*Ambrosia trifida*	C_3	Annual*
	Yellow foxtail	*Setaria viridis*	C_4	Annual
	Green foxtail	*Setaria lutescens*	C_4	Annual
	Quackgrass	*Elytrigia repens*	C_3	Perennial
	Common ragweed	*Ambrosia artemisiifolia*	C_3	Annual*
	Wild proso millet	*Panicum miliaceum*	C_4	Annual

Lists are generated by farmer surveys as reported by the Southern Weed Science Society (SWSS) for Louisiana (LA) and Missouri (MO) [37] and the University of Minnesota extension service for Minnesota (MN) (Frank Forcella, USDA-ARS Personal Communication). C_3/C_4 refers to photosynthetic pathway.
*Indicates herbicide resistant populations within that state.

the impacts of anthropogenic climate change on pest pressures and crop yields can be completely quantified, the current study suggests that farm management, through the increased use of chemical applications, may negate pest pressures on crop production that are incurred as a result of changing climate variables such as temperature. As such, pest-induced reductions in agricultural yield of soybean associated with climate change may be difficult to quantify where pesticides are widely used. At present, while the current analysis suggests that minimum daily temperatures could be used as a proxy to understand climate change impacts on pesticide use in soybean; critical information regarding the environmental and economic consequences of such impacts will require further study.

Materials and Methods

Soybean

For all seven states examined along the north-south transect, soybean is considered a major crop (>300,000 ha planted) [34]. Genetically modified (gmo) soybean (e.g., Round-Up Ready^{tm}) was introduced in 1995–1996, mainly as a means for blanket control of post-emergent weeds. Following its introduction it was quickly adopted by growers, with gmo soybean acreage >50% by 1999 [33] (**Table 3**). There is no evidence of different rates of adoption of gmo soybean over the period 1999–2012. For example, by 2001, the amount of soybean area plant to gmo was 63% for both Minnesota and Mississippi (**Table 3**). Since 2005, approximately 90% of all soybeans grown in the U.S. have been herbicide resistant gmo (**Table 3**).

Table 3. Genetically engineered soybean varieties for herbicide tolerance for states included in the north-south transect.

STATE	2000	2001	2002	2003	2004	2005	2006	2007	2008	2009	2010	2011	2012
Minnesota	46	63	71	79	82	83	88	92	91	92	93	95	91
Wisconsin	51	63	78	84	82	84	85	88	90	85	88	91	92
Iowa	59	73	75	84	89	91	91	94	95	94	96	97	93
Missouri	62	69	72	83	87	89	93	91	92	89	94	91	91
Arkansas	43	60	68	84	92	92	92	92	94	94	96	95	94
Mississippi	48	63	80	89	93	96	96	96	97	94	98	98	95
U.S.	54	68	75	81	85	87	89	91	92	91	93	94	93

Values are the percent of all soybeans planted. Data are available at:
www.usda.mannlib.cornell.edu/usda/nass/Acre/2010s/2011/Acre-06-30-2011.pdf#page=27.

Pesticide Use Rates

State level pesticide usage rate are available for 1999, 2000, 2001, 2002, 2004, 2006 and 2012 from the USDA National Agricultural Statistical Service Survey (NASS) data online (www.quickstats.nass.usda.gov/#5072A3CD-64E7-384A-8FF0-DDC931164C3B) [34] These data include chemical usage by compound for insecticides, fungicides and herbicides for soybean for the states included in the north-south transect. Although data are also available from 1991 through 1998 from this same source, the rapid rates of gmo adoption for herbicide resistant soybean, and the shift in herbicide usage during this time period was thought to, potentially, obfuscate any potential changes related to climate variables such as temperature. This is because application rates often vary by herbicide mode of action; hence, differential herbicide use would result in different amounts of active ingredient (ai) being applied. However, the large-scale shift to gmo soybean provided greater uniformity in herbicide type over a large regional area (**Table 3**). This is an important consideration as herbicides represent the largest component of total pesticide use (e.g. **Figure 2**). However, the temporal range of yearly averages (1999–2013) was considered insufficiently long to evaluate any potential increase in minimum daily temperatures associated with anthropogenic surface warming and pesticide usage (relative to year to year variability) for any one state.

Temperature Data

A software program developed by Texas A&M University [35] was used to identify at least four weather stations located within soybean growing regions of a given state used in the transect. Information on the lowest recorded (24 h) minimum temperature for a given year was determined and averaged for all weather stations within a state for the 1999–2013 period. These same stations and data sets were used to generate minimum daily temperature over a longer time period, from 1977 (i.e. the start of the rapid increase in global land-ocean temperature index) through 2013 [36].

Statistical Analysis

In examining pesticide use it is important to determine whether there may be fixed or random regional effects within the data. For example, are there state specific factors beyond temperature that would result in systematic shifts in pesticide usage? (e.g. one state using irrigation while another state does not, change in rotation rates, etc.) This analysis has been previously done by Chen and McCarl [21] who found with 99% confidence that a random state effect existed for pesticide use in corn, potatoes, soybeans and wheat [21].

A step-wise regression program (ver. 10.0 Statview, Cary NC, USA) was used to determine the best fit regression line for average minimum daily temperature and pesticide application rate (insecticides, fungicides and herbicides) for all states within the latitudinal transect. If the correlation coefficient was determined to be significant for a given pesticide category (e.g. insecticides); then a longer term minimum daily temperature data set (1977–2013) was generated for each state to quantify increases in minimum temperature by decade. The increase in minimum daily temperature by decade was then used to project potential short-term (2014–2023) pesticide use (by category and state) using the quadratic or linear functions provided in figures 1 and 2.

Author Contributions

Conceived and designed the experiments: LHZ. Performed the experiments: LHZ. Analyzed the data: LHZ. Contributed reagents/materials/analysis tools: LHZ. Wrote the paper: LHZ.

References

1. Adams RM, Rosenzweig C, Peart RM, Ritchie JT, McCarl BA, et al. (1990) Global climate change and U.S. agriculture. Nature 345: 219–224.
2. Reilly J, Tubiello F, McCarl B, Abler D, Darwin R, et al. (2003) U.S. Agriculture and climate change. Clim Change 57: 43–69.
3. Howden SM, Soussana JF, Tubiello FN, Cherti N, Dunlop M, et al. (2007) Adapting agriculture to climate change. Proc Nat Acad Sci (USA) 104: 19691–19696.
4. Lobell DB, Burke MB, Tebaldi C, Mastrandrea MD, Falcon WP, et al. (2008) Prioritizing climate change adaptation needs for food security in 2030. Science 319: 607–610.
5. Tao F, Yokozawa M, Xu Y, Hayashi Y, Zhang Z (2006) Climate changes and trends in phenology and yields of field crops in China, 1981–2000. Ag Forest Met 138: 82–92.
6. Schmidhuber J, Tubiello FN (2007) Global food security under climate change. Proc Nat Acad Sci 104: 19703–19708.
7. Schlenker W, Roberts MJ (2009) Nonlinear temperature effects indicate severe damages to US crop yields under climate change. Proc Nat Acad Sci 106: 15594–15598.
8. Ziska LH, Bunce JA, Shimono H, Gealy DR, Baker JT, et al. (2012) Food security and climate change: On the potential to adapt global crop production by active selection to rising atmospheric carbon dioxide. Proc Royal Soc (B) 279: 4097–4105.
9. Lobell DB, Gourdji SM (2012) The influence of climate change on global crop productivity. Plant Physiol 160: 1686–1697.
10. Scherm H (2004) Climate change: can we predict the impacts on plant pathology and pest management? Can J Plant Pathol 26: 267–273.
11. Gregory PJ, Johnson SN, Newton AC, Ingram JSI (2009) Integrating pests and pathogens into the climate change/food security debate. J of Exp Bot 60: 2827–2838.
12. Oerke E-C (2006) Crop losses to pests. J Ag Sci 144: 31–43.
13. Ziska LH, Runion GB (2007) Future Weed, Pest and Disease Problems for Plants. In: Newton PCD, Carran A, Edwards GR, Niklaus PA, Editors. Agroecosystems in a Changing Climate. CRC Press, Boston, MA, 262–279.
14. Patterson DT, Westbrook JK, Joyce RJV, Lingren PD, Rogasik J (1999) Weeds, insects, and diseases. Climatic Change 43: 711–727.
15. Cammel ME, Knight JD (1992) Effects of climatic change on the population dynamics of crop pests. Adv Ecol Res 22: 117–162.
16. Bradley BA, Blumenthal DM, Wilcove DS, Ziska LH (2010) Predicting plant invasions in an era of global change. TREE 25: 310–318.
17. McDonald AJ, Riha S, Ditomasso A, DeGaetano A (2009) Climate change and the geography of Weed damage: Analysis of US maize systems suggests the potential for significant range transformations. Ag Ecosys Environ 130: 131–140.
18. Vose RS, Easterling DR, Gleason B (2005) Maximum and minimum temperature trends for the globe: An update through 2004. Geophys Res Let 32: 1–5.
19. Intergovernmental Panel on Climate Change (2007) Summary for policymakers. In: Climate Change 2007: The Physical Science Basis. Contribution of Working Group I to the Fourth Assessment Report of the Intergovernmental Panel on Climate Change (Solomon, S. et al., eds), Cambridge University Press.
20. Lough JM, Wigley TML, Palutikof JP (1983) Climate and climate impact scenarios for Europe in a warmer world. J. Clim. Appl. Meteorol. 22: 1673–1684.
21. Chen C-C, McCarl BA (2001) An investigation of the relationship between pesticide usage and Climate change. Climatic Change 50: 475–487.
22. Intergovernmental Panel on Climate Change (2012) Managing the risks of extreme events and disasters to advance climate change adaptation. A special report of Working Groups I and II of the Intergovernmental Panel on Climate Change. Cambridge University Press, Cambridge.
23. Fuhrer J (2003) Agroecosystem responses to combinations of elevated CO_2, ozone, and global climate change. Ag Ecosys Environ 97: 1–20.
24. Malloch G, Highet F, Kasprowicz L, Pickup J, Beilson R, et al. (2006) Microsatellite marker analysis of peach-potato aphids (*Myzus persicae*, Homoptera: Aphididae) from Scottish suction traps. Bull Entomol Res 96: 573–582.
25. Tran JK, Yliqja T, Billings RF, Regniere J, Ayres MP (2007) Impact of minimum winter temperatures on the population dynamics of *Dendroctonus frontalis*. Ecol Appl 17: 882–889.
26. Treharne K (1989) The implications of the 'greenhouse effect' for fertilizers and agrochemicals. In: Bennet RC editor, Ministry of Agriculture, Fisheries and Food, London, UK 67–78.
27. Porter JH, Parry ML, Carter TR (1991) The potential effects of climatic change on agricultural insect pests. Ag For Meteor 57: 221–240.
28. Delye CD, Jasieniuk M, Le Core V (2013) Deciphering the evolution of herbicide resistance in weeds. Trends in Gen 29: 649–655.
29. Zimdahl RL (2007) Weed-crop competition: a review. John Wiley & Sons, New York, NY, 220 pages.
30. Ziska LH (2010) Global Climate Change and Carbon Dioxide: Assessing Weed Biology and Management. In: Rosenzweig C, Hillel D, Ediotrs. Handbook of Climate Change and Agro-ecosystems: Impacts, adaptation and mitigation. World Scientific Publishing, Hackensack, NJ, 191–208.
31. Robinson EA, Ryan GD, Newman JA (2012) A meta-analytical review of the effects of elevated CO_2 on plant-arthropod interactions highlights the importance of interacting environmental and biological variables. New Phytologist 194: 321–336.
32. Hansen J, Sato M, Ruedy R (2012). Perception of climate change. Proc Nat Acad Sci 109: E2415–E2423.
33. Carpenter J, Gianessi L (1999) Herbicide tolerant soybeans: Why growers are adopting Round-up Ready varieties. AgBioForum 2: 65–72.
34. National Agricultural Statistical Service, (NASS) (www.quickstats.nass.usda.gov/#5072A3CD-64E7-384A-8FF0-DDC931164C3B) (Accessed, January, 2014).
35. Yang Y, Wilson LT, Wang J (2010) Development of an Automated Climatic Data Scraping, Filtering and Display System. Comp Elect Ag 71: 77–87.
36. Hansen J, Ruedy R, Glascoe J, Sato M (1999). GISS analysis of surface temperature change. J Geophysical Res: Atmos (1984–2012), 104: 30997–31022.
37. Southern Weed Science Society (SWSS) (2013) SWSS Proceedings. 66: 275–287.

PERMISSIONS

All chapters in this book were first published in PLOS ONE, by The Public Library of Science; hereby published with permission under the Creative Commons Attribution License or equivalent. Every chapter published in this book has been scrutinized by our experts. Their significance has been extensively debated. The topics covered herein carry significant findings which will fuel the growth of the discipline. They may even be implemented as practical applications or may be referred to as a beginning point for another development.

The contributors of this book come from diverse backgrounds, making this book a truly international effort. This book will bring forth new frontiers with its revolutionizing research information and detailed analysis of the nascent developments around the world.

We would like to thank all the contributing authors for lending their expertise to make the book truly unique. They have played a crucial role in the development of this book. Without their invaluable contributions this book wouldn't have been possible. They have made vital efforts to compile up to date information on the varied aspects of this subject to make this book a valuable addition to the collection of many professionals and students.

This book was conceptualized with the vision of imparting up-to-date information and advanced data in this field. To ensure the same, a matchless editorial board was set up. Every individual on the board went through rigorous rounds of assessment to prove their worth. After which they invested a large part of their time researching and compiling the most relevant data for our readers.

The editorial board has been involved in producing this book since its inception. They have spent rigorous hours researching and exploring the diverse topics which have resulted in the successful publishing of this book. They have passed on their knowledge of decades through this book. To expedite this challenging task, the publisher supported the team at every step. A small team of assistant editors was also appointed to further simplify the editing procedure and attain best results for the readers.

Apart from the editorial board, the designing team has also invested a significant amount of their time in understanding the subject and creating the most relevant covers. They scrutinized every image to scout for the most suitable representation of the subject and create an appropriate cover for the book.

The publishing team has been an ardent support to the editorial, designing and production team. Their endless efforts to recruit the best for this project, has resulted in the accomplishment of this book. They are a veteran in the field of academics and their pool of knowledge is as vast as their experience in printing. Their expertise and guidance has proved useful at every step. Their uncompromising quality standards have made this book an exceptional effort. Their encouragement from time to time has been an inspiration for everyone.

The publisher and the editorial board hope that this book will prove to be a valuable piece of knowledge for researchers, students, practitioners and scholars across the globe.

LIST OF CONTRIBUTORS

Nicky McCreesh and Mark Booth
School of Medicine, Pharmacy and Health, Durham University, Durham, United Kingdom

Susan K. Skagen
United States Geological Survey, Fort Collins Science Center, Fort Collins, Colorado, United States of America

Valerie Steen
United States Geological Survey, Fort Collins Science Center, Fort Collins, Colorado, United States of America
Department of Fish, Wildlife, and Conservation Biology, Colorado State University, Fort Collins, Colorado, United States of America, Graduate Degree Program in Ecology, Colorado State University, Fort Collins, Colorado, United States of America

Barry R. Noon
Department of Fish, Wildlife, and Conservation Biology, Colorado State University, Fort Collins, Colorado, United States of America, Graduate Degree Program in Ecology, Colorado State University, Fort Collins, Colorado, United States of America

Hao Wang, Fan Lu, Yang Fang, Mei Yang1, Qianhong Wu, Jiakuan Chen and Wenju Zhang
Institute of Biodiversity Science, School of Life Sciences, Fudan University, Shanghai, China

Lu Xu
Institute of Biodiversity Science, School of Life Sciences, Fudan University, Shanghai, China
College of Life Sciences, Northwest Normal University, Lanzhou, China

Qiong La and Yang Zhong
Institute of Biodiversity Science, School of Life Sciences, Fudan University, Shanghai, China
Department of Biology, Tibet University, Lhasa, China

Kun Sun
College of Life Sciences, Northwest Normal University, Lanzhou, China

H. John B. Birks
Department of Biology, University of Bergen, Bergen, Norway
Environmental Change Research Centre, University College London, London, United Kingdom
School of Geography and the Environment, University of Oxford, Oxford, United Kingdom

Junjun Zhi, Shengpan Lin, Cao Zhang and Qiankun Liu
College of Environmental and Resource Sciences, Zhejiang University, Hangzhou, China

Changwei Jing and Jiaping Wu
Ocean College, Zhejiang University, Hangzhou, China

Stephen D. DeGloria
Department of Crop and Soil Sciences, Cornell University, Ithaca, New York, United States of America

Jose Luis Gonzalez-Andujar
Departamento Protección de Cultivos, Instituto de Agricultura Sostenible Consejo Superior de Investigaciones Científicos (CSIC), Córdoba, Spain
Laboratorio Internacional de Cambio Global, LINCG (CSIC-PUC), Santiago, Chile

Rosana Ferrero
Departamento Protección de Cultivos, Instituto de Agricultura Sostenible, Consejo Superior de Investigaciones Científicos (CSIC), Córdoba, Spain
Center of Applied Ecology and Sustainability (CAPES), Santiago, Chile

Mauricio Lima
Departamento de Ecología, Pontificia Universidad Católica de Chile, Santiago, Chile
Laboratorio Internacional de Cambio Global, LINCG (CSIC-PUC), Santiago, Chile
Center of Applied Ecology and Sustainability (CAPES), Santiago, Chile

Frank A. La Sorte
Cornell Laboratory of Ornithology, Cornell University, Ithaca, New York, United States of America

Stuart H. M. Butchart
BirdLife International, Cambridge, United Kingdom

Walter Jetz
Department of Ecology and Evolutionary Biology, Yale University, New Haven, Connecticut, United States of America

Katrin Böhning-Gaese
Biodiversity and Climate Research Centre (BiK-F), Senckenberg Gesellschaft für Naturforschung, Frankfurt (Main), Germany

Department of Biological Sciences, Goethe Universität, Frankfurt (Main), Germany

Erin A. Urquhart, Benjamin F. Zaitchik and Darryn W. Waugh
Department of Earth and Planetary Sciences, Johns Hopkins University, Baltimore, Maryland, United States of America

Seth D. Guikema
Department of Geography and Environmental Engineering, Johns Hopkins University, Baltimore, Maryland, United States of America

Carlos E. Del Castillo
Ocean Ecology Laboratory, NASA Goddard Space Flight Center, Greenbelt, Maryland, United States of America

Heather A. Hager, Sarah E. Sinasac and Jonathan A. Newman
School of Environmental Sciences, University of Guelph, Guelph, Ontario, Canada

Zéev Gedalof
Department of Geography, University of Guelph, Guelph, Ontario, Canada

Ingeborg G. Hallanger and Ireen Vieweg
Department of Arctic and Marine Biology, UiT The Arctic University of Norway, Tromsø, Norway

Jasmine Nahrgang and Jørgen Berge
Department of Arctic and Marine Biology, UiT The Arctic University of Norway, Tromsø, Norway
University Centre in Svalbard, Longyearbyen, Norway

Ekaterina Korshunova
Department of Arctic and Marine Biology, UiT The Arctic University of Norway, Tromsø, Norway
Akvaplan-niva, Fram Centre, Tromsø, Norway

Øystein Varpe
University Centre in Svalbard, Longyearbyen, Norway
Akvaplan-niva, Fram Centre, Tromsø, Norway

Svetlana Murzina
Institute of Biology of the Karelian Research Centre, Russian Academy of Science, Petrozavodsk, Russia

Prashant Goswami and Swathi Trithala Krishnan
CSIR Fourth Paradigm Institute, Bengaluru, India

Upadhayula Suryanarayana Murty and Srinivasa Rao Mutheneni
CSIR Indian Institute of Chemical Technology, Hyderabad, India

Michael M. Fuller
Department of Biology, University of New Mexico, Albuquerque, New Mexico United States of America

Mason J. Ryan, Norman J. Scott, Joseph A. Cook and Steven Poe
Department of Biology, University of New Mexico, Albuquerque, New Mexico United States of America
Museum of Southwestern Biology, University of New Mexico, Albuquerque, New Mexico United States of America

Beatriz Willink, Gerardo Chaves and Federico Bolaños
Escuela de Biología, Universidad de Costa Rica, San Pedro, San José, Costa Rica

Jan Ohlberger, Lauren A. Rogers and Nils Chr. Stenseth
Centre for Ecological and Evolutionary Synthesis (CEES), Department of Biosciences, University of Oslo, Oslo, Norway

Carlo Giupponi
Department of Economics, CáFoscari University of Venice, Venice, Italy

Claudio Biscaro
Department of Economics, CáFoscari University of Venice, Venice, Italy
Department of Management, CáFoscari University of Venice, Venice, Italy
Institut für Organization und Globale Managementstudien, Johannes Kepler Universität, Linz, Austria

Matjaž Kuntner
Institute of Biology, Scientific Research Centre, Slovenian Academy of Sciences and Arts, Ljubljana, Slovenia
Centre for Behavioural Ecology and Evolution, College of Life Sciences, Hubei University, Wuhan, Hubei, China
National Museum of Natural History, Smithsonian Institution, Washington, D. C., United States of America

Daiqin Li
Centre for Behavioural Ecology and Evolution, College of Life Sciences, Hubei University, Wuhan, Hubei, China
Department of Biological Sciences, National University of Singapore, Singapore, Singapore

Jonathan A. Coddington
National Museum of Natural History, Smithsonian Institution, Washington, D. C., United States of America

Magdalena Năpăruş
Centre of Landscape–Territory–Information Systems - CeLTIS, University of Bucharest, Bucharest, Romania, 5 Tular Cave Laboratory, Kranj, Slovenia

Lisa A. Eby and Olga Helmy
University of Montana, Wildlife Biology Program and Department of Ecosystem and Conservation Sciences, College of Forestry and Conservation, Missoula, Montana, United States of America

Lisa M. Holsinger
U.S. Department of Agriculture Forest Service, Rocky Mountain Research Station, Fire Sciences Laboratory, Missoula, Montana, United States of America

Michael K. Young
U.S. Department of Agriculture Forest Service, Rocky Mountain Research Station, Forestry Sciences Laboratory, Missoula, Montana, United States of America

Esther R. Frei, Jaboury Ghazoul and Andrea R. Pluess
Department of Environmental Systems Science, Ecosystem Management, ETH Zurich, Zurich, Switzerland

Neal D. Niemuth and Ronald E. Reynolds
United States Fish and Wildlife Service Habitat and Population Evaluation Team, Bismarck, North Dakota, United States of America

Kathleen K. Fleming
United States Fish and Wildlife Service Division of Migratory Bird Management, Laurel, Maryland, United States of America

Tim J. Smyth, Icarus Allen, Angus Atkinson, John T. Bruun, Rachel A. Harmer, Robin D. Pingree, Claire E. Widdicombe and Paul J. Somerfield
Plymouth Marine Laboratory, Plymouth, Devon, United Kingdom

Simon G. Haberle
Department of Archaeology and Natural History, College of Asia and the Pacific, Australian National University, Canberra, Australian Capital Territory, Australia

David M. J. S. Bowman and Grant J. Williamson
School of Plant Science, University of Tasmania, Hobart, Tasmania, Australia

Rewi M. Newnham
School of Geography, Environment and Earth Sciences, Victoria University of Wellington, Wellington, New Zealand

Fay H. Johnston
Menzies Research Institute Tasmania, University of Tasmania, Hobart, Tasmania, Australia

Paul J. Beggs and Alison K. Jaggard
Department of Environment and Geography, Faculty of Science, Macquarie University, Sydney, New South Wales, Australia

Jeroen Buters
Center for Allergy and Environment, Technical University of Munich, Munich, Germany

Bradley Campbell and Ian Godwin
School of Agriculture and Food Science, The University of Queensland, St Lucia, Queensland, Australia

Bircan Erbas and Don Vicendese
School of Public Health and Human Biosciences, La Trobe University, Bundoora, Victoria, Australia

Brett J. Green
Allergy and Clinical Immunology Branch, Health Effects Laboratory Division, National Institute for Occupational Safety and Health, Centre for Disease Control and Prevention, Morgantown, West Virginia, United States of America

Alfredo Huete
Plant Functional Biology and Climate Change, University of Technology, Sydney, New South Wales, Australia

Danielle Medek
School of Medicine, Australian National University, Canberra, Australian Capital Territory, Australia

Frank Murray
School of Environmental Science, Murdoch University, Murdoch, Western Australia, Australia

Ed Newbigin
School of Botany, University of Melbourne, Melbourne, Victoria, Australia

Michel Thibaudon
European Aerobiology Society, Réseau National de Surveillance Aérobiologique, Lyon, Rhône-Alpes, France

Janet M. Davies
Lung and Allergy Research Centre, School of Medicine, and Translational Research Institute, The University of Queensland, Woolloongabba, Queensland, Australia

Sean A. Parks and Carol Miller
Aldo Leopold Wilderness Research Institute, Rocky Mountain Research Station, USDA Forest Service, Missoula, Montana, United States of America

Marc-André Parisien
Northern Forestry Centre, Canadian Forest Service, Natural Resources Canada, Edmonton, Alberta, Canada
Solomon Z. Dobrowski
Department of Forest Management, College of Forestry and Conservation,
University of Montana, Missoula, Montana, United States of America

Clara L. Mackenzie
Centre for Marine Biodiversity and Biotechnology, School of Life Sciences, Heriot-Watt University, Edinburgh, United Kingdom

Sharon A. Lynch and Sarah C. Culloty
Aquaculture and Fisheries Development Centre, School of Biological, Earth and Environmental Sciences, University College Cork, Cork, Ireland

Shelagh K. Malham
Centre for Applied Marine Sciences, Bangor University, Menai Bridge, Anglesey, United Kingdom

Xiaofei Yu, Yuanchun Zou, Xianguo Lu and Guoping Wang
Key Lab of Wetland Ecology and Environment, Northeast Institute of Geography and Agroecology, Chinese Academy of Sciences, Changchun, China

Michael Grace
Water Studies Centre and School of Chemistry, Monash University, Clayton, VIC, Australia

Xuefeng Yu
State Key Laboratory of Loess and Quaternary Geology, Institute of Earth Environment, Chinese Academy of Sciences, Xían, China

Lewis H. Ziska
Crop Systems and Global Change Laboratory, United States Department of Agriculture, Agricultural Research Service, Beltsville, Maryland, United States of America

Index

S

T

U

V

W

Y